THE GREAT BARRIER REEF

AF757155

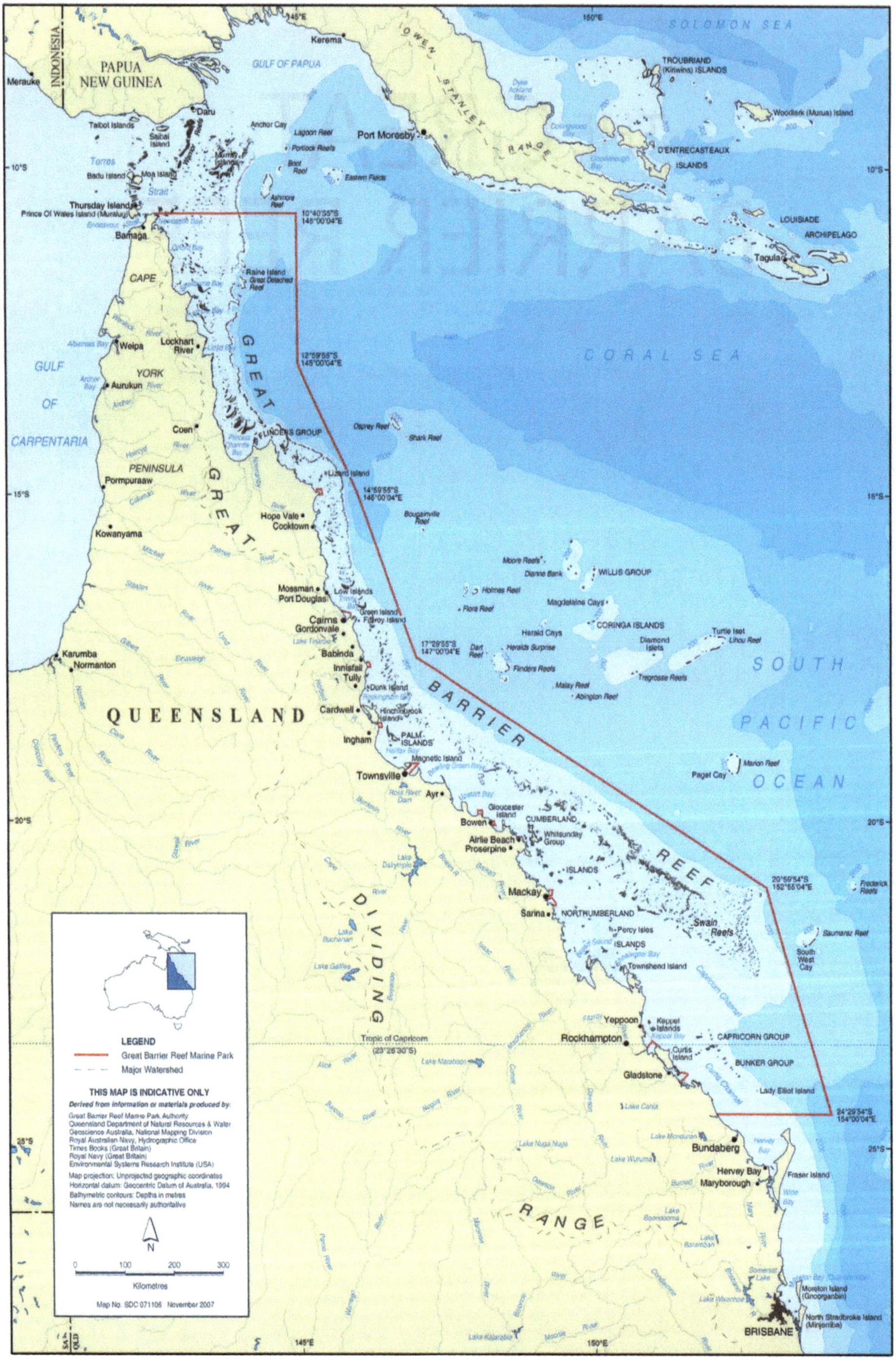

PAPUA NEW GUINEA
INDONESIA
SOLOMON SEA
GULF OF PAPUA
OWEN STANLEY RANGE
Kerema
Merauke
Daru
Port Moresby
TROUBRIAND (Kiriwina) ISLANDS
Woodlark (Murua) Island
D'ENTRECASTEAUX ISLANDS
LOUISIADE ARCHIPELAGO
Tagula
Talbot Islands
Saibai Island
Anchor Cay
Lagoon Reef
Portlock Reefs
Boot Reef
Eastern Fields
Ashmore Reef
Torres Strait
Badu Island
Moa Island
Thursday Island
Prince Of Wales Island (Muralug)
Bamaga
10°40'55"S 145°00'04"E
Raine Island
Great Detached Reef
CAPE YORK PENINSULA
GULF OF CARPENTARIA
Weipa
Lockhart River
Aurukun
Coen
Pormpuraaw
Kowanyama
12°59'55"S 145°00'04"E
CORAL SEA
GREAT BARRIER REEF
GREAT DIVIDING RANGE
FLINDERS GROUP
Osprey Reef
Shark Reef
Lizard Island
14°59'55"S 146°00'04"E
Bougainville Reef
Hope Vale
Cooktown
Moore Reefs
Diane Bank
WILLIS GROUP
Holmes Reef
Flora Reef
Magdelaine Cays
CORINGA ISLANDS
Herald Cays
Heralds Surprise
Diamond Islets
Turtle Islet
Lihou Reef
Dart Reef
Flinders Reefs
Malay Reef
Abington Reef
Tregrosse Reefs
Mossman
Low Islands
Port Douglas
Cairns
Green Island
Fitzroy Island
Gordonvale
17°29'55"S 147°00'04"E
Karumba
Normanton
Babinda
Innisfail
Tully
Dunk Island
Cardwell
Hinchinbrook Island
QUEENSLAND
Ingham
PALM ISLANDS
Magnetic Island
Townsville
Ayr
SOUTH PACIFIC OCEAN
Marion Reef
Paget Cay
Gloucester Island
Bowen
CUMBERLAND ISLANDS
Whitsunday Group
Airlie Beach
Proserpine
Mackay
Sarina
NORTHUMBERLAND ISLANDS
Percy Isles
Swain Reefs
20°59'54"S 152°55'04"E
Frederick Reefs
Saumarez Reef
South West Cay
Townshend Island
Capricorn Channel
Yeppoon
Keppel Islands
Rockhampton
Tropic of Capricorn (23°26'30"S)
CAPRICORN GROUP
Curtis Island
BUNKER GROUP
Gladstone
Lady Elliot Island
24°29'54"S 154°00'04"E
Bundaberg
Hervey Bay
Maryborough
Fraser Island
Moreton Island (Gnoorganbin)
North Stradbroke Island (Minjerriba)
BRISBANE
10°S
15°S
20°S
25°S
145°E
150°E
SA
QLD
LEGEND
Great Barrier Reef Marine Park
Major Watershed
THIS MAP IS INDICATIVE ONLY
Derived from information or materials produced by:
Great Barrier Reef Marine Park Authority
Queensland Department of Natural Resources & Water
Geoscience Australia, National Mapping Division
Royal Australian Navy, Hydrographic Office
Times Books (Great Britain)
Royal Navy (Great Britain)
Environmental Systems Research Institute (USA)
Map projection: Unprojected geographic coordinates
Horizontal datum: Geocentric Datum of Australia, 1994
Bathymetric contours: Depths in metres
Names are not necessarily authoritative
N
0
100
200
300
Kilometres
Map No. SDC 071106 November 2007

THE GREAT BARRIER REEF

Biology, Environment and Management

Editors

Pat Hutchings, Australian Museum

Mike Kingsford, James Cook University

Ove Hoegh-Guldberg, The University of Queensland

Australian Coral Reef Society

Australian museum nature culture discover

JCU James Cook University

The University of Queensland Australia

CSIRO PUBLISHING

Springer

© ACRS 2009
Softcover reprint of the hardcover 1st edition 2009

No part of this work may be reproduced, stored in a retrieval system, or transmitted in any form or by any means, electronic, mechanical, photocopying, microfilming, recording or otherwise, without written permission from the Publisher, with the exception of any material supplied specifically for the purpose of being entered and executed on a computer system, for exclusive use by the purchaser of the work.

Co-published by Springer Science + Business Media B.V., Dordrecht, The Netherlands and
CSIRO PUBLISHING, Collingwood, Australia

ISBN: 978-90-481-8034-9

Sold and distributed:
In the Americas, Europe and Rest of the World by Springer Science + Business Media B.V.,
with
springer.com

In Australia and New Zealand by **CSIRO** PUBLISHING, with
www.publish.csiro.au

Front cover: Main photo by R. Steene.
Other images (clockwise from top right): by R. Steene,
O.Hoegh-Guldberg, D. Wachenfeld, R. Steene, K. Fabricius.

Set in 10/14, Palatino
Cover design by James Kelly
Edited by Amanda Reid
Index by Russell Brooks

Every effort has been made to contact copyright holders for their permission to reprint/reproduce material in this book. The publishers would be grateful to hear from any copyright holder who is not acknowledged here and will undertake to rectify any errors or omissions in future editions of this book.

Contents

Preface

Hon Virginia Chadwick AO

Formerly Chairman of the Great Barrier Reef Marine Park Authority.

In 1922 the Australian Coral Reef Society was established as the first association in the world specifically concerned with the study, research and conservation of coral reefs. Over decades the Society's work has contributed significantly to our understanding of the Great Barrier Reef.

Increased knowledge and understanding is critical if students are to feel encouraged and inspired to follow careers in research or management of coral reef ecosystems.

For those in government, or in organisations such as the Great Barrier Reef Marine Park Authority, the results of scientific research provide both a catalyst and a foundation for policy responses to identified threats and challenges.

Additionally, through conferences, lectures, and the media, the Society provides the broad Australian community with independent and expert information about the issues impacting on a great Australian icon.

This book, 'The Great Barrier Reef: Biology, Environment and Management', published by CSIRO Publishing, continues to explain, inform and educate. It will be valued by non experts, students and researchers.

Those who have contributed are recognised as Australian experts in their field and each chapter has been peer reviewed. The end of each chapter lists 'Additional reading' relevant to the chapter. More references are available at at http://www.australiancoralreefsociety.org/GBR_book.htm [Verified 21 March 2008].

I commend the Australian Coral Reef Society for this initiative that provides such a wealth of information in an accessible form.

For those who care for the Great Barrier Reef this book provides confirmation of the continuing need for sound science and informed management if the reef, as we know it, is to survive current and future challenges.

Acronyms

ABRS	Australian Biological Resources Study, Canberra, ACT
ACRS	Australian Coral Reef Society
AIMS	Australian Institute of Marine Science, Townsville, Qld
ARC	Australian Research Council
CRC	Co-operative Research Centre
CSIRO	Commonwealth Scientific and Industrial Research Organisation
DDM	Day-to-day Management Program
DPI & FISH	Department of Primary Industries and Fisheries
ENCORE	Enrichment of Nutrients on Coral Reefs
EPICA	European Project for Ice Coring in Antarctica
GBR	Great Barrier Reef
GBRCA	Great Barrier Reef Catchment Area
GBRMP	Great Barrier Reef Marine Park
GBRMPA	Great Barrier Reef Marine Park Authority, Townsville, Qld
GBRWHA	Great Barrier Reef World Heritage Area
IPCC	Intergovernmental Panel on Climate Change
JCU	James Cook University, Townsville, Queensland
MODIS	Moderate Resolution Image Spectroradiometer on the Terra and Aqua satellites, available at http://modis.gsfc.nasa.gov/ [Verified 21 March 2008].
MST	Marine Science and Technology
NGOs	Non Government Organisations
NIWA	National Institute of Water and Atmospheric Research, Wellington, New Zealand
NQAIF	North Queensland Algal Identification/Culturing Facility
NTM	Museum and Art Gallery of the Northern Territory, Darwin, NT, Australia
QDNR&M	Queensland Department of Natural Resources and Management
QDPI	Queensland Department of Primary Industries
QPWS	Queensland Parks and Wildlife Service
OSCAR-NOAA	A NOAA project mapping, using satellite altimetry, the water surface elevation of the ocean, from which the near-surface water currents are calculated.
RAP	Representative Areas Program
ROV	Remote Operated Vehicle
RWQPP	Reef Water Quality Protection Plan
SeaWiFS	Sea-viewing Wide Field-of-view Sensor, see http://oceancolor.gsfc.nasa.gov/SeaWiFS [Verified 21 March 2008].

Author Biographies

S. T. Ahyong

Shane Ahyong is a Research Scientist at the National Institute of Water and Atmospheric Research (NIWA), in Wellington, New Zealand. He is an international authority on the stomatopod and decapod crustaceans. Research interests include crustacean phylogeny and systematics invasive species and biosecurity, and deep-water faunas.

P. Alderslade

Phil Alderslade has been researching the identification of octocorals since the early 1970's and for the last 25 years has been the Curator of Coelenterates at the Northern Territory Art Gallery and Museum (NTM. With Katharina Fabricius he has co-authored 'Soft Corals and Sea Fans'—the only comprehensive field guide to the shallow water octocorals of the Red Sea, Indo-Pacific and central west Pacific regions of the world.

C. Alexander

Christopher Alexander was appointed lecturer in Marine Biology at the University College of Townsville (later James Cook University (JCU)) in 1968 and retired in 2003. During that period, he taught a wide range of subjects, researched and supervised honours and graduate students in plankton, invertebrates (especially Crustacea) and Neurophysiology. He is currently Adjunct Associate Professor and regularly tutors in laboratory classes in plankton and crustaceans and gives guest lectures in these subjects and introductory neurophysiology. Degrees: Zoology (University of Wales), M.Sc. Plankton (University of Southampton), Ph.D. crustacean neurophysiology (University of Wales).

T. J. Anderson

Tara Anderson is a Research Scientist (Benthic Ecologist), Geosciences Australia. Her research over 15 years has focused primarily on the relationships between benthic habitats and biota, including biophysical sea floor mapping, in a broad range of coastal and offshore marine ecosystems in New Zealand, Australia, and along the west coast of the United States.

P. Bock

Philip Bock became fascinated by bryozoans while geological mapping in south-west Victoria, where the sediments are often packed with their skeletons. He took up study of the living bryozoans of southern Australia in order to gain an understanding of their environmental variation. After working at RMIT (Royal Melbourne Institute of Technology) University for over 30 years, he retired in 1997, and maintains an active interest, including keeping the Bryozoa Home Page website available at http://www.bryozoa.net/ [Verified 21 March 2008].

J. Brodie

Jon Brodie is a research scientist with the Australian Centre for Tropical Freshwater Research at James Cook University. He has held positions as Director, Institute of Applied Science, University of the South Pacific, and Director, Water Quality and Coastal Development Group, GBRMPA (Great Barrier Reef Marine Park Authority). Jon has collaborated with a wide range of research colleagues from AIMS (Australian Institute of Marine Science), GBRMPA, CSIRO (Commonwealth Scientific and Industrial Research Organisation), QDNR&M (Queensland Department of Natural Resources and Management), QDPI (Queensland Department of Primary Industries), JCU (James Cook University) and the Reef CRC (Co-operative Research Centre) to establish the effects of changed terrestrial runoff on GBR (Great Barrier Reef) ecosystems.

M. Byrne

Maria Byrne's research on the biology of invertebrates involves echinoderms and molluscs. She received her

B.Sc. from Galway University, Ireland, and Ph.D. from University of Victoria, Canada. At the Smithsonian Institution she investigated Caribbean echinoderms and returned to Ireland working on sea urchin aquaculture and fisheries. Her research on life history evolution uses Australian echinoderms with divergent life histories to investigate the role of evolution of development in generating larval diversity and speciation in the sea. Maria is Director of One Tree Island Research Station, southern Great Barrier Reef. She has published over 130 refereed articles and book chapters.

J. H. Choat

Howard Choat is an Adjunct Professor of Marine Biology at James Cook University and has worked extensively on coral reef fishes over the Indian, Pacific, and Atlantic Oceans.

B. C. Congdon

Brad Congdon is a Reader in Ecology at James Cook University, Cairns. He is a field ecologist with a special interest in seabird conservation and evolution and over 25 years experience working with seabirds both in Australia and overseas. His research group has recently demonstrated that seabirds are sensitive indicators of multiple, previously indistinguishable, climate change impacts at upper trophic levels and were the first to clearly establish rising sea-surface temperatures as a major conservation issue for seabirds of the GBR.

J. Day

Jon Day is currently the Director, Outlook Report Taskforce, within GBRMPA. He has been closely associated with the Great Barrier Reef for 22 years, having worked initially in the park management and planning sections in GBRMPA. Jon was responsible for commencing and co-ordinating the Representative Areas Program, the major rezoning program undertaken for the Marine Park, between 1998–2003. This marine planning approach is widely considered as 'world's best practice', and has received eleven national and international awards, including the Banksia and Eureka Awards, and the UNESCO/MAB Environmental Prize (2005 Sultan Qaboos Prize for Environmental Preservation).

G. Diaz-Pulido

Guillermo Diaz-Pulido grew up in Colombia. He completed his B.Sc. (Hons) in Marine Biology in Colombia in 1995 and his Ph.D. in Marine Botany at James Cook University in 2002. He has done pioneering work on the ecology and diversity of reef algae from the Caribbean Sea and the Great Barrier Reef. His current research focuses on the dynamics of algae after coral disturbances, coral-algal interactions, and impacts of climate change on macroalgae. He is associated with the Universidad del Magdalena in Colombia and is currently a Research Fellow at the Centre for Marine Studies at the University of Queensland.

P. J. Doherty

Peter Doherty, currently Research Director of the Australian Institute of Marine Science (AIMS), gained his Ph.D. in 1980 by describing the population dynamics of damselfishes at One Tree Island. In 1989 he joined AIMS and led a research group in tropical fisheries ecology. In 1998, he joined the Co-operative Research Centre for the Great Barrier Reef World Heritage Area. One of his achievements was to facilitate the $9 million research collaboration known as the 'GBR Seabed Biodiversity Project'. His fondest memories are more than 100 days spent at sea on the RV Lady Basten working the back deck from midnight to midday.

S. Dove

Sophie Dove obtained an undergraduate degree in Mathematics and Philosophy from the University of Edinburgh, and a Ph.D. in Biological Sciences from the University of Sydney. She is presently a Senior Lecturer at the Centre for Marine Studies at the University of Queensland. Her research interests predominantly lie in the area of dinoflagellate-coral symbiosis, especially with regard to their photobiology, clonal nature, and their ability to deposit calcium carbonate. The goal of this research is to understand how these basic features of scleractinian corals respond to the environmental change inclusive of anomalous atmospheric CO_2 concentrations.

N. C. Duke

Norm Duke is a Mangrove Ecologist of more than 30 years standing, specialising in mangrove floristics, biogeography, genetics, climate change ecology, vegetation mapping, plant-animal relationships, pollution and habitat restoration. As Principal Research Fellow he currently leads an active research and teaching group on marine tidal wetlands at the University of Queensland Centre for Marine Studies. With his detailed knowledge and understanding of tidal wetland processes he regularly advises on effective management and mitigation of disturbed and damaged ecosystems. He has published more than 130 peer-reviewed articles and technical reports, including his recent authoritative popular book 'Australia's Mangroves'.

K. Fabricius

Katharina Fabricius is a coral reef ecologist at the Australian Institute of Marine Science. She has worked on coral reefs around the world since 1988. Her main research interest is to better understand the roles of environmental conditions and disturbances, including changing water quality, for the biodiversity of nearshore coral reefs. Katharina was awarded a Ph.D. in 1995 for her work on octocoral ecology, and presently holds the position of a Principal Research Scientist. She has published over 70 journal articles, book chapters, and a book on Indo-Pacific octocorals, jointly produced with Phil Alderslade.

L. Gershwin

Lisa-ann Gershwin, National Marine Stinger Advisor, is an international authority on medusae and ctenophores. Her research interests are primarily on taxonomy, systematics, biogeography, and biodiversity, with recent work on the prediction of the presence of marine stingers and prevention and treatment of their stings. She has worked on medusae and ctenophores around the world since 1992, with particular focus on Australian species since 1998. While working in Australia she has collected many thousands of specimens, with at least 151 species new to science, including 14 new genera and four new families.

D. P. Gordon

Dennis Gordon FLS has been studying bryozoans for 40 years. He is a past President of the International Bryozoology Association and a Principal Scientist at the National Institute of Water and Atmospheric Research (NIWA), Wellington, New Zealand.

H. Heatwole

Harold Heatwole is an ecologist and herpetologist, who in the past 47 years successively held academic posts at the University of Puerto Rico, University of New England, and North Carolina State University. His main research interests are sea snakes, island ecology, ants and tardigrades.

He is the author of over 300 scientific articles and 11 books. He also produces videos for educational purposes. He holds a D.Sc. and Ph.D's in Zoology and Botany. The last dealt with vegetation dynamics on the small cays of the Great Barrier Reef. He is a Fellow of the Explorer's Club.

K. Heimann

Kirsten Heimann is a phycologist in the School of Marine and Tropical Biology at James Cook University (JCU) and has 20 years experience in this research area. She established the North Queensland Algal Identification/Culturing Facility (NQAIF) within the School of Marine and Tropical Biology. The research facility has established tropical microalgal cultures from the Great Barrier Reef, thereby creating the first systematic microalgal record for the region. She is especially interested in algal bloom development, the impact of phosphate/nitrate eutrophication on, for example, dinoflagellate/diatom reproduction dynamics/levels of toxicity, determinants of invasiveness, and the use of microalgae as indicators of water quality.

O. Hoegh-Guldberg

Ove Hoegh-Guldberg is Director of the Centre for Marine Studies at the University of Queensland. After completing his B.Sc. (Hons) Ove travelled to the United States to complete his Ph.D. After postdoctoral work and lectureship Ove returned to Sydney in 1992. In 2000 he moved to the University of Queensland where he leads a group focused on the physiological ecology of coral reefs, particularly regarding global warming and ocean acidification. He has produced over 120 peer reviewed scientific articles and runs the blog:

http://www.climateshifts.com [Verified 21 March 2008]. In 1999 he was awarded the Eureka prize for scientific research into coral bleaching and climate change. In May 2008, he became the 2008 Queensland Smart State Premier's Fellow.

J. N. A. Hooper

John Hooper, Head of the Biodiversity & Geosciences Programs, Queensland Museum, is an international authority on sponges (Phylum Porifera) with specific research interests in taxonomy, systematics, biogeography, biodiversity and conservation biology, and collaborating with 'biodiscovery' agencies over the past two decades in the search for new therapeutic pharmaceutical compounds (and discovering thousands of new species along the way).

D. Hopley

David Hopley is a coastal geomorphologist, holding the position of Professor Emeritus in the School of Earth and Environmental Sciences at James Cook University. He held a personal chair in marine science and has a 43 year association with the University. He worked on coral reef evolution, and changing sea levels, especially on the Great Barrier Reef. He has more than 150 scientific papers on this and related topics including two major books, 'The Geomorphology of the Great Barrier Reef: Quaternary Evolution of Coral Reefs' (Wiley Interscience, 1982) and, with Scott Smithers and Kevin Parnell, 'The Geomorphology of the Great Barrier Reef: Development, Diversity and Change' (Cambridge University Press, 2007).

T. P. Hughes

Terry Hughes has written over 90 research articles on coral reef science, including 21 in the journals *Science* and *Nature*. He has received numerous prizes and awards, and was elected to the Australian Academy of Sciences in 2002, for his contribution to reef science. His publications focus mostly on the dynamics of coral reefs, and issues relating to managing fisheries, pollution and climate change. Terry is Director of the Australian Research Council's Centre of Excellence for Coral Reef Studies, where he leads a $50 million research program. He provides frequent advice to governments and NGOs on marine science, coral reef management and policy.

P. A. Hutchings

Pat Hutchings is a Senior Principal Research Scientist at the Australian Museum and has spent her research career working on the systematics and ecology of polychaetes. In addition, she has been studying the process of bioerosion not only on the GBR but also in French Polynesia, collaborating with French researchers. As well as publishing extensively she has been actively involved in the Australian Coral Reef Society (ACRS) for many years and commenting on management and zoning plans for Australian coral reefs.

R. Kelley

Russell Kelley is a science communication consultant specialising in invisible or time dependent processes through animation and multimedia techniques. His printed work includes a series of popular publications visualising the biological and physical connections between the Great Barrier Reef and its catchments.

M. J. Kingsford

Michael Kingsford is currently Head of the School of Marine and Tropical Biology at James Cook University. The School is a recognised world leader in tropical marine studies. He is also co-ordinator of the Area of Research Strength, Marine Science at James Cook University (JCU), member of the International Advisory Committee of the Great Barrier Reef Research Foundation, Immediate Past President of the Australian Coral Reef Society, and the former Director of One Tree Island Research Station in the southern Great Barrier Reef. He has published extensively on the ecology of reef fishes, jellyfishes, and biological oceanography. His projects have encompassed a range of latitudes and include a well respected book on temperate marine environments. A major focus of his research has been on connectivity of reef fish populations and how the findings can assist managers of marine parks. In addition to research and leadership, he teaches undergraduate and postgraduate students and supervises many postgraduate students.

P. Kott

Patricia Mather (née Kott) AO, FMLS was, before retirement, a senior curator in the Queensland Museum. She has published on the history and roles of museums, taxonomy, conservation, and the Great Barrier Reef and is an internationally recognised authority on the taxonomy of the Ascidiacea. Her appointments include: membership of the MST Grants Committee, ABRS (Australian Biological Resources Study) Consultative Committee, Australian Research Council (ARC) Biological Panel, and a nine-year term on the Consultative Committee of the Great Barrier Reef Marine Park Authority that she had helped to establish. She is a past president of the Great Barrier Reef Committee and senior editor of the 'Coral Reef Handbook'.

A. W. D. Larkum

Anthony Larkum has worked in many fields—from molecules to ecosystems. His early interests were in the way plants absorb nutrients. However, an interest in SCUBA diving stimulated an interest in algae and in how algae are adapted to light fields underwater. This led to a lifelong interest in the physiology and ecology of algae and seagrasses. He has edited two books on the biology of seagrasses. He was instrumental in setting up the University of Sydney's One Tree Island Research Station and has been fascinated with the various roles of algae in the coral reef ecosystem. He initiated the ENCORE (Enrichment of Nutrients on Coral Reefs) Project at One Tree Island that looked at the effect of raising the local levels of nitrogen and phosphorus on coral reef organisms. He is also currently working on the potential effects of global climate change.

V. Lukoschek

Vimoksalehi Lukoschek completed her Ph.D. in 2007 at James Cook University where she studied the evolutionary and conservation genetics of hydrophiine sea snakes, with a focus on species occurring in Australian waters. She obtained her B.Sc. in marine and terrestrial biology with an emphasis on marine conservation, and a First Class Honours investigating the foraging ecology of benthic, carnivorous coral reef fishes. She has also worked on a variety of projects in molecular biology and marine mammal ecology.

H. Marsh

Helene Marsh is Professor of Environmental Science and Dean of Graduate Research Studies at James Cook University, Townsville. Her research interests include marine mammalian population ecology with an emphasis on life history, reproductive ecology, population dynamics, diet, distribution, abundance and movements of dugongs. She has also supervised several PhD students working on coastal cetaceans. Helene has produced some 200 scientific publications including one book, about 100 articles in refereed journals, chapters in books and encyclopaedia, conference proceedings and technical reports. She was awarded a Pew Charitable Trust Fellowship in Marine Conservation, the most prestigious international scientific award in marine conservation in 1998 and a Distinguished Service Award by the Society of Conservation Biology in 2007.

D. McKinnon

David McKinnon has over 25 years experience as a biological oceanographer, and has published extensively on copepod dynamics and systematics. David completed his undergraduate education at the University of Otago, New Zealand before moving to Melbourne, where he completed an M.Sc. His Ph.D., on environmental regulation of copepod production, was completed at the University of Queensland. As part of the Australian Institute of Marine Science Water Quality and Ecosystem Health team he leads research into the environmental impacts of tropical aquaculture, and on the biological oceanography of Australia's tropical seas.

J. M. Pandolfi

John Pandolfi is Professor in the Centre for Marine Studies and the Department of Earth Sciences at the University of Queensland. His research integrates palaeoecological, ecological, historical, and climate data to provide critical insights into how marine communities are assembled and structured in the face of environmental variability and human impacts over extended periods of time.

C. R. Pitcher

Roland Pitcher, Principal Research Scientist, CSIRO Marine and Atmospheric Research, and Principal Investigator of the Great Barrier Reef Seabed Biodiversity Project. His research over 25 years on seabed habitats and biota, including distribution and abundance mapping, effects of prawn trawling, recovery and dynamics, population modelling and assessments, provides an objective foundation to assist management in achieving sustainability of the seabed environment.

B. C. Russell

Barry Russell is Principal Scientist (Marine Biodiversity) with the Northern Territory Department of Natural Resources, Environment and the Arts. He has over 30 years research experience on the systematics, ecology, and behaviour of tropical demersal fishes of the Indo-West Pacific. His current research interests include the taxonomy and phylogenetics of wrasses (Labridae), groupers (Serranidae), threadfin breams (Nemipteridae) and lizardfishes (Synodontidae).

C. Syms

Craig Syms is a lecturer at the School of Marine and Tropical Biology at James Cook University. He has published a range of papers on the relationships between reef fishes and their habitats. His current research examines the role of different spatial and temporal scales of habitat variability in structuring communities. In addition to research, he has also advised extensively on marine resource management and evaluation of marine reserves in California. He teaches postgraduate sampling and experimental design and statistics, and supervises postgraduate students in a range of different marine projects.

C. C. Wallace

Carden Wallace is the Principal Scientist of the Queensland Museum, based at the Museum of Tropical Queensland in Townsville. She has researched coral biodiversity and evolution on reefs around the world and is author of the monograph *Staghorn Corals of the World*. Her current research focus is on taxonomy, biogeography, fossil history, evolution and post-bleaching recruitment of family Acroporidae. She is also reviewing the sea anemone fauna of Australia, with Dr Daphne Fautin and others.

R. C. Willan

Richard Willan is a molluscan taxonomist, presently Senior Curator of Molluscs at the Museum and Art Gallery of the Northern Territory (MAGNT) in Darwin. Formerly he was on the staff of the Zoology Department at the University of Queensland in Brisbane, from where he studied the molluscs of the Great Barrier Reef. During that time he visited research stations on the reef, studying opisthobranchs and bivalves. He is an authority on invasive marine molluscs in Australia. As the result of many visits to the Heron Island Research Station, he collaborated with Julie Marshall to write 'Nudibranchs of Heron Island, Great Barrier Reef'.

E. Wolanski

Eric Wolanski is a coastal oceanographer and holds positions at James Cook University (JCU) and the Australian Institute of Marine Science (AIMS). He has more than 300 publications. He is a fellow of the Australian Academy of Technological Sciences and Engineering, the Institution of Engineers Australia, and l'Académie Royale des Sciences d'Outre-Mer. He was awarded an Australian Centenary medal, a Doctorate Honoris Causa, and a Queensland Information Technology and Telecommunication award for excellence. He is the chief editor of *Estuarine, Coastal and Shelf Science* and *Wetlands Ecology and Management*. He is an Erasmus Mundus scholar and is listed in Australia's *Who's Who*.

1. Introduction to the Great Barrier Reef

P. A. Hutchings, M. J. Kingsford & O. Hoegh-Guldberg

The Great Barrier Reef is one of the world's most spectacular natural features and one of the few biological structures visible from space (Fig. 1.1). The sheer size of the GBR Marine Park (over 360 000 km^2) as well its beauty and biodiversity draw people from all over the world. The reef stretches over 2200 km from subtropical waters (~27°S) to the tropical waters of Torres Strait (8°S) and as far as 400 km from the coast to the outer shelf slope. For Australians, the reef is a source of much pride and enjoyment. The GBR is one of the most prominent icons of Australia, with the majority of visitors coming specifically to Australia to see it. This draw underpins substantial industries such as tourism and commercial and recreational fisheries. What is perhaps a surprise is how much more we still need to know about coral reefs like the GBR. We are still struggling to describe the myriad of species and processes that define the GBR—all with an urgency now that is heightened by the unprecedented local and global pressures that currently face the reef.

The underlying concept behind this book is to describe the patterns, processes, human interactions and organisms that underpin large reef ecosystems like the GBR. Although much of the content of this book is focused on the GBR, we consider it highly relevant to coral reefs in other parts of Australia and the rest of the world. There has been no other comprehensive introduction to the biology, environment, and management of the GBR, especially with regard to the major processes that underpin it or how issues such as deteriorating coastal water quality and climate change affect it. Extending our knowledge and understanding of these processes is vital if we are to sustainably manage the Reef, especially during the coming century of climate change. Only by understanding and managing it wisely do we have a chance of keeping the GBR 'great'.

This book is aimed at undergraduate and postgraduate students and the informed public, as well as researchers and managers who would like to familiarise themselves with the complexity of coral reefs like the GBR. The project arose out of an advanced undergraduate course that has been held on the GBR over the past decade and extensive discussions on the need for a book like this at the ACRS. Through this society and the course, we were able invite the appropriate international experts to contribute to this book. The ACRS (which started out as the Great Barrier Reef Committee) is the oldest coral reef society in the world and most of the authors are members and

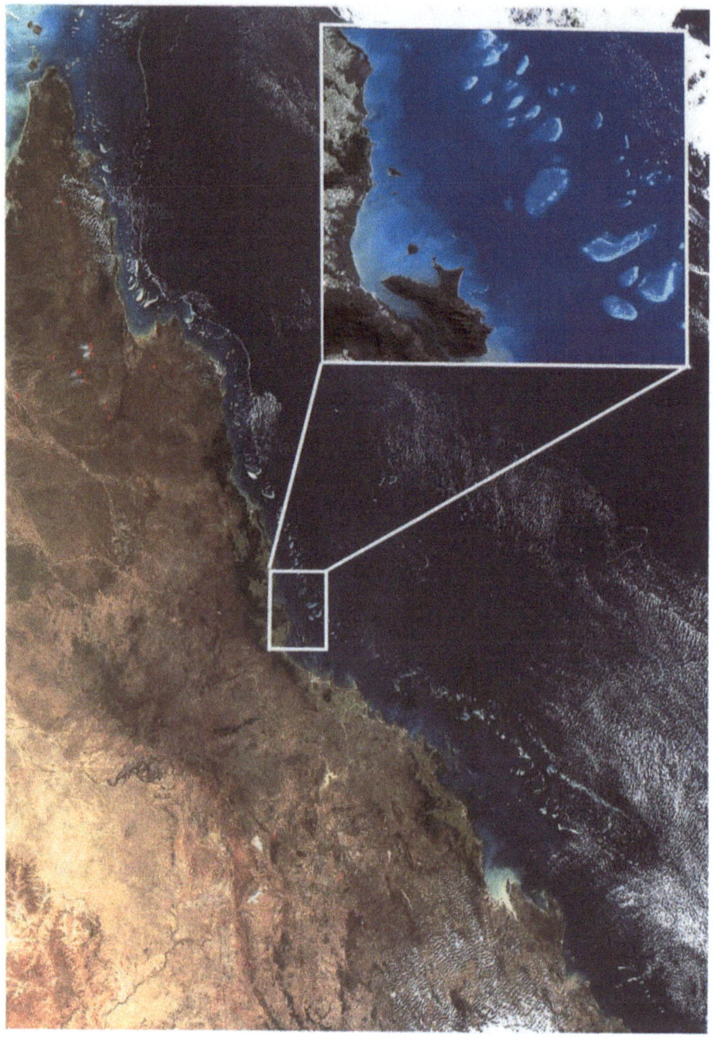

Figure 1.1 Image of the entire GBR from space showing the mainland from Cape York to Gladstone (approximately 2000 km). A mosaic of reefs can be seen parallel to the mainland and extending hundreds of kilometres across the GBR. Quasi-true colour image showing the Great Barrier Reef along the continental shelf of north-eastern Australia. (Image generated from the Moderate Resolution Imaging Spectroradiometer (MODIS), data courtesy of NASA/GSFC, image courtesy of Dr Scarla Weeks, Centre for Marine Studies at The University of Queensland. Insert shows a closer cross-shelf view of the reefs out from Hitchinbrook, near Townsville.

associates of this society. It was also intimately involved in the establishment of the world's largest marine park, the GBR Marine Park in 1975, which was recently enlarged and rezoned based on our much increased understanding of coral reefs.

The book is divided into three sections. In the first section, the geomorphology and paleobiology of the GBR are discussed along with its oceanography. The various habitats of the GBR are discussed, not only from the point of view of coral dominated ecosystems, but also includes the important associated inter-reefal areas. These components of the GBR, along with catchments and offshore deeper waters, are highly interconnected (Fig. 1.2). The second section focuses on the major processes that are affecting the reef and includes a review of photosynthesis and primary production, as well as the flow of energy and nutrients within the reef ecosystem. Other chapters deal with the major

Figure 1.2 Crossing the Blue Highway. Designed and written by science communicator Russell Kelley and published by the ACRS, the Blue Highway poster portrays the reefs of the Great Barrier Reef as part of a larger supporting system that includes the coastal catchments. The poster illustrates how natural nutrient loads from runoff and ocean upwelling fuel a connected mosaic of ecosystems and the role inter-reef habitats play in supporting migrating species as they move from inshore nursery grounds to the outer reefs. The model species of fish is *Lutjanus sebae* (red emperor snapper) that spawn near the shelf edge and recruit to estuaries as larvae. Juveniles move from recruitment habitat to reefs and inter-reefal habitats, before they mature. Printed copies of this poster are available from the GBRMPA. (Artwork: G. Ryan.)

forces within and around the reef, illustrating its inherent dynamic personality. This section also reviews our current understanding of how local (declining water quality and over exploitation of fisheries) and global factors (ocean warming and acidification) have changed the circumstances under which coral reefs have otherwise prospered over the last millennia.

The third and last section of this book, which is perhaps the most important, deals with the diversity of organisms that live in and around coral reefs (Fig. 1.3). In this section, the reader is introduced to the basic taxonomy of the major groups, as well as their biology and ecology. This is a fascinating journey through the unique and wonderful creatures of coral reefs. By weaving the basic taxonomy of these groups together with fascinating details of their lives, it is hoped that the interest of the reader will be inspired to explore this incredible diversity.

Throughout this book we come back to the major challenges that reefs face in our changing world. For this reason, our book is unique is in that it reviews the past, current, and future trajectories and management of

Figure 1.3 Coral reef biodiversity: a snapshot of corals, fishes and gorgonians on the Great Barrier Reef. (Photo: D. Wachenfeld.)

the GBR. Reefs of the world are at risk and knowledge-based management is critical. The GBR is at the forefront of this. It is our hope that this book will help develop a better understanding of coral reefs and to assist in maintaining their ecological resilience in order to allow them to survive the challenges of the future.

This said, we hope that you will enjoy using this book to discover the intricacies of the world's most diverse marine ecosystem. As editors, we would like to thank the generous contributions from the many authors that have contributed to this book, the production of which would have not been possible otherwise.

2. Geomorphology of Coral Reefs with Special Reference to the Great Barrier Reef

D. Hopley

Coral reefs are found in shallow waters throughout the tropics. Although corals contribute the largest amount to reef-building (up to 10 kg $CaCO_3$ m^2 yr^{-1}) many other organisms that secrete limestone-like forms of calcium carbonate such as coralline algae, molluscs, bryozoans, and the green alga *Halimeda* also play a part. Simultaneously, other plants and animals such as chitons, clionid sponges, clams and parrotfish will be destroying the reef through bioerosion at rates that can equal those of production. The reef structure is thus the net result of construction and removal, so that not everywhere that corals grow do significant reefal structures develop.

However, corals remain the most important contributors to reef growth and have some strict environmental requirements, many of which are related to the symbiotic zooxanthellae that require light for photosynthesis. Thus, corals in turbid waters may be limited to a depth of only 8 m but in the clear waters of the outer shelf of the GBR on Myrmidon Reef off Townsville living corals have been observed to depths of 100 m and *Halimeda* to 125 m.

Temperature is also important. The optimum temperature is about 26–27°C with an absolute range of 18–36°C; extremes being tolerable for short periods, for example, at low tide in reef flat pools. Individual corals appear to adapt to their ambient environmental conditions, generally also through the tolerance of their zooxanthellae. Temperature excursions, either higher or lower from the normal range, even if withstood by the same species elsewhere, may locally result in expulsion of the zooxanthellae, coral bleaching and possible mortality.

Salinity affects most reef builders and optimal conditions of about 36‰ or about the level of the open ocean are ideal, with the accepted range being between 30‰ and 40‰. Lower salinity produced by heavy rainfall or runoff can only be tolerated for short periods and is generally the reason for the lack of reefs close to the mouths of major rivers. As the freshwater floats on the salt water, lowering of the low salinity surface water onto a reef flat at low tide can be devastating.

Turbidity is also regarded as a major controlling factor on reef growth. Muddy substrata are unsuitable for larval settlement and reefs are not found where fine sediments have accumulated over time. Turbidity, the result of suspension of sediment in the water column, is more transient and surprisingly high levels may be tolerated. The study of fringing reefs at Cape Tribulation and Magnetic Island have shown that much of the turbidity is caused by resuspension of sediments during rough weather and, as long as the sediment does not settle on the reef and is kept moving by wave action, turbidity levels of more than 100 mg L^{-1} can be withstood. Further, corals from inshore environments are far more tolerant of high suspended sediment levels than exactly the same species on mid and outer shelves. Great diversity can be maintained on these reefs, for example, 141 species from 50 genera of hard corals have been recorded from the Cape Tribulation reefs. Thus, although the total amount of sediment input to GBR waters has increased by more than three times in the last 150 years this is insignificant compared to the billions of tonnes delivered to the nearshore zone and resuspended by wave and current activity over the last 6000 years. Only where new sediment is reaching existing reefs does it have detrimental effects. For example, the fringing reef on High Island, south of Cairns, extends down to 20 m depth and new sediment settling on the lower part of this reef is causing damage as it is below the level of wave action and cannot be resuspended. Increased nutrients may also be associated with increased sediment yield, further adding to reef decline.

GEOMORPHOLOGICAL ZONATION

Coral reefs generally display a strong zonational pattern parallel to the reef front. Ecological zones are clearly related to these geomorphological zones and the higher the energy conditions, the stronger and more distinct the zonation:

- The windward coral-covered reef front is generally steeper in the high energy areas such as the ribbon reefs. Near the top of the reef front may be a distinctive 'spur and groove' zone of coral covered buttresses and intervening channels within which sediment may move.
- The reef crest commences with a living coral zone just below the low tide level. Then follows an intertidal algal pavement, the exact composition of which varies but on the GBR is rarely similar to the prominent coralline algal ridge of mid oceanic atolls. Coralline algae are present, but this zone is dominated by turf algae, which hold together sediment such as coral shingle. On this surface there may be large reef blocks up to 4 m in diameter, torn from the reef front in cyclones or by tsunamis.
- The highest part of the reef is a rubble zone of coral shingle and larger fragments, forming a shingle rampart. It is formed of material from the reef front and deposited by waves as they lose energy passing over the reef flat.
- The aligned coral zone forms the rearmost part of the reef flat, with coral colonies growing in lines perpendicular to the reef front and separated by narrow sandy channels.
- The sheltered back-reef area, often in the form of a sand slope, contains the most fragile branching colonies.

Within this zonation are distinctive changes to coral colonial morphology (Fig. 2.1). Light, wave energy, and emersion are the major controls on both ecomorphology and diversity. At greatest depth on the reef front, light is the most limiting control and corals may adopt a globose or plate-like shape. Higher up, in the optimal photic zone, but below wave base, colonies can be intricately branching, but moving up into the wave zone stronger colonial structures dominate with encrusting forms on the reef crest.

On the reef flat a variety of forms may be found, with exposure the limiting factor. The upper limit to coral growth is about low tide level but varies with species and with local environmental conditions related, for example, to wave action. However, in some instances the low tide level may be controlled by moating behind shingle ramparts or algal ridges and under these conditions, the same level is achieved on every low tide. Corals respond by growing to these levels and form distinctive microatolls: flat discs usually circular and up to half a metre thick. Both head and branching corals form microatolls, though the majority are head corals, the most common being *Porites* sp. They can be up to 10 m in diameter, and, as their upper surface is determined by the moated low tide level, any change in this level, for example as the result of a cyclone lowering a shingle rampart, will produce a change to the surface morphology. Annual growth rings can determine the age of these environmental modifications.

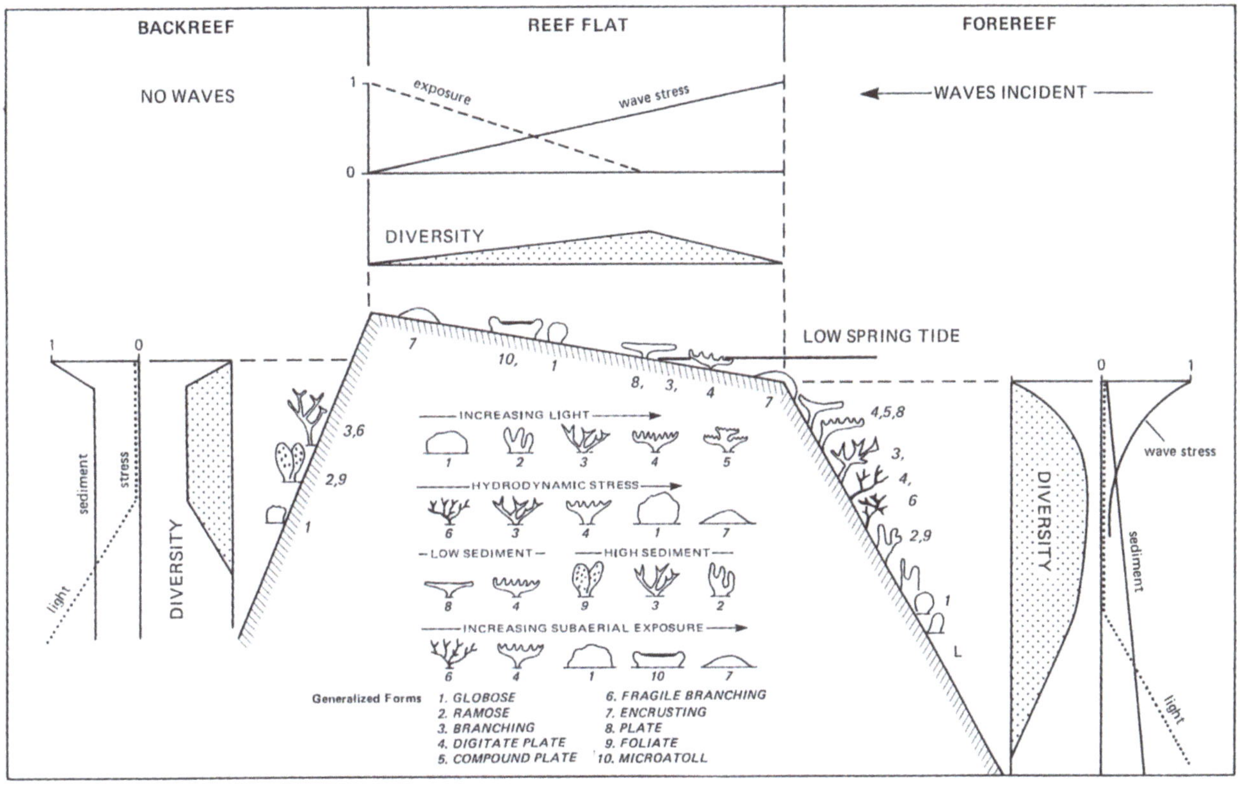

Figure 2.1 Geomorphological and ecomorphological zonation on a reef in response to light, sediment and wave stress, and subaerial exposure (after Chappell 1980). Reprinted by permission from Macmillan Publishers Ltd from: Chappell, J. (1980: 250).

Light may become a factor in determining the morphology in nearshore turbid environments. In turbid waters the whole ecomorphological zonation may be vertically compressed and colonial morphology that encourages sediment shedding dominates.

THE GREAT BARRIER REEF

The GBR (Fig. 2.2) is the largest marine province in the world and has great geomorphological diversity, lacking only open ocean atolls. From 9°15′S the GBR extends southwards off the coast of Queensland to Lady Elliott Island at 24°07′S, a latitudinal spread of 14°52′. The outer perimeter of the reef province is about 2300 km. The outer reefs vary greatly in distance from the mainland coast, approaching within 23 km at 14°S to a maximum of 260 km at 21°S. On the wider shelf, reefs are established on the outer third with a wide area of reefless sea floor at depths of between 20 m and 40 m but with a very gentle slope. In the north, reefs occupy a much greater proportion of the shelf but a narrow reefless 'shipping channel' can still be recognised. The outer edge of the GBR lies on the shoulder of the continental shelf in water depths of up to 100 m, beyond which the sea floor falls away almost vertically in the north, more gently in the central and southernmost GBR (Bunker-Capricorn Group) and in a step-like fashion off the southern-central GBR in the vicinity of the Pompey Reefs.

Dimensions of the GBR are often related to those of the Great Barrier Reef Marine Park (345 500 km^2) and the GBR World Heritage Area (348 000 km^2) both of which end at 10°45′S and thus do not include the reefs of Torres Strait. Within the GBRMP the area of continental shelf is approximately 224 000 km^2, the area of reefs and shoals of 20 055 km^2 making up about 9% of the total area. Although it is sometimes difficult to

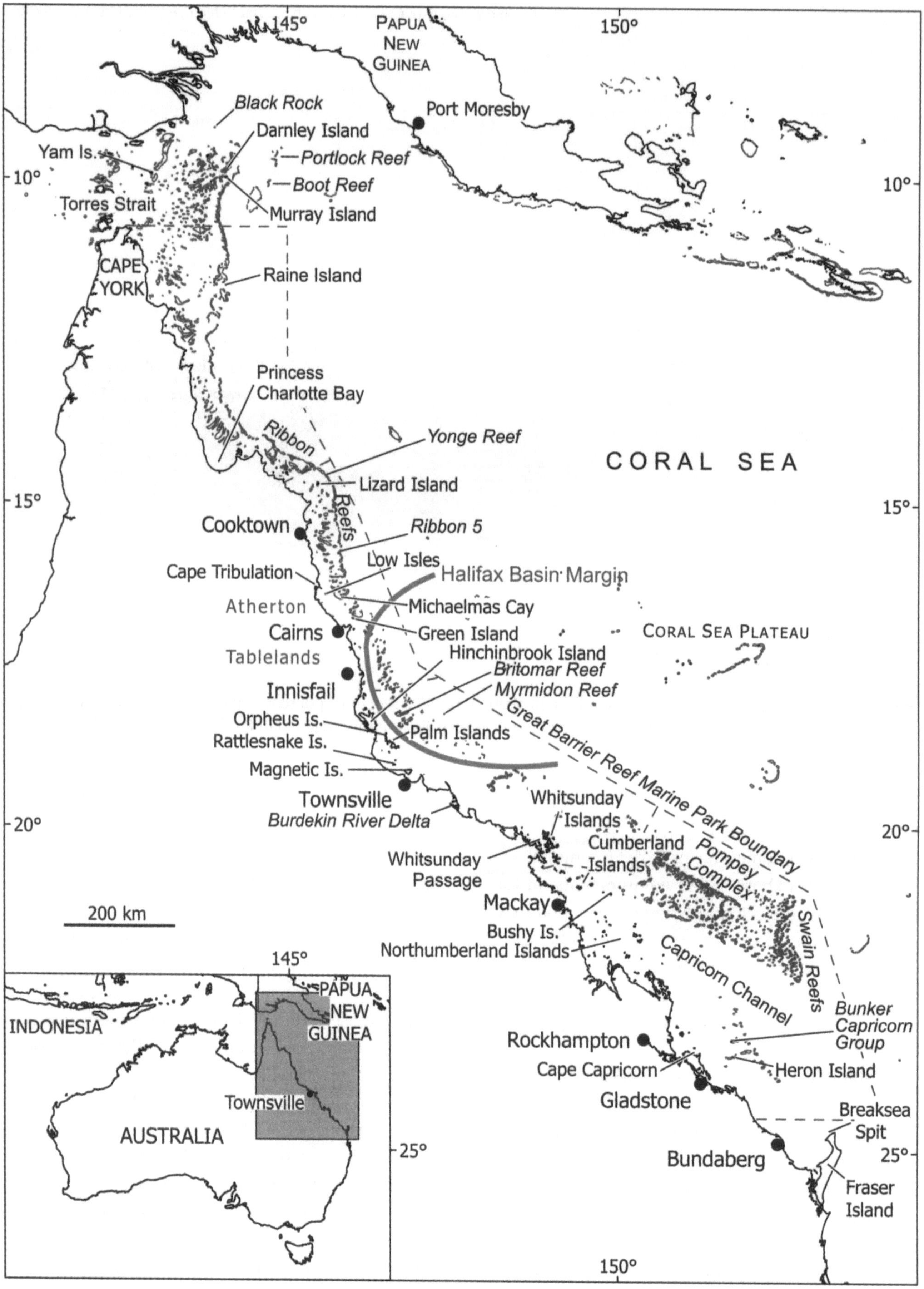

Figure 2.2 The Great Barrier Reef – locations.

count separate reefs in complex parts of the GBR and to fully recognise submerged shoals, a figure of 2900 reefs is generally accepted for the Marine Park. For Torres Strait there is an additional 37 000 km^2 of shelf area and a further 750 reefs covering an area of about 6000 km^2.

THE MOST RECENT EVOLUTION OF THE GBR

The early establishment of the GBR is discussed in Chapter 3 where the importance of Pleistocene sea level changes is indicated. The last time sea level was at or above its present level was about 125 000 years ago at which time the GBR was at least as extensive as it is today. Subsequently, a series of interstadial high sea levels between 60 000 and 70 000 yrs BP(Before Present) and only 15 m to 25 m below present level were capable of adding further growth to the reefs of the outer shelf. Sea level reached its lowest point of at least −125 m around 20 000 years ago. At this time the whole of the GBR, including the shoulder of the continental shelf, was exposed and rivers such as the Burdekin flowed across it to the shelf edge.

The modern GBR developed as the postglacial marine transgression flooded the continental shelf. This took place about 10 000 years ago although the first flooding may not have favoured coral growth as Pleistocene soils and regolith were reworked, producing turbid, eutrophic conditions especially in the lee of the ribbon reefs where extensive meadows of *Halimeda* were the first carbonate structures to develop. The *Halimeda* bioherms, or mounds, up to 18.5 m thick are found inside the ribbon reefs from Cooktown to Torres Strait. They have been maintained to the present day by the jetting of deep nutrient rich water through the passes that exist between the ribbon reefs.

By 9000 years ago sea level was about −20 m below the present position and the older Pleistocene reefs were being inundated. The oldest radiocarbon ages from the Holocene reefs are just over 9000 years and, once inundated, recolonisation and vertical accretion was rapid. The bulk of the Holocene reef was laid down between 8500 and 5500 yrs BP at rates of up to 13 m ka^{-1} (mean about 6 m ka^{-1}). Effectively the transgression was complete by 6500 yrs BP, but shelf warping as the result of hydro-isostasy (depression of the outer shelf and compensatory upwarping of the inner shelf due to weight of the water) produced some cross shelf contrasts. Thus, whilst modern sea level had been achieved by 6500 yrs BP on the inner shelf (documented by raised reefs of up to 1.5 m on inner fringing reefs and the inner shelf low wooded island reefs of the northern GBR), the age becomes progressively younger towards the outer shelf where no emergence took place.

There is also evidence of tectonic shelf-warping in the central GBR where the Halifax Basin of the Coral Sea impinges upon the shelf. Down-warping is suggested by the extensive line of submerged reefs on the outer shelf south of Cairns. There are also regional patterns in the depths down to the Pleistocene foundations of the modern reefs. Although some variation may be expected due to original morphology or subsequent erosion, the pattern as shown by reef drilling also supports the downwarping on the outer half of the continental shelf. In Torres Strait and the far northern region south to Cooktown there is a gradual increase in minimal depth from –5 m in the north to between −15 m and −17 m in the south. South of Cairns the depth to the Pleistocene foundation increases by more than −5 m and everywhere, except Britomart Reef (where it is at −19.8 m), is at depths greater than 20 m. On a cross shelf transect just south of Mackay the variation is from 0 m at Digby Island, between −8.0 m and −13 m on mid-shelf reefs and −17.5 m beneath Cockatoo Reef in the Pompey Complex.

REEF TYPES OF THE GBR

Charles Darwin's reef classification that shows fringing reefs around subsiding volcanic islands that evolve into barrier reefs and finally atolls is applicable only to open ocean 'hot spot' volcanoes and reefs and is not appropriate for the relatively stable, continental shelf of the GBR that has no atolls. Nor is there any great latitudinal contrast on the GBR except for that related to shelf width. Instead cross shelf environmental gradients divide the GBR into three zones:

Fringing and nearshore reefs

Some 545 fringing reefs with recognisable reef flats and 213 incipient fringing reefs (shore attached reefs lacking reef flats) have been recognised in the GBR

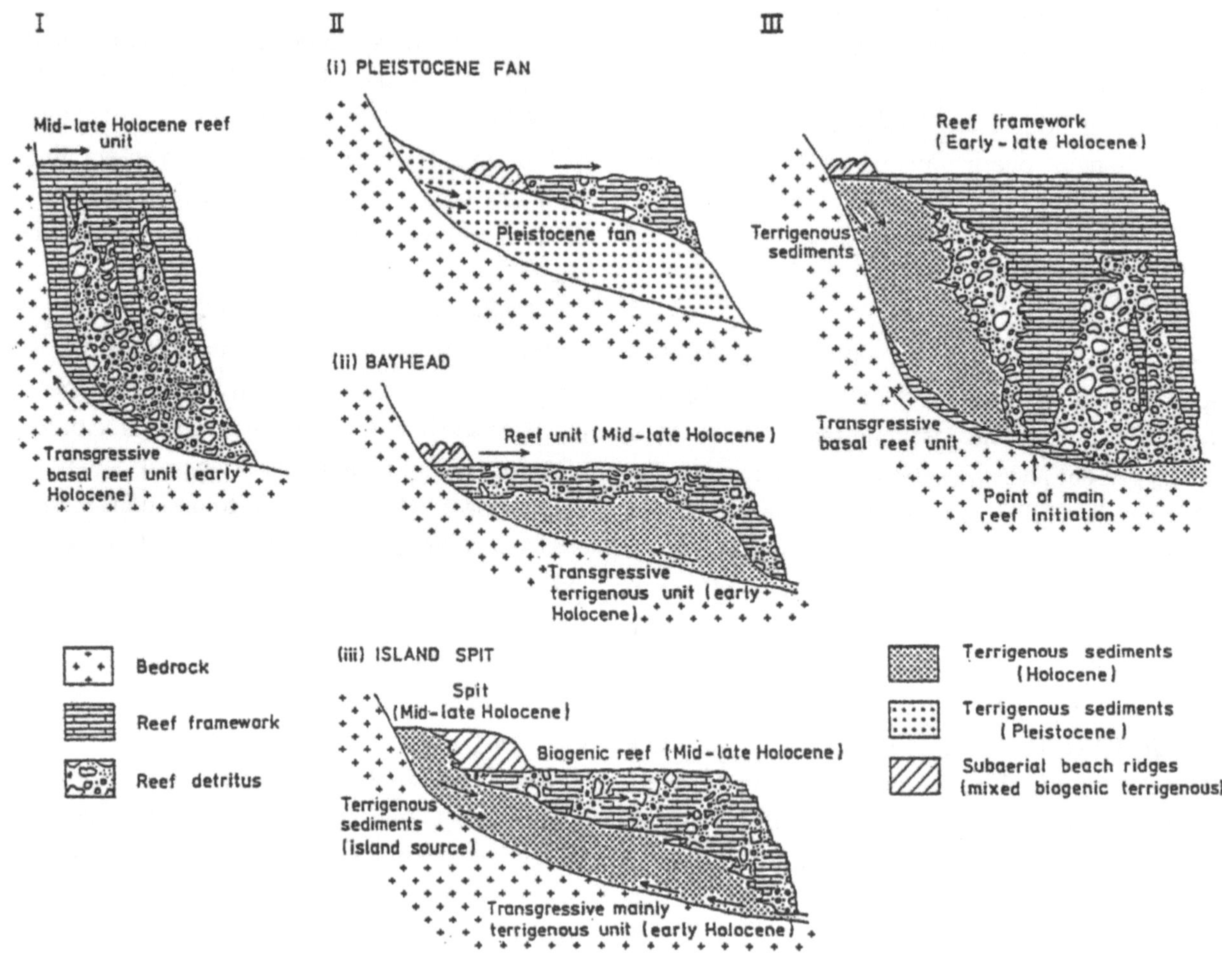

Figure 2.3 Fringing reef classification based on foundation type. (After Partain and Hopley 1987: 39; see also Hopley *et al.* 2007, fig. 7.7.)

Marine Park. Most are small (average 1 km^2) and in total they cover only 350 km^2. In addition there may be up to 100 small inshore reefs and shoals, only a few of which have been mapped because of the turbidity of the waters in which they are found. Fringing reefs attached to the mainland occur in the Whitsunday area and north of Cairns, probably the best known being those of Cape Tribulation. However, the majority are attached to the high continental islands of the inner shelf and their location puts them within the zero isobase of hydro-isostatic uplift. Some of the oldest reef flat dates on the GBR are found on these reefs (>6000 yrs BP) and inner flats are frequently composed of raised microatolls indicative of a 1.5 m higher sea level. Similar to the outer reefs, the fringing reefs had commenced Holocene growth by 9000 yrs BP and accumulated much of their structure by 5500 yrs BP.

Fringing reefs have a wide range of substrata over which they have developed (Fig. 2.3):

(i) Rocky foreshore – formed as sea level rose against the rocky slopes of offshore islands, especially on the windward side. These reefs are generally narrow and not as common as might be expected.

(ii) Pleistocene reefal substratum – unlike mid and outer shelf reefs, older Pleistocene fringing reef foundations are not common. The large fringing reef of Hayman Island is the best example. At Digby Island, Pleistocene reef is exposed at the surface.

(iii) Preexisting positive sedimentary structures such as alluvial fans (Magnetic and Great Palm Is), transgressionary sedimentary accumulations (Pioneer Bay, Orpheus Island), leeside island sandspits (Rattlesnake Island), deltaic gravels (Myall Reef, Cape Tribulation) and boulder spits (Iris Point, Orpheus Island).

Various modes of reef growth have been recognised, including the growth of offshore structure and subsequent backfilling to form the reef flat (e.g. Hayman Island and Yam Island in Torres Strait) or the development of a nearshore reef subsequently attached to the mainland by shoreline progradation. However, the most common form is progradation, usually episodically, due to sea level fall or other environmental factors, from shoreline outwards to reef front.

Mid-shelf reefs

Except on the narrower shelf north of Cairns, the mid-shelf reefs lie outside the hydro-isostatic zero isobase and therefore do not record the higher mid-Holocene sea level, and many did not reach present sea level until after the highstand. They can be described via an evolutionary classification (Figs 2.4, 2.5*A–E*), commencing with all these reefs growing over older Pleistocene reefal foundations at depths ranging from −5 to more than −25 m. The younger stages generally grow from the deepest foundations and include:

A Juvenile Stage consisting of:

(i) the drowned Pleistocene reef with possibly only *Halimeda* growth;

(ii) submerged shoal reefs, usually with coral growth over the higher parts of the older reef foundation, and

(iii) irregular reef patches, formed when coral growth first reaches sea level.

A Mature Stage consisting of:

(i) crescentic reefs formed when reef patches coalesce into more extensive reef flat on the windward margins, with a hardline reef front;

(ii) lagoonal reefs that form as the outer reef flat extends around the margins enclosing a lagoon that is slowly infilled by patch reef growth and by transport of sediment from the windward margins into the lagoon.

A final Senile Stage consisting of:

(iii) planar reefs, with the lagoon infilled and reef flat extending across the entire reef. Sediment and seagrass beds or mangroves may dominate the reef flat and coral cays are common on such reefs, including low wooded islands.

Two dominant factors determine the stage a mid-shelf reef has reached. The first is the depth to the Pleistocene antecedent reef (that influences the depth of lagoon to be infilled), with reefs growing from shallow foundations most quickly reaching modern sea level and progressing to a mature or senile stage. The second is the size of the reef, as this determines the ratio of highly productive perimeter to the volume of lagoon to be infilled. Larger reefs, once they reach sea level, progress much more slowly. It was shown earlier that the depth of Pleistocene foundations on the GBR was not random, with shallow depths in the north and far south, a cross shelf gradient in the southern-central GBR, and generally depths greater than 20 m on the central GBR shelf. Reef types reflect this variation with submerged reefs, patches and crescentic reefs dominating the central GBR and most planar reefs found north of Cairns or south of Mackay.

Outer shelf reefs

Most reefs on the shoulder of the continental shelf (with the exception of the detached reefs in the north) are linear, parallel to the shelf edge and have originated as fringing reefs at earlier low sea levels, though they are certainly multicyclic. Morphology of the shelf edge and tidal range are the greatest influences on reef type.

Northern detached reefs. These reefs appear to grow on isolated pinnacles of fragmented continental crust. Ribbon-like reefs occur on the windward side of some, with large *Halimeda* meadows to the lee. On Raine Island, the top of the Pleistocene is 11 m below reef flat level, with the reef flat developing about 6000 years ago, and the cay (a significant turtle nesting site) initially accumulating about 4500 years ago.

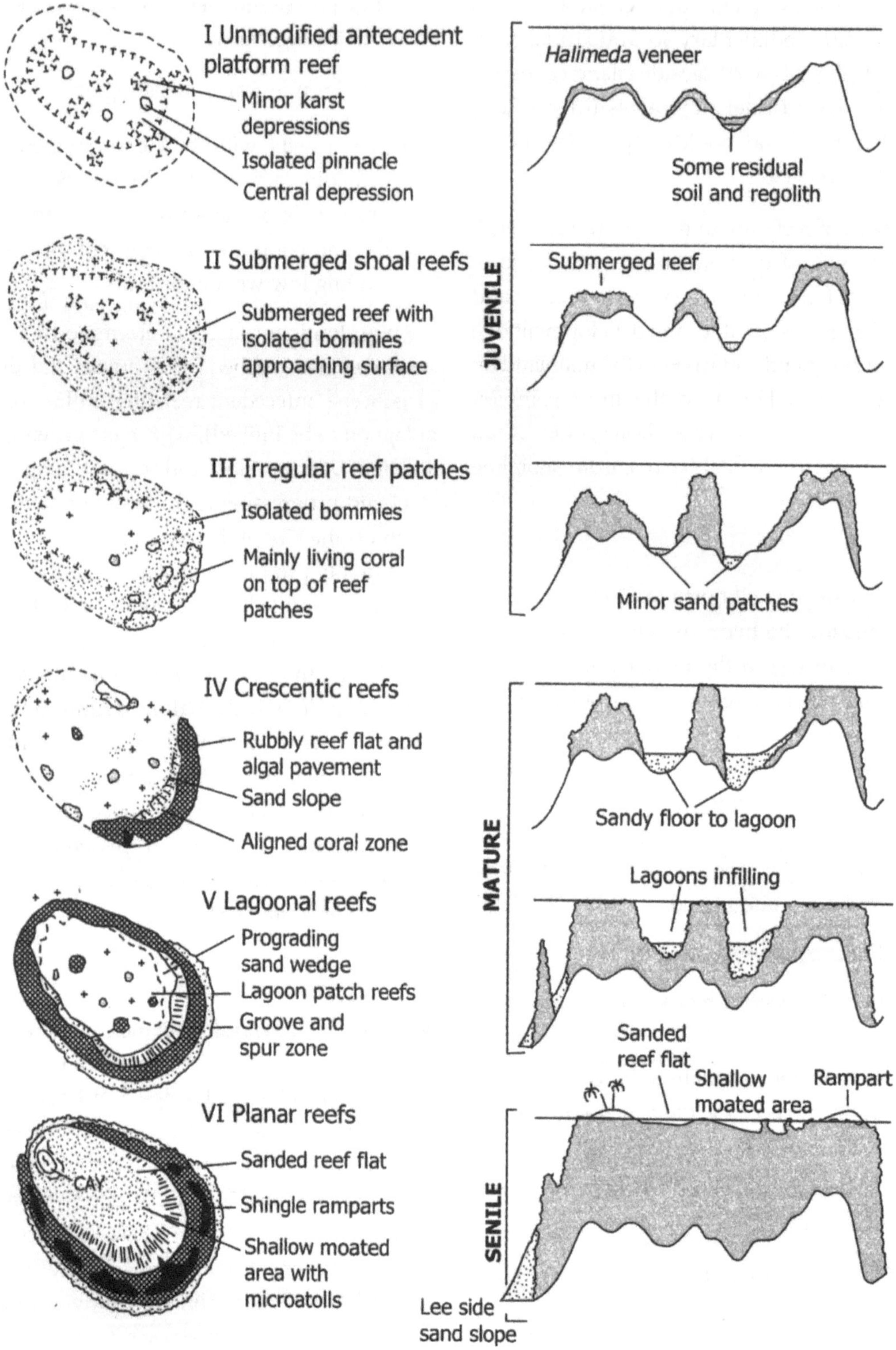

Figure 2.4 An evolutionary classification for mid-shelf reefs growing from Pleistocene reefal foundations (from Hopley, 1982).

Northern deltaic reefs. The northernmost 96 km of shelf edge of the GBR south of the Gulf of Papua consists of short (up to 4 km) narrow reefs parallel to the shelf edge, separated by passages up to 200 m wide. Water depths at the shelf edge are only 30 m to 40 m but on the ocean side is a steep drop-off down to >700 m. Complex delta-like lobes have formed on the inner western side of each passage, deposited as sedimentary structures in response to strong flood tide currents. These have formed the foundation for reef growth.

Ribbon reefs (Fig. 2.5*E*). These form a classic shelf edge barrier reef system between 11°S and 17°S (about 700 km). Individual reefs are up to 28 km long, separated by narrow passages most of which are less than 1 km wide. Currents through the passes are weaker than opposite Torres Strait and although the ends of the reefs curve inwards there are no deltaic structures. However, tidal jets carrying nutrient rich water from the front of the reefs (that have a steep drop-off down to >1000 m) surge through the channels and maintain the extensive *Halimeda* bioherms described above. The outside of Ribbon 5 has been explored by submersible and shows dead reefs at −50 m and −70 m above an almost vertical limestone cliff down to 200 m. The Pleistocene is at −15 m on Ribbon 5 and −19 m on Yonge Reef. Holocene growth commenced about 8000 years ago and the ribbon reefs appear to have reached modern sea level between 6000 and 5000 yrs BP. Because of the high energy conditions, the ribbon reefs display a strong zonation.

The submerged reefs of the central GBR. South of Cairns, where the shelf widens, reefs are set back from the shelf edge, which is in 70 m to 80 m of water, beyond which is a more gently inclined slope reaching only 200 m within 3 km. Some reefs and shoals are found in this area. Where examined, all outer shelf reefs show the Pleistocene antecedent surface at depths below 20 m, and, although colonised quickly after they were drowned by the rising Holocene sea level (basal dates greater than 8500 yrs BP), they did not reach modern sea level until ~4000 yrs BP, even though they maintained vertical accretion rates of >6 m ka^{-1}. The most significant feature of this 800 km section of the GBR is the presence of up to three lines of submerged reefs rising from water depths down to −100 m. In places these reefs may be as much as 70 m deep but elsewhere approach the surface though many remain unmapped. These features are composite, formed during many transgressive and regressive episodes. Many appear to be dead, perhaps drowned by rapid sea level rise, sediment shedding from upslope or decline in water quality when the shelf margin was first drowned. However, some dense stands of coral have been seen on ROV dives. This area of the shelf has experienced hydro-isostatic subsidence but the impingement of the active Halifax Basin onto the shelf may be the most important influence on the distribution of the submerged reefs.

The Pompey Complex. The unique reefs of the Pompey Complex extend over a shelf distance of 140 km and include some of the largest reefs of the GBR, many between 50 km^2 and 100 km^2. The entire reef tract, which parallels the shelf edge, is between 10 and 15 km wide. However, the reefs themselves are not on the shelf edge, which lies a further 20 km seaward, separated from the Pompey Reefs by a series of probably fault controlled steps, most prominently at −70 m and −80 m, the edges of which are the location of further submerged reefs some of which come to within 10 m of the surface. Narrow channels cut through the Pompey Reefs reaching depths of over 100 m, up to 40 m deeper than the shelf on either side. These reefs are on the outer edge of the widest part of the GBR shelf, in the area of highest tidal range that reaches 10 m on the mainland and exceeds 4 m within the reefs. Tidal currents greater than 4 m s^{-1} have been recorded and are responsible for both the scouring of the channels and the deltaic reefs at both ends of them. Unlike the northern deltaic reefs, water depths have been sufficiently shallow for ebb tide deltas to develop on the seaward side of the reefs as well as the western side, producing extremely complex morphology. Intricate lagoons, with blue hole sinkholes (to 90 m deep) formed at low sea levels also typify the Pompey Reefs.

The Swain Reefs. These form a contrasting area of numerous small planar and lagoonal reefs to the south

Figure 2.5 Reef types of the Great Barrier Reef. *A*, Reef patches, the youngest stage of reef development, Barnett Patches off Hinchinbrook Island. *B*, Crescentic reef with hardline growth on the windward side, Eagle Reef, off Bowen. *C*, Enclosed lagoonal reefs, infilling from internal patch reef growth and transport of sediment from the productive margins, Heron and Wistari Reefs, Bunker-Capricorn Group. *D*, A small senile planar reef with flat top (still with a living coral veneer) and no lagoon, Wheeler Reef off Townsville (note the small unvegetated sand cay on the leeward side). *E*, The continuous line of outer shelf ribbon reefs, Yonge Reef near Lizard Island. *F*, Typical high island fringing reef, Great Palm Island. (Photos: D. Hopley.)

of the Pompeys. A steep drop-off occurs around the northern margin, but to the south and west adjacent to the Capricorn Channel the slope is far gentler.

The Bunker-Capricorn Group. The southernmost reefs of the GBR are set back from the shelf margin, which is poorly defined. The reefs grow over shallow Pleistocene foundations and are either planar or lagoonal with many cays. Corals are not found in the shelf sediments south of 24°S and it is possible that there was no reef growth at this latitude at the height of the glacial periods.

Coral islands

Wave action accumulates sediments produced by coral reefs in particular areas of the reef flat to form various types of coral islands or cays. Reef flat is required and as some can be up to 6000 years old, reef islands are all younger than this. As reefs of the central GBR are relatively immature, no vegetated reef island exists between Green Island (16°45′S) and Bushy Island (25°57′S). Centripetal action of refracted waves moves sediment over the reef, deposition occurring when the waves no longer have the capacity to carry it further. Capacity to carry shingle is quickly lost in normal weather conditions and shingle cays and ramparts are found on the windward side of reefs. Sand may be carried greater distances and sand cays are usually found towards the lee of the reef, the shape depending on the wave refraction pattern. On oval reefs the depositional area may be fairly compact and the cay is oval. On elongate reefs the depositional area where wave trains from each side of the reef meet is linear and the resulting island is longer and narrower, such a shape being susceptible to changing location and erosion in response to changes in weather patterns. On stable, older cays, mature soils, perhaps with guano deposits, have had time to form and a climax woodland vegetation, for example, *Pisonia grandis*, may exist but because of constant changes all stages of the vegetational succession may be present on a single island.

Classification and description of reef islands is based on sediment type (sand or shingle) and vegetation cover (from unvegetated to highly complex including reef flat mangroves). Island shape and size are also important as there is a threshold island width of about 120 m to support a freshwater lens on which mature vegetation depends. Some reefs have more than one island, a windward shingle cay and a leeward sand cay but the most complex are the low wooded islands of the northern Reef. These consist of windward shingle ridges and cemented ramparts that give protection to the reef flat over which mangroves can become established (in some instances over almost the entire reef top). Sand cays are usually found on the leeside of these reefs.

There is a distinctive pattern in the distribution of island types. Unvegetated islands occur throughout the GBR but are least common on the central GBR, where vegetated cays are totally absent. Vegetated cays are most numerous in the far south and far north and occur on six of the Swain Reefs. Low wooded islands are limited to the inner shelf north of Cairns. Within the GBRMP are 213 unvegetated cays, 43 vegetated cays and 44 low wooded islands (total 300) though changes are continually taking place, especially to those with ephemeral vegetation.

■ GEOMORPHOLOGY AND THE FUTURE OF THE GREAT BARRIER REEF

The GBR has changed dramatically over the last 10 000 years and, in terms of coral cover and ecological variety, was probably at its peak about 6000 years ago. Since then reefs have progressed towards more senile stages. Global climate change may accelerate this trend. Even a simple rise in sea level may see a renewal of carbonate productivity over reef flats that have been at sea level for 5000 years or more, but with more efficient movement of sediment as water levels deepen. However, increases in coral mortality from bleaching, diseases, and fragility of many surviving corals as oceans become more acidic are all going to accelerate sediment production and infilling of lagoons. These same processes may counteract the rise of sea level against reef islands. Geomorphological studies show that far from disappearing as some have suggested, islands may increase in size. Tropical cyclones, which may increase in frequency and magnitude, can both erode and build island structures largely dependant on tidal conditions. All these changes will tend to remove and rework the oldest part of the islands containing mature soils and vegetation returning them to an early stage of their vegetational succession. Changing climate will have major effects on geomorphological processes that in turn will impact on the ecology of the GBR.

ADDITIONAL READING

There is a large database on the geomorphology of the GBR that has been used in this digest but is too large to acknowledge here. Two major references by the author include over 1000 references on processes, morphology, and evolution and are a source for further information. They are:

Hopley, D. (1982). 'Geomorphology of the Great Barrier Reef: Quaternary Development of Coral Reefs.' (John Wiley, Interscience: New York.)

Hopley, D., Smithers, S. G., and Parnell, K. E. (2007). 'The Geomorphology of the Great Barrier Reef: Development, Diversity and Change.' (Cambridge University Press: Cambridge.)

Figures were sourced from:

Chappell, J. (1980: 250). Coral morphology diversity and reef growth. *Nature* **286**: 249–252.

Partian, B. R., and Hopley, D. (1989: 39). Morphology and development of the Cape Tribulation fringing reefs, Great Barrier Reef, Australia. Technical Memoir No. 21, Great Barrier Reef Marine Park Authority, Townsville.

3. The Great Barrier Reef in Time and Space: Geology and Palaeobiology

J. M. Pandolfi & R. Kelley

INTRODUCTION

Reefs in their many forms are found throughout the fossil record and represent some of the earliest structure-forming ecosystems on Earth. Since the explosion of metazoans in the Cambrian around 540 My (million years ago) many groups of organisms have formed 'reef like' features on the sea floor, making reef communities difficult to singularly characterise. Following the greatest extinction of all time, the Permo-Triassic event (251 My), scleractinian corals, bivalve molluscs, and crustose coralline algae have dominated the construction of wave resistant organic carbonate structures on the planet—commonly called reefs.

While the definition of just what is a 'reef' has a long and tortuous history of debate in the scientific literature, there is no dispute about the importance of the reef ecosystems of the coastlines, continental shelves and ocean provinces of the tropical realm. In this chapter we take a broad spatial and temporal view of the largest of the world's platform reef provinces—the Great Barrier Reef (GBR). In particular we look at the boundary conditions and mechanisms that underpin the perpetuation of the GBR province in space and time and how environmental change has influenced the reef biota.

The last decade has seen an increasing appreciation of the importance of understanding the GBR from a total system perspective. It has also seen the recognition of the need for a temporal perspective in every aspect of ecology, especially where it seeks to relate to natural resource management. Understanding the ecosystems we live in and exploit over medium term time scales allows natural resource managers to plan for ecological resilience that is key to sustainability.

But what about the long term view? Geological evidence accumulated over the last 30 years shows that the 'reef' part of the GBR is a relatively young feature—less than 1 million years old. Because reefs are built mainly during rising sea levels, and the highest sea levels are associated with interglacial periods, much of our attention is attracted towards GBR reef growth during the interglacial high sea level episodes of the last 500 000 years. But the GBR, like reefs elsewhere in the tropics, survived during lower sea level stands as well. In this chapter we explore the life and times of the GBR and the dynamic ecological response of coral reefs to constant environmental change.

ORIGINS OF THE GREAT BARRIER REEF

The history of the GBR is influenced by the post-Gondwanan continental drift history of the Australian continent and repeated episodes of global environmental change associated with the late Tertiary and Pleistocene ice ages. The 'reefal' GBR is a relatively young geological structure that was slow to respond to favourable environmental conditions early on. In fact, the central Queensland continental shelf has enjoyed warm tropical waters that could well have supported coral growth for the past 15 million years (My). However, the best evidence indicates that the initiation of the GBR did not occur until around 600 thousand years ago (ka), and the regional province of reef systems as we now know them probably did not occur until around 365–452 ka. This is coincident with Marine Isotope Stage (MIS) 11, perhaps the warmest interglacial of the past 450 thousand years (ky), and one with climatic conditions most similar to those we are now experiencing. Some workers believe that the 'switching-on' of the GBR was related to the mid-Pleistocene transition from 41 ky to 100 ky-long climatic cycles, and to the development during MIS 11 of a marked highstand that enabled sustenance of both a cyclone corridor and a reef tract along a relatively wide and deeper water continental shelf (Fig. 3.1*A*).

Cores drilled through Ribbon Reef 5 have shown that the GBR has been able to reestablish itself repeatedly during high sea level episodes associated with major environmental fluctuations in sea level, temperature and CO_2 over the past several hundred thousand years. Moreover, these reefs have maintained a similar coral and algal species composition during their repeated formation (see section on Palaeoecology below) (Fig. 3.1*B*, *C*).

REEF GROWTH AND GLOBAL SEA LEVEL CHANGE

The growth and decay of ice sheets in the northern hemisphere were controlled by 10^4- to 10^5-year scale climate changes forced by natural cyclic changes in several parameters of Earth's orbit (so called Milankovitch cycles). These cycles influence the amount of sun energy received by the Earth. They include obliquity (changes in the angle of Earth's axis of rotation with respect to the sun), eccentricity (changes in the circularity of Earth's orbit around the sun), and precession of the equinoxes (changes in the position of the Earth in its orbit around the sun at the time of the equinox). The cycles are 41 000, 100 000, and 23 000 years respectively. During the last 500 000 years, global sea level underwent at least 17 such cycles of rise and fall. Average rates of sea level change between glacial and interglacial intervals approached 5 m per thousand years with the possibility of greater rates associated with Heinrich events (abrupt climatic episodes associated with ice-rafted detritus during the last glacial). The magnitude of sea level change from one interglacial to the next is on the order of 120 m, a major repetitive 100 m$^+$ rhythm to the late Pleistocene ice ages with which all marine life contends (Fig. 3.2*A*).

The GBR is very similar to other reefs around the world in having grown during rises in sea level, or transgressions, associated with the deglaciation part of the cycle. One of the best examples of transgressive reef growth that has been clearly related to the oxygen isotope record for the late Pleistocene occurs at the Huon Peninsula, Papua New Guinea (PNG) (Fig. 3.2*B*, *C*). In this remarkable tectonically active locality, ongoing uplift during the last several hundred thousand years has left a record of transgressive reef terraces like 'bath rings' along over 80 km of coast. Here, nine transgressive reef growth phases are recorded between 125 ka and 30 ka. Overall the record of dated transgressive reef growth episodes extends back to at least 340 ka.

During rising seas, reefs can accumulate at rates exceeding 10 m per thousand years. This involves a huge bulk of cemented biological framework, principally coral and coralline algae, and even larger quantities of associated sediments. However, once reefs reach sea level, or sea level rise slows and stabilises, this growth slows. From here the interplay between the growth of the bound biological framework, the production of reef associated skeletal sediment and their destruction by bioerosion and physical forces becomes of critical importance to the maintenance of reef growth.

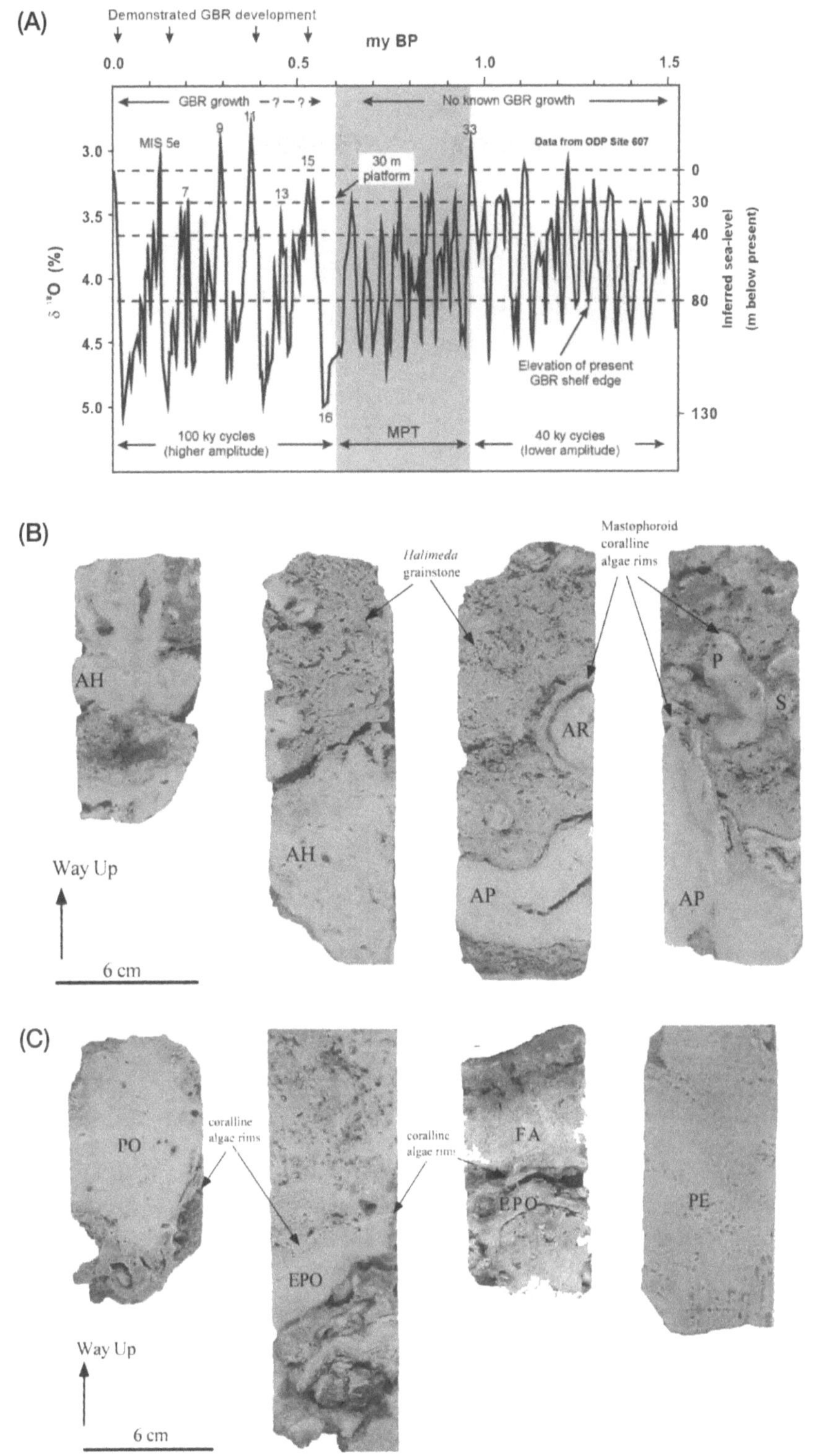

Figure 3.1 *A*, Curve for the past 1.5 million years showing the change in frequency and amplitude of the climatic, and by inference, sea level fluctuations after the mid-Pleistocene transition (MPT) at 0.9–0.6 My (MPT, a shift in the periodicity of radiative forcing by atmospheric carbon dioxide that caused higher amplitude climate periodicities). Growth of the GBR occurred after the MPT (after Larcombe and Carter 2004.) *B*, Photographs from the Ribbon Reef 5 core showing the major coral components of 'Assemblage A'. These include robust branching corals of species from the *Acropora humilis* group (AH), the *Acropora robusta* group (AR), *Acropora palifera* (AP), *Stylophora pistillata* (S), and *Pocillopora* (P). Much of the coral framework is encrusted with coralline algae. *C*, Photographs from the Ribbon Reef 5 core showing the major coral components of 'Assemblage B1' and 'B2'. These include massive *Porites* (e.g. *Porites* cf *lutea*) (PO), encrusting *Porites* (EPO), and massive faviids such as *Favites* (FA) and *Plesiastrea versipora* (PE). Again, extensive coralline algal rims encrust much of the coral framework. (*B* and *C* from Webster and Davies 2003.)

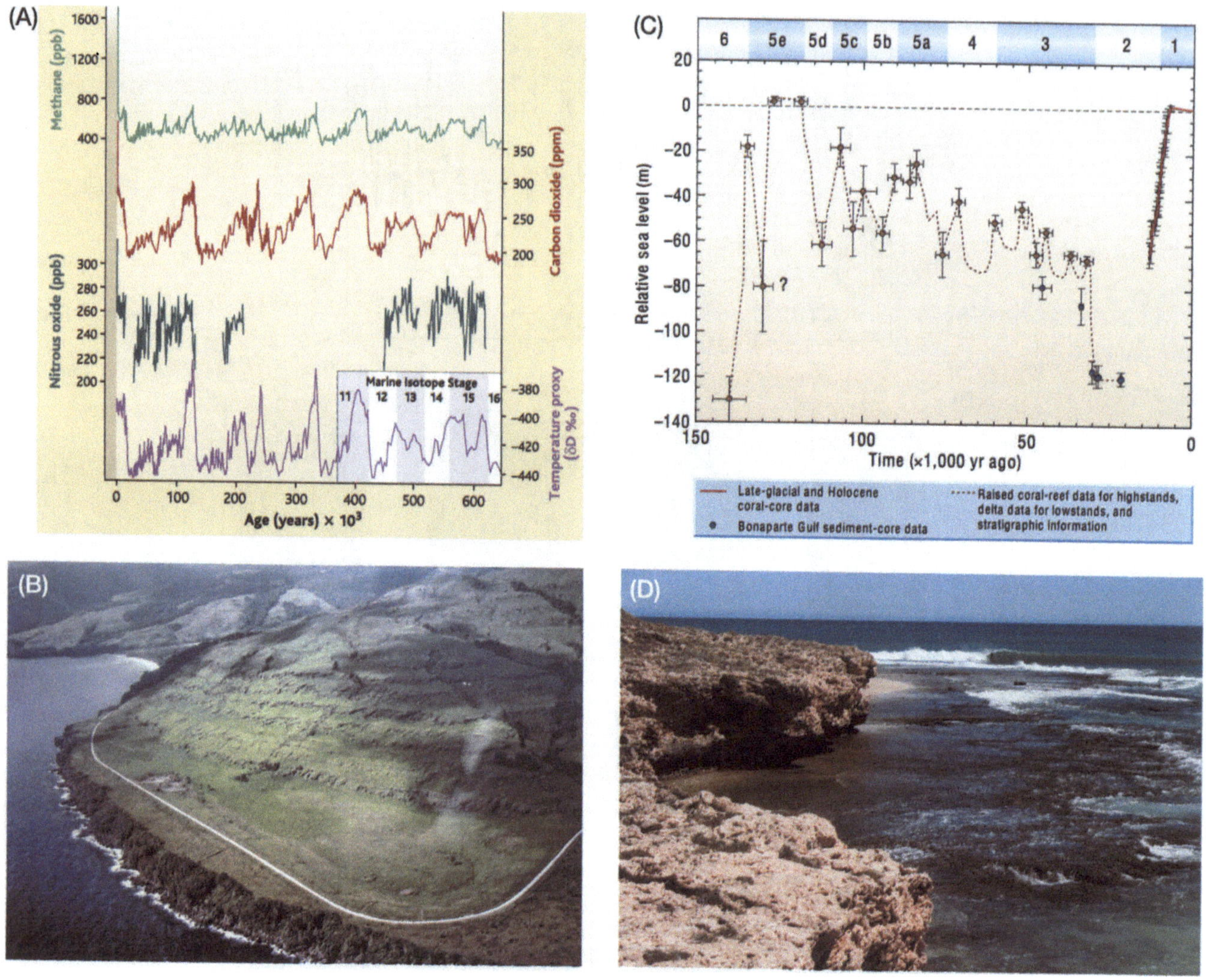

Figure 3.2 *A*, Sea level, temperature, and greenhouse gas fluctuations over the past 650 ky from the EPICA ice core from Antarctica. *B*, View of the Pleistocene and Holocene raised reef terraces at Huon Peninsula, PNG. (Photo: R. Kelley.) *C*, Sea level curve for the past 150 ky derived from Huon Peninsula, supplemented with observations from Bonaparte Gulf, Australia (from Lambeck *et al.* 2002). *D*, Pleistocene reef terrace from the 125 ka reef at Exmouth, Ningaloo, Western Australia. (Photo: R. Kelley.)

THE GREAT BARRIER REEF AND GLOBAL ENVIRONMENTAL CHANGE

The geology, geomorphology and age structure of the GBR is described in detail in Chapter 2. While there is evidence of Pleistocene age reef growth older than 140 ka, we will focus on what the more recent evidence can tell us about the GBR ecosystem in time and space. Here we discuss the GBR during its most recent 'life cycle'—from the previous to the current interglacial cycle and spanning the last ice age.

The superbly exposed and documented record from the Huon Peninsula, PNG, provides a template for expected expressions of transgression within the physical GBR province. Specifically, we should find evidence of reef growth leading to a still stand (i.e. when sea level has ceased to rise or fall) in the previous (128–118 ka: −10 to +5 m a.s.l. (above sea level)) and the present (10 ka to present: −15 to 0 m a.s.l.) interglacials (Fig. 3.2C). There is extensive physical evidence from drill cores of reef growth leading into both of these interglacials. Chapter 2 discusses the dating literature associated with the

Holocene transgression from the Last Glacial Maximum (18 ka) to the current high sea level stand. There is also radiometric evidence from the GBR that the Holocene reef growth was superimposed upon relic Pleistocene reef topography from the last interglacial age.

In other tectonically stable parts of the world this last interglacial reef is well documented at about +2–6 m a.s.l. For example, the last interglacial reef (125 ka, ~5 m a.s.l.) is emergent along the West Australian Ningaloo coast, where it is extensively preserved in near desert conditions (Fig. 3.2*D*). Whereas the north Queensland coast is well endowed with evidence of Pleistocene shorelines in the form of beach rock, dunes, dune foundations, and beach ridges (e.g. Cowley beach near Innisfail), there are relatively few occurrences providing surficial expression of last interglacial reef framework from the GBR (e.g. Stradbroke Island, Evan's Head, Lord Howe Island, Saibai Island in the Torres Strait, Digby Island). One possible explanation for this is that the northeast coast's moist airflow from onshore tradewinds has weathered and eroded the emergent 125 ka GBR reef below the current high sea level such that they now only exist as a base for Holocene reef growth. But perhaps the last interglacial reef did not everywhere grow to the high sea level. One recent explanation is that vertical movements in the form of hydro-isostasy or tectonic lowering are the major factors.

Because of the Milankovitch cycles discussed earlier, sea level fluctuations are not confined to 'glacial' (i.e. Last Glacial Maximum [LGM] 18 ka) and 'interglacial' (i.e. Last Interglacial [LI] 125 ka) periods; smaller scale fluctuations are referred to as 'stadial' (temporary ice advance) and 'interstadial' (temporary ice retreat) times. Abundant studies carried out in tropical seas correlate the growth of 'wave resistant organic structures' such as coral reefs with sea level transgressions. We therefore might expect to see transgressive reef deposits developed during the smaller scale sea level changes between the high stand LI and the low stand LGM (Fig. 3.2*C*). But is there any evidence for GBR reef-building during these lower sea levels?

The nature of low sea-stand reefs has been studied in the Huon Gulf in Papua New Guinea. Here are found similar rates of accumulations and coral communities that were not unlike their high sea level stand counterparts in the adjacent raised reef terraces of the Huon Peninsula, PNG. Regardless, reefs must be seen as dynamic and fluid, reacting to sea level throughout the major and minor Pleistocene fluctuations in sea level (Fig. 3.2*A*).

On the GBR, there is a history of investigation of the terraces and positive-relief features on the continental shelf and margin for evidence of lower sea levels. 'Wave-cut' terraces have been recorded in the southern GBR at −175 m and in the central GBR at −113 m, −88 m and −75 m, where they were interpreted to correspond to postglacial shorelines. Submerged reefs, terraces and notches have been consistently recorded on single beam echo sounder transects across the southern GBR shelf edge but so far there is insufficient evidence on the spatial distribution of these features to make accurate comparisons against sea level curves.

Recent investigations by marine geologists using multibeam echo sounders have revealed that drowned reefs extend for hundreds of kilometres along the GBR outer shelf edge in −40 m to −70 m depth. They appear to be submerged 'barrier reefs' approximately 200 m wide and are comprised of two parallel ridges of eroded limestone pinnacles (Fig. 3.3*A*). These drowned shelf-edge reefs might be an important archive of past climate and sea level changes, and potentially provide predictive tools for GBR coral community response to future climate changes. It is now also possible to map shelf depth palaeo-drainage in greater detail than ever before (Fig. 3.3*B*). Very recent work has extended the occurrence of these submerged shelf-edge reefs as far south as the northern end of the Swain Reefs.

In previous decades the inherent difficulty of remote underwater exploration has restricted the usefulness of this work. A submarine terrace might represent a constructional feature—an interstadial reef—but it might also represent an erosional feature—a wave cut cliff or bench. Modern acoustic techniques involving multibeam sonar hold great promise for finally illuminating the inter-reef and shelf-edge stories by combining high resolution 3-dimensional structure with an ability to map its regional extent.

Despite the limitations of technology, four decades of exploration combined with the new hydrographic charts do tell us one thing. The interstadial GBR does

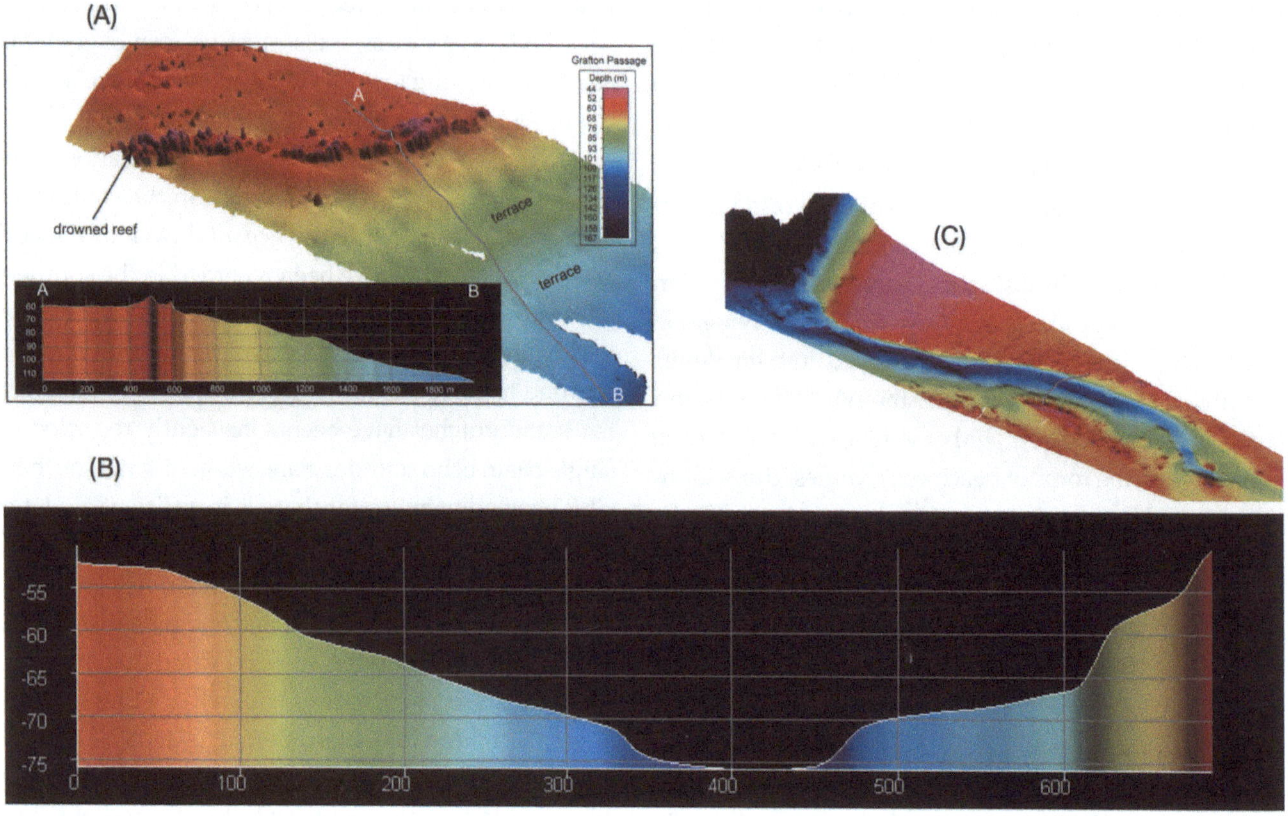

Figure 3.3 *A*, Drowned shelf-edge reef at Grafton Passage. Recent investigations by marine geologists at the James Cook University School of Earth and Environmental Sciences using multi-beam echo sounders have revealed drowned reefs that extend for hundreds of kilometres along the GBR outer shelf edge in −40 m to −70 m depth. This submerged 'barrier reef' near Grafton Passage is approximately 200 m wide. (Image: R. Beaman.) *B*, *C*, Palaeochannel near Cruiser Passage, North Queensland. During the last glacial maximum, sea level was over 100 m lower than today. During these times, rivers deposited floodplain and channel sediments on the continental shelf and upper slope. (Image: R. Beaman.)

not have as grand or extensive an expression as its modern interglacial sibling in terms of accumulated organic carbonate features. There are some obvious factors that may ultimately explain this. The first is the sea floor slope. Worldwide, continental shelves typically have very shallow gradients from the coast to the shelf-slope break where the gradient markedly increases. Here, a one metre rise in sea level can result in kilometres of shoreline displacement. During times of rapidly rising sea level, rates of reef growth from 4 m ky^{-1} to 10 m ky^{-1} are common and environmental gradients are shallow, broad and dynamic. This means that on a shallow continental shelf environmental conditions have the potential to change very rapidly both in a 'turn on' (increased oceanic circulation/reduced shoreline terrigenous influence), and 'turn off' (decreasing circulation, water depth, increasing sedimentation) mode. By contrast, steeper gradients, seen in the steep drop-offs on the GBR Ribbon Reefs and in atoll settings, are more like dipsticks, where the shoreline recedes little during sea level rise and environmental gradients are steep, narrow and less dynamic.

A further consideration influencing reef development is the effect of sea level when still stand is achieved. Rivers that flow across the continental shelf during ice ages have their floodplain sediments remobilised during the next transgression. These materials are moved inshore by the wave climate and end up, in the eastern Australian case, coming onshore in spectacular dune fields. Geological studies of a dune island barrier system enclosing Moreton Bay, southern Queensland, showed that when sea level rise stops, the onshore movement of sediments into the near-shore sediment profile slows and coastal dune building decreases in size and extent. In the Moreton Bay example a few thousand years of still stand also led to the development of sedimentary deposits (coastal plains and tidal deltas) and their inshore environmental correlates (mangroves, seagrasses etc.). This restricted back-barrier circulation increased the estuarine nature of these environments with negative consequences for mid-Holocene back-barrier coral communities.

In the Moreton Bay example many millions of tonnes of coral carbonate was deposited throughout the entire bay in sequences up to 8 m thick immediately after sea level stabilised between 6 ka and 4 ka. So extensive were these deposits they supported a dredge mining operation for over six decades. While corals are still found in Moreton Bay today the reduced circulation and increasingly estuarine conditions experienced after 4 ky has reduced their extent, growth and diversity.

These corals are of interest to the GBR context for what they are not as much as for what they are. They did not form 'reefs' according to the 'wave resistant structure of organic origin' definition. Rather, the corals were flourishing mounds or banks of corals in a back-barrier setting with open circulation. Unpublished radiocarbon dates from Moreton Bay show these *Acropora* dominated coral communities (~40 spp.) grew and accumulated carbonate at rates of up to 5 m ky^{-1}, similar to those known from GBR reefs. The Moreton Bay back-barrier model for coral communities is therefore a diverse, fast acting and geologically significant vehicle for corals over time. A significant difference between these communities and true reefs is that if sea level had continued to rise, these uncemented carbonate deposits would most likely have been eroded away.

Further instances of coral communities from turbid environments forming detrital mounds (as opposed to 'true reefs') have been documented from near-shore reefs of the GBR, including Broad Sound and Paluma Shoals. We feel it is helpful to differentiate these 'coral communities' from 'coral reefs' because they provide an alternative phase during the 'life cycle' of the GBR. As sea levels rise and fall, barrier islands will form in response to still stands providing for Moreton Bay-style opportunities again and again. From a palaeoecological perspective they are a useful alternative to the high sea level reef paradigm.

Having discussed the geological boundary conditions and some of the processes that frame the GBR in space and time we can now better understand our notion of a single interglacial to interglacial 'life cycle' of the GBR and also better scrutinise some of our assumptions about the system. A review of the drilling data shows that the majority of framework growth associated with the current interglacial GBR grew between 9 ka and 4 ka—5 ka window at depths shallower than 30 m. If we assume a similar window for the previous interglacial GBR then the extensive matrix of high sea level platform reefs we know as the GBR was probably active for about 10% of the ecological time between the last two interglacials. Moreover, it may only have been actively growing for about 5% of that time.

A model of the GBR in time and space also needs to account for environmental gradients. Today, there is no single locality that supports all of the roughly 400 coral species found in the GBR region. Richest reefs are in the far northern to central region. Areas like Princess Charlotte Bay, the Palm and Whitsunday Islands provide important ecological space for 'inshore' or turbid water coral communities. These communities collectively contain most species present in the GBR coral fauna, with only a small pool of species apparently restricted to offshore reefs. There are, nevertheless, substantial differences in species' abundance in respect of the major environmental gradients, resulting in more or less characteristic community types across and along the GBR. In particular, there are major differences in species composition between the wave washed, clear water reef crest communities of the seaward slopes of outer barrier reefs and their highly sheltered, turbid water, inshore counterparts, most notably those of the deeper reef slopes of leeward sides of continental islands. These communities are at opposite ends of the physico-chemical spectrum and environmental gradients for the GBR.

So do wave resistant high-stand reefs adequately represent a model for the ecological and geological propagation of the GBR in time and space? Clearly, only partially. If sea level were to fall by 10 m, 20 m and then 30 m would the corals of the GBR, and the thousands of coral connected species, go charging out to the Queensland Plateau to form a clear water 'reef'? Again, probably not. Just as understanding the workings of the modern GBR benefits from a wider whole-of-system approach, grappling with the GBR in time and space requires a broader conception of coral communities than just the clear water platform reefs where we might prefer to go diving.

WHAT CAN WE LEARN FROM PALAEOECOLOGICAL RESPONSES IN ANCIENT REEFS?

Like tropical marine communities throughout the world, the corals, coral communities and coral reefs of the GBR are fundamentally influenced by their response to climate change and associated environmental parameters, including magnitude and rates of sea level change, CO_2, temperature, and turbidity. There is a growing recognition that the integration of palaeoecological and climate data on the GBR provides essential insight into how natural communities are assembled and structured in the face of environmental variability over extended periods of time. Given our ability to discriminate among the various kinds of coral and reef development on the GBR, we next consider some examples of the ecological dynamics of coral reef communities over long time frames.

Although less true for the GBR, many reef organisms are sufficiently preserved in fossil sequences around the world as to provide generic or even species level information on community structure, including corals, molluscs, echinoderms, coralline algae, and foraminifera. We will discuss these examples to help illustrate what the long term ecology of GBR reefs might have been.

In the Indo-Pacific, recent evolution of corals has been rather slow, with less than 20% of new taxa appearing in the past 2–3 My. In the Caribbean, only two species have gone extinct in the past 125 ky As such, palaeoecological patterns from Quaternary reefs (past 1.8 to 2.6 My) can be investigated from what are essentially modern faunas. For example, during the last interglacial (128–118 ka), sea level was two to six metres higher than present levels. This has left a fossilised remnant reef in a large number of locations through the tropics (Fig. 3.2*D*), giving global insight into the ecological nature of reefs in the recent geological past.

One of the best archives for understanding the ecological effects of sea level fluctuations on coral reefs is contained in Pleistocene reef sequences from several tectonically active sites around the world, the most famous of which is the Huon Peninsula in PNG, where nine such reefs were developed between 125 ka and 30 ka (Fig. 3.2*B*). This series of coral reef terraces, formed by the interaction between Quaternary sea level fluctuations and local tectonic uplift, allows investigation of the assembly of coral reefs during successive sea level rises. Here, ecological trends over millennial time scales point to high levels of persistence in community structure, regardless of the magnitudes of change in environmental variables.

In the Caribbean, similar coral community structure was noted among four reef-building episodes ranging in age from 104 ka to 220 ka on Barbados. Remarkably, the high similarity in community composition derived from surveys of common species was also characteristic of separate surveys targeting rare taxa. These studies point to persistence in coral community structure over successive high sea level stand reefs that grew optimally during rising sea level, and are consistent with the rare glimpses we have of the GBR that also show that recurrent associations of coral reef communities are the norm (Fig. 3.4*A*).

Current concern over the deteriorating condition of coral reefs worldwide has focussed intense attention upon the relationship between past 'natural' levels of disturbance and community change versus modern human-induced agents of decline. To understand the impact of humans our only recourse is to study the fossil record. The uplifted Holocene reef at the Huon Peninsula, PNG, age-equivalent to the GBR, has been studied to determine the frequency of disturbance in fossil sequences with little or no human impacts. Rates of mass coral mortality were far lower (averaging one in 500 years) than are presently being experienced in living reefs (multiple events per decade) (Fig. 3.4*B*). Recovery from disturbance was swift and complete, and the history of communities provides predictive power for the nature of their recovery. The stark contrast between living and fossil reefs provides novel insight to the abnormally high disturbance frequencies now occurring.

But what happens when sea level falls or stands still, and how do reefs respond to habitat reduction caused by lowered or lowering sea level? Some spectacular

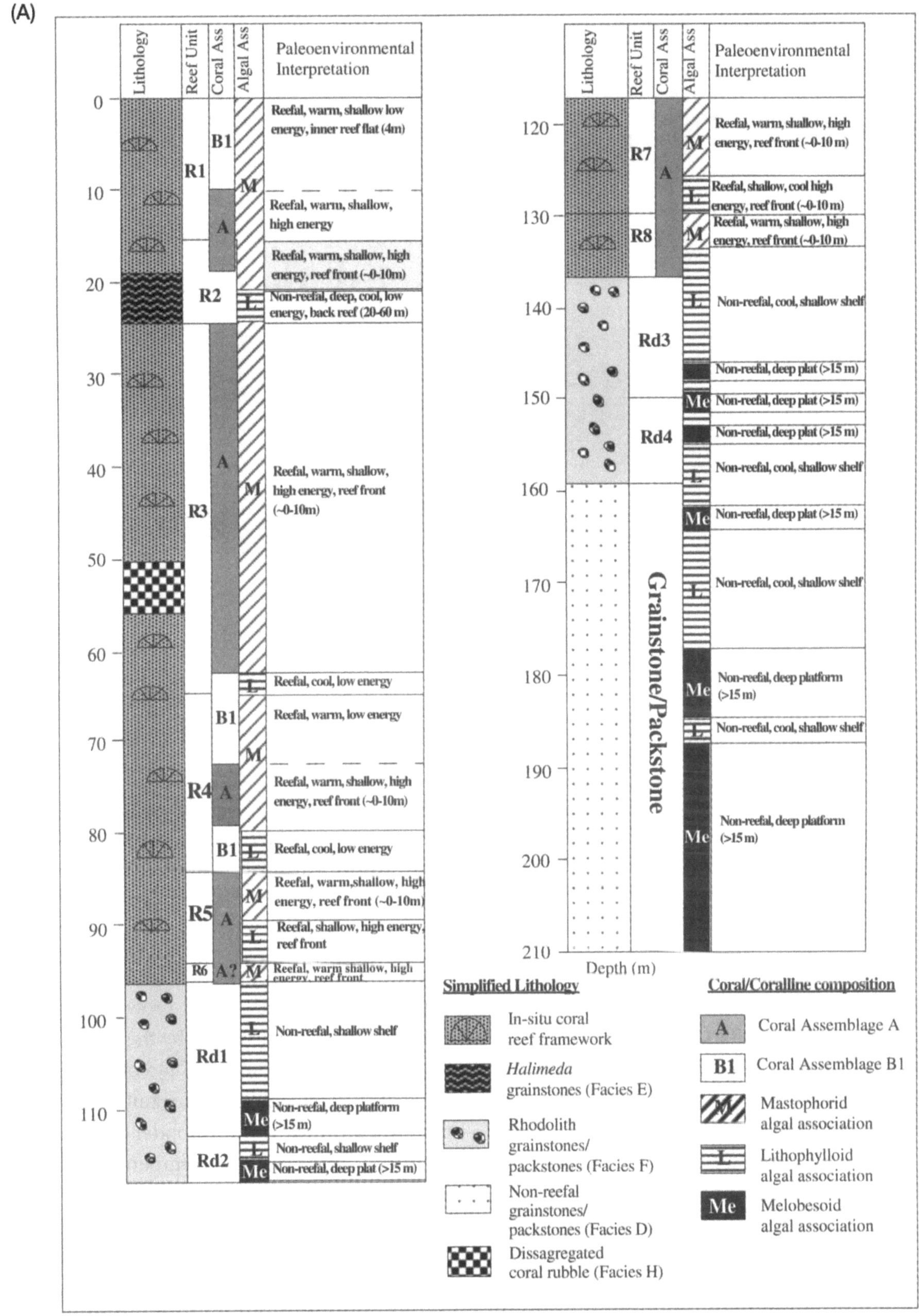

Figure 3.4 *A*, Lithologic and biologic variation in the Ribbon Reef 5 core through the past ~600 ky Ancient environments and coral assemblages remained constant through a number of cycles during the growth and development of the GBR (from Webster and Davies 2003). *(Continued)*

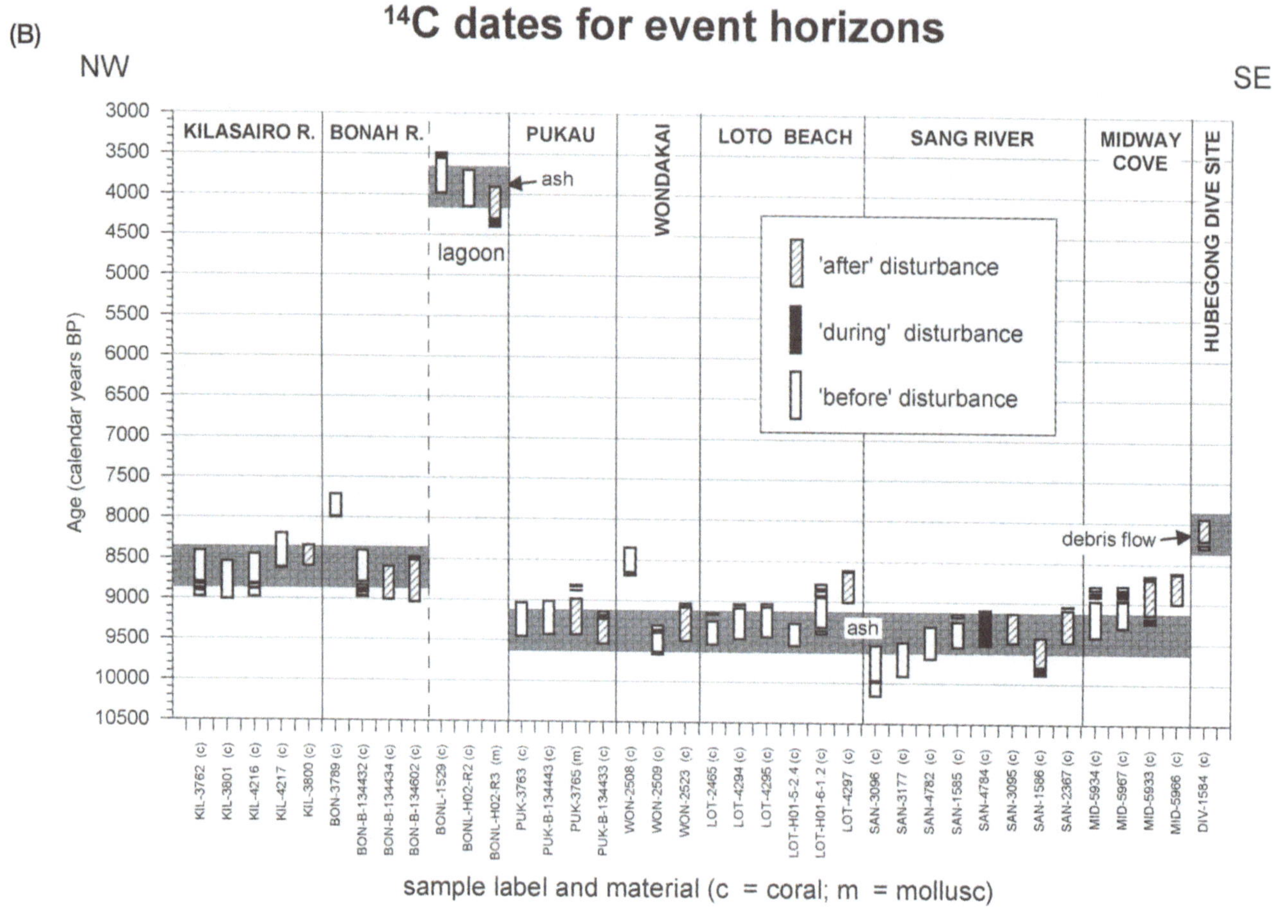

Figure 3.4 *(Continued)* *B*, ^{14}C dates of coral mass mortality along 27 km of the Holocene raised reef terrace from the Huon Peninsula, PNG. Two widespread disturbance events were dated at ~9100–9400 years BP (before present), and ~8500 years BP. Isolated examples of coral mortality were observed in the Bonah River lagoon, and the Hubegong dive site. Mortality events are shaded, and labelled with their likely cause where this has been deduced ('ash', associated with a volcanic ash layer, or, 'debris flow', associated with a submarine debris flow). Shown here are the 2 sigma age ranges in calendar years BP (from Pandolfi *et al.* 2006).

sequences of drowned coral reefs occur at significant water depths in Hawaii and the Gulf of Papua. Although the scale of resolution is much less for these drowned reefs, the overall picture is one of community similarity through large intervals of geological time. When sea level falls and a new reef grows, there is a high degree of predictability in the coral composition of the reefs.

■ EPILOGUE

Global environmental change has had a profound influence over the development of tropical coral reefs since time immemorial, and their effects are no less profound for the GBR. Coral growth culminating in today's GBR has been shaped by the natural variations in sea level, temperature, CO_2 and other climate variables that control light levels, rainfall, turbidity and ocean acidification. It is no wonder then that the ecology of coral and coral reef communities must be seen as dynamic and fluid in their response to Pleistocene sea level and climatic fluctuations.

During the last interglacial to interglacial cycle the GBR has experienced large scale platform reef accretion during two short intervals near the peak of each transgression. However, for more extensive periods the GBR has been doing 'something else', including intervals of restricted shelf edge fringing reef development on the eastern slopes coupled with other non-reefal modes of coral community exploitation of the southern regions of the GBR province during the regressive and glacial intervals. Further work into the comparative response of coral communities and reef growth to these two end member phases should provide important insight into prediction of reef response to future climatic changes.

Now the GBR, like many coral reefs around the world, is changing dramatically in response to human interaction. But living ecological systems provide few clues as to the extent of their degradation. The only recourse into understanding the natural state of living reefs is the fossil record. Our knowledge of past ecosystems on the GBR contributes to formulating sound approaches to the conservation and sustainability of the GBR; specifically in ensuring that policy makers and managers use geological contexts and perspectives in setting realistic goals and measuring their success.

ADDITIONAL READING

Sea level change

Beaman, R. J., Webster, J. M., and Wust, R. J. A. (2008). New evidence for drowned shelf edge reefs in the Great Barrier Reef, Australia. *Marine Geology* **247**, 17–34.

Chappell, J., Omura, A., Esat, T., McCulloch M., Pandolfi, J., Ota, Y., and Pillans, B. (1996). Reconciliation of late Quaternary sea levels derived from coral terraces at Huon Peninsula with deep sea oxygen isotope records. *Earth and Planetary Science Letters* **141**, 227–236.

Lambeck, K., Esat, T. M., and Potter, E.-K. (2002). Links between climate and sea levels for the past three million years. *Nature* **419**, 199–206.

Reef development

Larcombe, P., and Carter, R. M. (2004). Cyclone pumping, sediment partitioning and the development of the Great Barrier Reef shelf system: a review. *Quaternary Science Reviews* **23**, 107–135.

Pickett, J. W., Ku, T. L., Thompson, C. H., Roman, D., Kelley, R. A., and Huang, Y. P. (1989). A review of age determinations on Pleistocene corals in eastern Australia. *Quaternary Research* **31**, 392–395.

Webster, J. A., and Davies, P. J. (2003). Coral variation in two deep drill cores: significance for the Pleistocene development of the Great Barrier Reef. *Sedimentary Geology* **159**, 61–80.

Reef systems

Cappo, M., and Kelley, R. (2001). Connectivity in the Great Barrier Reef World Heritage Area – an overview of pathways and processes. In 'Oceanographic Processes of Coral Reefs: Physical and Biological Links in the Great Barrier Reef'. (Ed. E. Wolanski.) pp. 161–187. (CRC Press: New York.)

Hopley, D., Smithers, S., and Parnell, K. (2007). 'The Geomorphology of the Great Barrier Reef: Development, Diversity and Change.' (Cambridge University Press: Cambridge.)

Pandolfi, J. M., Tudhope, A., Burr, G., Chappell, J., Edinger, E., Frey, M., Steneck, R., Sharma, C., Yeates, A., Jennions, M., Lescinsky, H., and Newton, A. (2006). Mass mortality following disturbance in Holocene coral reefs from Papua New Guinea. *Geology* **34**, 949–952.

4. Oceanography

M. J. Kingsford & E. Wolanski

■ REEFS AND OCEANOGRAPHY

The coral reefs that form the GBR are scattered over the continental shelf, which is shallow and fringed by the deep water of the Coral Sea (Fig. 4.1). Oceanography affects in many ways the organisms and the nature of contemporary geological processes. Seawater erodes and shapes reefs; it influences the transport of sediment and the deposition of material to the substratum. Organisms of all sizes are affected by oceanography. The storm generated seas of cyclones destroy reef structures, kill organisms and alter the nature of habitats. Changes in habitats can in turn affect organisms that typically 'respond' to different habitat types and the influence of oceanography on habitats can influence broad scale patterns of biogeography. The richness of inter-reefal and reef based flora and fauna are strongly influenced by nutrient input from rivers and upwelling over the shelf break. In the pelagic environment, plankton have limited control of their horizontal movements and, therefore, transport and dispersion will be influenced by currents.

Currents have a great affect on highly mobile pelagic organisms. Nekton are attracted to oceanographic features such as convergences for the purposes of feeding and reproduction. For example, flying fish lay their demersal eggs on flotsam in convergences and this may provide suitable conditions for larvae. Concentrations of prey and variation in sea water temperature can also be critical for the survival of larval phases through to their settlement on reefs or other habitats. Some plankton will only spawn while aggregated. Many jellyfish and larvaceans, for example, primarily spawn when concentrated in convergence zones so that the chances of fertilisation are greatest. Currents can move plankton to favourable or unfavourable environments and therefore influence the connectivity of populations. This is particularly important for larval forms that must settle on coral reefs; unfavourable currents may expatriate them from

Figure 4.1 3-D rendering of the bathymetry of the GBR and the adjoining Coral Sea. The South Equatorial Current transports waters that are 4000 m deep to the GBR on a continental shelf 30 m to 100 m deep. (After Wolanski 2001.)

BOX 4.1 GLOSSARY

Eckman transport. Wind-related transport where the upper water layer drags the layer beneath. The rotation of the Earth (= Coriolis effect) results in a slight change in current direction with depth. In the southern hemisphere, average transport in the Eckman layer is to the left (opposite for northern hemisphere).

ENSO. El Nino Southern Oscillation (defined in the text)

Halocline. The region of transition between low salinity waters and high salinity waters.

Internal wave. The result of a deformation or perturbation of the thermocline. Internal waves can be solitary (solitons), or often occur in groups (packets). Internal waves can channel cool water to reefs and influence transport of organisms in slicks at the surface.

Plankton. All organisms that are considered 'wanderers' or 'drifters'. Plankton includes viruses, autotrophs and heterotrophs, phytoplankton and zooplankton.

Pycnocline. The region of transition between low density waters and high density waters.

Sigma-T. A measure of water density. Water density is explained by water temperature (about 65% of the relationship) and salinity.

sv. Volumetric measure of flow (1 million cubic metres per second)

Thermocline. The region of transition between warmer surface waters and colder deep oceanic waters.

Upwelling. A process whereby cold, generally nutrient rich waters from the ocean rise to the surface.

the GBR and favourable currents may sweep them towards reefs, either other coral reefs (meaning the reefs are then connected) or their natal reefs (meaning the reef is self-seeded).

Oceanography affects the GBR at large and small spatial scales. At the largest spatial scales (thousands of kilometres) major oceanic currents of the Coral Sea bath the GBR, affecting patterns of connectivity between reefs and the likelihood of coral bleaching. At very small spatial scales (centimetres to metres) small scale turbulence can affect the settlement patterns of organisms such as corals. The oceanography of the GBR is affected by oceanic currents, tides that flow on and off the shelf, complex topography that generates jets, eddies, stagnation, and convergences, and wind that varies in direction and strength with time of year. Variation in vertical physical structure (i.e. thermoclines and haloclines) and the input of freshwater and mud from rivers also affect water density and related currents. Across the shelf, near-shore waters are often impacted by freshwater whereas outer reefs are more affected by upwelling.

■ CURRENTS IN THE CORAL SEA

South Equatorial Current and south flowing tropical currents

The strong, westward flowing, South Equatorial Current (SEC) prevails in the Coral Sea and is directed towards the GBR (Fig. 4.2*A*, *B*). The strength of this flow is indicated by the Southern Oscillation Index (SOI), or, the difference in atmospheric pressure between French Polynesia and Darwin. The large scale features of this oceanic circulation (Fig. 4.2*A*) are readily measured by satellite-mounted altimeters and scatterometers. Shipborn CTD data are necessary to measure and map smaller scale circulation features (Fig. 4.2*B*) such as the bifurcation of the SEC that generates the southward

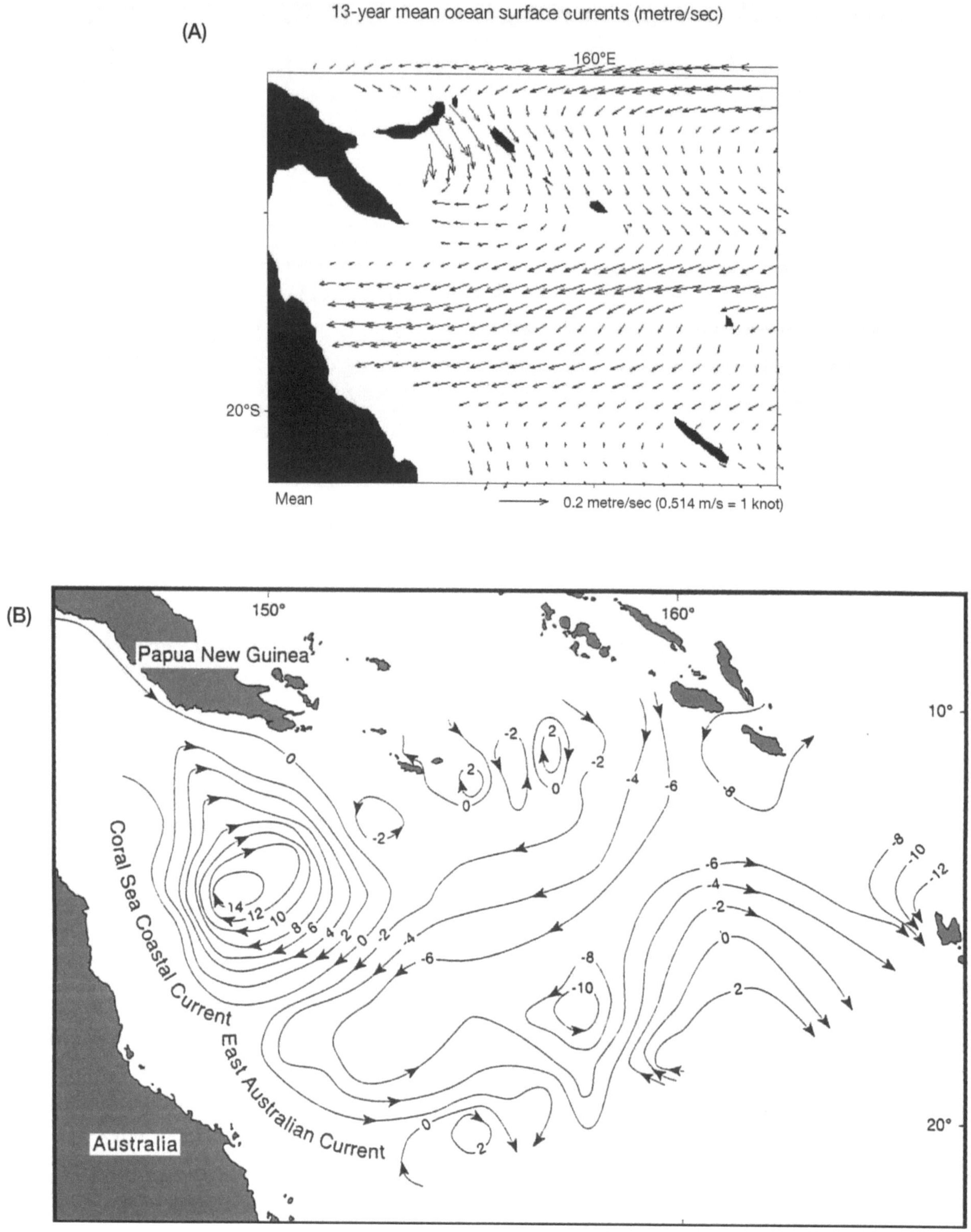

Figure 4.2 *A*, Mean surface currents in the Coral Sea as derived by OSCAR-NOAA satellite data. (Source: http://www.oscar.noaa.gov/). *B*, Details of the circulation derived from ship-born CTD data (plotted as contours of volume transport (in Sv; 1 Sv, 1 million $m^3 s^{-1}$) for the top 1000 m (positive for clockwise flow, negative for counter-clockwise flow). (Adapted from Andrews and Clegg 1989.)

flowing East Australian Current (EAC) south of the bifurcation point and the northward flowing Coral Sea Coastal Current (CSCC) north of the bifurcation point.

The bifurcation point fluctuates between Townsville and Lizard Island. The EAC does enter the reef mosaic of the GBR and generates a coastal GBR lagoon flow from the Central GBR to the Swains (Fig. 4.3). This flow is weak and episodic compared to the flow on the outside of the reef. In southern Queensland this boundary current can reach speeds of over 0.5 m s^{-1}. The influence of the EAC extends over a 1000 km south of the GBR and veers eastward, forming a temperature discontinuity called the Tasman Front.

A warm, southward flowing tropical boundary current is also found on the west coast of Australia. This is the Leeuwin Current (LC) that is the continuation of the Indonesian Throughflow from the Indonesian seas, and flows south down the coast of Western Australia

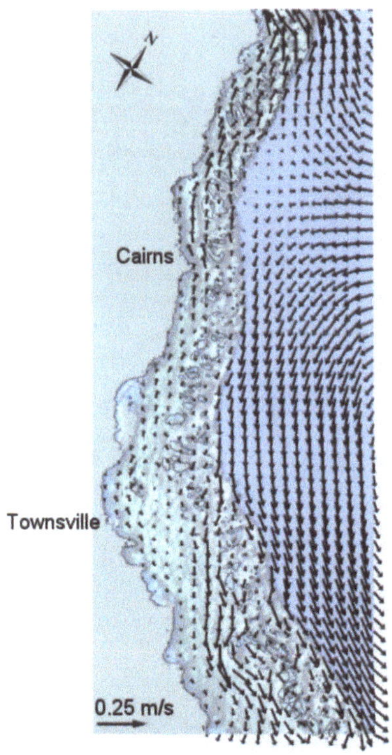

Figure 4.3 Synoptic views of model-predicted mean currents in the austral winter of 1981 under the influence of wind, tides and the SEC. B, bifurcation; CSCC, Coral Sea Coastal Current that flows north to become the Hiri Current; CSLC, Coral Sea Lagoonal Current; EAC, East Australian Current. (Modified from Brinkman *et al.* 2002.)

(Fig. 4.4). It turns east at Cape Leeuwin, where it becomes the South Australian Current, and flows eastwards below most of South Australia to Tasmania, where it is known as the Zeehan Current. The LC creeps closest to the coast near Ningaloo, about 1200 km north of Perth, where the current may be only 10 km off the coast. The thermal signal of the LC can be readily detected at Port Lincoln.

The EAC and LC both carry warm and clear waters that are nutrient poor (oligotrophic). The direction of transport and the oligotrophic waters have a great influence on the biogeography of the east and west coasts. Corals are found as far south as Rottnest Island in the west and near Port Stephens in the east. On both coasts, the larvae of tropical species are advected south. A host of newly settled tropical fishes and invertebrates often arrive at high latitude toward the end of the summer, but few survive the cold of winter. The northern extent of macroalgae (e.g. *Macrocystis* and *Ecklonia*) is also affected by warm currents, as these algae cannot persist in warm and nutrient poor waters.

El Niño Southern Oscillation (ENSO)

The intensity of the SEC and resultant EAC on the east coast of Australia, and the LC on the west coast, depends on the southern oscillation index (SOI). La Nina is sometimes referred to as 'super-normal conditions' as the westward flow of water from Peru is exaggerated. This results in high and positive values of SOI. The SEC is the largest during long periods of steady SOI and smallest during periods of rapidly variable SOI. During El Niño conditions (negative values of SOI), the upwelling fails off Peru. There is a 'decadal oscillation' that corresponds to periods of time when SOI is more regularly in the positive than the negative and visa-versa. Variation in SOI has a great impact on biology; the failure of upwelling off Peru correlates with movements of whales and pelagic fishes and death of sea birds that can no longer find prey. Galapagos iguanas that rely on cool upwelled waters to support biomass of their algal food die *enmass* as the algae disappear. The number of lobsters that recruit to the west coast of Australia vary with strength of the LC. When flow is weak during El Niño, recruitment of puerelus larvae to limestone reefs is poor. Coral bleaching on the GBR and other parts of the

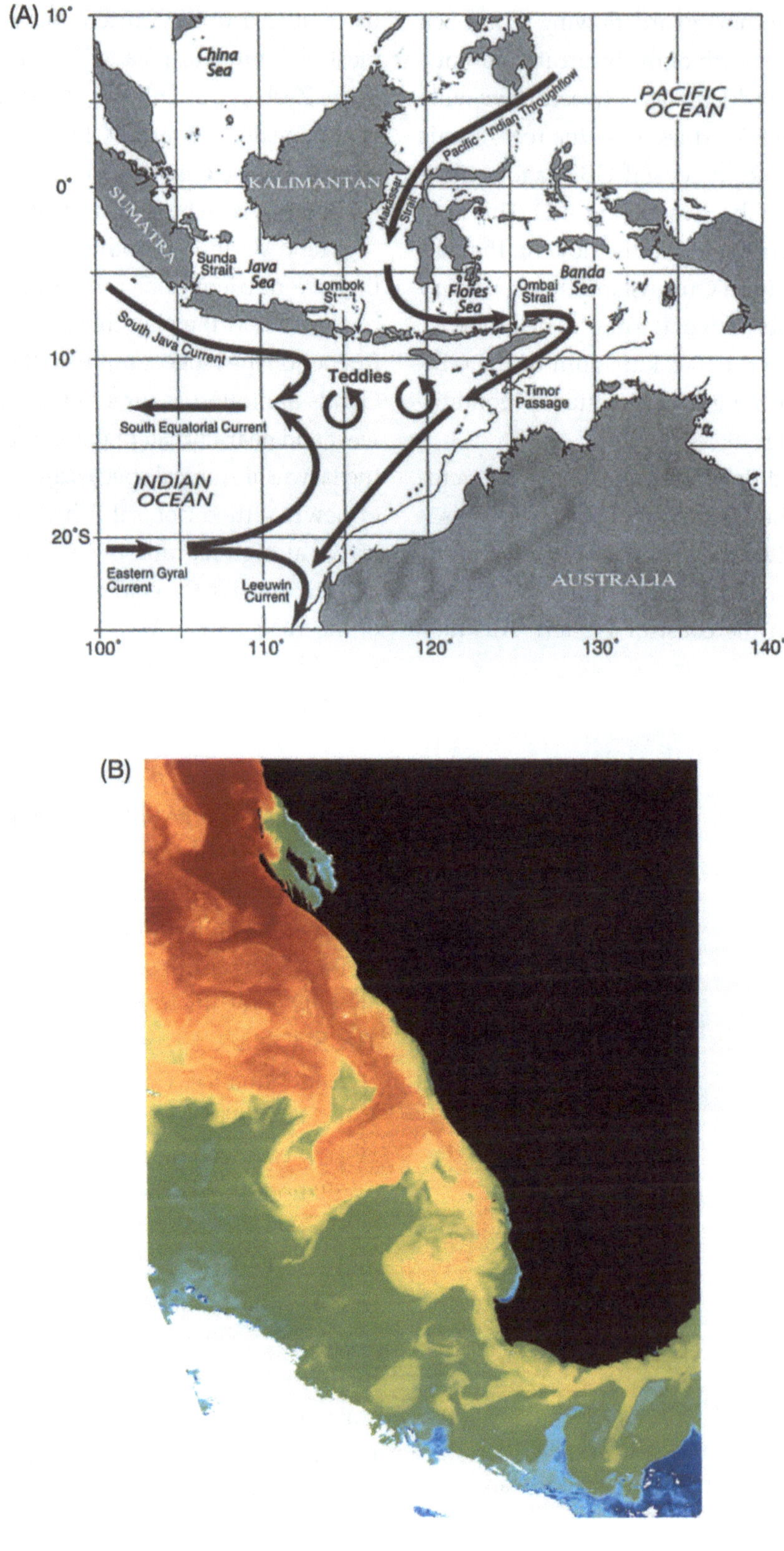

Figure 4.4 *A,* Sketch of the Indonesian Throughflow, the 'teddies' (eddies imbedded within that current), and the origin of the Leeuwin Current (reproduced from Nof *et al.* 2002). *B,* satellite image of Sea Surface Temperature showing the warm Leeuwin Current flowing southward and into the Great Australian Bight. (Images: CSIRO.)

BOX 4.2 CONSEQUENCES OF GLOBAL WARMING ON GBR OCEANOGRAPHY

Global warming will cause changes in the physical and chemical oceanography of the GBR as well as related biological change. There is great speculation on exactly what will happen, but predictions include the following: global warming is likely to cause increases in the frequency and duration of bleaching events on the GBR through warm water intrusions. Storm frequency and intensity may also increase and this will have a direct physical affect on reefs. Increased freshwater and nutrient input from cyclones would facilitate phase shifts from coral-dominated to algal-dominated inshore reefs and the nature of the pelagic environment would change as would its suitability to the survival of larvae. For example, the survival of crown-of-thorns starfish (COTS) larvae may increase, causing greater frequency of COTS outbreaks. Other forms of plankton, however, may struggle with major changes in the abundance and species richness of the plankton. An increase in atmospheric CO_2 will lower the pH of the sea so making it more acidic (0.3–0.5 of a pH unit by 2100). This is critical for corals, shellfish and some plankton (e.g. coccolithophores) because a reduction in pH affects carbonate concentration ions, making it more difficult to precipitate calcium carbonate.

world have been recorded during La Nina and El Niño, but the biggest bleaching events have been recorded during El Niño conditions (1998 and 2002). Patches of water that generally exceed 31°C stagnate on the GBR and when they remain in areas for too long the symbiotic algae of corals (zooxanthellae) are lost and in severe circumstances the coral will bleach and die.

TIDES

Tides have a great influence on currents within the GBR lagoon. The tides flow on and off the shelf, resulting in a considerable east-west movement of water. Maximum amplitudes of tides, by area, on the GBR range from 2.5 m to 7 m. The tides are generally semidiurnal (twice per day) and spring tides generate greatest currents (cf. neap tides). In regions that include the southern GBR and northern regions, tidal currents of 1 m s^{-1} are generated through channels between coral reefs. Numerical models of particle transport indicate that particles can travel about 15 km westward on incoming tides and 15 km eastward on outgoing tides. Tide and bathymetry are critical components of oceanographic models.

The interaction of tidal currents with the EAC depends on the region on the reef and distance across the shelf. Where the EAC enters the GBR in the central region and forms the lagoonal current that flows south, particles move on an east-west axis with the tide and there is some movement to the south as a result of the EAC. The influence of the CSCC (flowing north) on inter-reefal waters of the northern GBR is lower and the residence time of water masses tends to be greater in this area, therefore, the effect of tide generally dominates here.

TOPOGRAPHY

Topography has a great influence on currents of the GBR and over 2500 reefs generate great complexities in flow. Complex bathymetry and channels generate water jets, eddies and convergence zones such as thermal fronts and internal waves. All of these features have a great influence on the transport and aggregation of plankton, connectivity of larvae among reefs and the aggregated plankton is often an attractant for nekton.

■ EDDIES

Eddies vary greatly in spatial scale from small scale features that are tens to hundreds of metres wide to those that are tens of kilometres wide. Eddies form on the down current side of reefs and will influence the trajectory of particles. Eddies are three-dimensional structures with a shape like a doughnut (Figs 4.5, 4.6*A*). Particles are subducted at convergence zones around the edge of the eddy. Rotation of the eddy facilitates upwelling up through the core and sea level is lowest in the core of the eddy. These eddies are common around reefs. It takes an hour or two for them to generate after a change of the tide and particles may only do a few orbits of the doughnut before the eddy moves off with a change of the tide and dissipates. Some reefs that are in the Coral Sea have topographically stable eddies for periods of much greater than a day because the current does not change direction. Large scale eddies are also found on or near the GBR (Fig. 4.5), perhaps the most conspicuous is the eddy that is driven by the EAC and which forms in the lee of the Swains (Fig. 4.5). The EAC forms many topographically unconstrained eddies that are transported south. Eddies of all sizes are of biological important because they can retain particles (including larvae) near individual reefs or regions through cyclonic transport or through aggregation in convergence zones and they can influence local upwelling of nutrient rich water.

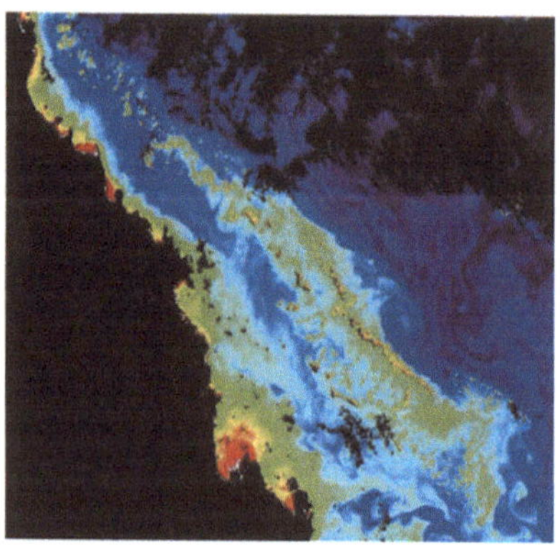

Figure 4.5 SeaWiFS satellite image of chlorophyll a concentrations on August 7, 2002, in the central and southern Great Barrier Reef, showing large eddies on the continental slope. The blue flow of the CSLC can be seen between the reef edge and the mainland (Source: NASA for MODIS ocean colour and AIMS image enhancement.)

■ JETS

Water jets through the Ribbon Reefs of the GBR on the outgoing tide and generates complex three dimensional structures that extend hundreds of metres to over a kilometre from the edge of the reef (Fig. 4.6*A–B*). Floating particles are captured and aggregated along slicks by these circulations at the convergence zone or along the leading edge of the jet. At rising tide, a bottom-tagging, cold, nutrient rich water mass is formed by a Bernouilli-effect upwelling water on the oceanic side (Fig. 4.6*C*); the primary beneficiary may be the *Halimeda* algae that form large meadows near these passages. At falling tide, the water leaving the continental shelf forms a buoyant jet that lifts off the bottom and vertically entrains deeper oceanic water (Fig. 4.6*D*); this upwelling may explain the aggregation of black marlin in front of these passages.

Buoyancy-driven flows

The flow of GBR rivers is stochastic with respect to time of year, but particularly large rain events coincide with cyclones and major weather fronts from the south that occur occasionally in some years, while other years can miss out on significant rainfall. The Burdekin River has a huge catchment area of 128 860 km^2. The river is often dry, but over short periods of time it can be Australia's largest river, in terms of volume. In some flood events up to 1690 million cubic metres per day have been recorded. These huge volumes of freshwater, and associated nutrients from the land, are transported in a plume that can extend northward for over 200 km. Plumes are significant turbidity signatures that can be viewed from land and air (Fig. 4.7). Plumes set up a three-dimensional structure that will have some influence on local flow and the transport of particles in the vicinity of plumes. These plumes are highly stratified in salinity, with fresher water on top, and the buoyancy, together with the Coriolis effect, make the plume drift northward at velocity of 0.1–0.3 m s^{-1} even against a prevailing

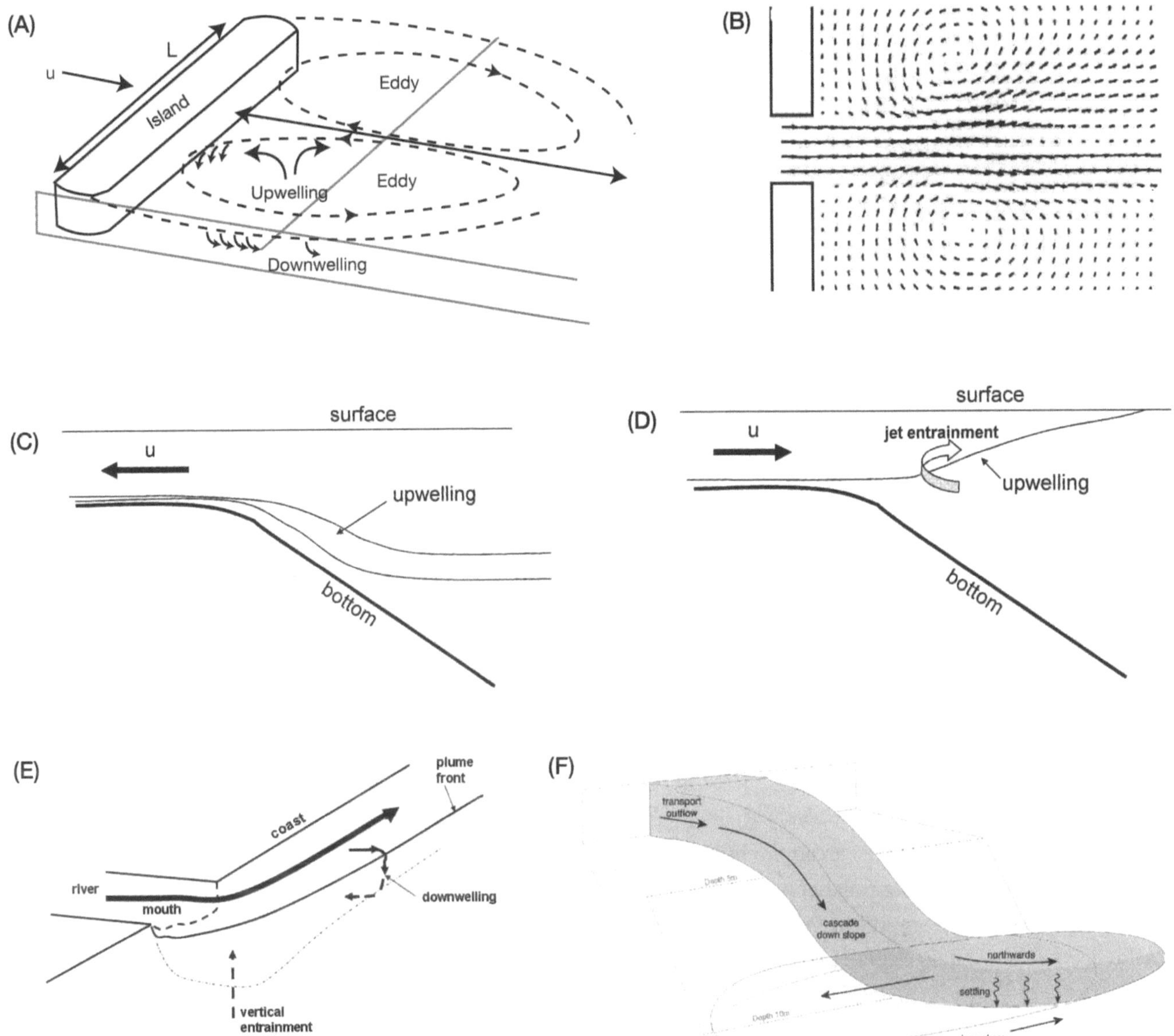

Figure 4.6 Topographic influences and density driven flows. *A*, Sketch of the 3-D circulation in an island wake (u is the undisturbed current speed upstream of the island, L is the current speed in the eddy, and W is the width of the island). (Modified from Wolanski *et al.* 1988). *B*, Model-predicted currents near a reef passage showing the formation of a mushroom jet at rising tide. (Modified from Wolanski 2001). *C*, Internal structure within the jet at flood tide showing the Bernouilli-effect upwelling of deeper, colder, nutrient-rich oceanic water into the reef passage and the jet-driven landward movement of that passage (u is the water speed in the passage). *D*, Internal structure at ebb tide showing upwelling by jet entrainment seaward of the reef passage (u is the water speed in the passage). *E*, Internal structure of a riverine plume exiting from the mouth of the river and advecting along the coast. Downwelling occurs at the front and entrainment of coastal waters under the plume. (Modified from Wolanski 2007). *F*, Density currents generated by muddy water escaping from Trinity Bay off Cairns, cascading down to deeper water and being entrained by the prevailing currents on the shelf. (Modified from Brinkman *et al.* 2004).

Figure 4.7 Turbidity plume from the Burdekin River, flood of February 2007. The plume was visible from space and was 10 km to 30 km wide from Cairns and into the Coral Sea. (Photo by M. J. Kingsford.)

southward mean oceanic current. By turbulent mixing, the width of the plume grows as the water moves northward away from the river mouth.

Strong winds entrain muddy sediment in suspension in embayments, such as Trinity Bay in the Cairns region of the GBR. When the wind calms, this water is negatively buoyant because of the sediment load; the density-driven flows makes this water cascade downward into deeper water as a bottom-tagging nepheloid layer a few metres thick, carrying the mud towards coastal and inner shelf reefs that it will help degrade (Fig. 4.6*F*).

■ CONVERGENCES

Convergences are caused by a variety of physical features that include, eddies, jets, wind, fronts, internal waves and freshwater plumes. Convergence zones accumulate plankton and floating objects such as flotsam, drift algae (see Chapter 14) and aggregations of cells (e.g. *Trichodesmium*). The convergence zones of windrows that form parallel to the direction of the wind are Langmuir cells and aggregate floatsam and organisms such as plankton, coral eggs and jellyfish. Tidally induced fronts are generated through differences in the density of waters masses, differences in depth (i.e. water will travel relatively faster in deeper water creating a shear zone or convergence), and the edge of eddies. When lagoons heat up, sea water becomes less dense and on the outgoing tide a thermal front will form where the warm water meets relatively cool inter-reefal waters. The differences in temperature can range from 0.2 °C to 1.5 °C and a convergence will form at the front. These and other tidal fronts generally dissipate with a change of the tide.

Internal waves result when there is a density difference with depth and on the GBR this is usually restricted to reefs on the outer shelf and in the Coral Sea, such as Raine Island. Internal waves commonly also form near the shelf break from tidal motions. The outgoing tide deforms the thermocline downward and as the tide changes to flood it perturbs the thermocline and propagates a packet of internal waves shoreward. Parallel convergences form at the surface over the rear of each wave and aggregated material is transported shoreward. The largest internal waves in the world (amplitude of 270 m) have been found around isolated oceanic coral reefs. The deformation of the thermocline upward over a reef combined with channeling by local topography will result in benthic organisms being bathed in cool water and this can constitute significant upwelling events (enhancing productivity) and thermal shocks (stressing organisms).

■ WIND AND UPWELLING

Wind has great influence on currents. Although tides and the EAC are important, the direction and intensity of wind also greatly affects water movement. Although transport by the wind is downwind at the surface, this is not the case through the whole water column. Transport of particles in the wind affected layer (Eckman layer) deviates to the left in the southern hemisphere with depth and more to the right in the northern hemisphere as a result of the rotation of the Earth. On average, therefore, particles will be transport toward shore in the Eckman layer with a wind from the south and offshore with a wind from the north. When surface waters are transported away from the shore, a sea level low is created which is replaced by cool water from beneath. This is one of the major mechanisms for generating upwelling of nutrient rich waters. Upwelling in the clear waters of the GBR appears to an oxymoron as tropical waters are generally thought to be oligotrophic.

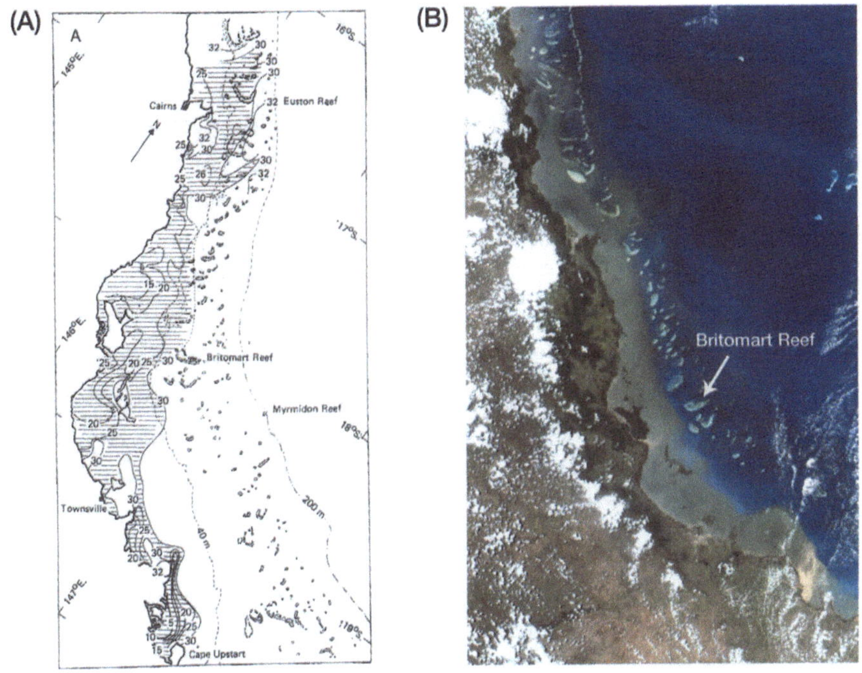

Figure 4.8 *A*, The distribution of measured surface salinity on 26–27 January, 1981, during a flood event. The flood plume moved northward from the mouth of the Burdekin River, was restricted to a coastal strip off Townsville, and extended over the entire width of the continental shelf off Cairns. A similar pattern was found in other floods in other years. (Modified from Wolanski and van Senden 1982). *B*, Satellite image of the plume from the Burdekin River (lower right) in February 2007. The plume flowed up the coast and joined the plume of the Herbert River (opposite Brittomart Reef), before it was advected about 50 km into the Coral Sea near Cairns. The 'marine snow' flocs from these plumes are very sticky and readily smother coral polyps and cause mortality of juvenile corals. (Image: CSIRO.)

However, upwelling has been documented along the shelf break of the GBR and the resultant mixing of nutrients with planktonic assemblages results in clear chlorophyll signatures that can be detected from space (Fig. 4.5). The cool water will often not reach the surface, it typically slides over the shelf as a high density water mass that remains close to the substratum; in some cases these intrusions can make it most of the way across the shelf. Although data on the periodicity of upwelling are sparse, we would predict that upwelling would be most common from November to March when winds from the north are most common.

Wind and associated wind-driven swell often has a great influence on circulation patterns in lagoons. Because waters of lagoons are often shallow (1–4 m deep) a combination of the wave set and wind will facilitate lagoon flushing as found at One Tree Island.

CONNECTIVITY

The fish, corals and other invertebrates spawn eggs and sperm at particular times of the year. The eggs are fertilised and the larvae that develop will drift with the water. The level to which reefs are self-seeded in these larvae (i.e. larvae that return to their natal reefs) and to which reefs are connected (i.e. the exchange of larvae between reefs) partly depends on the water currents during the drift period. Since these currents vary with the wind and the influence of the SEC—itself controlled by large scale oceanic features within the Coral Sea—oceanography predicts that the level of self-seeding and connectivity may vary seasonally and interannually from reef to reef. In areas where the reefs are scattered at relatively low density on the shelf, a reef is usually both self-seeded and connected with other reefs located 'upstream'. In areas where the reefs are scattered at high density, the currents are steered around the dense reef matrix instead of flowing through the reef matrix, akin to a sticky water effect; reefs within the dense matrix are highly self-seeded and little connected with reefs outside. Recent evidence indicates that self seeding by recruit fishes can be as high as 30% to 60% and it is likely that the behavioural abilities of presettlement fishes as well as current influence connectivity.

FRESHWATER INPUT, CYCLONES AND MUD

Riverine input to the GBR is substantial and it has its greatest impact near-shore (within 10 km of the mainland) and in and around what would be considered 'inner' reefs of the GBR. The river plumes also impact mid-shelf and outer reefs in the Cairns area (Fig. 4.7 and 4.8). Patterns of rainfall vary along the GBR. Largest rainfall occurs north of Hinchinbrook Island (near Tully). There is a distinct seasonal pattern to rainfall where there is a rainy season that extends from *c.* December to April. Major perennial rivers such as the Normandy and Daintree are in this area (see Chapter 11).

Muddy marine snow flocs as large as several millimetres are formed and carried in suspension by the aggregation of mucus and mud, and they support intense biological activity. The mucus is formed by bacteria and plankton, and this is enhanced by nutrients from land runoff. The mud is brought in from river plumes, wave resuspension and bottom-tagging nepheloid layers. These aggregates settle in quiescent areas in calm weather, as controlled by the oceanography, and being very sticky can readily smother coral polyps and organisms living on coral reefs. This mud can later be resuspended by waves and is carried away but the reef damage has been done in the meantime.

Freshwater input has a great influence on the biology of inshore waters. In many parts of the world it has been demonstrated that catch and recruitment rates of prawns and fishes vary with freshwater input. Recruitment rates usually go up with input of freshwater. Changes in freshwater runoff may happen through anthropogenic alteration of catchments and through changes in rainfall that relate to global warming (see Box 4.2). There is concern that an increase in runoff and nutrients could affect inner shelf reefs through phase shifts (i.e. from coral to algae) and the survival of crown-of-thorns starfish (COTS) larvae.

The wet season is the same as the cyclone season and river floods often result from cyclones. Cyclones can thus be very destructive because of the huge input of sediment and freshwater input to the GBR and also of wave height on reefs. Physical destruction can be considerable and the swath of damage is asymmetrical with respect to the position of the eye of the cyclone. Cyclones rotate clockwise in the southern hemisphere and anticlockwise in the northern hemisphere as typhoons (Asia) or hurricanes (America). The impact of cyclones on the GBR is greater on the southern side of the eye due to the greater fetch from the open ocean and the larger wave height. The destruction of reef habitat and death of organisms can be substantial. Storm swell can blast large chunks of coral (including large *Porites*) onto the reef flat. Damage will usually only be to the windward side of reefs, but life on coastal fringing reefs can be all but obliterated. Great changes in habitat type (e.g. from a species rich assemblage of live coral to coral rubble) will have a great influence on local species diversity of most taxa (see Chapter 5).

ADDITIONAL READING

Biological oceanography

Burgess, S. C., Kingsford, M. J., and Black, K. P. (2007). Influence of tidal eddies and wind on the distribution of presettlement fishes around One Tree Island, Great Barrier Reef. *Marine Ecology Progress Series* **341**, 233–242.

Glynn, W. (1988). El Nino-Southern Oscillation 1982–1983: Nearshore population, and ecosystem responses. *Annual Review of Ecology and Systematics* **19**, 309–346.

Kingsford, M. J., Wolanski, E., and Choat, J. H. (1991). Influence of tidally induced fronts and Langmuir circulations on distribution and movements of presettlement fishes around a coral reef. *Marine Biology* **109**, 167–180.

Mann, K. H., and Lazier, J. R. N. (1991). 'Dynamics of Marine Ecosystems: Biological-physical Interactions in the Ocean.' (Blackwell: Oxford.)

Suthers, I., Taggart, C. T., Kelley, D., Rissik, D., and Middleton, J. H. (2004). Entrainment and advection in an island's tidal wake, as revealed by light attenuance, zooplankton, and ichthyoplankton. *Limnology and Oceanography* **49**(1), 283–296.

Great Barrier Reef

Andrews, J. C., and Gentien, P. (1982). Upwelling as a source of nutrients for the Great Barrier Reef ecosystems: A solution to Darwin's question. *Marine Ecology Progress Series* **8**, 257–269.

Brinkman, R., Wolanski, E., Deleersnijder, E., McAllister, F., and Skirving, W. (2002). Mass flux from the Coral Sea into the Great Barrier Reef. *Estuarine, Coastal and Shelf Science* **54**, 655–668.

Burrage, D., Steinberg, C. R., Skirvig, W. J., and Kleypas, J. A. (1996). Mesoscale circulation features of the Great Barrier Reef Region Inferred from NOAA Satellite Imagery. *Remote Sensing of Environment* **56**, 21–41.

Furnas, M. J. (2003). 'Catchments and Corals: Terrestrial Runoff to the Great Barrier Reef.' (Australian Institute of Marine Science: Townsville.)

Nof, D., Pichevin, T., and Sprintall, J. (2002). Teddies and the origin of the Leeuwin Current. *Journal of Physical Oceanography* **32**, 2571–2588.

Wolanski, E. (2001). 'Oceanographic Processes of Coral Reefs: Physical and Biological Links in the Great Barrier Reef.' (Boca Raton: CRC Press.)

See website for an extended list, with updates: http://www.australiancoralreefsociety.org

5. Coral Reef Habitats and Assemblages

C. Syms & M. J. Kingsford

CORAL REEFS AS HABITATS

Coral reefs often characterise the immediate subtidal environment in tropical regions. The definition of a coral reef is a loose one. In one sense, it is simply a hard substratum that supports corals. Many geologists, however, restrict the term to large subtidal hard bottoms ('reefs') in which the hard base (that may be 1–2 km thick on some reefs) consists mostly of dead corals, generated by the death and assimilation of the carbonates fixed by coral polyps into the hard substratum, usually beneath a living veneer of live corals. This process is unique to reef-building or hermatypic corals, which contain symbiotic algae or zooxanthellae, and typically occur in tropical regions in which the average winter water temperature is greater than 18°C. However, corals themselves, and the organisms associated with them, are not constrained to reefs generated in this way. On the Great Barrier Reef (GBR) corals occur both on carbonate reefs, as geologists define coral reefs, and on geologically-formed rock bases such as granite. These rock bases occur on the mainland and on islands in the centre of the GBR itself. In these areas corals simply form a structural veneer, contributing physical and biological structure to the underlying hard rock. On volcanic-origin islands, particularly in the south (to about 30°S), scleractinian corals may also be found mixed with tufting and laminarian algae (groups more characteristic of temperate reefs). The organisms that occupy carbonate and non-carbonate reefs are the same, and interact with the reef and other organisms in the same way, consequently we include both carbonate and non-carbonate reefs under the broad umbrella of 'coral reefs'. This imprecise definition is unfortunate, but necessary. Geologists have a clear definition of a 'coral reef' as a geomorphological structure built by corals. It can also be argued that the structural veneer—the coral communities themselves—are the central component of coral reefs. However, not all organisms on a coral reef are positively associated with corals. A reef community is more than its physiographic origin and dominant space-occupiers. A key to the diversity of a coral reef lies in its habitat heterogeneity, which of course includes areas that are not occupied by corals themselves.

Although corals have an important geological and environmental role in fixing carbonates and generating hard structure in tropical marine systems, in this chapter we will discuss the role of coral reefs as habitats for other organisms and their importance within the broader tropical subtidal landscape (or maybe, more appropriately, seascape). We will not restrict our discussion to particular taxa, but rather use occasional examples to illustrate general points about coral reef habitats and habitat associations. Although some authors have lumped the entire range of environmental factors such as temperature, water quality and so forth into the definition of an organism's 'habitat', we instead will use a more literal definition of habitat from the latin *habitare* 'the place where something lives' (see Box 5.1). So 'habitat' under this definition becomes a spatial concept, and reflects an emphasis on habitat as

BOX 5.1 GLOSSARY

Assemblage. A set of the organisms that occur at any defined level of habitat. Occasionally used interchangeably with 'community', the complete set of ecologically interacting species.

Facultative habitat associate. A species usually found in a particular habitat patch type, but which can survive in other habitats, such as the damselfish *Pomacentrus moluccensis*.

Habitat. In a strict sense, the place where an organism lives. Usually considered at one of several spatial scales.

Habitat determiner. An organism that interacts with, and alters the impact of, a habitat former. Crown-of-thorns starfish are habitat determiners. Their predation on corals may exert strong effects on those species reliant on the habitat structure or heterogeneity provided by the corals.

Habitat former. An organism that generates a physical structure that forms a habitat for another organism. Tabulate corals such as *Acropora hyacinthus* are habitat-formers.

Habitat generalist. A species that occurs across a wide range of habitat patch types.

Habitat homogeneity. The degree to which constituent areas forming a habitat mosaic, habitat zone, or landscape are similar in type or distribution.

Habitat interface. The junction between different adjacent habitat patches.

Habitat mosaic. A spatially cohesive collection of habitat patches.

Habitat patch. A discrete, internally homogenous uniform area with a characteristic physical structure or biotic composition.

Habitat responder. An organism that simply occupies habitat, without forming habitat for others or altering the action of a habitat former.

Habitat zone. A habitat mosaic with a recognisable physiographic and environmental setting.

Landscape/seascape. A spatially cohesive collection of mosaics or habitat zones. Landscapes (or seascapes) are not easily defined and are perceived at a wide range of spatial and temporal scales by marine organisms. While many coral reef organisms are sedentary and restricted in their ability to move between habitat patches, during their larval phase they may sample the environment at much larger spatial scales.

Microhabitat. A characteristic position within a habitat patch.

Obligative habitat associate. A species that is only found in a particular habitat patch type, such as gobies of the genus *Gobiodon*.

being what an organism perceives within its range of movement and response. We will start by considering organisms in the context of their interactions with their habitat then introduce the concepts of temporal and spatial variability in generating habitat heterogeneity. Then we shall consider how defining habitat independently of a particular organism or set of organisms is rather challenging, and conclude with some general comments on the importance of habitat heterogeneity in maintaining biodiversity on the GBR.

HABITAT FORMERS, DETERMINERS, AND RESPONDERS

An important distinction between temperate rocky reefs and coral reefs is that living organisms (e.g. hermatypic corals, and octocorals) generate the primary hard superficial benthic structure in coral reefs. While temperate macroalgae such as giant kelps, like *Macrocystis pyrifera*, also generate habitat structure by extending physical complexity into the water column, corals contribute to complexity whether dead or alive. Corals can be considered as *habitat formers*. This biological structuring has several important consequences for the ecology of coral reefs. Coral growth forms vary greatly from one species to another, and even within the same species under different environmental conditions. Growth forms range from featureless encrusting forms, to massive corals with large physical mass but little small scale structure, through to branching forms with complex structure and interstices that can provide shelter to a large number of species, and large numbers of individuals within a species (Fig. 5.1). Consequently, the coral species composition and environmental conditions at a site can, to a large extent, determine the habitat structure. Corals also exist in a range of assemblage types. Monospecific stands of corals will generate habitats with a very similar structure across their extent, whereas mixed-species stands will generate more variable habitats. In addition, not all hard bottom in a coral reef is covered by corals, so further habitat heterogeneity is generated by rock, rubble, and sand substrata, and other habitat formers such as soft corals.

Because corals are living organisms, they are subject to a range of ecological processes that alter their ability to survive and grow at a site. Corals may be outcompeted by other species, their reproductive success is temporally variable, and they may be stressed by poor environmental conditions. Corals are also the prey for other organisms. Crown-of-thorns starfish (COTS) and *Drupella* whelks can kill entire colonies and, in the case of COTS, entire communities across large swaths of reefs. Following death, the physical structure provided by the corals quickly erodes and this reduces the complexity of the physical habitat. Species like crown-of thorns starfish are *habitat determiners*. Their ecological role alters the physical structure of the habitat. In this way, a feedback loop of ecological interactions occurs. Habitat determiners are

Figure 5.1 Hard corals are important habitat formers. Coral growth structures range from featureless encrusting forms, to massive corals with large physical mass but little small-scale structure (*A*) through to tabular (*B*) and branching (*C*) forms. Even within the branching forms, the degree of branching may vary, generating a range of different potential shelter sites. This structure remains intact even after death, until eroded away by biological and physical processes. Eroded corals then become carbonate sand, which in turn eventually provides habitat for a new range of species. (Photos: Australian Institute of Marine Science.)

attracted to habitats of a certain type, but then alter that habitat type and may move on. Habitat determiners, however, need not necessarily be destructive. For example, urchins, such as *Diadema*, graze algae, and surgeonfishes, such as *Ctenochaetus striatus*, remove detritus from the reef top and by doing so both species help maintain space that is suitable for coral recruitment.

Coral reefs are important because of the organisms or *habitat responders* that live on them. Although fishes are the most obvious occupants of coral reefs, a vast array of organisms from a wide range of phyla including molluscs, worms and so forth make up tropical reef biodiversity. Associations between organisms and the coral reef habitat are complex. Some organisms, such as coral gobies (*Gobiodon*, *Paragobiodon*) are very specialised in their habitat association and will only be found in a few species of hard corals (Fig. 5.2*A*, *B*). They are also obligate specialists—they never occur on a reef if their preferred coral is not found there. Most coral reef habitat responders are not obligate specialists, however. Many fishes, for example the lemon damsel *Pomacentrus moluccensis*, are commonly associated with particular habitats such as branching hard corals (Fig. 5.2*C*, *D*), but they are not restricted to particular species of coral, and indeed may be found in habitats that are devoid of live coral. Other fish species may be found in reef habitats irrespective of live coral, such as many wrasses (e.g. *Cheilinus digramma*, Fig. 5.2*E*, *F*). Measuring the predictability of the composition of habitat-responding communities, given habitat availability, is fraught with difficulties. Various studies have reported a difference of anywhere between 5–70% coral cover is required to detect changes in abundance of habitat responders such as fishes. A large amount of the variability in estimates may be due to methodological differences. This is an issue that needs to be addressed, but it appears that coral cover alone does not sufficiently explain patterns of variability in many habitat responders.

Figure 5.2 Habitat specialisation occurs at a range of spatial scales. Obligate specialists such as the goby *Gobiodon citrinus* (*A*) occur on a specific range of hard branching coral species (*B*). They are never found in the absence of these coral species. Facultative specialists such as the damselfish *Pomacentrus moluccensis* (*C*) are found on a range of hard coral species and growth forms (*D*), and may even be found in areas devoid of hard corals. Generalists such as the wrasse *Cheilinus digramma* (*E*) are not typically associated with any particular coral type, but are generally found in areas that contain a range of different structural habitats (*F*). (Photos: P. Munday (A); Great Barrier Reef Marine Park Authority (C–E); Australian Institute of Marine Science (F).)

■ SPATIAL VARIABILITY IN CORAL REEF HABITATS

It is important to realise that coral reefs are not simply homogenous areas of corals on which organisms live. Spatial variability occurs at a range of scales (Fig. 5.3). At the finest scale, relatively homogenous areas such as single coral heads, rubble patches, or algal covered rock, form *habitat patches* at scales from centimetres to metres (Box 5.1). These habitat patches may indeed form the entire habitat of extreme specialists. Coral gobies, for example, are restricted to a limited species range of live coral heads and they will not move from a single coral head. For many organisms though, the habitat patch is not necessarily the smallest habitat unit. In a patch of staghorn coral, for example, some species may live at the base of the coral whereas others may live near the top. In this way, habitat patches can also be subdivided in a range of *microhabitats*. The microhabitat need not be completely contained within a single habitat patch. Some species use microhabitats at the transition of different patches. Physical, temporal, and biological variability generates heterogeneity in the environment at larger scales of tens to hundreds of metres. On coral reefs this is often termed a *habitat zone*, that is, a collection or *mosaic* of different habitat patches, and is usually characterised by a combination of biotic, physical, and physiographic factors. On the GBR the seaward side of reefs (facing the continental shelf break) typically have a shallow reef crest, which may also be present or absent on the leeward side of reefs. This is a shallow zone in which coral growth rates are high, but the corals are also subjected to wave action and currents (Fig. 5.3). The reef crest will consist of stands of different coral species, interspersed with rock covered with encrusting algae.

The reef crest sharply breaks to a reef slope, which is generally steeper, with a stronger depth gradient on seaward sides of reefs that are exposed to the prevailing winds, and hence have much more coral growth with well developed reef crests and slopes (but also subject to more wave disturbance). This reef slope is not just a continuous band of corals, however. On exposed reefs in particular, the reef slope is punctuated by grooves in which coral rubble accumulates as a result of wave disturbance and rips. On the leeward side of the reef crest, a shallow reef-flat zone often containing rubble, sand, and fleshy macroalgae such as *Padina* or *Sargassum* species might occur. This zone is in very shallow water, exposed to wave action, with little vertical physical structure and heterogeneity. On reefs with lagoons, the reef-flat zone will give way to a lagoonal habitat zone, relatively sheltered from wave disturbance. The lagoon may often be very coral-rich in clear lagoons, but possibly devoid of corals in turbid lagoons. In the centre of the lagoon fine sediments occur, providing yet another set of microhabitats. Large *Porites* colonies (metres wide and high) are rare on the exposed sides of reefs, but are often common on the leeward sides of reefs and in lagoons. On the leeward side of a typical reef, the lagoon will give way to a back-reef habitat zone. The back-reef is often rich in corals because visibility is good; not as good as the exposed side, but wave disturbance is much lower. This collection of habitat zones forms a *landscape* or *seascape* within which biodiversity exists on a single coral reef and while there is considerable small scale spatial variability in habitats, at the spatial scale of 100's of metres to kilometres the physical structure of the reef is relatively unchanging and predictable at ecological time scales and the spatial distribution of habitat responders across a reef landscape becomes very predictable.

Clearly the integration of habitat patches into zones in the reef landscape generates an enormous range of different types of places for different types of organisms to live, even taxonomically related species. And while these organisms can be characterised as 'coral reef' species, they may not actually be associated with corals themselves. However, the reefs themselves form part of a larger seascape of multiple reefs that constitute the GBR (Fig. 5.3).

At this scale, geological, oceanographic, geographic, environmental, and biogeographic patterns exert additional effects on the individual coral reefs that make up the GBR (see Chapter 2). *Cross shelf* variability at scales of tens of kilometres are associated with large differences in turbidity, wave exposure, and nutrient load. Physiographic, oceanographic, and current regimes will be different at different latitudes at scales of hundreds of kilometres due to a combination of continent

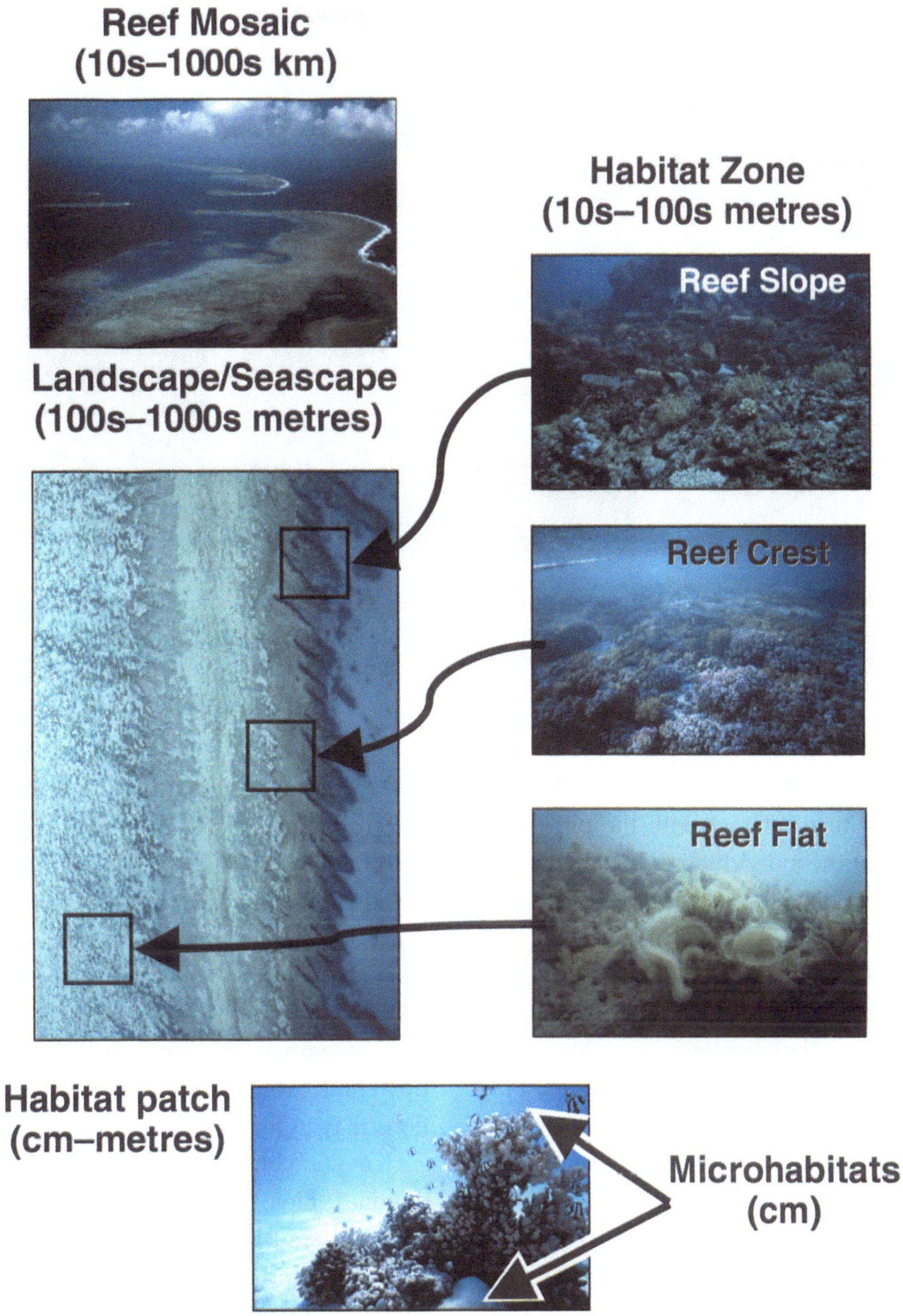

Figure 5.3 Habitat heterogeneity occurs at a range of spatial scales. Individual coral stands, rubble areas and sandy areas, among others, form habitat patches. Within patches, however, the physical structure may not be homogenous so finer-scale microhabitats such as the base versus the top of a coral head may exist. These habitat patches themselves form mosaics that typically occur in physiographic zones with characteristic physical conditions such as aspect to wave exposure. Depth is an important factor in zonation, and even with the same aspect over a vertical range of a few metres, reef tops and reef crests may have a very different appearance and species composition than reef slopes. These habitat zones are part of a wider landscape of different coral and non-coral habitat types at the scale of 100's of metres, and generate heterogeneity at the scale of a single reef. The reefs themselves form a landscape or seascape that varies at scales of 10's of kilometres across the shelf, and 100's to 1000's of kilometres along the GBR. (Photos: Great Barrier Reef Marine Park Authority.)

Figure 5.4 Different scales of coral disturbance. Parrotfishes (*A*) can generate local damage to hard corals (*B*) but rarely at large scales. Crown-of-thorns starfish outbreaks (*C*) can generate widespread damage to hard corals (*D*), leaving dead corals that eventually erode to algal-covered rubble. Cyclones (*E*) can generate widespread damage to large areas of multiple reefs (*F*). (Photos: Australian Institute of Marine Science (A–C); Great Barrier Reef Marine Park Authority (D–F).)

scale current patterns (see Chapter 4), and smaller scale currents, which in turn interact with the profile and relative positions of the individual reefs along the continental shelf. These have important consequences for larval transport, and ultimately biological connectivity between the organisms on different reefs. The GBR crosses a large latitudinal range of approximately 14°52′, which means that temperature tolerances of many species may exert large effects on the assemblage composition of southern *v.* northern reefs, just as the large width of the continental shelf exerts effects due to sedimentation and exposure regimes. This wide range of spatial variability in habitat availability has profound effects on the habitat formers, determiners, and responders that live on the reefs. The 'typical' zonation patterns described previously may not occur on all GBR reefs, so there is also an additional level of heterogeneity at the reef level.

TEMPORAL CHANGES IN CORAL REEF HABITATS

Coral reefs, as with most habitats, change through time. Coral reefs are subject to a wide range of natural and anthropogenic perturbations, which can alter their structure and heterogeneity (Fig. 5.4). At large spatial scales, the Pacific Decadal Oscillation can generate oceanwide changes in temperature and primary productivity regimes, which may operate for many decades. At slightly shorter temporal scales, El Niño Southern Oscillations may generate oceanwide fluctuations in temperature, wind strength, and oceanographic upwelling over time periods of 3–7 years. These fluctuations may in turn contribute to large scale disturbances such as coral bleaching during warm water events, and cyclones. Biological disturbances may also operate at large scales, such as parrotfish feeding-scars (Fig. 5.4*A*, *B*). Crown-of-thorns starfish (COTS) outbreaks can kill corals on vast areas of reef (Fig. 5.4*C*, *D*). Large scale perturbations typically take a long time to recover, and if they occur together (e.g. coral bleaching and COTS outbreaks) their combined effect may further increase recovery time. Physical conditions will also facilitate biological disturbances. Particular combinations of suitable oceanographic conditions may result in increased survival rates of COTS through increased larval nutrition and condition and dispersal to suitable habitat. Conversely, some environmental conditions may favour survival of pathogens, which may affect habitat determiners such as *Diadema* urchins. The interaction between physical and biological processes is important because of the different scales of each; physical processes, which usually occur at a large scale, may effectively force biological processes to occur at a similar scale. This spatial autocorrelation (i.e. the abundance or ecological interactions of organisms are more similar at locations closer to each other) has been termed the Moran Effect. The consequence of this is that minor biological effects that would appear only to exert local effects can, if generated by physical forcing, occur over a much larger spatial extent than would be predicted.

Small scale disturbances may also have important and immediate consequences for habitat structure. Storm damage can range from localised damage to individual coral heads, up to complete coral removal across 100 m swaths within windward sides of reefs, and across several reefs (Fig. 5.4*E, F*). These types of disturbances are not necessarily detrimental to a reef's overall health. Removal of a coral colony by storm damage frees up space for recolonisation by other species. It also generates heterogeneity in the substratum, so habitat responders that prefer bare rock or rubble can occupy the space that is cleared by the disturbance. In this way, local disturbances within a coral reef can actually increase species diversity. In low disturbance regimes competitively dominant species, which may be fast growing in disturbed habitats, large, and can overgrow other species, will form large monospecific stands. The diversity of corals in these stands, and habitat responders will be low. In contrast, highly disturbed areas will have a low coral diversity, and the substratum will consist primarily of encrusting algal and coral forms. These organisms provide little structural heterogeneity, and consequently species diversity in highly disturbed habitats will be low. However, at intermediate levels of disturbance, the abundance and cover of competitively dominant species is reduced due to their higher susceptibility to disturbance, which in turn frees up space for other coral species. At intermediate disturbance levels, coral diversity increases. The relationship between coral diversity and disturbance is reflected in the diversity of many habitat responders (Fig. 5.5*D, E*). Fish diversity, for example, tracks coral diversity with the highest species richness found at intermediate coral densities that in turn are associated with higher levels of habitat heterogeneity. In contrast with the competitive-disturbance mechanism generating coral diversity, the tracking of coral diversity by habitat responders is usually not due to competition, but rather the provision of a range of habitat types, which enables a range of species with different habitat preferences to occupy the same area.

Although natural disturbance is a normal part of coral reef ecology, and an important process that generates habitat variability, anthropogenic disturbances also exert effects at a range of spatial and temporal scales. Boat anchors may generate local coral damage. Artisanal fishing practices such as netting, trampling, and rock-throwing (although not a major influence in the GBR) may generate local coral damage in reef systems. Commercial fish extraction practices such as cyanide and explosives, again not common on the GBR but widespread in other parts of the world, may generate more widespread physical damage. Extraction of certain fish species may itself have biologically mediated effects. For example extraction of herbivorous fishes such as parrotfishes (Scaridae) may lead to increased growth of macroalgae, which may in turn lead to lower coral cover and, importantly, reduce the ability of corals to recruit onto bare space. At larger scales, increased sedimentation and eutrophication may result from land-based human practices such as deforestation, agriculture, and sewage outfalls. The resulting changes in water quality are usually long-lived, and may alter the trophic regime, light levels, and increase stress to corals via sedimentation. These stresses may in turn lead to increased susceptibility of corals to environmental fluctuations such as temperature changes (see Chapter 9).

Natural and anthropogenic temporal changes generate habitat heterogeneity by generating space that is devoid of corals. The form of disturbance is important in determining what the reef mosaic will look like. Moderate disturbances, which fragment large coral colonies will reduce the amount of physical structure but as many corals can grow from fragments, this may not result in much loss of coral cover. Increased levels of physical disturbance may remove and kill entire colonies, and thus provide bare space. This space will

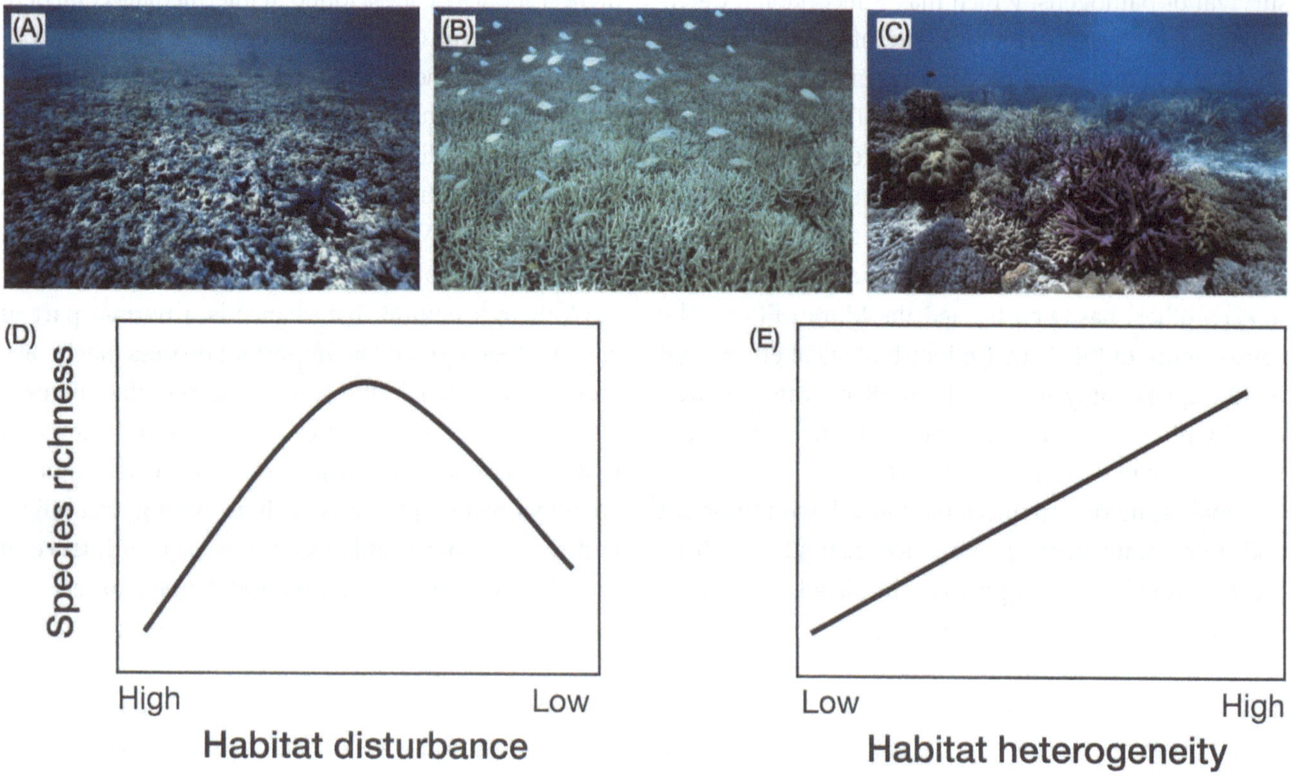

Figure 5.5 Coral cover alone does not determine diversity of a coral reef assemblage. *A*, Bare rock or rubble habitats may contain only a few species. Monospecific stands of coral cover (*B*) will contain only a few coral specialist and coral generalist species, but varied stands of coral interspersed with some bare space and rubble (*C*) will contain a range of coral specialist and generalist species, in addition to rubble/rock specialists, patch-interface specialists, and generalists. Species richness, therefore, is highest at intermediate levels of coral cover (*D*), where habitat heterogeneity is highest (*E*). (Photos: Great Barrier Reef Marine Park Authority.)

undergo a natural succession of community states in which the space is colonised by encrusting coralline algae, which then provides a suitable substratum for coral settlement. Rapidly growing forms, such as branching and plating corals, then occupy this space but may be supplanted over time by more wave resistant forms. Extreme physical disturbances or widespread coral death may result in a reef being covered by coral rubble. Whether the rubble forms a substratum for coral or algal communities may be dictated by the numbers of herbivores on the reef. Exclusion of herbivores can result in algal takeovers of areas previously occupied by corals and fleshy and tufting algae such as *Padina* can occupy space following COTS outbreaks or bleaching. It is important to realise that coral reefs are not static, stable, unchanging ecosystems. The abundances of coral reef organisms change at temporal scales ranging from days to multiple decades, and at spatial scales ranging from a single coral head to entire reefs in which their preferred habitat occurs. Species that are not as habitat-specific will be found on a wider range of reefs. So, a reef with a diverse range of corals will tend to have more species, both specialists and generalists. However, not all species are associated with corals. Some species prefer areas without corals, or the interfaces between different patches. Consequently, species diversity may actually be higher in areas where coral cover lies at an intermediate value. Pelagic species may occur around coral reefs simply because they form the only physical structure in the ocean, which in turn may provide shelter and alter water currents that, in combination may attract prey and provide suitable spawning sites. If alternative structures are available, these species may occupy them irrespective of whether they were generated by, or covered by corals. These different temporal and spatial scales of habitat variability combine synergistically to provide an enormous range of different types of potential habitat for a huge diversity of organisms on the GBR.

CONSEQUENCES OF HABITAT VARIATION AND HABITAT ASSOCIATION FOR CORAL REEF ASSEMBLAGES

We defined *habitat* as the place where an organism lives, yet discussed habitat patches and zones (e.g. reef slope, rubble grooves) relatively independently of any particular organism. This is because defining a habitat based on where an organism lives introduces a curious circularity in its definition. For example, the damselfish *Pomacentrus coelestis* is primarily found living on rubble. This introduces a problem because it implies there could be 'empty' habitats in the environment: if a patch of rubble doesn't contain *P. coelestis* then is the patch suitable habitat for the species? So the challenge lies in defining habitats even though they may be unoccupied. Of course in the example above, this is not too difficult if we view suitable habitat as a probability of occupation, rather than strict occupancy of any single particular piece of space. While there is spatial and temporal variability in habitats, there is also greater spatial and temporal variability in species abundance. In general, rubble patches on reefs contain *P. coelestis*, and conversely *P. coelestis* is usually found on rubble patches. The abundances of course may vary on leeward *v.* seaward sides of the reef, across the continental shelf, and latitudinally because of increasingly larger scales of environmental variation. This introduces another problem though: if habitat is the place where an organism generally lives, then we need to identify the scales at which an organism perceives and responds to the habitat characteristics, and also identify what the important characteristics are.

Determining what exactly is important about the habitat to an organism can be surprisingly difficult for some species. In an obligate species such as a coral goby or coral crab, it is easy—a particular species of coral needs to be present. For species with a wider range of movement such as wrasses, then simple correlations between species abundance and different types of coral cover can be misleading. For example, many fish species appear to be positively correlated with soft coral cover, but removal of soft corals at small scales may result in no change to the fish assemblage. The apparent correlation may simply be due to wrasses preferring habitats within a diverse range of substratum types, coupled with soft corals also being positively correlated with a diverse range of substratum types. Prior knowledge of the habitat characteristics that are actually important to organisms is essential for identifying and predicting responses of organisms to habitat changes.

Identifying coral reef habitats, environments and cross shelf and latitudinal gradients is important both for pure ecological studies, and management of the GBR. Many ecological questions and management solutions require the habitats of a range of organisms to be identified or protected, so measuring habitat diversity is an important task because it is closely linked to species diversity. This is often known as a *community*-level approach, although a community in a strict sense is a complete suite of interacting species. On coral reefs 'resident' species also interact with non-resident species. For example planktivores eat pelagic species, and pelagic fishes such as jacks (Carangidae) may eat resident fishes despite not being considered 'coral reef' fishes themselves. Additionally, some fish species such as some wrasses feed on hard reef and soft bottom substratum, thus linking the hard and soft bottom 'communities' energetically. In reality, the community-level approach is really an *assemblage*-level approach, in which we consider a co-occurring subset of the entire range of interacting species, so we will use the word assemblage in preference to community. Measuring or managing the habitats of a wide range of species raises some interesting problems. First, if habitat can only be defined as the place where an organism lives, do we need to identify the typical range at which each of the organisms in the community perceive their habitat? Obviously we could not do this easily with such a diverse flora and fauna, with each species interacting at different scales with their habitat and each other. While this may be an important nuisance to coral reef ecologists, for GBR managers the solution is rather easier. Simply recognising habitat as a spatial concept, and recognising that reefs vary along the GBR, a spatial management protocol can use an 'umbrella' approach. If large spatial units (e.g. individual reefs) can be managed or protected from some type of anthropogenic pressure, then all habitats contained within that spatial unit will be automatically protected as well. In addition, if these units can be placed along all axes of the main environmental gradients such as latitude and shelf position, then in all probability this protection will extend over the entire range of habitats without really needing to know exactly how to define habitats for each species, or even know which habitats occur on each reef. Such an approach was used in the zoning of the GBR in which large regional areas with clusters of similar habitats ('bioregions') were identified and different zoning levels applied to protect representative habitats across the reef (see Chapter 12).

To summarise: on coral reefs living organisms contribute to the habitat structure of a vast array of organisms, and so ecological processes such as competition, predation, and biological disturbance interact with physical processes such as water movement and physical disturbance to alter the physical structure of the habitat. In this way, local biological processes, perhaps correlated over large spatial and temporal scales, can exert very strong effects on those organisms that respond to the habitat-forming organisms. The diversity of habitat-responding organisms is not restricted to places in which the habitat-former is present; many 'coral reef' organisms are not associated with corals themselves, but rather the habitat heterogeneity they provide. Temporal and spatial variability in habitat structure generates a range of opportunities for habitat responding organisms to coexist at a range of spatial scales. Although defining habitats for individuals and assemblages is problematic, recognising that habitat is a spatial concept means that management of habitats can be a relatively simple process, as long as big enough areas of space are controlled. This diversity of places to live, in combination with the vast array of organisms that are available and able to live in these places undoubtedly combine to maintain biodiversity on the GBR.

ADDITIONAL READING

Connell, J. H. (1978). Diversity in tropical rain forests and coral reefs. *Science* **199**: 1302–1310.

Done, T. J. (1992). Effects of tropical cyclone waves on ecological and geomorphological structures on the Great Barrier Reef. *Continental Shelf Research* **12**: 859–872.

Halford, A., Cheal A. J., Ryan, D., and Williams D. M. (2004). Resilience to large-scale disturbance in coral and fish assemblages on the Great Barrier Reef. *Ecology* **85**: 1892–1905.

Hughes, T. P., Rodrigues, M. J., Bellwood, D. R., Ceccarelli, D., Hoegh-Guldberg, O., McCook, L., Steneck, R. S., and B. Willis. (2007). Phase shifts, herbivory, and the resilience of coral reefs to climate change. *Current Biology* **17**: 360–365.

Jones, G. P., and Andrew, N. L. (1993). Temperate reefs and the scope of seascape ecology. In 'Proceedings of the Second International Temperate Reef Symposium, NIWA Marine, Wellington, New Zealand'. (Eds C. N. Battershill, D. R. Schiel, G. P. Jones, R. G. Creese, and A. B. MacDiarmid.) pp. 63–76. (NIWA Marine: Wellington.)

Jones, G. P., and Syms, C. (1998). Disturbance, habitat structure and the ecology of fishes on coral reefs. *Australian Journal of Ecology* **23**: 287–297.

Jones, G. P., McCormick, M. I., Srinivasan, M., and Eagle, J. V. (2004). Coral decline threatens fish biodiversity in marine reserves. *Proceedings of the National Academy of Sciences of the United States of America* **101**: 8251–8253.

Syms, C., and Jones, G. P. (2001). Soft corals exert no direct effects on coral reef fish assemblages. *Oecologia* **127**: 560–571.

See website for an extended list, with updates: http://www.australiancoralreefsociety.org

6. Seabed Environments, Habitats and Biological Assemblages

C. R. Pitcher, P. J. Doherty & T. J. Anderson

The modern Great Barrier Reef rises from a shallow (0–100 m) continental shelf of about 210 000 km², between the north-east Australian coast and the Coral Sea. While shallow coral reefs are widely recognised as an iconic habitat, they occupy only five percent of the region. This chapter is about the diversity of benthic habitats and biota of the deeper lagoon and inter-reefal seabed areas of the remainder of the region.

Only a small fraction of this vast deeper seabed had been examined by marine biologists during the 20th century. Extensive new information was provided during 2003–2006 by the GBR Seabed Biodiversity Project, which mapped the distribution of seabed habitats (including mud, sand and gravel flats; algae and seagrass beds, sponge and gorgonian gardens, hard and shoal grounds) and their associated biological diversity of more than 7000 species, many new to science, across the length and breadth of the GBR Marine Park.

The GBR seabed has alternated between terrestrial and marine environments, due to repeated glacial periods over geological history (Chapters 2 and 3). Today's seabed habitats and biodiversity still partly reflect this history and are broadly influenced by coastal processes (Chapter 11) on one side and oceanic processes (Chapter 4) on the other, with numerous subpatterns due to regional differences in several physical factors. Overlaid on the broad patterns, the majority of variation in the biota results from a multitude of biological and stochastic processes.

GEOPHYSICAL CONTEXT AND PROCESSES

Offshore from Townsville in the central GBR (about 18.2–19.2°S), where most previous studies have been conducted, the continental shelf is about 120 km wide. On the outer 40–50 km of the shelf, an open matrix of coral reefs rises from 50–70 m depth. Between the coast and the reef matrix, a 75 km wide stretch of water known as the GBR Lagoon is almost devoid of reefs possibly due to the action of frequent cyclones.

North of 18°S, the shelf narrows (to about 50 km wide) and becomes shallower (typically 30 m, rarely >50 m). The outer shelf becomes delimited by a nearly continuous barrier of ribbon-like reefs; with steep descents to the abyss beyond. Platform reefs are common, often quite densely packed across the mid- and outer-shelf, and the GBR Lagoon becomes narrower (30 km to 20 km).

South of 19°S, to 22°S, the shelf broadens to as much as 250 km and generally becomes much deeper with substantial areas >70 m. Two lines of hard-packed reefs, the Pompeys, with narrow channels between, emerge from a shallow (30–50 m) relic limestone platform that dominates along much of the outer shelf. The Lagoon also broadens to >100 km and deepens (50–70 m) to the southeast, becoming the Capricorn Channel with depths in excess of 100 m at its mouth.

At the very southern end of the GBR (23.0–24.5°S), the Capricorn–Bunker Reefs rise from the outer areas of an 80 km wide sandy shelf of 30–40 m depth.

Coastal influences

Along much of the coast, rivers export terrigenous sediments to inshore seas. After heavy rainfall, turbid flood plumes carry suspended solids and nutrients along the coastline and into the GBR Lagoon. Over the last 8000 years, this has resulted in a ~15 km wide inshore deposit of muddy sediments, the Holocene Wedge, particularly between 12–21°S. Elsewhere, the inshore sediments are largely silica sands—especially at the southern end, where the shelf is extensively covered by silica sands from the Great Sandy Region further south.

On exposed coasts, wave action from trade winds regularly turns over the sediments to depths of 20 m, creating unstable habitat and redistributing finer particles to less exposed areas. Cyclones are particularly frequent in the open central GBR, and their disturbing effects reach much greater depths, transporting fine particles north and inshore, leaving behind a thin veneer of coarse particles over much of the shelf in this region.

Oceanic influences

In the Coral Sea, the westerly South Equatorial Current bifurcates at the continental margin between 14–16°S, to produce a northward off-shelf current and the southerly East Australia Current (EAC). The ribbon barrier reefs in the north limit the exchange of oceanic water onto the shallow shelf; here the trade winds drive transient northward flows of shelf water. The Pompeys hard-line has a similar barrier effect in the south. In contrast, the open reef matrix of the deeper central GBR shelf allows episodic inflows of water from the EAC. Occasionally, upwellings of cool nutrient rich water intrude up and across the shelf.

Oceanic tides exert a strong influence on parts of the shelf. The topography of the southern GBR creates a tidal node, which causes very large amplitude tides in the Broad Sound and Shoalwater Bay region and extreme tidal currents. The same processes also cause very strong currents through the narrow passages between the Pompeys hard-line reefs offshore. Such currents scour away sediments, progressively depositing them in less energetic areas.

In the far northern GBR, the out-of-phase tides of the Coral and Timor Seas also cause very strong tidal currents, with similar effects. In addition, tidal jets flowing into passages between some ribbon reefs can also pump nutrient rich Coral Sea water onto localised areas of the outer shelf.

BIOPHYSICAL RELATIONSHIPS AND ASSEMBLAGE PATTERNS

The Seabed Biodiversity Project sought to examine the relationships between the major driving factors outlined above and the biology observed by video and collected in epibenthic sled and scientific trawl samples at almost 1400 sites. These relationships are important in order to deliver integrated landscape maps of seabed assemblages, which are needed for management in the Marine Park. Such maps were produced by predicting species distributions and assemblage patterns in areas not sampled, on the basis of modelling their relationships with 28 more broadly available geo-physico-chemical variables. While this surrogate approach has limitations, it is the only feasible option given the vast size of the GBR, which means that despite considerable effort, sampling sites were relatively sparsely (average ~12 km apart) distributed over the shelf seabed.

The biophysical modelling indicated the major environmental factors that appeared to affect the distribution patterns of seabed habitats and assemblages in the GBR, including: sediment grain size (particularly the percentage of mud); force of water currents on the seabed (benthic stress); chlorophyll and/or turbidity; and, to a lesser extent, depth and some nutrients. While the correlations between the 28 physical factors in the GBR are complex and multidimensional, they have been simplified to just two dimensions (Fig. 6.1), after rescaling them in proportion to their importance for biological patterns. The most common environments

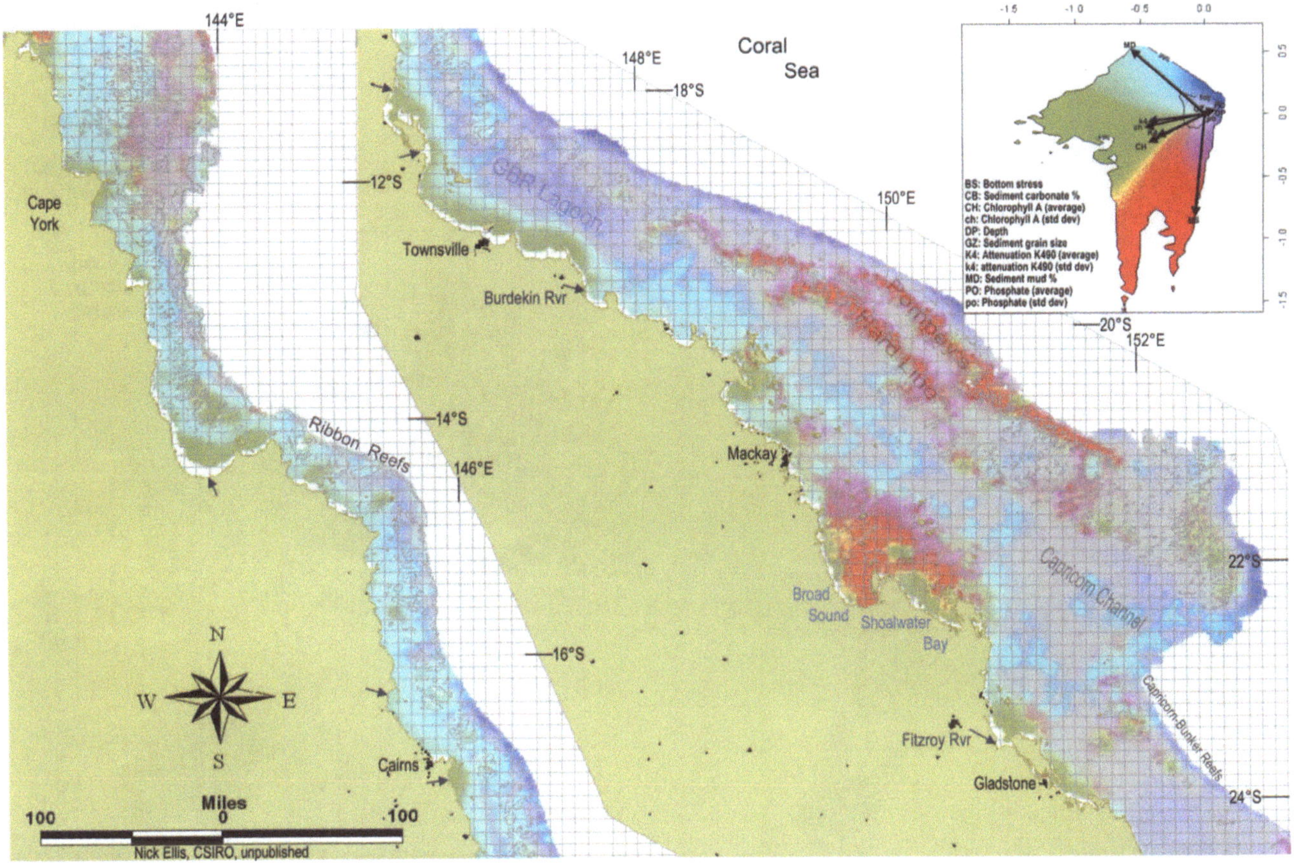

Figure 6.1 Biophysical map of the GBR continental shelf. Inset: colour key showing distribution of >170 000 seabed 0.01° grid cells on the first two principal dimensions of variation; contours show 99th and 50th percentiles of grid cell density; labelled arrows indicate direction of major physical factors. (Image: N. Ellis, CSIRO.)

(50% of the seabed) are coloured light grey in the centre of the plot and represent mostly coarse sands, usually carbonate and often with sparse algae and benthos; the corresponding map shows the location of such areas. Less common and more extreme environments are shown in higher intensity colours and the arrows indicate the axes of the major physical factors, with the length indicating their relative influence on biological variation. By relating these axes to the colour key, the distribution of the dominant factors can be identified in the map. The colour key of Fig. 6.1 has been chosen to maximise the representation of the major physical environments of the GBR so that similar environments have a similar colour; nevertheless, the region is so varied that differences will occur among areas having the same colour. Further, only 65% of the environmental variation can be represented in the two dimensions of the key.

Major biological patterns and important physical factors

The primary factor influencing biological assemblages is mud (MD vector, Fig. 6.1). The terrestrial muddy inshore areas along much of the length of the GBR (Fig. 6.4*A*) are often also very turbid (K4) and/or have high levels of chlorophyll (CH; green areas Fig. 6.1). These inshore (green–blue) areas tend to be devoid of visible biological habitat attached to the seabed (Fig. 6.2 (white), Fig. 6.4*A*), but typically are about 60–80% bioturbated (Fig. 6.2 (grey), Fig. 6.4*B*) indicating the presence of burrowing animals such as worms, shrimps and bivalves. The mobile fauna has low diversity and low biomass (Fig. 6.3) but still includes many small fishes, prawns, crabs, mantis shrimps, gastropods and heart urchins—many of which feed on the deposits. A few sessile filter feeders, such as sea pens, survive in

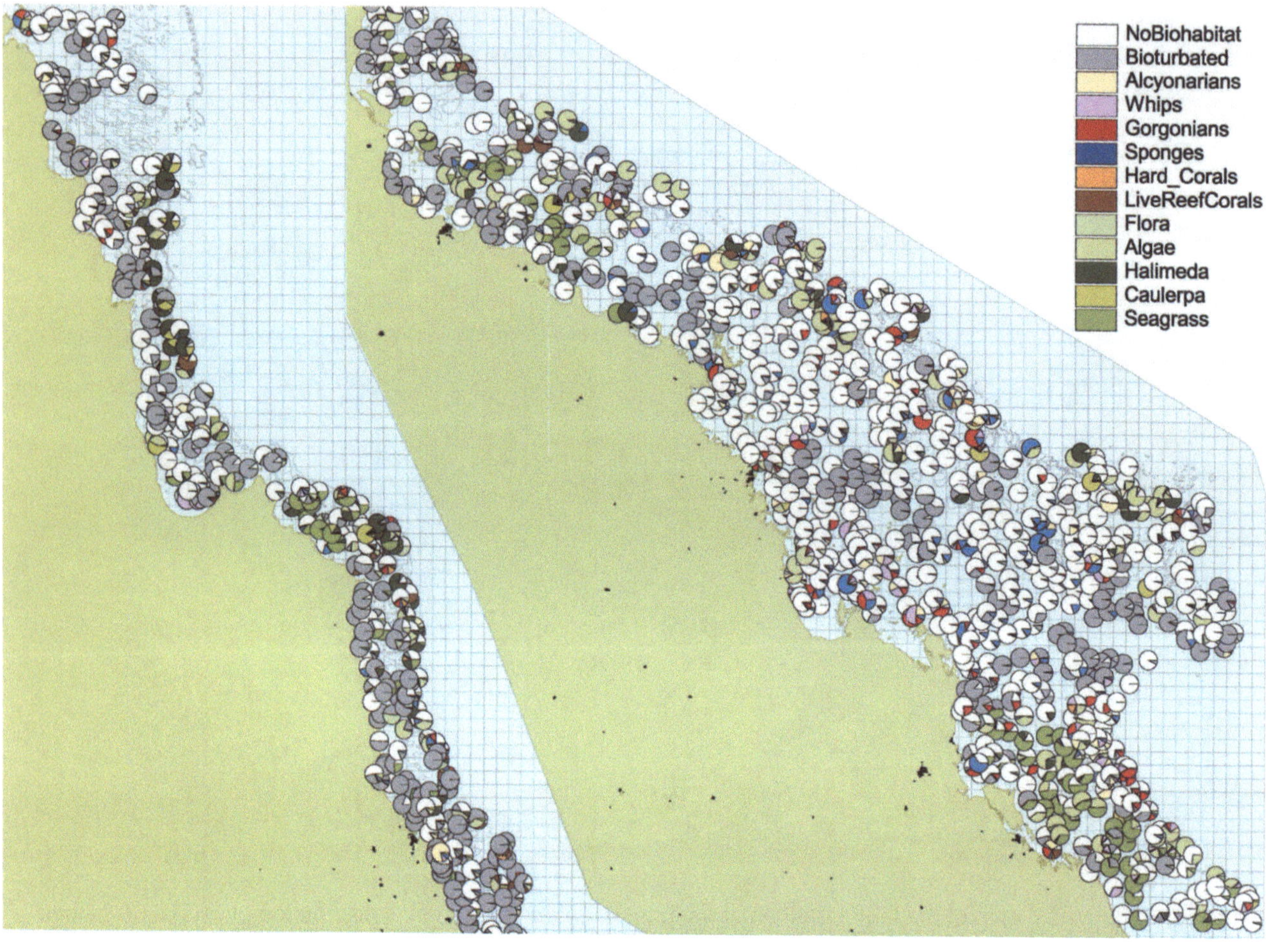

Figure 6.2 Map of the distribution of broad biological seabed habitat types observed during towed video camera transects. (Image: CSIRO.)

these soft sediments. Further offshore where the turbidity is lower the muddy sediments (blue areas, Fig. 6.1) are of carbonate, indicating their biological origin. Typically these habitats are about 40% bioturbated (Fig. 6.2) and have a fauna similar to the inshore muds, but with some stalked and cryptic sponges and surface dwelling bryozoans (lace corals) (Fig. 6.3). In the north, carbonate muddy habitats extend across most of the shelf (Fig. 6.1) given the protection of the ribbon reefs. In the deeper Capricorn Channel, the carbonate muddy habitats are among the most barren (Fig. 6.2), with about 20% bioturbation, and have the lowest diversity and biomass (Fig. 6.3) comprising a few crabs and fishes. Typically, with greater distance across the shelf, the substratum is sandier or even coarser (Fig. 6.4C), comprising entirely of biogenic carbonate and may support a variety of biohabitats (Fig. 6.2), sometimes even bioturbating animals (Fig. 6.4*D*), though more commonly surface biota such as starfish, crabs, gastropods, fish, algae, and filter-feeding alcyonarians and sponges—sometimes in great abundance (Fig. 6.3).

At the opposite extreme to mud, benthic stress (BS) is one of the strongest biophysical forces (red–orange areas, Fig. 6.1). The inshore vicinity of Broad Sound and Shoalwater Bay (orange) has the largest tidal range in the GBR and is accompanied by extreme currents (lower left orange branch of key) that scour sediments away, exposing rubble, stones and rock. Offshore, these tidal forces also cause extreme currents of clearer water (lower right red branch of key) to surge through the narrow channels of the Pompeys hard-line (red), again scouring the seabed to the limestone base (Fig. 6.4*E*). Strong benthic stress also occurs in the far northern GBR and Torres Strait, and in some

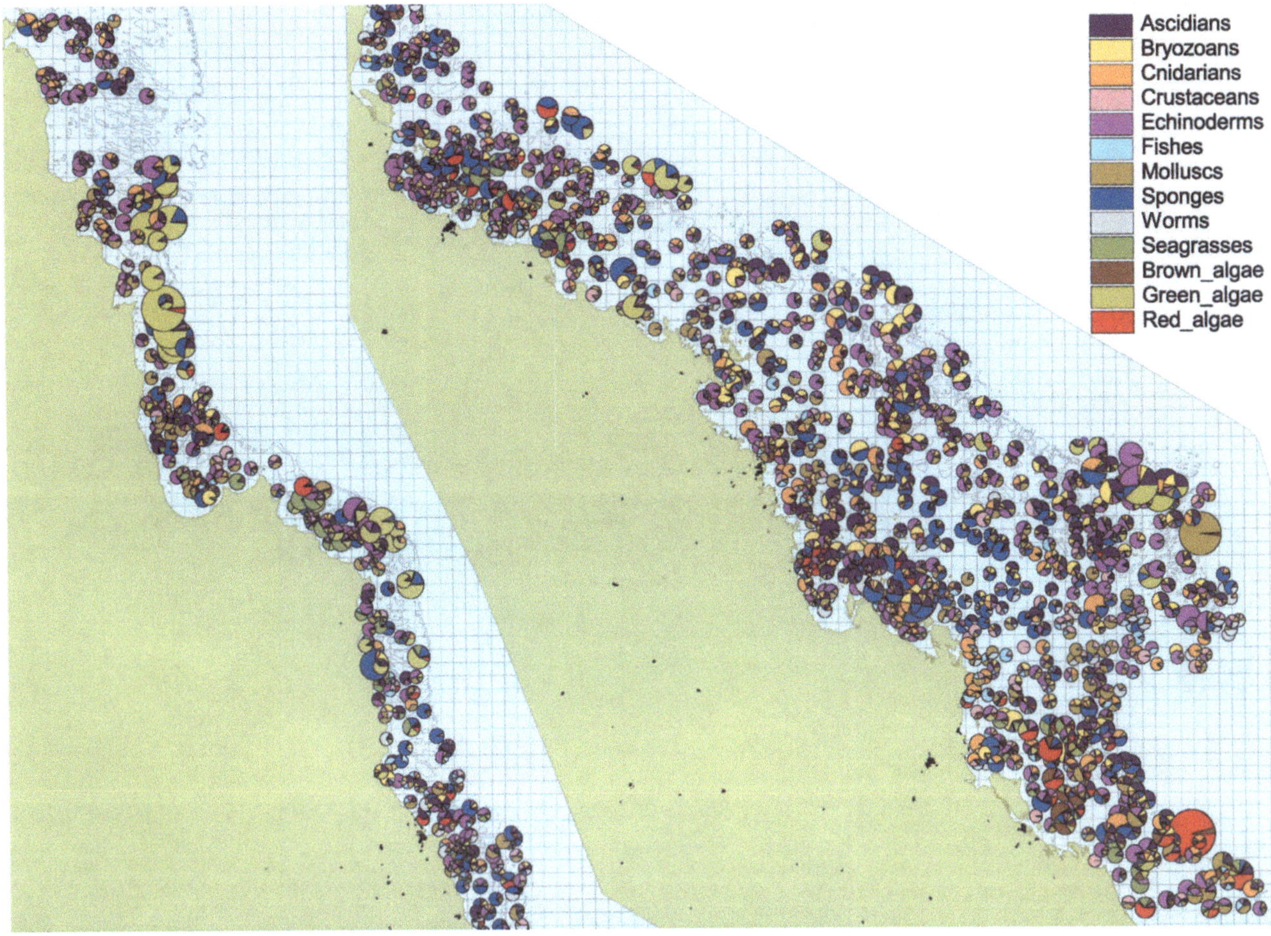

Figure 6.3 Map of the distribution of biomass of the major biological groups sampled by an epibenthic sled. (Image: CSIRO.)

local areas such as the Whitsunday Passage. In many of these areas, bare seabed is often interspersed with gardens of colourful filter-feeding sea fans, whips and sponges (Figs 6.2, 6.4*F*, *G*), which benefit from the stable hard seabed and suspended food particles—in the offshore passages, encrusting bryozoans form extensive biogenic rubble habitat (Fig. 6.4*H*). The fauna has very high diversity and moderate biomass (Fig. 6.3) and includes many species of bryozoans, sponges, and gorgonians, as well as fishes, crustaceans, bivalves, starfish, urchins and corals. The scoured sediments typically are deposited in ripples, waves and dunes (Fig. 6.4*I*) on the fringes of these high stress (red) areas. In offshore sandy areas with medium currents, crinoid feather stars may be extremely abundant on the seabed (Fig. 6.4*J*); also, relic coralline outcrops and shoals with a rich sessile biota, including living hard corals, may occur in deep areas between emergent coral reefs (e.g. Fig. 6.4*K*).

The deeper clear waters near the outer edge of the continental shelf may be influenced by upwelled nutrients (e.g. PO, indigo areas, Fig. 6.1). Despite the depth, algae (including crustose coralline algae (Fig. 6.4*L*) that are adapted to low-light conditions) may be prolific in these clear nutrient rich waters (Fig. 6.2) and contribute to habitats of moderate diversity and biomass (Figs 6.3, 6.4*M*). Areas where upwellings intrude onto the shelf can also be seen in Fig. 6.1 (faint indigo on grey), particularly offshore from Townsville. Here, along the outer margins of the inshore turbid/muddy areas, where the water is clear enough to allow sufficient light to reach the seabed, a >200 km long band of mixed algae (including *Caulerpa*, Fig. 6.4*N*) and patchy seagrass (primarily *Halophila spinulosa*) proliferates (Fig. 6.2). Similarly,

Figure 6.4 Photographs of some example seabed habitat types observed during towed video camera transects. *A*, turbid muddy inshore seabed, with filefish; *B*, bioturbated silty inner shelf seabed; *C*, coarse outer shelf sediment with sparse biota; *D*, large bioturbation mounds in offshore sand; *E*, scoured rocky seabed in extreme current area; *F*, soft corals in strong current channel; *G*, gorgonian garden on hard ground; *H*, bryozoan rubble in strong current channel; *I*, rippled sand in inshore high current area; *J*, crinoids on sand in offshore strong current area; *K*, shoal ground in deep water; *L*, *Ulva* growing on patches of coralline algae at shelf edge; *M*, diverse algae and coral near shelf edge; *N*, dense algal bed (*Caulerpa*); *O*, seagrass (*Halophila spinulosa*) bed; *P*, *Halimeda* algae bank. (Photos: CSIRO.)

around the Turtle and Howick Islands in the central northern GBR (~14.5°S), meadows of dense *H. spinulosa* occur (Fig. 6.2). Algae and *H. spinulosa* habitats also occur over much of the shelf in the Capricorn region (Figs 6.1, 6.2, 6.4*O*). On the shallow shelf just inside the outer barrier ribbon reefs in the northern GBR, the signature of nutrients pumped in by tidal jets can be seen faintly in Fig. 6.1. These have encouraged the development of vast banks of *Halimeda* algae (Figs 6.2, 6.4*P*) up to 15 m thick, comprised of the deposited carbonate skeletons of these algae. *Halimeda* also occurs elsewhere (Fig. 6.2) but has not formed such banks. These varied marine plant communities form substantial areas of habitat for a plethora of biota, including numerous species of green, brown and red algae, small fishes, gastropods, sea slugs, crustaceans, starfish, sea cucumbers, urchins and corals (Fig. 6.3), which may attain moderate to high biomasses and, particularly in the case of mixed algal–seagrass meadows, very high species diversity. The moderately diverse biota of *Halimeda* banks includes numerous species of green algae, small fishes, gastropods, urchins, corals, sponges and alcyonarians.

General patterns of abundance and diversity of the major phyla

The GBR seabed biota are represented by more than a dozen major phyla, which can be quantified in terms of biomass abundance, frequency of occurrence and numbers of species. Our understanding of these biota depends on the devices used to sample them. Animals living down in the sediments such as small worms, crustaceans and molluscs (infauna), usually are collected by a grab or core—a device not used by the Seabed Project. An epibenthic sled samples sessile biota and slow moving invertebrates (Fig. 6.3) living on the seabed or in the top few centimetres of the sediment, whereas a research trawl typically samples the more mobile fauna living just above the seabed, such as fishes and crustaceans.

Together, the sled and trawl confirmed the dominant biomasses of algae, particularly green algae, in the areas outlined above (Figs 6.2, 6.3). After algae, sponges are the next most abundant group; encrusting and massive morphotypes may reach high biomasses in the higher current areas, stalked and cryptic types are sparsely distributed in sedimentary areas. Ascidians are the next most abundant sessile group and have a similar pattern of distribution, followed by cnidarians, which tend to be more restricted to the higher current areas and harder ground, as are bryozoans. Echinoderms are the most abundant and widespread of the mobile invertebrates, with overall biomass between that of sponges and ascidians. Molluscs, while widespread in softer sediments, appear to be about half as abundant. Fishes, better sampled by trawl, are next in abundance—inshore and muddy areas in particular tend to have high relative proportions of fishes. Crustaceans were much less abundant. These were followed by worms, elasmobranchs and minor phyla, none of which were well sampled by either device.

The ordering of these groups by frequency of species occurrence differed from that of biomass. Fish species occurred most frequently, followed by crustaceans, molluscs, echinoderms, sponges, corals, algae and ascidians. In terms of numbers of species, sponges were the richest with more than 1100 taxa, followed by molluscs (>1000), fishes (>850), crustaceans (almost 600), echinoderms (>500), algae (>400), corals (almost 400), bryozoans (>300) and ascidians (>300). These statistics indicate a very high diversity for the GBR seabed, yet the true diversity is much greater, given the infauna are not included and some phyla were not fully sorted. More detailed identifications of these samples, all of which are lodged with the Queensland Museum, will continue to reveal this diversity.

■ HUMAN INFLUENCES ON SEABED HABITAT AND ASSEMBLAGES

Terrestrial runoff has been reported to be elevated by human activities and to have implications for coastal coral reefs (see Chapters 9, 11). Coastal processes influence the composition of seabed biota as outlined above, but it is unclear whether possible anthropogenic increases in turbidity and sedimentation have caused any changes to the benthos in recent decades.

A widespread activity on the seabed in the GBR is trawling for prawns. Previous research showed that trawling can have direct impacts, particularly on easily removed and/or slow to recover biota, but stressed the importance of assessing the results in the context of the regional distribution of vulnerable biota in relation to the distribution of trawl effort. The Seabed Project has now completed this risk assessment, which will contribute to legislative requirements that all fishing in the Marine Park be sustainable. Trawlers now use devices to reduce the incidental catch in their nets of nontarget marine life (bycatch). These devices are effective for elasmobranchs and some fishes, which is beneficial as the distributions of these groups have the greatest overlap with trawl effort. The risk assessments have indicated that the majority of vulnerable fauna have distributions that overlap little with trawl effort. Combined with recent management that has reduced trawl effort substantially, this means that these fauna are unlikely to be at significant ongoing risk. The marine park zoning plan (Chapter 12) contributes to conserving benthic biodiversity by protecting 20%–100% of the populations of all species assessed, and the 2004 rezoning increased levels of protection by approximately 30%. These management actions provide greater assurance for the future of the diversity of habitats and biota on the seabed in the GBR.

ADDITIONAL READING

Pitcher, C. R., Doherty, P., Arnold, P., Hooper, J., Gribble, N., Bartlett, C., Browne, M., Campbell, N., Cannard, T., Cappo, M., Carini, G., Chalmers, S., Cheers, S., Chetwynd, D., Colefax, A., Coles, R., Cook, S., Davie, P., De'ath, G., Devereux, D., Done, B., Donovan, T., Ehrke, B., Ellis, N., Ericson, G., Fellegara, I., Forcey, K., Furey, M., Gledhill, D., Good, N., Gordon, S., Haywood, M., Hendriks, P., Jacobsen, I., Johnson, J., Jones, M., Kinninmoth, S., Kistle, S., Last, P., Leite, A., Marks, S., McLeod, I., Oczkowicz, S., Rose, C., Seabright, D., Sheils, J., Sherlock, M., Skelton, P., Smith, D., Smith, G., Speare, P., Stowar, M., Strickland, C., Van der Geest, C., Venables, W., Walsh, C., Wassenberg, T., Welna, A., and Yearsley, G. (2007). Seabed Biodiversity on the Continental Shelf of the Great Barrier Reef World Heritage Area. CRC Reef Research Final Report, AIMS/CSIRO/QM/QDPI, Brisbane.

Poiner, I. R., Glaister, J., Pitcher C. R., Burridge, C., Wassenberg, T., Gribble N., Hill, B., Blaber, S. J. M., Milton, D. A., Brewer D., and Ellis, N. (1998). The environmental effects of prawn trawling in the far northern section of the Great Barrier Reef Marine Park: 1991–1996. Final Report to GBRMPA and FRDC, CSIRO Division of Marine Research, Queensland Department of Primary Industries, Brisbane.

7. Primary Production, Nutrient Recycling and Energy Flow through Coral Reef Ecosystems

O. Hoegh-Guldberg & S. Dove

Corals reefs provide a spectacular contrast to the oceanic ecosystem that surrounds them. While coral reefs are ablaze with thousands of species living within a highly productive ecosystem, the surrounding ocean is usually devoid of particles and has a low rate of primary production. Whereas coral reefs are often likened to the 'rainforests of the sea', the surrounding waters are often referred to as 'nutrient deserts' due to the low quantities of essential inorganic nutrients such as ammonium and phosphate ions. How a marine 'rainforest' exists in an oceanic 'desert' was one of the grand puzzles that faced the early workers on coral reefs, including Charles Darwin.

In this chapter, the various drivers of the energy flow through coral reefs are explored. After investigating which physical and chemical factors define corals and the reefs that they build, we describe some of the fundamental biological relationships that lead to the capture of energy and nutrients by coral reefs, and how these essential requirements of life flow through coral reef ecosystems. As we will see, while primary production is similar to that of other shallow marine ecosystems, coral reefs have a high proportion of pathways that involve the recycling of nutrients between closely associated primary producers and consumers. We will also see that the highly productive nature of coral reefs is deceptive in that it is accompanied by the rapid and efficient recycling of nutrients and energy such that overall rates of accumulation of organic carbon as reef growth are low. This theme lies at the heart of the puzzle of how highly productive and diverse coral reefs can exist in the 'nutrient deserts' of tropical and subtropical oceans.

ENVIRONMENTAL FACTORS DRIVING THE DEVELOPMENT OF CORAL REEFS

Coral reefs grow in shallow, sunlit waters of tropical and subtropical oceans and are built from the activities of many types of calcifying organisms. Globally, coral reefs occupy 284 300 square kilometres, which is less than 1% of the global ocean. They dot oceans and line coasts within a band between 30° north and south of the equator. In Australia, coral reefs stretch northwards along continental shelves, from northern NSW on the east coast, from Rottnest Island in the south-west to the north-west coast of Australia and along the western edge of the Gulf of Carpentaria. Throughout this range there is a tremendous variability in the structure of coral reefs. Coral reefs range from poorly developed reefs that fringe continental islands and the Australian coastline to extensive carbonate barrier reefs. At latitudes from 22–30°S, extensive coral reefs form but fail to accumulate limestone (i.e. reef erosion exceeds calcification) and are referred to as non-carbonate reef systems (see Chapter 2).

Central to the existence of coral reefs are reef-building corals (Order Scleractinia, Class Anthozoa, Phylum Cnidaria, see Chapter 20). Reef-building corals are often referred to as the framework builders of coral reefs, but they are not the only calcifiers on coral reefs. Others organisms such as giant clams, foraminifera and calcareous algae contribute substantial quantities of calcium carbonate to reef structures and sediment. Calcareous algae (see Chapter 15) play a particularly important role in reef frameworks, some by coating and 'gluing' the coral framework together, and others by building thick edifices of their own. The framework provided by corals and other marine calcifiers forms the three-dimensional structure into which hundreds of thousands of species of animal, plant, fungi and bacteria live.

Several authors have explored the conditions under which coral reefs thrive globally. Explanations have frequently focused on the fact that coral reefs form in warm seas, leading to the 'rule of thumb' that coral reefs are limited to waters that do not decrease below 18°C in the winter. This principal is often incorrectly perceived as the ultimate limit to the development of coral reefs that ignores the many other variables that also change at higher latitudes. While reefs are adapted to their local temperature regime, the amount of light and the concentration of carbonate ions are at least as important in limiting reef development. An exhaustive study of the environmental factors associated with coral reefs using data from close to 1000 coral reef locations found that light availability and the concentration of carbonate ions (that is ultimately determined by temperature) are as potentially important as temperature in defining where limestone coral reefs are found (Table 7.1).

The conditions that are associated with the distribution of coral reefs vary across spatial and temporal scales. Variability across the year can be substantial, while diurnal variability in temperature is usually small (except in areas such as shallow intertidal reef crests). The seasonal variability of conditions becomes important on coral reefs at high latitudes where extremes in both winter (that can be too cold and dark) and summer (too hot and bright) can cause stress on coral reef organisms. At these sites in Australia, inter-annual variability such as that associated with the El Niño cycle along the east coast of Australia can play a large influence on coral reefs through warm (e.g. 2002) and colder years (e.g. 2003). In coming decades, due to rising background sea temperatures, warmer years can exceed the tolerance of symbiotic organisms giving rise to mass coral bleaching and associated mortality events (see Chapter 10).

The global distribution of carbonate and noncarbonate coral reefs is strongly (and perhaps not surprisingly) correlated with the concentration of carbonate ions, which is ultimately determined by ocean temperature, salinity, and factors such as the atmospheric carbon dioxide concentration (see Chapter 10). The concentration of carbonate ions is highest at the equator and decreases at high latitudes due to the effect of sea temperature on the solubility of CO_2 (see Chapter 10). Coral reefs do not exist at carbonate concentrations below 200 µmol kg^{-1}; this has significance in terms of the problem of ocean acidification for coral reefs as discussed in Chapter 10.

The last variable that is a major determinant of where coral reefs are found is light. Because of the dependence of primary production and calcification on light, coral reefs are limited to clear tropical and subtropical waters where depths are less than 100 m. Both light quantity and quality (wavelength) are important, driving the primary step of photosynthesis of the symbionts within corals and many other photosynthetic organisms. Coral reefs only form where the average irradiance is at least 250 µmol m^{-2} s^{-1} (roughly 10% of surface irradiances in tropical and sub-tropical regions).

Several variables affect the light available for coral reefs. Light enters the outer atmosphere of the Earth (Fig. 7.1) and is selectively filtered such that some wavelengths (ultraviolet, infrared) are largely removed by the ozone layer and water vapour (e.g. clouds). The penetration of Photosynthetically Active Radiation (PAR, 400–700 nm) is also reduced by dust and clouds. Then, at the surface of the ocean, more light is reflected,

Table 7.1 Environmental factors identified by Kleypas *et al.* (1999) associated with more than 1000 reef locations worldwide

Variable	Minimum	Maximum	Mean	Standard Deviation
Temperature (°C); based on NOAA AVHRR-based sea temperature records				
average	21	29.5	27.6	1.1
minimum	16	28.2	24.8	1.8
maximum	24.7	34.4	30.2	0.6
Salinity (ppt)				
minimum	23.3	40	34.3	1.2
maximum	31.2	41.8	35.3	0.9
Nutrients (μmol L^{-1})				
NO3	0	3.34	0.25	0.28
PO4	0	0.54	0.13	0.08
Aragonite saturation (Ω-arag)				
average	3.28	4.06	3.83	0.09
Maximum depth of light penetration (m) calculated from the monthly average depth at which average light decreased below the perceived minimum for reef development of 250 μmol m^{-2} s^{-1}.				
average	−9	−81	−53	13.5
minimum	−7	−72	−40	13.5
maximum	−10	−91	−65	13.4

with the amount reflected decreasing as the height of the waves increases. The light that enters the ocean is absorbed by water molecules, or is scattered and absorbed by dissolved compounds, plankton, and suspended sediments. These interactions are wavelength dependent such that its spectral breadth and intensity decrease with depth. In a uniform water column, the intensity of light decreases exponentially as described by Beer's Law (Fig. 7.1).

Different water columns vary with respect to the amounts of dissolved substances and particles they contain. Inshore sea waters, that receive fresh waters from rivers and land runoff, often have large quantities of sediments, tannins and phytoplankton (often referred to as 'Gelbstoff' or yellow substances). These waters have spectra that are green-yellow shifted and light attenuation coefficients (k, Fig. 7.1) that may range up to 0.5 m^{-1}, which means that it gets very dim at quite shallow depths. Waters that are offshore or are located away from rivers, have far less scattering and absorption. These waters may have light attenuation coefficients as low 0.01 m^{-1}. The effect of these differences in light at depth is substantial, causing the depth limits of corals in typical inshore regions to be around 5 m as compared with offshore sites where the depth limits may be in excess of 50 m. Light at depth in offshore sites is blue shifted due to a relatively small influence of Gelbstoff substances and the greater relative influence of the water itself as a source of scattering and absorption. One of the key reasons why coral reefs are not found near major rivers is because the light environment deteriorates due to heavy sediments blocking light transmission through the water column, in addition to the problems of the low availability of stable

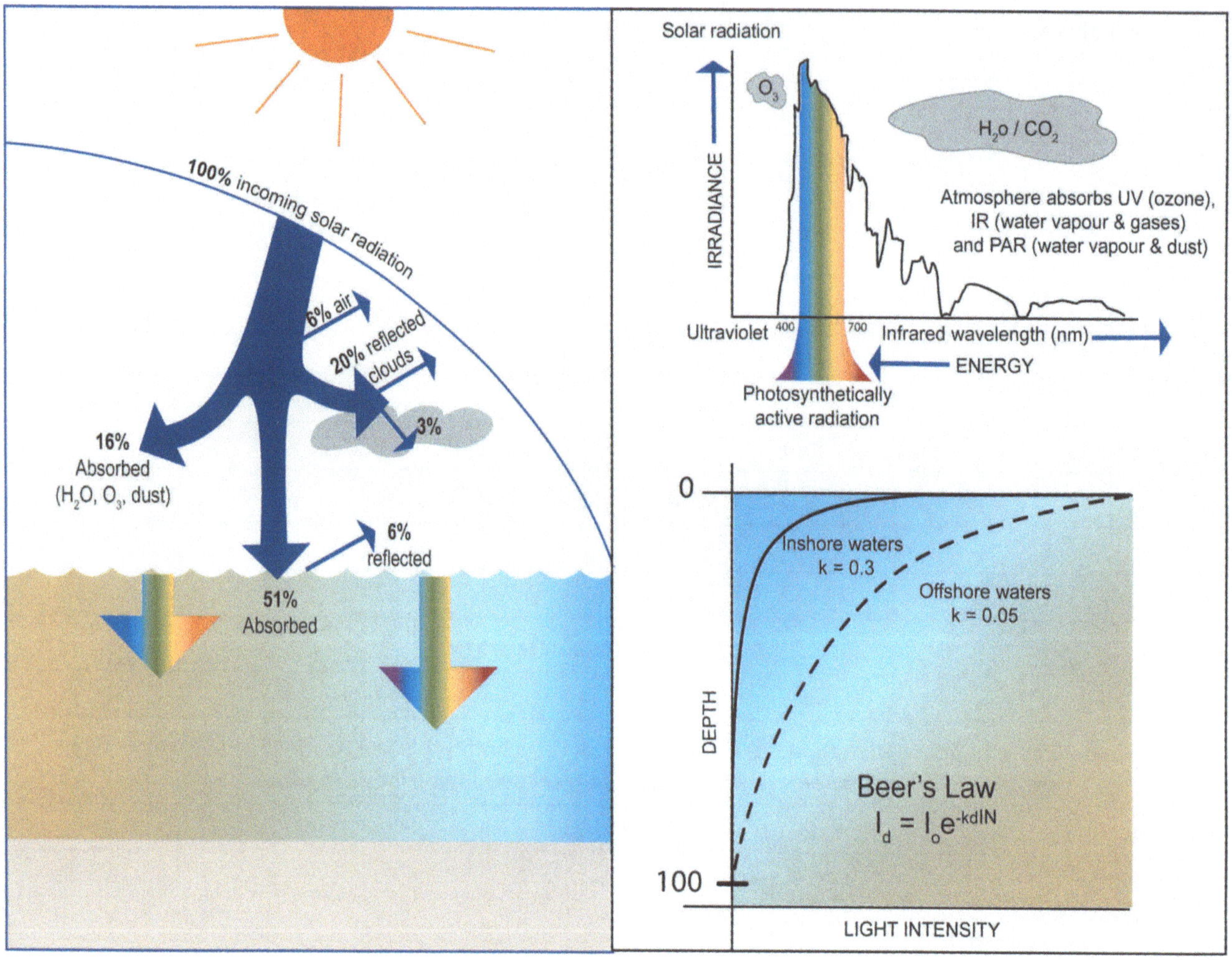

Figure 7.1 Depiction of the pathway for solar radiation entering the ocean from the outer atmosphere. Radiation interacts with atmosphere components such as clouds, dust and specific gases such as carbon dioxide and water vapour. Light is reflected as it crosses the surface interface of the ocean. Once it has entered the water it is scattered and absorbed by water and components such as suspended sediments. Beer's Law relates the passage of light through the water column to the amount of suspended material in the water column. I_o, intensity of light at the surface, while I_d is the intensity at depth, d. The attenuation Coefficient, k, is a measure of the extent to which light is absorbed and scattered by the water column and its internal constituents. (Figure: D. Kleine and O. Hoegh-Guldberg.)

clean substrata for corals to settle and grow due to high rates of sedimentation.

The light environment of coral reefs also varies temporally. Seasonal changes in solar flux are significant at higher latitudes, and hourly light environments at any single point on a reef vary according to the angle of the sun. Coral reefs growing close to islands or patches of coral reef growing in crevasses experience light environments that can fluctuate considerably. Tidal variations in depth can also have significant influences on the short term light environment as will variations in cloud cover over periods of minutes. At even finer time scales, effects such as the focusing of light through the lens effect of wave surfaces (sub flecks) can increase the intensity of light over very short (millisecond) time scales.

■ PROCESSES UNDERPINNING PRIMARY PRODUCTIVITY OF CORAL REEFS

The capture of the sun's energy is a fundamental step in the energy and nutrient cycles of biological systems on Earth. Only chemosynthetic organisms that derive energy from the geothermal sources of reduced compounds such as sulfide stand apart from the overwhelming majority of organisms that are dependent ultimately on solar energy trapped by photosynthetic organisms. The myriad of photosynthetic organisms on coral reefs provide the basis for the vigorous energy and nutrient cycles that typify coral reefs.

Many organisms participate in photosynthetic activities on coral reefs. Prominent among these are blue-green bacteria (Cyanophyta), macroalgae (seaweeds), microalgae (phytoplankton) and photosynthetic protists such as dinoflagellates and diatoms. The reef habitats they occupy are also diverse, with some of the most significant activity occurring on uncharismatic 'mossy' substrata such as rocks and sediments. Importantly, their densities and actions are profuse within invertebrate host organisms such as corals, clams and foraminifera. Fundamentally, photosynthesis involves the capture of carbon dioxide using the energy of sunlight trapped using the green pigment chlorophyll. In a balanced equation, six molecules of carbon dioxide are incorporated and six molecules of water split to produce one molecule of sugar (glucose) and six molecules of oxygen.

$$6CO_2 + 6H_2O + Energy_{sunlight} \rightarrow C_6H_{12}O_6 + 6O_2$$

The matching process, respiration, acts in the reverse, and oxidises organic molecules (here glucose) to produce carbon dioxide, water and energy.

$$C_6H_{12}O_6 + 6O_2 \rightarrow 6CO_2 + 6H_2O + Energy_{metabolic}$$

There are two parts to the photosynthetic process. One part is referred to the 'light' reactions (where light energy is trapped by chlorophyll and is converted in the chemical energy of ATP and other molecules). The other part involves the 'dark' reactions (where the chemical energy that is trapped during the light reactions is used to fix carbon dioxide to generate organic molecules). The dark reactions start with the fixation of CO_2 by the abundant enzyme ribulose bisphosphate carboxylase/oxygenase (or Rubisco) and involve the set of reactions comprising the Calvin-Benson Cycle. Whereas the light reactions are powered by the sun's energy, the dark reactions do not need light if the appropriate levels of ATP and other reduced molecules are made available to power the enzymatically catalysed reactions involved.

Chlorophyll is the central pigment involved in the transduction from light to chemical energy, but there is a range of accessory pigments to assist the process. Accessory pigments interact with light in a variety of ways and consequently add colours from blue-green (Cyanophyta) to red (Rhodophyta) to the organisms that contain them. As light increases, so does the rate of gross photosynthesis (P_G). All organisms (whether photosynthetic or not) respire and release the energy of carbon-carbon bonds in organic substrata. In photosynthetic organisms, the rate of respiration (R, usually measured in the dark when no photosynthesis can occur) is subtracted from P_G to calculate the net photosynthesis (P_N). This is essentially a measure of the rate at which organic carbon molecules (and the associated energy) accumulates during photosynthesis over and above those consumed during respiration. When measured per square metre of coral reef, values of P_N can be used as a measure of the Net Primary Productivity. This is an important number as it defines the extent of energy being added to an ecosystem like a coral reef.

The relationship between P_N and light has a characteristic shape (Fig. 7.2*A*). In the dark, P_N is negative and equals the rate of respiration (R). As light increases, however, P_N also increases until it equals zero, the co-called Compensation Irradiance (I_c). At this point, the rate at which organic carbon (energy) is being produced by photosynthesis is just balanced by the consumption of organic carbon by respiration. No net accumulation of organic carbon (or energy) occurs at this point. I_c is also the point at which the flux of oxygen into the organism just balances the rate at which oxygen is consumed by respiration. The reverse is true of carbon dioxide, which travels in the reverse direction to oxygen. The net rate of photosynthesis continues to increase as the light levels increase, with a net accumulation of organic carbon and production of oxygen.

The relationship between photosynthesis and light is linear at first and has a characteristic slope ('α') that is a measure of the efficiency of photosynthesis. Essentially α is measure of the rate at which photosynthetic activity increases with an increase in light (quanta). α varies according to the type of organism, their light

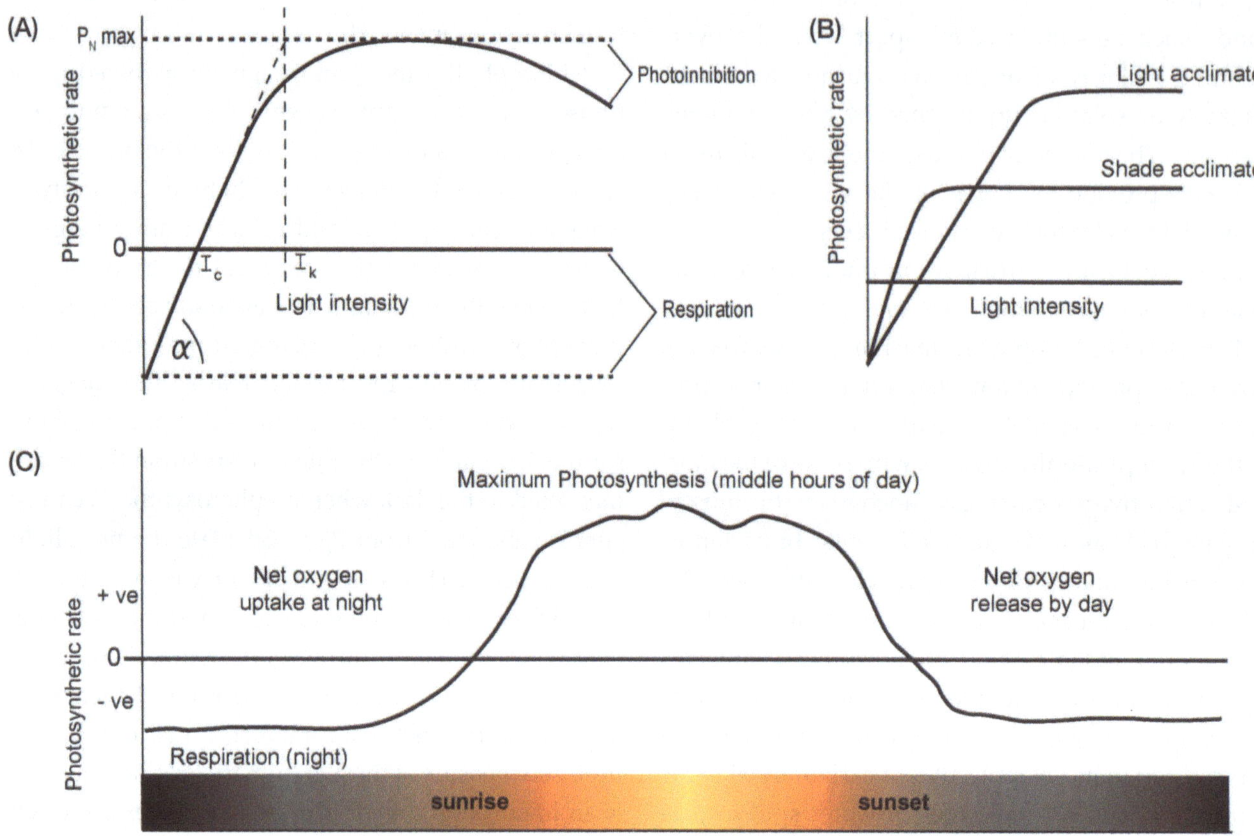

Figure 7.2 Typical characteristic of photosynthetic processes. *A*, Typical photosynthetic curve showing how the rate of photosynthesis increases with light intensity until it matches the respiratory rate at the compensation irradiance (Ic) and eventually saturates at the net maximum rate of photosynthesis ($P_{N\,max}$). The Saturation Irradiance (I_k) is a measure of the irradiance at which the photosynthetic rate is maximised. The efficiency with which light is converted into photosynthetic production is measured by the photosynthetic efficiency (α). If light levels continue to increase, the net rate of photosynthesis may decrease due to photoinhibition. *B*, Organisms may acclimate or adjust physiologically to different light environments. In light acclimated organisms, maximum rates of photosynthesis will be higher while photosynthetic efficiencies will be lower. The opposite is true in organisms that have acclimated to living in low light conditions. *C*, A typical pattern over time associated with the activity of a photosynthetic organism on a reef (net photosynthetic rate per hour). At night time, respiration dominates. At dawn, photosynthetic activity increases as light levels rise to achieve maximum levels of both light and photosynthetic activity at the middle of the day. (Figure: D. Kleine and O. Hoegh-Guldberg.)

trapping processes and their acclimation state (Fig. 7.2*A*). As light increases, however, the slope begins to decline to zero at which point the photosynthesis versus irradiance curve has reached a maximum value (referred to as $P_{N\,max}$). Further increases in light do not result in any further increases in photosynthetic production due to the saturation of the dark reactions of photosynthesis. That is, the capacity of these reactions to convert energy and carbon dioxide into organic carbon molecules has been exceeded. At very high light levels, P_N may decline below $P_{N\,max}$ due to a detrimental process called *photoinhibition*. The latter occurs because high light levels cause the light reactions to over-produce trapped energy from light capture resulting in oxygen being converted into highly reduced molecules of 'active oxygen'. Oxygen receives the excitation energy (essentially becoming supercharged) and, if these resulting supercharged molecules of active oxygen are not neutralised by various antioxidant enzyme pathways, serious damage to cells and tissues can eventuate. This damage is an important step in 'coral bleaching' (see below).

The characteristics of photosynthetic versus irradiance curves, while being driven by light, can change to some extent through the process of 'photosynthetic acclimation' (often mistakenly referred to as 'photoadaptation', which implies wrongly that evolutionary 'adaptation' is involved). To optimise light capture in low light while reducing the risk of photoinhibition, photosynthetic organisms actively manipulate the efficiency of light capture (α) by adding or subtracting chlorophyll and other pigments to or from the photosynthetic components responsible for light capture. This ability to process captured light (that reduces the relative risk of photoinhibition) is high in organisms that grow in high light habitats, and low in organisms that grow in low light. Organisms can change their ability to process captured light by increasing the capacity of their dark reactions (Calvin-Benson cycle) to increase $P_{N\,max}$. These changes define many of the differences between shade and light-acclimated photosynthetic organisms (Fig. 7.2*B*).

Over day-night cycles, the photosynthetic activity of reef communities fluctuates between periods of net consumption (i.e. at night, Fig. 7.2C) and periods of net production of organic molecules (i.e. during the day, Fig. 7.2C). On the way between these two extremes, photosynthetic reef organisms pass through periods (usually a few hours after sunrise or before sunset) in which the fluxes of carbon dioxide, oxygen and energy are zero (I_c, as described above). As the sun rises, the rate of photosynthetic activity becomes positive, increasing toward maximum values in the middle of the day. At this point, P_G approaches $P_{G\,max}$ that may be several times the absolute value of respiration. The daily photosynthesis to respiration ratio (P:R ratio) can be derived from the integrated photosynthetic activity over a day ($P_{G\,24\,h}$) divided by the respiration that occurred over the same period ($R_{24\,h}$) and is used by many physiological ecologists to examine how dependent an organism is on energy derived directly from light. P:R ratios can be calculated for short (hourly) or long (daily, yearly) periods. Organisms that have P:R ratios that are greater than 1.0 are referred to as autotrophic ('self feeding'; also referred to as phototrophic or 'light feeding') while organisms that have ratios of less than 1.0 are referred to as heterotrophic ('feeding on others'). Organisms on coral reefs have a wide range of P:R ratios. Photosynthesis to respiration ratios can be applied to communities as well, giving important insight into how much organic carbon or energy is entering a particular patch of land or seascape. In the latter case, these measurements are easily integrated into measurements of the net accumulation of organic carbon per unit time (primary production). These types of measurements provide an important basis from which to answer key questions such as whether coral reefs are a net sink or source CO_2.

Productivity varies across the globe, with some of the highest values being found in tropical rainforests (700–800 g C m^{-2} y^{-1}. The highest values that occur in the ocean range from 200 to 500 g C m^{-2} y^{-1} and are associated with nutrient rich coastal areas where upwelling occurs (e.g. off the west coasts of South America and Africa). The oceans in which coral reefs occur generally have low primary production rates of around 0.1–1.0 g C m^{-2} y^{-1}. Coral reefs, on the other hand, show average rates of primary production that range from 3–100 g C m^{-2} y^{-1} (averaged over all components including sand and rocky substrata) with some components (e.g. algal turfs, macroalgae, symbiotic corals) having rates that match those of the highest productivities in terrestrial systems).

There is no single value for the primary production of all coral reefs. Primary productivity varies with such characteristics as whether a reef is inshore or offshore, or whether it is a high or low latitude coral reef (Fig. 7.3). Values for primary productivity also vary according to location within a reef. The well flushed conditions of the fore reef have the highest primary production values (up to 1500 g C m^{-2} y^{-1}) while the poorly flushed and more variable conditions (in terms of temperature and light) of back-reef areas can have very low primary production values (up to 300 g C m^{-2} y^{-1}). Primary productivity also varies with the component of reef ecosystem considered. Benthic microalgae (see Chapter 15) and corals (and other symbiotic invertebrates) can have primary productivities of up to 2000 g C m^{-2} y^{-1}.

In addition to sunlight and CO_2, photosynthetic organisms require inorganic nutrients, particularly

Figure 7.3 Variation in primary productivity (red numbers: units = g C m^{-2} y^{-1}) and reef accretion (calcium carbonate) from the tropical water column to the shore of a typical coral reef coastline. Open ocean conditions are usually constant although productivity is low due to a paucity of nutrients. Productivity and reef accretion is significant on the reef slope and crest due to good conditions (temperatures, light and water movement are optimal). Going shoreward, conditions eventually become more extreme due to the ponding of water behind the reef crest. (Figure D. Kleine and O. Hoegh-Guldberg.)

inorganic nitrogen (usually as ammonium or nitrate ions) and phosphorus (as phosphate). They get these compounds from the breakdown (waste) products of animals that have consumed other organisms, or by the breakdown of debris generated by predators (e.g. fish) or scavengers (crabs, starfish) and use them to build new organic molecules. A range of organisms are involved in these natural cycles with different types of bacteria playing a dominant role in degradation. Huge populations of bacteria inhabit the sediments associated with coral reefs, which provide the ideal microenvironment for processing these compounds. One of the best known of these process pathways is the nitrogen cycle (Fig. 7.4), in which organic material settles on the surface of the sediments and is quickly buried by scavengers such as worms, molluscs and crustaceans.

In the upper few millimetres of the sediment, where oxygen levels are relatively high, a diverse range of microorganisms (bacteria like *Vibrio*, as well as many actinomycetes and fungi) use proteinases to strip amine groups off proteins and release ammonium ions (a process called 'ammonification'). Some ammonium ions escape from the sediments into the water column for use by photosynthetic organisms to build amino acids and proteins. Some ammonium ions continue down a pathway of 'nitrification', in which case the ammonium is oxidised to produce nitrate. This occurs in the O_2 rich upper sediment layers by a set of bacteria that must have oxygen to survive (Fig. 7.4). Some nitrate released by this process leaves the sediment also for use by marine photosynthesisers: some stays in the sediments and undergoes further denitrification. The next step involves the conversion of nitrate to nitrite that has to occur without oxygen and consequently occurs within the sediments, below the upper oxygen rich layers (first few millimetres). Here, oxygen levels decrease to zero due to the lack of light and the abundance of metabolising fauna that cause the oxygen concentration to decrease. The processes of ammonium oxidation and nitrate reduction are coupled. Aerobic nitrification sits adjacent to anaerobic denitrification (via the sediment gradient in oxygen availability), which is ideal for this coupling of the two processes together. The eventual outcome is that nitrate is reduced to nitrite, and nitrite is denitrified to nitrous oxide (NO) and/or nitrogen (N_2) gas that is lost to the atmosphere.

Before leaving the topic of the nitrogen cycle, it is important to consider where inorganic nitrogen and hence

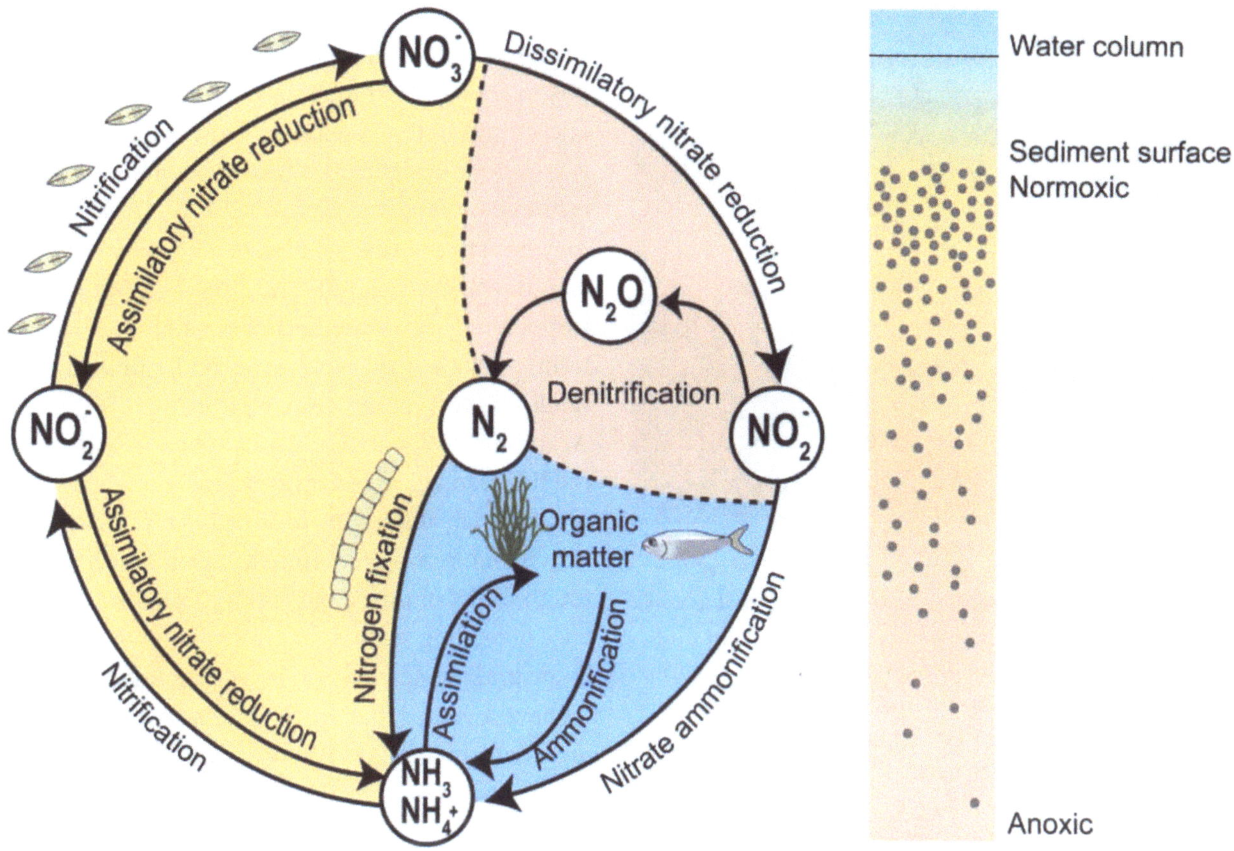

Figure 7.4 Nitrogen cycle associated with coral reefs. The fixation of nitrogen into organic compounds occurs in the water column (above the sediments). Plants take up ammonium and may be eaten by herbivores, with ammonium being released and recycled after burial of some organic material. At this point, released ammonium may participate in assimilatory nitrate reduction in the sediment layers that have relatively high oxygen levels. Nitrate may then be denitrified in the underlying sediments that are low in oxygen. The colours associated with the left-hand diagram indicate when the majority of each part of the cycle occurs within the water and sediment profile (indicated in the right-hand figure). (Figure: D. Kleine and O. Hoegh-Guldberg.)

nitrogen comes from originally. Cyanobacteria (so called 'blue-green algae') are important parts of the production line. One of their key characteristics is that they contain an enzyme called nitrogenase that can split the powerful triple bonds of atmospheric nitrogen gas (N_2) in a process called 'nitrogen fixation'. Such nitrogen fixing bacteria occur in a large number of habitats and ecosystems, including microbial mats and symbioses such as with sponges and legume plants. Recent work has revealed that nitrogen fixation is prolific in coral reefs and may play a critical role in supplying nitrogen to coral reef ecosystems. It appears that nitrogen fixers are abundant on benthic surfaces (sediments, rocks) and in the water column of coral reefs as blooms of the cyanobacterium *Trichodesmium* (commonly called 'sea sawdust', Fig. 7.5).

Coral reefs are an exception to the general rule that high primary productivity is strongly dependent on the availability of a high standing stock of inorganic nutrients (particularly N and P). Coral reef primary production is indeed high, yet the nutrient levels in the surrounding waters are exceeding low. What could be going on? To find the answer to this problem, we need to explore the ways that energy (contained in carbon compounds) and more particularly nutrients (N and P compounds) move through coral reef ecosystems after the initials stages of primary production.

Figure 7.5 Cyanobacteria are prominent members of the communities that cover most surfaces on coral reef as well as forming large slicks in the waters surrounding coral reefs. Slicks such as the one shown off the central portion of the Great Barrier Reef (genus *Trichodesmium*) eventually sink to the bottom of the ocean and may contribute significantly to the nitrogen cycle of coral reefs. (Photo: O. Hoegh-Guldberg.)

ENERGY FLOW THROUGH CORAL REEF ECOSYSTEMS

Organic carbon that accumulates through primary production is eaten, degraded and recycled by other organisms called 'consumers'. The mass of primary producers far exceeds that of primary consumers. This is due to inefficiencies in the transfer of energy and nutrients between trophic levels and to entropic considerations. The 'higher' plant and animal life of coral reefs forms a trophic pyramid, with a successive decrease in the mass of each successively higher trophic level (Fig. 7.6*A*) as discussed above, however, microscopic life and intracellular processes on coral reefs support this pyramid by mediating energy and nutrient flows in unique and noteworthy ways.

Herbivory

As discussed elsewhere (Chapters 7 and 28), herbivory is an important process on coral reefs. An adequate level of herbivory is necessary for maintenance of the balance between reef-building corals and macroalgae, and hence many of the typical characteristics of a coral reef. Any visit to an intact coral reef will yield countless observations of animals moving across the substrate and cropping the vegetative growth of marine plants. Key herbivores include gastropod molluscs, chitons, echinoids, crabs and fish (Fig. 7.6). These organisms may be active at different times of the day or light level (e.g. dawn, dusk, night) and tidal heights. Many gastropods (e.g. abalone) and chitons set out from their hiding places at night to graze on benthic microalgae while many grazing fish (e.g. Scarids, Siganids and Acanthurids) are active only by day and will wait until the tide floods the reef crest before moving in to graze on these rich and highly productive reef areas. Grazing activity is often evident from scrape marks made by the tough beaks of grazing fish (Fig. 7.6*B*), or by the presence of feeding scars from grazing molluscs (Fig. 7.6*C*). Some grazing activity can be very subtle. Recent studies have revealed that porcellanid crabs (see Chapter 23) are significant grazers on coral reefs, cropping the benthic microalgae from the substrate using their chelae that are like tiny scissors. In this case, there is little visual evidence of the grazing activities of these organisms despite their importance in terms of the mass of material removed each day from a coral reef.

Predation

Undisturbed coral reefs are visually exciting for many reasons, one of which is the presence of predators of all sizes and shapes (see Chapters 28, 29, 30, 31). Predation is a dominant driving process, from the tiniest predators such as the chaetognaths and fish larvae that eat copepods, to sharks that eat large organisms such as rays, turtles, groupers and smaller shark species. In coral reef ecosystems, sharks are often referred to as apex predators (Fig 7.6*A*). As with herbivores, some predators are nocturnal and have evolved to hunt in almost no light at all. Others prey on diurnal grazers and time their visits to the reef with the tidal cycle.

Particle feeding

Many of the organisms that live on coral reefs feed on particles such as plankton or small pieces of debris suspended in the water column, or deposited on submerged

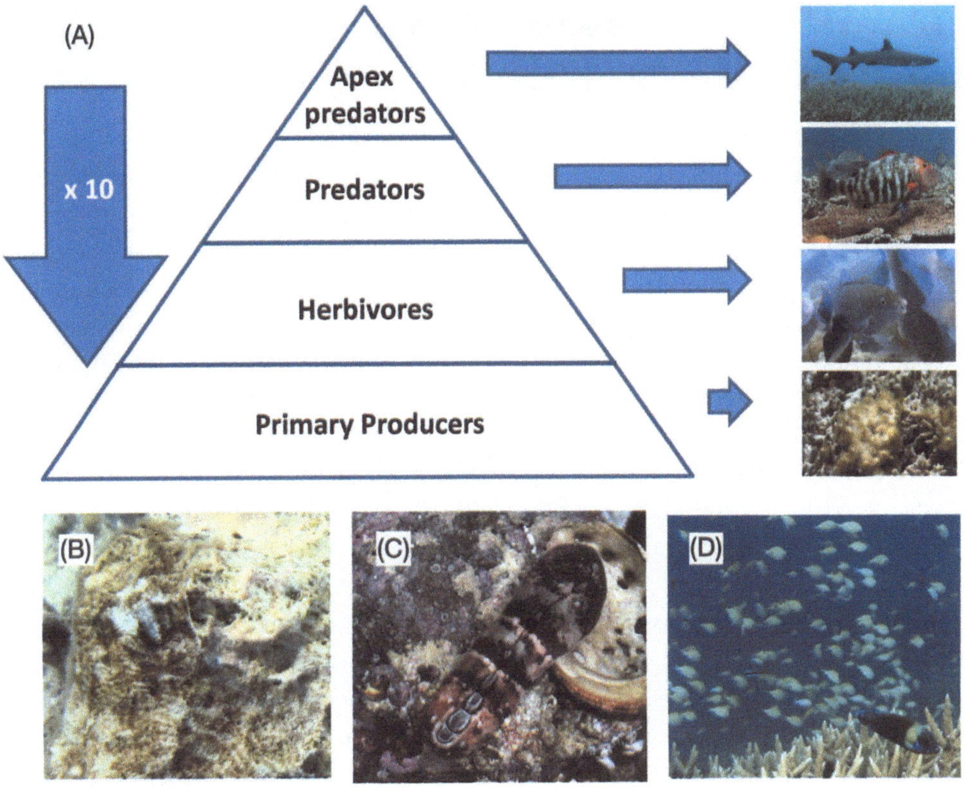

Figure 7.6 Trophic interactions on coral reefs. *A*, Trophic pyramid showing the flow of energy from primary producers such as the benthic algae to apex predators such as the White-tip reef shark shown. *B*, Herbivores such as parrotfish leave characteristic scrape marks. *C*, Other invertebrates such as chitons graze the surfaces in between coral colonies and may be significant reef grazers. *D*, Another important input of energy to the reef system occurs through the large numbers of particle feeders that line the seaward edges of coral reefs forming a 'wall of mouths'. (Photos: O. Hoegh-Guldberg; Figure: D. Kleine and O. Hoegh-Guldberg.)

surfaces. Particle feeders are part of an important energy and nutrient loop (see next section). Particle feeders range from the largest (whale sharks) to the smallest organisms (copepods) of reef associated species. They include sponges, which use special cells called choanocytes to drive water through small channels and trap particles. Many simple (hydroids, anemones) and advanced invertebrates (polychaetes, bivalve molluscs, ascidians) use cilia to drive water past their feeding structures and use sticky mucous to trap them prior to ingesting them. Other organisms such as crinoids use tube feet to pick particles out of the water while orienting feeding structures into the currents. Polychaete worms and sea cucumbers have specialised feeding structures that allow them to feed on particles deposited on sediment surfaces. Some fish actively pick particles out of the water column, while others sift sediments using their gill rakers to find small pieces of detritus for ingesting.

The assemblage of particle feeders along the reef crest has been termed the 'wall of mouths'. The 'wall' plays a crucial role in the accumulation of energy and nutrients on coral reefs. Few large zooplankton from the open ocean survive contact when they float by the cloud of particle feeding fishes at the edges of coral reefs (Fig. 7.6*D*). The amount eaten (and hence the energy acquired by the reef ecosystem) is highly significant, as shown when feeding rates are calculated per unit time per metre of reef-ocean interface (in one study, approximately 0.5 kg m^{-1} d^{-1} wet weight zooplankton, mostly larvaceans and copepods, entered the reef economy). It appears that the 'wall of mouths' represents

BOX 7.1 THE MUTUALISTIC ENDOSYMBIOSIS OF CORALS AND DINOFLAGELLATES

Reef-building corals and invertebrates from at least five invertebrate phyla form close associations with dinoflagellates from the genus *Symbiodinium*. Often referred to as zooxanthellae (a loose, non-taxonomic term), these single-celled plant-like organisms live within the endodermal cells (gastroderm) of reef-building corals. Here, they photosynthesise like other phototrophs, but instead of retaining the organic carbon that they make, *Symbiodinium* releases up to 95% to the host. This energy is used by the coral to grow, reproduce and produce copious amounts of calcium carbonate, which forms the framework of coral reefs. Only corals that have a symbiosis with *Symbiodinium* are able to calcify at the high rates that are typical of reef-building corals.

In return for this copious energy, *Symbiodinium* receives inorganic nutrients from the waste metabolism of the animal host (Fig. 7.7). Given the shortage of inorganic nutrients such as ammonium and phosphate ions in tropical and subtropical water columns, the provision of these nutrients is critical to the high rates of photosynthesis and energy production of *Symbiodinium*. Because there are benefits for both partners in this symbiosis and one cell lives inside the cells of another, this symbiosis is referred to as a 'mutualistic endosymbiosis'. The tight recycling of energy and nutrients between the primary producer and consumer avoids the problem of the low concentrations of these materials in typical tropical seas. This is thought to be one of the key reasons why coral reefs are able to prosper in the otherwise nutrient 'deserts' of tropical seas.

yet another important part of the puzzle for how coral reefs maintain themselves in the nutrient-poor (oligotrophic) conditions of tropical seas.

One of the hallmarks of coral reefs is the high number of mutualistic symbiotic relationships across a large range of organisms. These relationships number in the thousands and involve all sorts of interactions, from those between gobies and burrowing shrimp to the cellular symbioses between sponges and bacteria. One of the central hypotheses surrounding coral reefs is that the large proportion of mutualistic symbioses have arisen due to the low nutrient conditions that dictate the advantages of a close association of primary producer and consumer. The ultimate outcome of these close associations is that the inorganic nutrients required by the primary producer are obtained directly from the animal consumer. This avoids the dilution that would otherwise happen if the nutrients and organic matter were to enter the water column. There is no better example of the ultimate close association than that of reef-building corals.

Reef-building corals form a mutualistic symbiosis with single-celled dinoflagellate protists (genus *Symbiodinium*) that live inside the gastrodermal cells of corals where they photosynthesise, passing large amounts of captured energy to the coral host (see Fig. 7.7 and Box 7.1). In return for the energy contributed to the coral host, the symbiotic dinoflagellates receive access to inorganic nutrients arising from animal metabolism. The advantages of the close coupling of coral and *Symbiodinium* spp. are enormous, resulting in large photosynthetic rates that power the metabolically expensive process of calcification. Significantly, only animals that are symbiotic with *Symbiodinium* calcify at rates that are significant enough to contribute significant amounts of energy to reef accretion. The close relationship between corals and symbiotic dinoflagellates has been in existence for at least 220 million years and is largely responsible for the huge reserves of limestone found in the

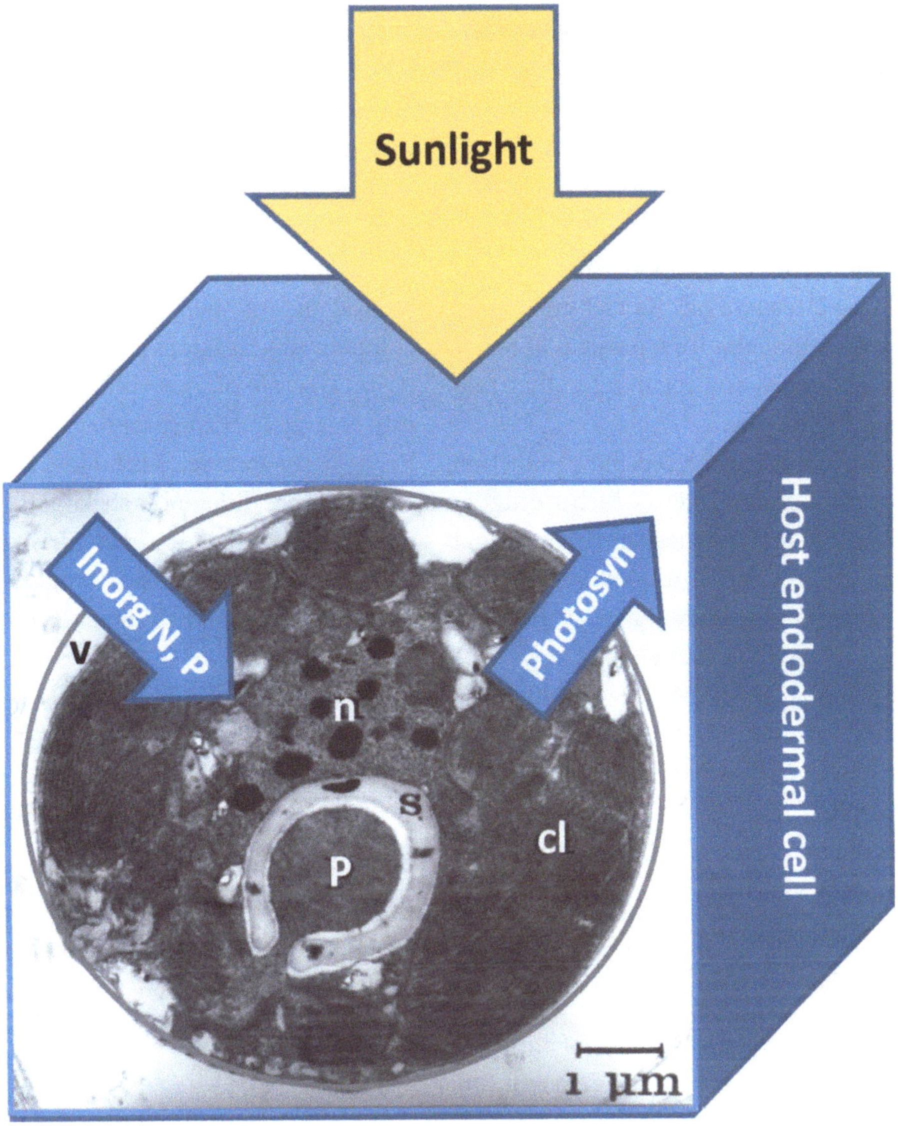

Figure 7.7 Illustration of the relationship between *Symbiodinium* (transmission electron micrograph, scaled 1 µm) and the host endodermal cells of reef building corals. Sunlight illuminates the transparent host cells driving photosynthesis of *Symbiodinium*. *Symbiodinium* passes copious amounts of resulting photosynthetic products (labelled 'photosyn') to the host cell that in return allows *Symbiodinium* access to inorganic nutrients such as ammonium and phosphate arising from host catabolism (labelled 'Inorg N, P'). n, nucleus; s, starch cap; cl, chloroplast; p, pyrenoid; v, vacuole space between host vacuole membrane and out plasmalemma of the enclosed *Symbiodinium* cell. (TEM image: O. Hoegh-Guldberg.)

upper layers of the Earth's crust. As pointed out elsewhere, the limestone structures generated by corals and other organisms generate the habitat for over a million species of plant, animal, fungi and bacteria worldwide.

Studies of the food webs on coral reefs have identified a major role for the mucus generated by reef-building corals. Mucus is considered to be relatively cheap to produce due to the abundant energy available for corals in shallow habitats (hence the concept of 'junk carbon'). It is primarily produced to prevent the surfaces of corals from being colonised by fouling organisms and may have a role in protecting corals from excessive light (PAR and Ultra-Violet Radiation, UVR). It tends to be sloughed of corals at the end of the day (after extensive photosynthesis has occurred). Corals on the intertidal reef flat at Heron Island exude up to 4.8 litres of mucus per square metre of reef area per day, and, of that, up to 80% dissolves in the reef water. While the dissolved component stimulates a burst of metabolic activity in the sediments where it is largely metabolised, the remaining particulate proportion is eaten by fish and other particle feeders on the reef crest. This transfer of energy is thought to represent a major trophic exchange of energy, and relative to other marine food webs is fairly unique.

In summary, this chapter explored the production and flow of energy through coral reefs, which are highly productive ecosystems that prosper in the nutrient poor waters of the tropics. While tropical oceans that surround most coral reefs, they have a primary productivity that is close to zero, the coral reefs that they bathe often have levels of primary productivity that are among the highest in the ocean. This productivity is a manifestation of the efficient photosynthetic processes and recycling that occurs within the warm and sunlit setting of coral reefs. We also examined one of the key nutrient cycles of coral reefs, that of nitrogen, observing that nitrogen is regenerated by nitrogen fixation and that it cycles between the different organisms within the food web of coral reefs along with the energy of organic carbon bonds. The 'wall of mouths' clouds of small fish and other particle feeders that forage at the interface of coral reefs and the open ocean play an important role in acquisition of energy. As well, the efficiencies of mutualistic symbioses like those seen between corals and symbiotic dinoflagellates have huge benefits to a wide range of organisms in the dilute nutrient conditions of tropical seas.

Probably no single factor can explain why coral reefs are so productive. The answer probably lies in combinations of characteristics and mechanisms that generate and recycle nutrients. There is one important take-home message that may not be obvious at first: it is a mistake to think that much of the energy generated can be harvested as a net product of the system. Tight recycling of nutrients and energy means that the majority of primary production is rapidly recycled back into the ecosystem by the many pathways elucidated in this chapter. This has been likened to a 'beggar's banquet' where the table is set for a feast that looks at first glance to be generous and abundant yet very little can be eaten or taken away from the table. This situation appears to be fundamentally different than that seen in marine ecosystems such as kelp forests where the rate of primary production can be large seasonally and substantial amounts of energy and organic carbon are exported out of the kelp ecosystem. Another way of understanding this is to compare the measures of P_G and P_N on a community basis. The rate of gross photosynthesis of the community is large in both coral reefs and in kelp forests, but $P_{N\,community}$ of coral reefs is far less than $P_{N\,community}$ of kelp forests. These differences strike at the heart of the unique nature of coral reefs and may drive other emergent features such as the sensitivity of coral reefs to small changes in environment that surrounds them.

ADDITIONAL READING

Anthony, K. R. N., and Hoegh-Guldberg, O. (2003). Variation in coral photosynthesis, respiration and growth characteristics in contrasting light microhabitats: an analogue to plants in forest gaps and understoreys? *Functional Ecology* **17**, 895–899.

Benson, A., and Muscatine, L. (1974). Wax in coral mucus – energy transfer from corals to reef fishes. *Limnology and Oceanography* **19**, 810–814.

Darwin, C. R. (1842). 'The Structure and Distribution of Coral Reefs.' (Smith Elder and Company: London.)

Hamner, W. M., Jones, M. S., Carleton, J. H., Hauri, I. R., and Williams, D. (1988). Zooplankton, planktivorous fish, and water currents on a windward reef face: Great Barrier Reef, Australia. *Bulletin of Marine Science* **42**, 459–479.

Hatcher, B. G. (1988). Coral reef primary productivity: a beggar's banquet. *Trends in Ecology and Evolution* **3**, 106–111.

Hoegh-Guldberg, O. (1999). Coral bleaching, climate change and the future of the world's coral reefs. *Marine and Freshwater Research* **50**, 839–866.

Hughes, T. P., Baiard, A. H., Bellwood, D. R., Card, M., Connolly, S. R., Folke, C., Grosberg, R., Hoegh-Guldberg, O., Jackson, J. B. C., Kleypas, J., Lough,

J. M., Marshall, P., Nyström, M., Palumbi, S. R., Pandolfi, J. M., Rosen, B., Roughgarden, J. (2003). Climate change, human impacts, and the resilience of coral reefs. *Science* **301**, 929–933.

Kleypas, J. A., McManus, J., and Menez, L. (1999). Using environmental data to define reef habitat: Where do we draw the line? *American Zoologist* **39**, 146–159.

Muscatine, L. (1990). The role of symbiotic algae in carbon and energy flux in reef corals. In 'Coral Reefs'. (Ed. Z. Dubinsky.) pp. 75–84. (Amsterdam: Elsevier.)

Wild, C., Huettel, M., Klueter, A., Kremb, S. G., Rasheed, M., and Jørgensen, B. B. (2004). Coral mucus functions as an energy carrier and particle trap in the reef ecosystem. *Nature* **428**, 66–70.

8. Calcification, Erosion and the Establishment of the Framework of Coral Reefs

P. A. Hutchings & O. Hoegh-Guldberg

Much like a city represents an equilibrium between processes of building construction and demolition, coral reefs are the net result of processes that form calcium carbonate (calcification) and those that take it away (physical and biological erosion). The resulting reef framework is quintessential to coral reefs, forming the habitat for tens of thousands of reef species. This chapter examines the major forces that control this equilibrium state, examining the physical, chemical and biological forces that are involved. It also investigates the role that humans have in influencing these processes, highlighting the large scale effects of both local (eutrophication, land runoff) and global (global warming, ocean acidification) factors on reef structures. Some of these issues will also be discussed further in Chapter 10.

CALCIFICATION

Calcification is highest in the warm, sunlit waters of the tropics and subtropics where the concentrations of calcium and carbonate ions are highest. As discussed in Chapter 2, these three factors are considered to be the major determinants for where coral reefs grow, with waters becoming the too cold, dim and low in the abundance of these crucial ions as one goes poleward. This association of calcification with low latitude environments appears to have held for many hundreds of millions of years, resulting in huge deposits of calcium carbonate within the Earth's geological structure.

The organisms that calcify on coral reefs are diverse and include a large number of phyla including cnidarians, molluscs, crustaceans and foraminifera as well as green and red algae. Many of the organisms that calcify at high rates are also symbiotic with dinoflagellates such as *Symbiodinium* (see Chapter 7). This association with organisms having high photosynthetic capabilities is considered to be indicative of the high energy requirements of the calcification process. These organisms take up calcium and carbonate ion from the super-saturated concentrations typical of tropical and subtropical waters, depositing either calcite or aragonite (two forms of calcium carbonate crystals). The form of calcium carbonate deposited depends mostly on the organism involved. Corals, for example, deposit aragonite while red coralline algae deposit Mg-calcite.

Organisms use calcium carbonate to create skeletons that either provide rigid support structures and/ or protective shells or cases. These functions are likely to be crucial in the highly dynamic coral reef ecosystem, where competition and predation can be high.

The mechanism by which organisms produce calcium carbonate skeletons has yet to be conclusively determined. Within the water column of tropical and subtropical oceans that is saturated with respect to both calcium and carbonate ions, there appears to be three possible ways in which a high rate of precipitation can be fostered. The first depends on bulk metabolic energy to concentrate calcium and carbonate ions within confined spaces and subsequently to cause a rapid precipitation of aragonite or calcite. The second is that the symbionts assist by removing so-called 'crystal poisons' such as phosphate that otherwise retard the formation of crystals. The third is the production of specialised proteins that are often referred to as 'skeletal matrix proteins'. These particular proteins tend to be highly anionic (covered in negative charges) and contain regions that are associated with enzymes such as carbonic anhydrase (that catalyses the rapid conversion of carbon dioxide to bicarbonate and protons, a reaction that occurs rather slowly in the absence of a catalyst). Although there is some debate over which is more important, there is good evidence that all of these processes may play roles of differing importance within the variety of organisms that calcify within coral reef environments.

There are a number of ways calcium carbonate deposition is measured. These are outlined in Box 8.1. The rates of calcification on coral reefs can be extremely high in equatorial or low latitude areas of the planet. On a more regional scale, the deposition of calcium carbonate varies with the presence or absence of rivers, where high nutrients and sedimentation may slow the deposition of calcium carbonate. In this respect, inshore coral reefs on the GBR do not deposit calcium carbonate as fast as those reefs that are in more offshore positions. Again, this is a consequence of changes in factors such as light, temperature and nutrients. At the scale of a reef, calcium carbonate deposition can be quite dynamic and will vary between the slope, crest and back-reef areas as discussed already in Chapter 7. Coral reefs are also dynamic in geological time frames, with the shape of deposited calcium carbonate varying over time in response to prevailing winds and currents. These aspects of reef construction are discussed in Chapter 2.

The skeletons of calcifying organisms build up and construct the accumulated calcium carbonate debris that constitutes the solid component of the framework of coral reefs. Calcifiers on coral reefs can have different roles, for example the massive and branching structures provided by corals require the activities of encrusting red algae to essentially glue them into a consolidated framework (Fig. 8.1*A*). The rate of calcification normally greatly exceeds the rate of erosion on carbonate coral reefs. Estimates of calcification suggest that rates vary from 1–2 m per century while rates of reef growth are about 1–2 m per millennium. Based on these rough figures, this would suggest that rates of calcification are between 3–10 times higher than the rate at which calcium carbonate is removed by physical and biological erosion. As we will see later in this chapter, the balance between the two forces (calcification versus erosion) is critical to understanding the impacts of global change, such as ocean acidification.

PHYSICAL AND BIOLOGICAL EROSION

The removal of calcium carbonate from coral reefs (erosion) is a key process on coral reefs that involves a number of elements including dissolution, physical breakage and the activities of a number of so-called bioeroders. These elements are intertwined and it is difficult to separate them. They are a feature of recent as well as fossil coral reefs.

Wave action erodes the reef slowly over time by physical action and chemical dissolution of the reef substratum. During storms, however, this rate will increase and large boulders may be dislodged. As they roll down the reef slope they may physically remove many more coral colonies (Fig. 8.1*B*, *C*). These forces can have significant impacts on the shape of coral reefs (e.g. spur and groove formations such as those seen on Wistari Reef (Fig. 8.1*D*)). Depending on the wave energy and the relative hardness of coral skeletons (that itself is affected by chemical and biological factors), the impacts of storms can be substantial. After a cyclone has passed through an area it appears as if the living veneer of the reef has just been peeled off and shed.

Biological erosion consists of the loss of reef substratum by boring and by grazing. A suite of organisms

BOX 8.1 HOW TO MEASURE CALCIFICATION

Accurate measurements of coral calcification are critical to any in-depth understanding of the rate that it plays in reef processes. There are a number of methods that have been used over the years to measure the rate at which calcium carbonate is deposited by corals and other calcifying invertebrates. A number of these methods are outlined below, with some comments on their ease of use and efficacy.

Dye to measure linear extension

Corals build their skeletons by incrementally adding layers of calcium carbonate. If the thickness of added calcium carbonate is known, then it is possible to obtain a relative measure of the rate of calcification. Many studies have used the fluorescent dye Alizarin, which binds effectively to proteinaceous elements of the coral skeleton. To undertake measurements with marker dyes, corals are incubated in non-toxic levels of the dye for several hours before being placed back in the field. Following exposure to the dye, which essentially marks the beginning of the period over which skeletal growth will be measured, corals are allowed to grow under field conditions before being harvested several months to years later. Back at the laboratory, the corals are killed and their skeletons cut into thin sections to show the annual banding patterns (exposed using X-ray photography). By measuring the distance between the skeletal surface (the site of the latest deposition of skeleton) and the fluorescent band (representing the beginning point), the linear extension of skeleton can be calculated as a function of time. While this method is easy to use in the field, it can only be used as a relative measure as growth varies across the colony surface, and hence is not a reliable measure of the volume of skeletal material that has been formed during the experimental period.

Using radioisotopes to follow the deposition of calcium atoms

Calcium is the cation that is deposited along with carbonate ions as the coral forms a skeleton. If $^{45}Ca^{2+}$ (a radioactive isotope of Ca^{2+}) is mixed into the water surrounding a coral, it immediately gets taken up by the coral as calcification occurs. By measuring the ratio of radioactive atoms to non-radioactive atoms of Ca^{2+}, the rate at which all Ca^{2+} atoms are deposited can be calculated. This technique is very precise and can yield an accurate measure of the rate of calcification of corals. The downside of this technique, however, is that it must be done in the laboratory, is potentially hazardous if the right precautions are not taken, and may not accurately depict calcification as it might occur on a coral reef (due to its requirements for lab conditions).

Buoyant weight method

From the difference in density between calcium carbonate and seawater (known entities), it is possible to convert the weight of an object that has been measured in sea water into an absolute measure of calcium carbonate deposition. Assumptions made in applying this technique are that the only negatively buoyant part of the coral colony is its calcium carbonate, which is largely true for the tissues of corals. This technique has the advantage of producing an accurate measurement of the total calcium carbonate in an experimental coral as well as having the advantage that corals do not have to be killed during the measurement process. Many studies use the buoyant weight technique on corals that have been grown on tiles that can be brought in from the field and periodically weighed.

Figure 8.1 Wave action is a key process on coral reefs and underpins the physical erosion of reefs. *A*, section through the reef crest at Heron I. (L, living corals; C, consolidated framework of dead corals stuck together by calcareous red algae); *B*, reef crest at the Low Isles; *C*, reef crest at Heron I., showing the reef break where physical eroding forces are maximum; *D*, Heron I., spur and groove. (Photos: O. Hoegh-Guldberg.)

including polychaetes, molluscs, sponges, barnacles, sipunculans, and various micro-organisms such as bacteria and algae bore into coral substrata. Endolithic algae colonise the skeletons of corals (Fig. 8.2*A*) that colonise the surface layers of the substratum together with the turf algae (Fig. 8.2*B*). Algae is grazed by fish (Fig. 8.2*C*, *D*) as well as invertebrate grazers such as chitons (Fig. 8.2*E*), echinoids (Figs 8.2*F*, *G*; 26.8*A*–*F*), and gastropods (Fig. 8.2*H*). These organisms physically bite or scrape the substratum to collect the algae (Fig. 8.3*A*) and with it they take particles of the substratum that has become honeycombed by the action of the borers (Fig. 8.3*B*). This calcium carbonate matrix, together with the algae, passes through the gut of the grazers and is ground up, separating the algae from the calcium carbonate. Cellulose enzymes break down the plant cells, the nutrients are then absorbed, and then the calcium carbonate is defecated as a fine powder. Swimming behind schools of large schools of parrotfish (scarids, Fig. 8.3*C*) one often sees the water column becoming cloudy as this fine powder is ejected (Fig. 8.3*D*). Similarly the faecal pellets of grazing echinoids consist largely of compacted finely ground calcium carbonate. The lagoonal sediments, especially those offshore, are largely composed of these products of bioerosion and physical erosion—along with mollusc

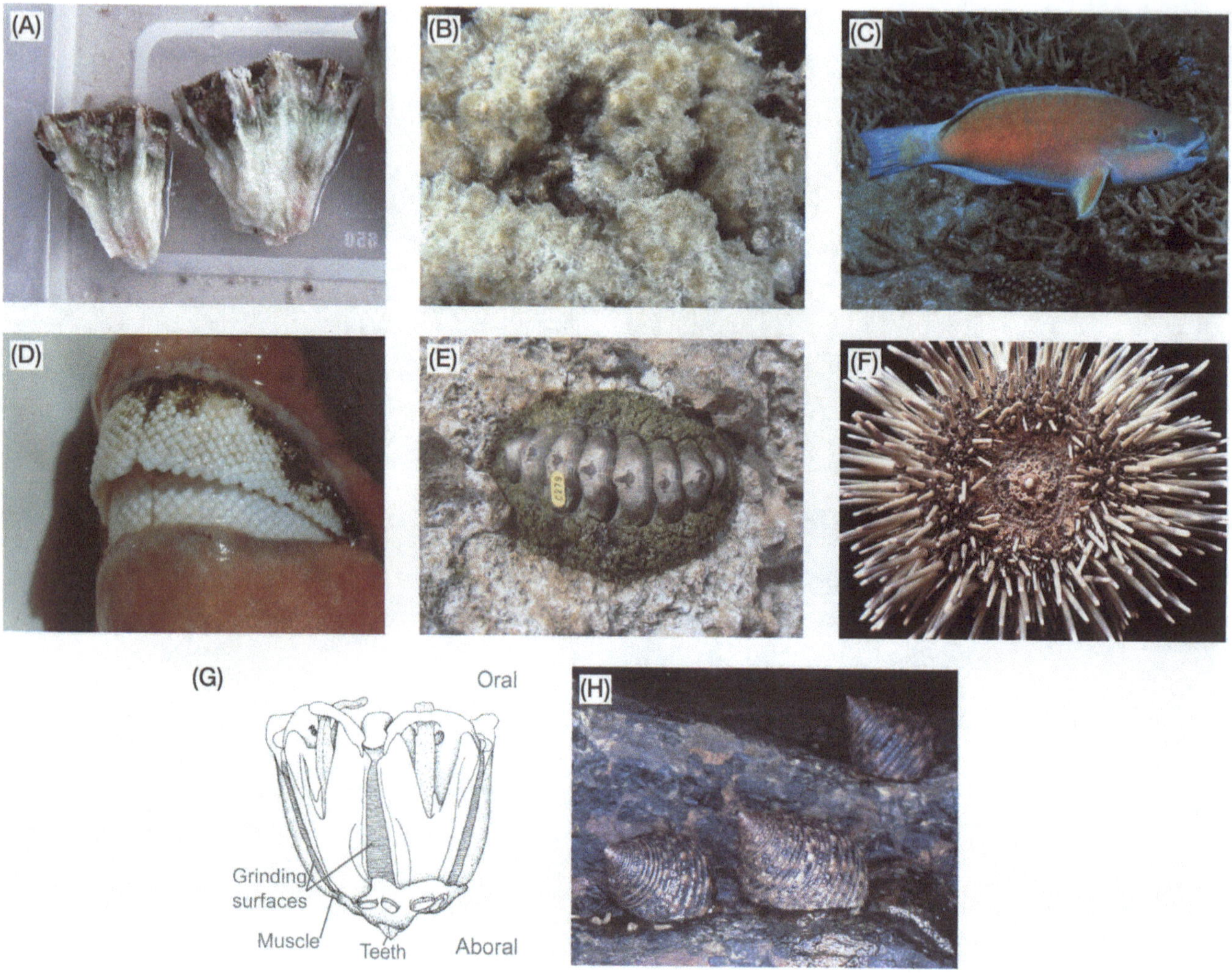

Figure 8.2 *A*, Endolithic algae inhabit the skeletons of corals, living amongst the crystals and over time weakening the skeleton. *B*, Dead coral substratum covered by turf algae. (Photo: O. Hoegh-Guldberg.) *C*, The parrot fish *Scarus* sp. With well developed jaws about to take a lump of dead coral substratum full of endolithic algae. (Photo: O. Hoegh-Guldberg.) *D*, Jaws of *Bolbometopon muricatum* on the outer barrier near Lizard I. (Photo: D. Bellwood.) *E*, Close-up of *Acanthopleura gemmata* from One Tree I., nestled onto its home scar. (Photo: B. Kelaher.) *F*, The grazing echinoid *Echinometra mathaei*, oral surface showing Aristotle's lantern partially protruding from the mouth that it uses to actually scrape off the surface of the coral. (Photo: A. Miskelly.) *G*, Diagram of Aristotle's lantern. (Illustration after Anderson, 1996.) *H, Monodonta labio* (Trochidae) feeding. (Photo: K. Gowlett-Holmes.)

shells, carapaces of crustaceans, foraminifera tests and sponge spicules. Only sediments adjacent to the coast or large islands have a component of terrestrially derived sediments.

Storm activity will dislodge coral colonies both live and dead (Fig. 8.1*B*), which have often been weakened by borers attacking the base and branches of corals. These coral colonies and broken off branches of the staghorn corals (*Acropora*) can be washed down to the bottom of the reef slope or thrown up onto sand cays. This band of coral rubble, which is often well developed on the windward side of cays, forms an important coral reef habitat for a wide range of organisms. Coral rubble is itself subjected to further bioerosion, although often it develops a protective coat of coralline algae that provides some protection from the colonisation of the substratum by endolithic algae. Such surfaces lacking endolithic algae are therefore not grazed by parrotfish

Figure 8.3 *A*, bite marks of a scarid (f), and a boring barnacle embedded in *Porites lutea* (b) (photo: O. Hoegh-Guldberg); *B*, in situ dead coral habitat split open to reveal boring sipunculans and bivalves, burrow of boring bivalve (t) (photo: P. Hutchings); *C*, schools of *Bolbometopon muricatum* at Osprey Reef, Coral Sea (photo: P. Hutchings); *D*, defaecation by parrotfish, fine sediment produced by the grinding of the ingested coral fragments (photo: D. Bellwood).

and echinoids. Dead coral substrata adjacent to river mouths, where large plumes of sediment loaded water flow out onto the reef during the wet season (see Chapter 11), tend to be covered in a thick layer of silt that again protects the substratum from endolithic algal colonisation. This has been documented on reefs adjacent to Low Isles and those at the mouth of the Daintree River in North Queensland. Live coral colonies have mechanisms to eject the sediment as it settles on the coral polyps, whereas it can accumulate on dead substratum.

While live coral colonies are typically unbored, once parts of the colony die the substratum is rapidly colonised by borers. As all borers have pelagic larvae it is difficult for them to settle on the living veneer of a coral colony without being eaten by the coral polyps. In cases where borers have settled in living colonies it is presumed that larvae have settled on damaged polyps allowing them to metamorphose and rapidly bore into the substratum before being eaten by a neighbouring polyp. Once coral substratum becomes available for colonisation by borers, a distinct succession occurs, with the early settlers being bacteria, fungi and endolithic algae that appear to condition the substratum and facilitate the next suite of colonisers, primarily polychaete worms, and later sponges, sipunculans and molluscs including bivalves and boring barnacles. Many of these early colonisers are short-lived and create burrows for a suite of non boring organisms or nestlers to colonise. In contrast, the sipunculans, sponges, bivalves and some of the larger polychaete borers are long-lived and once they have created their burrow they are entombed in the substratum and may then live for several years (Fig. 8.3*A*, *B*). These organisms bore by either physically eroding the substratum or chemically dissolving it, or by a combination of these methods.

Rates of boring decrease once the borers are established, and subsequent rates are just sufficient to allow those organisms to grow. This is particularly true for sponge colonies. Borers must obviously retain a link to the outside of the substratum in order to obtain their food, for respiration and for discharging their gametes. Once established in the substratum they are effectively entombed, often in flask-shaped burrows (Fig. 8.3*B*). They cannot leave their habitat, although some species of molluscs, primarily *Conus* spp., search out for particular boring species of polychaetes and sipunculans and insert their proboscis into the burrow and then proceed to suck out the worm.

Experimental studies have shown that recruitment of boring organisms is seasonal with maximum recruitment occurring during the summer months; however, some recruitment of borers occurs throughout the year. This means that within weeks of substratum becoming available it is already being colonised by borers. Recruitment also varies between years and this is presumably related to the availability of larvae, the supply of which will be influenced by weather patterns at the time of spawning. Net rates of bioerosion (losses due to grazing and boring plus gains from accretion from coralline algae and encrusting organisms, plus physical and chemical erosion) vary between sites on an individual reef as well as between reefs (Fig. 8.4), and differences occur between oceans. Factors such as water quality and sediment load influence not only the rates, but the agents responsible for grazing and boring.

On the Great Barrier Reef, various groups of scarids (parrotfish) are important grazers. Scarids can be divided into three distinct functional groups depending on the osteology and muscle development on the oral

BOX 8.2 HOW ARE RATES OF BIOEROSION MEASURED?

While a piece of dead coral substratum can be cut open and the amount of calcium carbonate that has been removed calculated, and borers identified, this does not give you any idea of erosion *rates*. A better way to study bioerosion is to use experimental blocks of freshly killed coral that show no sign of boring and to lay them out on the reef (Fig. 8.5C) for fixed time periods. As the original dimensions and density of the blocks are known, losses and gains can be calculated and assigned to the various organisms. Obviously, replicates need to be collected for each time period in order to determine the variation within a site before considering variation between sites. Thin sectioning of the blocks allows the distribution and density of the various microborers to be determined. Part of the block can be dissolved in order to extract the borers that can then be counted and identified to species level. While net rates of grazing can be determined from changes to the dimensions of the block (See Fig. 8.5*D*), potential grazers in the region need to be identified, their densities calculated and the amount of calcium carbonate in their faecal pellets measured. In the case of scarids, the depth and dimensions of the feeding scars (Fig. 8.3*A*) can be measured and for echinoids, their population density can be measured and faecal pellets collected over a 24 hr period to estimate the amount of calcium carbonate they contain in order to calculate rates of grazing. Knowing rates of calcification in the area, a balance sheet of losses and gains for the area can be constructed. However, it must remembered that these rates may vary considerably within a reef, so numerous replicates must be used in order to gain reliable data.

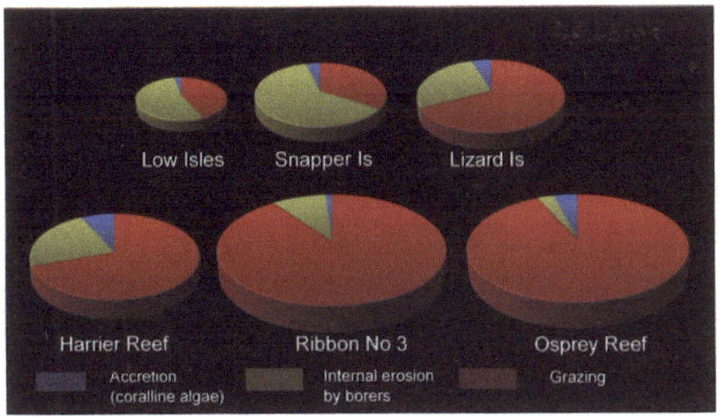

Figure 8.4 Variations in rates of grazing, accretion and internal erosion by borers across the Great Barrier Reef from the Daintree River out into the Coral Sea along a gradient of increasing water clarity determined from experimental blocks as illustrated in Figure 8.5*C* (after Hutchings *et al.* 2005). Loss (grazing and boring) + gain (accretion) = net rates of bioerosion. (Drawing: K. Attwood.)

and pharyngeal jaws. While all scarids have well developed scraping plates or a beak (Fig. 8.2*D*), the actual development of these plates and associated muscles determines what they can feed on. One type, the croppers, remove only algae and associated epiphytic material, whereas scrapers and excavators remove pieces of substratum together with the algae and leave distinctive feeding scars (Fig. 8.3*A*). The only difference between scrapers and excavators is the depth to which they can bite. These species are able to break down the calcium carbonate using their pharyngeal jaws. These latter two categories are feeding on the surface layers of dead coral substratum containing endolithic algae (Fig. 8.2*A*). The large double header *Bolbometopon muricatum* (Fig. 8.3C) feeds almost exclusively on live coral, often on the faster growing species such as *Pocillopora* and the tabulate acroporids. On the GBR, many species of scarids occur and their distribution and functional roles such as grazing, erosion, coral predation and sediment reworking vary across the reef. Inner shelf reefs support large numbers of scarids, although their biomass is low, and they exhibit high rates of grazing and sediment reworking. In contrast, the outer shelf reefs have much lower densities of scarids but there the biomass is much higher and they are responsible for higher rates of erosion by grazing and coral predation. Mid-shelf reefs have intermediate values. In areas of overfishing, loss of scarids and other herbivorous species of fish such as surgeonfishes, rabbitfishes and drummers can have significant impacts on the reef, as the algae are not being removed by grazing leading to intense competition for space between corals and benthic macroalgae. This can lead to a shift from a coral dominated environment to one dominated by macroalgae.

In the Caribbean and on some French Polynesian reefs echinoids are important grazers, especially *Diadema setosum* (Fig. 8.5*A*), *D. savignyi* and *Echinometra mathaei* (Fig. 8.2*F*, *G*). These species are relatively uncommon on the GBR. These echinoids graze on the algal covered reefal substratum using their Aristotle's lantern, a complex series of calcareous plates that scrape the surface (Figs 8.2*G*, 26.8*C*). The densities of these species are influenced by water quality and high levels of nutrients encouraging algal growth. A recent study in Papeete, Tahiti, showed that overfishing has reduced fish populations that graze on juveniles of echinoids and this has allowed high densities of *Echinometra mathaei* (Figs 8.2*F*, 8.6) to develop, which thrive on the excessive algal growth that is being driven by high levels of nutrients in the water column. These high levels of nutrients are being washed down from nearby rivers where untreated sewage is being discharged. Excessive algal growth restricts coral recruitment and over time if water quality is not improved, rates of grazing and associated boring will far exceed rates of calcification and this is already leading to substantial loss of reef framework on this reef (Fig. 8.6). This will have massive flow-on effects, including loss of protection from storm activity on nearby low lying areas, loss of coral reefs leading to reduced fish landings and the loss of tourism.

On the GBR, species of *Echinostrephus* (Fig. 8.5*B*) are present and can often be seen nestling in their home depression that they have eroded in the reef substratum. They feed on plankton and suspended matter and basically stay in these depressions.

The chiton *Acanthopleura gemmata* is common on the GBR in the intertidal zone (Fig. 8.2*E*). It forms deep depressions to which it returns after foraging on algae during low tides. It is has been suggested that feeding at low tide may decrease the predation risk from fishes and sharks that move over the area as the tide rises.

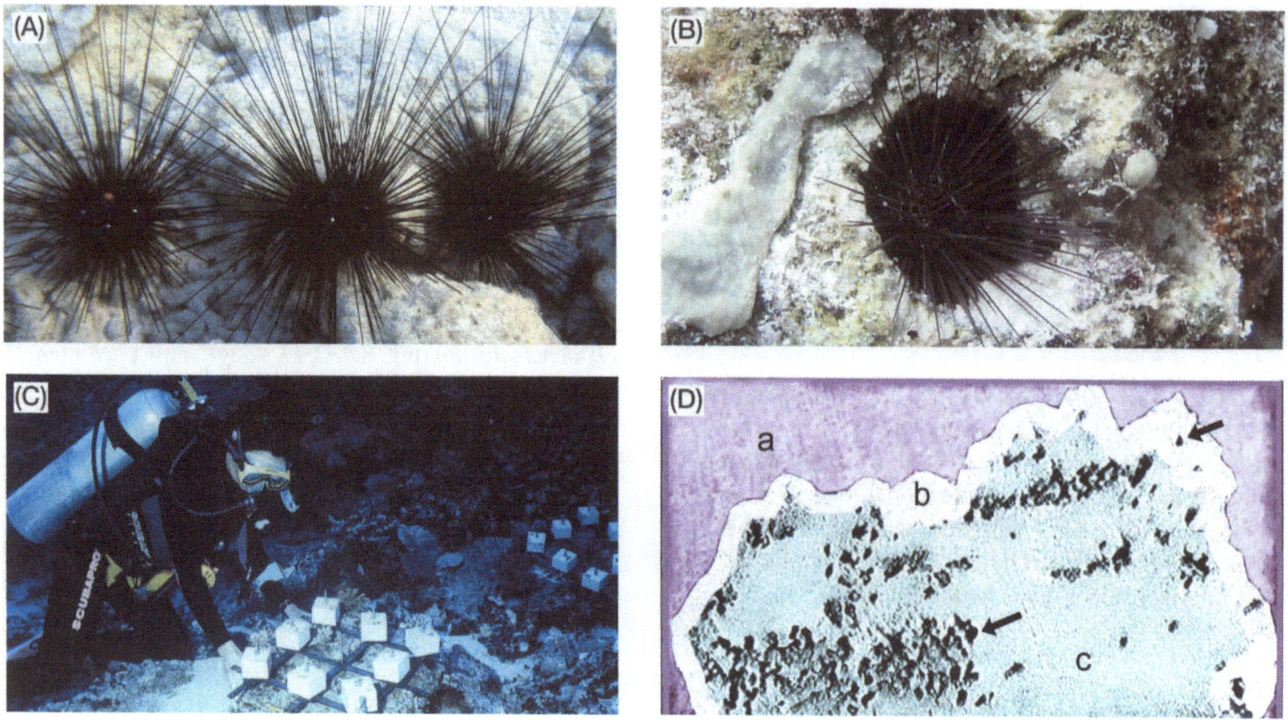

Figure 8.5 *A, Diadema setosum* a grazing echinoid linked to major erosion of western Indian Ocean reefs (photo: O. Hoegh-Guldberg); *B, Echinostrephus* sp., sitting in its home scar that it has eroded (photo: O. Hoegh-Guldberg); *C*, experimental study of bioerosion at Osprey Reef, Coral Sea, two replicate grids with newly laid coral blocks to be exposed for varying lengths of time (photo: J. Johnson). *D*, Diagrammatic representation of coral block illustrating how the various components of bioerosion (i.e. grazing, accretion and boring) are determined from a series of sections through each block. Knowing the density of the coral block, these measurements can then be scaled up to rates per square metre and then net rates of bioerosion calculated. a, original block; b, accretion; c, block remaining after grazing and boring (image: K. Attwood).

Figure 8.6 Experimental blocks after six months showing extensive grazing by *Echinometra mathaei* at Faaa, Tahiti. (Photo: M. Peyrot-Clausade.)

These chitons use their radulae to scrape the surface of the boulders, collecting the algae attached to the substratum. Examination of the extruded pellets of these chitons allows an estimate of the rate of sediment production to be calculated; in this restricted environment, rates are high.

A recent study of bioerosion across a cross shelf transect from the northern GBR out into the Coral Sea found that levels of sediment flowing down the Daintree River heavily influenced rates and agents of bioerosion. Experimental substrata at sites at the mouth of the river and Low Isles, after short periods of exposure are covered in a thick layer of silt that restricts development of the endolithic algae and hence levels of grazing are low. The silt settles out from the turbid water column. In contrast, at sites out in the Coral Sea where water is clear (Fig. 8.3*D*), substrata are heavily bored with endolithic algae, encouraging high rates of grazing by scarids. Boring communities vary between inshore and offshore sites with deposit feeding polychaete species dominant at inshore sites and filter and surface deposit feeders at offshore sites. Boring sponges are most abundant at inshore sites and boring bivalves at offshore sites. Net rates of erosion vary between sites and the relative importance of the components of erosion change markedly along the cross-shelf transect, supporting the data on the distribution, abundance and species composition of scarids across a similar transect (Fig. 8.4).

■ TIPPING POINT: HUMAN INFLUENCES ON CALCIFICATION AND EROSION

As human populations have expanded in the coastal areas of tropical and subtropical oceans their influence on the environment that surrounds coral reefs has increased dramatically. Changes to the nutrient and sediment concentration of the waters surrounding corals have impacted the growth and calcification of a wide range of organisms. In the last two decades, these local impacts have been joined by global factors such as global warming and ocean acidification (see Chapters 9 and 10). Together, local and global factors have decreased the growth in calcification of reefs while at the same time probably increasing the rate of dissolution and/or bioerosion. These changes are complex and interactive, and have far-reaching consequences for both natural ecosystems and the human societies that depend on them.

The impact of coral bleaching, crown-of-thorns starfish (see Chapters 5 and 26) and a wide array of other factors has decreased the proportion of reefs covered by living coral, which normally maintains the carbonate reef substrates against infestation by bioeroders. The loss of corals has in turn provided an increase in supply of suitable substrata for bioerosion, so rates will increase across the reef after a bleaching event. These rates will either remain high or decline to prebleaching levels, depending on other factors such as water quality and supply of coral recruits. Clearly, if the growth and survival of coral reefs is to continue to decline under the rapid changes in global climate that are projected for this century, then there will be an increasing proportion of reefs that will be no longer growing and will be in net erosion. How fast a reef matrix can disappear is probably dependent on a number of factors. Some studies have suggested that accumulated calcium carbonate structures typical of many reefs can disappear quite quickly.

An additional problem has arisen from the buildup of carbon dioxide in the atmosphere as explained in Chapter 10. In this particular case, roughly 40% of the carbon dioxide that has entered the atmosphere has been absorbed by the ocean. In the ocean, carbon dioxide reacts with water to create a dilute acid called carbonic acid. This acid releases protons which, combined with carbonate, convert it to bicarbonate. The net effect is that the carbonate ion concentration has been declining and will decline further as carbon dioxide builds up in the atmosphere. Decreasing carbonate ion concentrations will decrease the ease with which calcification can occur and will increase the tendency for calcium carbonate crystals to dissolve. This has many people concerned about whether projected increases in carbon dioxide in the atmosphere will tip the balance of coral reefs away from the accumulation of calcium carbonate and towards the erosion of this important resource. The implications that reefs are eroding are highly significant and may involve a loss of the three-dimensional structure of reefs. This structure is important habitat for many thousands of species worldwide as well as being the 'front line' defence along coastlines throughout the world. The prospect of reef barriers disappearing as they erode in a warm and acidic sea may mean increased exposure of other ecosystems such as mangroves and seagrasses, which generally shelter behind the reef crests from the full force of ocean waves. These changes in wave energy, especially when combined with sea level rise, could have dire implications for the extensive human infrastructure that often lines tropical coastline, and which also shelters behind these crucial reef barriers.

In summary, the factors controlling rates and agents of bioerosion are complex and interrelated as are those controlling reef growth, and superimposed on these are location and regional factors. Anthropogenic impacts seriously modify rates and agents of bioerosion, and commonly these are cumulative. For example, a reef can recover from a single bleaching event but if such events become more regular, and if this is combined with a crown-of-thorns starfish plague, the reef has little chance to recover before it is hit by another bleaching event. In the meantime, rates of bioerosion are increasing and are causing physical loss of reef structure. When combined with yet other aspects such as the overfishing of key functional groups, then once healthy coral reefs can become rapidly algal dominated reefs.

ADDITIONAL READING

Bellwood, D. R., and Choat, J. H. (1990). A functional analysis of grazing in parrotfishes (family Scaridae) on the Great Barrier Reef, Australia. *Environmental Biology of Fishes* **28**, 189–214.

Hoegh-Guldberg, O., Mumby, P. J., Hooten, A. J., Steneck, R. S., Greenfield, P., Gomez, E., Harvell D. R, Sale, P. F., Edwards, A. J., Caldeira, K., Knowlton, N., Eakin, C. M., Iglesias-Prieto, R., Muthiga, N., Bradbury, R. H., Dubi, A., and Hatziolos, M. E. (2007). Coral reefs under rapid climate change and ocean acidification. *Science* **318**, 1737–1742.

Hutchings, P. A. (1986). Biological destruction of coral reefs—A review. *Coral Reefs* **4**(4), 239–252.

Hutchings, P. A., Peyrot-Clausade, M., and Osnorno, A. (2005). Influence of land runoff on rates and agents of bioerosion of coral substrates. *Marine Pollution Bulletin* **51**, 438–447.

Kleypas, J. A., and Langdon, C. (2006). Coral reefs and changing seawater chemistry. In 'Coral Reefs and Climate Change'. (Eds. J. Phinney, O. Hoegh-Guldberg, J. Kleypas, and W. Skirving.) pp. 73–110. *Coastal and Estuarine Studies* **61**(73), 73–110.

Kleypas, J. A., McManus, J., and Menez, L. (1999). Using environmental data to define reef habitat: Where do we draw the line? *American Zoologist* **39**, 146–159.

Peyrot-Clausade, M., Chabanet, P., Conand, C., Fontaine, M. F., Letourneur Y., and Harmelin-Vivien, M. (2000). Sea urchin and fish bioerosion on La Reunion and Moorea Reefs. *Bulletin Marine Sciences* **66**, 477–485.

Raven, J., Caldeira, K., Elderfield, H., Hoegh-Guldberg, O., Liss, P., Riebesell, U., Shepherd, J., Turley, C., and Watson, A. (2005). Ocean acidification due to increasing atmospheric carbon dioxide. Special Report, Royal Society, London.

9. Human Impact on Coral Reefs

T. P. Hughes

Coral reefs provide important ecosystem goods and services, such as fisheries and tourism, and have great aesthetic and cultural value. Until recently, the direct and indirect effects of overfishing and pollution from agriculture and land development have been the most significant causes of the accelerating degradation of coral reefs in many places, particularly in the Caribbean. These human impacts have caused ecological shifts away from the original dominance by corals to a preponderance of fleshy seaweed or other weedy non-coral species. Importantly, these changes to reefs are compounded by the more recently superimposed impacts of global climate change, including coral bleaching and the emergence of disease. Even otherwise lightly impacted reefs, such as the northern and outer Great Barrier Reef, are increasingly vulnerable to climate change. Coral reefs are in serious decline globally; an estimated 30% are already severely damaged, and close to 60% may be lost by 2030.

Coral reefs are often described (inaccurately) as fragile ecosystems in delicate balance with nature; this notion goes hand in hand with the outmoded idea from visiting colonial scientists that the tropics are benign and stable environments (Fig. 9.1). According to this perspective, humans are typically portrayed as the disrupter of nature's delicate balance. But are coral reefs stable, fragile ecosystems? The answer is no, especially at the scales most relevant to human interaction with reefs. Coral reefs are subject to a high frequency of recurrent disturbances, and they have evolved and thrive in a dynamic environment.

From a demographic perspective, a decline in coral or a fish population is the result of births (or recruitment) being exceeded by mortality. As well as increasing the mortality of corals and other organisms, human activities also have significant effects on the regenerative processes of coral reef species, such as fecundity, fertilisation success, larval development and rates of settlement and recruitment. The scale of dispersal of larvae is only beginning to be understood, and is crucial to understanding patterns of larval recruitment and recovery from large scale disturbances. Human impacts on reproduction and dispersal

Figure 9.1 An upper reef slope at Lizard I. (3–4 m depth), showing a vibrant coral assemblage, dominated by tabular and branching species of *Acropora*. (Photo: T. P. Hughes.)

are much less obvious and harder to measure than catastrophic mortality, but they nonetheless play a crucial role in the long term dynamics of reefs.

Two categories of stress, acute and chronic, are useful for assessing human impacts and natural disturbances. Acute disturbances act suddenly, and usually for a short time, although their impacts may have long term repercussions. Examples include a ship grounding, an oil spill, or a nuclear bomb test. Chronic impacts occur over an extended period and are often difficult to stop. For instance, subsistence overfishing in densely populated developing countries, deforestation leading to coastal runoff of nutrients and sediment, or discharge of sewage from a coastal city are ongoing, chronic disturbances. There is some evidence that recovery from some types of human impacts is more difficult or slower than recovery from natural disturbances. According to a recent review, coral assemblages suffering from chronic (usually human) impacts recovered in only 27% of cases, compared to 69% for acute impacts.

Another useful way to view impacts on reefs is to consider how human activities affect the structure of food webs. The removal of species near the top of a food chain by fishing can lead to an increase in abundance of their prey (called a top-down effect). Many reefs worldwide have been severely overfished. Megafauna such as sharks and turtles are increasingly rare worldwide, and fisheries have moved lower down the food web targeting increasing numbers of herbivores such as parrotfish. Similarly, the addition of nutrients can stimulate growth of species at the bottom of the food web (primary producers such as phytoplankton and fleshy algae). This bottom-up effect can propagate upwards in a food web by providing more food for herbivores and their predators. Top-down and bottom-up distortions of food webs typically happen simultaneously.

Can we identify reefs that are most at risk? Anticipating and preventing damage is likely to be more effective than restoration afterwards (Box 9.1). Obviously, the number of people near a reef is crucial, for example,

BOX 9.1 CORAL REEF RESTORATION

The number of restoration projects is increasing as governments and NGOs attempt to 'do something' about the worldwide decline of coral reefs. Excluding artificial reef projects, nearly two hundred coral reef restoration studies have been undertaken worldwide over the past three decades, at a combined cost exceeding US$200 million. Most of them are small scale transplant experiments, where one or two species of corals are removed from one reef and relocated at a damaged site (e.g. after a ship grounding, cyclone, or the Asian tsunami). The total area of all of these projects is less than one square kilometre, while globally the amount of reef that has been degraded in the past few decades is about 10^5 times greater.

Coral reefs are much more diverse and complex than other systems such as mangrove stands, grasslands, or lakes, where restoration has sometimes been possible. However, no one has been successful at artificially restoring the biodiversity or ecological functions of a coral reef at a meaningful scale. A better outcome for sustaining coral reefs will come from addressing the root causes of reef degradation and from targeted interventions that build resilience to phase-shifts (e.g. by improving land-use practices in reef catchments, by establishing alternative employment options to reduce fishing pressure, and by reducing greenhouse gas emissions).

International Coral Reef Initiative resolution on artificial coral reef restoration and rehabilitation. 2005: available at http://www.icriforum.org/library/ICRI_resolution_Restoration.pdf [Verified 21 February 2008].

the cities of Honolulu, Miami and Jakarta have all impacted very significantly on nearby coral reefs through sewage discharge, increased sedimentation, industrial pollution, overfishing, and so on. In comparison, the land area adjoining the GBR has relatively few people, and most of the reefs are tens of kilometres offshore. In contrast, the reefs of Jamaica and elsewhere in the Caribbean are typically 100 m or less from the beach, and millions of people live next to them. Arguably, reefs at risk should be the highest priority for conservation and management efforts.

Reefs closest to the mainland or fringing populated high islands are more likely to be at risk than reefs on outer continental shelves or unpopulated oceanic atolls. For example, on the GBR inshore reefs generally have more human impacts than elsewhere (see Chapter 11). Nutrient and sediment runoff from farming activities on land have impacted many of the reefs closest to the mainland. The chemical signals in annual growth bands of century-old coral skeletons reveal a sharp increase in coastal runoff following the arrival of cattle and sheep and large-scale land clearing in the 19th century. Coastal development and recreational fishing have also significantly impacted nearshore reefs. Historical photographs of mainland reefs show vibrant stands of corals along the Queensland coast that are increasingly degraded today.

■ TYPES OF HUMAN IMPACTS

Human impacts vary in scale (how big they are, and whether they are acute or chronic), and the response of reefs is correspondingly variable. Next is a brief overview of some of the major categories of human activities that affect reefs: overfishing, runoff from land, climate change, coral harvesting, and recreational impacts.

Overfishing

Coral reefs support highly productive, diverse and economically important fisheries. However, human population growth and growing market demands for seafood have depleted many targeted stocks. On the GBR, earlier export fisheries that flourished following European colonisation (e.g. for sea cucumbers, pearl shell, *Trochus* snails, and turtles) have collapsed or are no longer commercially viable. Harvested tropical megafauna (e.g. whales, dugongs, crocodiles, turtles, and sharks) are severely depleted worldwide. For example, the number of dugongs on the GBR has declined by more than 90% in the past 30 years. Larger species of carnivorous fishes (e.g. groupers and snappers), especially those with vulnerable spawning aggregations, have also been heavily overfished. Even on the GBR, where fishing pressure is relatively modest compared to most coral reefs, the biomass of carnivorous fishes has been reduced by 4–5 fold on fished reefs compared to adjacent reefs that are zoned as 'No-Take Area'.

Top-down effects. In a classic example of 'fishing down the food chain' many reef fisheries today rely heavily on herbivores and planktivores. Australia is unusual because it lacks a large human population reliant on subsistence fishing, and there is virtually no recreational or commercial fishing of tropical reef herbivorous fishes. Over-harvesting of herbivorous fishes and addition of nutrients both promote blooms of fleshy algae that outcompete adult corals and impede new recruitment. The depletion of herbivorous fishes on Caribbean reefs has caused ecosystem collapse, with unchecked algal blooms replacing corals. On some reefs, overfishing has reduced levels of predation and competition from fishes, triggering unsustainably high populations of grazing sea urchins. This top-down effect is unstable because of emergent diseases that cause mass mortalities of super-abundant sea urchins, and because bioerosion of the substrate by huge numbers of sea urchins can exceed the accretion rate of the reef (see Chapter 8).

The relative importance of herbivory and nutrient supply ('bottom-up versus top-down control') in regulating the biomass and composition of seaweed is an ongoing, sometimes contentious debate. Algal biomass is often highest where herbivores are naturally scarce, for example, on intertidal reef flats, in turbulent shallow water, or within the defended territories of pomacentrid damselfish. Reef substrates that are protected experimentally from grazing rapidly become colonised by macroalgae, which in turn inhibit recruitment and growth of corals. Previous experimental approaches to explore herbivore-algae-coral interactions have used a variety of

BOX 9.2 NO-TAKE AREAS

No-Take Areas (NTAs) are an important tool for protecting fish stocks and for maintaining the critical ecological functions of fishes. No-Take Areas also provide a refuge from harvesting and destructive fishing practices (e.g. damage from poisons, dynamite and fishing gear), and of course they allow targeted species to grow older and larger. These big individuals typically have disproportionately higher reproductive success, and their offspring spill over into adjoining areas that have lower levels of protection, influencing the seascape as a whole. Many tourist operations are located at NTAs, so that snorkelers and divers can view larger fish.

Once new NTAs are established, the recovery of severely depleted fish stocks continues for many years. In the Philippines, a long term study revealed that the biomass of predatory fishes increased exponentially (by 17-fold), and showed no sign of slowing after 18 years of continuous protection. On the GBR, researchers are also working on new NTAs (green zones) established in 2004 (see Chapter 12) where the size and number of coral trout and other targeted species is already increasing. NTAs are seen as important tools for managing the resilience of reefs, that is, their capacity to cope with natural disturbances and human impacts (Box 9.3). Grazing by herbivorous fish that are protected within NTAs can facilitate recruitment by corals after a disturbance, by preventing blooms of seaweed.

techniques, including algal transplantation (e.g. from areas of low to high herbivore density), herbivore enclosure and exclusion experiments, herbivore and algae removals, and so-called 'natural experiments' such as the die-off of the Caribbean herbivorous sea urchin, *Diadema antillarum*. All of these approaches have their strengths and weaknesses. It is clear that both top-down and bottom-up effects are important, and they are difficult to separate since overfishing of herbivorous fishes and declining water quality often go hand in hand. No-Take Areas (NTAs), where fishing is prohibited, are an increasingly common management tool for sustaining targeted species and maintaining their ecological roles (Box 9.2).

Runoff from land

Runoff from land causes elevated nutrient loads and increased turbidity from suspended sediments (see Chapter 11). Excessive levels of sedimentation are caused by activities such as soil erosion from agriculture, dredging, and drilling for oil and gas. The most widespread of these is soil erosion, due to changes in land use practices, into rivers that flow onto coastal reefs. Throughout the tropics, there has been widespread deforestation and land clearing for agriculture, aquaculture and urbanisation. Turbidity influences the physiology, growth and survival of corals in several ways (see also Chapter 7). First, corals are forced to expend energy cleaning themselves of sediment and repairing damaged tissues. Second, the amount of light reaching a coral colony is reduced by increased turbidity, slowing their growth. Third, turbid waters tend to be enriched in organic matter and nutrients, which can boost the energy intake of some hardy corals. However, in general, too much sediment is bad for reefs. It is especially damaging to juvenile corals that are easily smothered by silt (Fig. 9.2).

Turbidity and sedimentation on the GBR, particularly inshore, are high compared to many reefs elsewhere. Muddy sediment from land has built up along the inner part of the shallow continental shelf off Queensland since sea level stabilised close to its present level about 6500 years ago (Chapter 3), and it is easily resuspended in windy weather. Visibility (that depends on the amount of

Figure 9.2 A degraded reef with some alcyonacean soft corals and almost no hard coral cover. The substrate has a fine layer of algae and silt. (Photo: T. P. Hughes.)

suspended material) on inner reefs is typically 1–5 m, and about 10–20 m on average further offshore on mid-shelf reefs. This compares with 40 m or more in, for example, the Caribbean and on oceanic reefs.

Bottom-up effects. Inputs from sewage and runoff of fertilisers can potentially cause eutrophication on coral reefs. The iconic example of sewage effects on a coral reef come from Kaneohe Bay, Hawaii. The bay is very shallow, connected to the ocean by a narrow opening (i.e. it has a very low flushing rate compared to most reefs), and the land area surrounding it is densely populated. It has a long history of other impacts such as dredging and overfishing and has a high proportion of pest species introduced by shipping. From 1963–1977 secondary sewage was discharged into Kaneohe Bay from three outfalls at a rate of up to 20 000 m^3 of sewage each day. There are also several streams entering the bay, carrying urban and suburban runoff. The increased nutrients resulted in higher sediment loads and phytoplankton blooms. Coral patch reefs were colonised by benthic macroalgae and suspension feeders (bivalves and sponges), while coral cover declined sharply. These effects exhibited a gradient away from the sewage outfalls. In 1977 the sewage pipe was extended ('diverted') into deeper water. Water clarity improved, the filter feeders and algae declined, and the corals slowly increased.

Destructive population explosions of the coral-feeding crown-of-thorns starfish, *Acanthaster planci*, may also be related to nutrient enrichment (see Chapters 11 and 26). The outbreaks were first observed in the late 1950s and 1960s, when many corals reefs in Australia, Guam, Japan, the Red Sea and elsewhere were badly damaged by enormous densities of starfish. Since then, there have been repeated outbreaks throughout most of the starfish's geographic range, and they have become a chronic issue on many reefs, including the GBR (Fig. 9.3).

Climate change

For coral reefs, climate change due to enhanced greenhouse gasses is not some distant threat that might happen in the future. Recent climate has been affecting coral reefs since the mid 1980s, and many locations have by now already experienced multiple bouts of coral bleaching in the past 25 years or so, following periods of unusually high water temperatures (Fig. 9.4). Coral bleaching occurs when corals become physiologically stressed and lose most of their symbiotic zooxanthellae. Localised bleaching has been described in the older coral reef literature following extreme weather and floods. However, regional scale bleaching such as the 1998 El Niño event (see Chapter 10) is a new phenomenon driven by global warming. Mortality of bleached corals is often high over very large areas, and

Figure 9.3 Multiple crown of thorns feeding on hard coral seen during Tioman COTs. (Photo by Badrul Huzaimi, http://www.reefbase.org.)

Figure 9.4 *Acropora* corals, bleached by unusually warm water, growing at 5 m depth on a vertical wall at Lizard I. (Photo: T. P. Hughes.)

Figure 9.5 Corals provide critical habitats for adult and juvenile fishes, such as pomacentrid damselfishes. The bleached *Acropora* corals shown here may soon die, leaving dead skeletons that soon crumble away. (Photo: O. Hoegh-Guldberg.)

the frequency and intensity of warm water events is set to rise in coming decades. Unprecedented large scale coral bleaching occurred on the GBR in 1998 and 2002 (Chapter 10). On the GBR, researchers have documented a steady increase in the incidence of coral disease over recent years, which may relate to rising sea-surface temperatures. In the longer term, ocean acidification is also likely to affect the growth rate and skeletal composition of corals.

Coral bleaching, like most forms of disturbance, affects some species more than others. For example, branching and tabular *Acropora* species are usually much more susceptible to thermal stress than many faviids and *Porites* (Figs 9.5, 9.6). There are two important unknowns concerning the responses of corals and their symbiotic zooxanthellae to global warming: whether they can adapt quickly enough to rising temperatures, and the extent to which warm-adapted genotypes may be able to move. Corals typically have very large geographic ranges that usually straddle the equator and extend to 25–30° north and south. Consequently, local populations experience substantially different temperature regimes. Importantly, bleaching does not occur at the same absolute temperature everywhere. Rather, it happens when temperatures have risen by about 2°C for several weeks above the ambient local temperature regime. Clearly, local populations have adapted to their local thermal environment (see Chapter 10). However, nobody knows how long this local adaptation has taken to evolve, or the extent to which warm-adapted strains can migrate via larval dispersal to higher latitudes.

Coral harvesting

Corals are mined in many places to provide a source of cheap building materials and rubble for roads and air strips, especially in remote locations where transport costs are high. Massive corals like *Porites* are

Figure 9.6 Loss of corals from bleaching also impacts on fishes. Here, juvenile reef fishes stay close to bleached colonies of *Acropora* and *Pocillopora* from the 2006 bleaching event in the Keppel Islands (southern Great Barrier Reef). If the corals die, recruitment by fishes will diminish. (Photo: O. Hoegh-Guldberg.)

collected and cut into building blocks. Coral fragments are widely collected for lime preparation (for making cement and for the consumption of beetel nut). Coral mining was widespread on the GBR prior to World War II, particularly as a source of lime for sugar cane farming.

Live and dead coral collecting for the ornamental, souvenir and aquarium trade is widespread, and has led to significant reef damage in the Philippines and elsewhere. All scleractinian corals are classified by the UN Convention on International Trade in Endangered Species (CITES) as 'endangered', and in theory they cannot be traded internationally without import and export licenses. There is a very small coral fishery on the GBR that is sustainable and tightly regulated.

Recreational impacts

Recreational fishing is a major contributor to the depletion of targeted species in wealthier tropical countries (see Fisheries, above). For example, the catch by recreational hook and line fishing for predatory fishes on the GBR rivals commercial fishing. Tourism (Fig. 9.7) also contributes to runoff and nutrient loads on coral reefs through the development of infrastructure such as hotels, marinas and sewage outfalls (see Runoff from Land, above). Anchor damage to corals, accidental boat grounding, fin damage from divers, trampling and littering all have localised lethal and sublethal effects on susceptible species. Public education can alleviate most of these problems, and the reef tourism industry can and does play a major role in promoting responsible activities by visitors. Public

Figure 9.7 The incidence of coral disease is increasing on the Great Barrier Reef. Here, a diver records a diseased table coral (*Acropora cytherea*). (Photo: B. Willis.)

moorings at popular sites have sharply reduced anchor damage.

Longer term trends

The selectivity of both natural disturbances and human impacts is a key issue for predicting their longer-term consequences. If cyclones, overfishing or coral bleaching affected all species equally, they would have no direct impact on species composition. Overall abundance would decline, but there would be no direct change in the relative abundance of each species. Massive destruction (e.g. more than a 90% decline in coral cover) is exceptional, even at the relatively small spatial scale that ecologists normally study. A large number of studies demonstrate selective mortality among corals and other organisms—from cyclones, predators, diseases, and human impacts. As well as the immediate changes that selective mortality induces, natural disturbances and human impacts also promote longer-term changes. For example, delayed mortality from outbreaks of disease among injured survivors (Fig. 9.8), bioerosion of damaged coral skeleton, and altered predator-prey relationships (if one survives better than the other) may occur for years after a cyclone has struck. Some species also rebound faster than others after an acute disturbance, further changing the composition of coral reefs.

Increasingly, many coral reefs today have a diminished ability to recover from recurrent natural disasters, due to other chronic human impacts. For example, the recovery of Red Sea corals after catastrophic mortality from low tides were much lower on a reef flat that was chronically impacted by oil-spills, compared to a nearby location that was free of oil. This erosion of resilience (Box 9.3) is often characterised by major shifts in species composition, called phase-shifts. Overfishing typically leads to dominance by fleshy

Figure 9.8 Tourists snorkelling at Lizard I. Tourism on the Great Barrier Reef contributes more than $5 billion each year to the Australian economy. (Photo: T. P. Hughes.)

macroalgae (Fig. 9.9), due to the depletion of herbivorous fishes. Nutrient addition can also contribute to algal blooms, and promote suspension feeders such as molluscs and sponges at the expense of corals. Managing the resilience of reefs (Chapter 12), so that they can continue to absorb natural and human impacts without fundamentally changing and degrading, is the major challenge for the future.

ADDITIONAL READING

Bellwood, D. R., Hughes, T. P., Folke, C., and Nyström, M. (2004). Confronting the coral reef crisis. *Nature* **429**, 827–833.

Connell, J. H. (1997). Disturbance and recovery of coral assemblages. *Coral Reefs* **16**, 101–113.

Harvell, D., Aronson, R., Baron, N., Connell, J., Dobson, A., Ellner, S., Gerber, L., Kim, K., Kuris, A., McCallum, H., Lafferty, K., McKay, B., Porter, J., Pascual, M., Smith, G., Sutherland, K., and Ward, J. (2004). The rising tide of ocean diseases: unsolved problems

BOX 9.3 WHAT IS RESILIENCE?

Resilience is the ability of reefs to absorb and recover from recurrent disturbances (e.g. from cyclones, outbreaks of predators, or coral bleaching events). Loss of resilience can lead to a sudden switch, known as a phase-shift, to an alternate assemblage of species that is typically dominated by fleshy seaweeds. Many reefs around the world have lost their resilience to routine disturbances, and instead of recovering as before are becoming more and more degraded. A resilience-based approach to reef management focuses on avoiding thresholds that lead to undesirable phase-shifts.

The traditional approach to management of marine fisheries is based on a flawed concept: the 'optimal' harvesting of single species in systems that are assumed to be reasonably stable. An emerging alternative approach recognises that ecosystems are characterised by complex dynamics and thresholds, and that disturbance and change are inevitable. Ecosystem-based management is replacing earlier single-species approaches, emphasising the broader array of ecological processes that sustain the delivery of harvestable resources. The traditional view of No-Take Areas as being primarily for managing fisheries is waning, with an increasing emphasis on their broader utility for managing biodiversity, restoring the structure of foodwebs, maintaining ecological functions, and building ecosystem resilience to future shocks.

Resilience Alliance: available at http://www.resalliance.org [Verified 21 February 2008].

Figure 9.9 *Galaxea* surrounded by algae. Macroalgal blooms can prevent recruitment by corals and smother adult colonies. (Photo: L. Anderson.)

and research priorities. *Frontiers in Ecology and the Environment* **2**, 375–382.

Hughes, T. P. (1994). Catastrophes, phase shifts, and large scale degradation of a Caribbean reef. *Science* **265**, 1547–1551.

Hughes, T. P., Baird, A. H., Bellwood, D. R., Card, M., Connolly, S. R., Folke, C., Grosberg, R., Hoegh-Guldberg, O., Jackson, J. B. C., Kleypas, J., Lough, J. M., Marshall, P., Nyström, M., Palumbi, S. R., Pandolfi, J. M., Rosen, B., and Roughgarden, J. (2003). Climate change, human impacts, and the resilience of coral reefs. *Science* **301**, 929–933.

Jackson, J. B. C., Kirbym M. X., Berger, W. H., Bjorndal, K. A., Botsford, L. W., Bourque, B. J., Bradbury, R. H., Cooke, R., Erlandson, J., Estes, J. A., Hughes, T. P., Kidwell, S., Lange, C. B., Lenihan, H. S., Pandolfi, J. M., Peterson, C. H., Steneck, R. S., Tegner, M. J., and Warner, R. R. (2001). Historical overfishing and the recent collapse of coastal ecosystems. *Science* **293**, 629–638.

Nyström, M., Folke, C., and Moberg, F. (2000). Coral reef disturbance and resilience in a human-dominated environment. *Trends in Ecology and Evolution* **15**, 413–417.

Pandolfi, J. M., Bradbury, R. H., Sala, E., Hughes, T. P., Bjorndal, K. A., Cooke, R. G., McArdle, D., McClenachan, L., Newman, M. J. H., Paredes, G., Warner, R. R. and Jackson, J. B. C. (2003). Global trajectories of the long term decline of coral reef ecosystems. *Science* **301**, 955–958.

Scheffer, M., Carpenter, S., Foley, J. A., Folke, C., and Walker, B. (2001). Catastrophic shifts in ecosystems. *Nature* **413**, 591–596.

Wilkinson, C. R. (2004). (Ed.) Status of the coral reefs of the world: 2004. Global coral reef monitoring network and Australian Institute of Marine Science, Townsville.

10. The Future of Coral Reefs in a Rapidly Changing World

O. Hoegh-Guldberg

Coral reefs are renowned for their breathtaking beauty and the bounty of diverse life forms that live within their carbonate structures. This book is a testimony to this splendour and variety. Unfortunately, coral reefs are changing due to the influence of humans on the marine environment. These changes are coming about due to the exploitation of coral reef organisms and through insidious changes to the condition of the waters (increased sediments and nutrients) that surround them. While these changes are having a very large impact on coral reefs (Chapter 9), changes to important atmospheric components called greenhouse gases are driving even larger impacts due to rapid changes to global climate, including the seas in which coral reefs live. These effects of climate change have impacted coral reefs in a spectacular fashion, with thousands of square kilometres of often remote coral reefs becoming stressed, with often large numbers of reef-dwelling organisms dying. While the impacts on coral reefs are only a small subset of the ecosystem changes that are happening across the planet as a result of climate change, coral reefs have grabbed public attention. This is because they are the first iconic ecosystem to demonstrate that biological systems, rather than changing gradually, may change abruptly and substantially as a result of climate change.

In this chapter, the issue of climate change is outlined and discussed, particularly with references to past, current and future climate change. There are several major issues. The first is the unusual nature of the current changes in our climate system. The second is the response of coral reef organisms to our rapidly warming and acidifying ocean. The last concerns possible options we have for minimising the impacts of greenhouse driven change on coral reefs. Discussion of these issues leads us to several conclusions. First, coral reefs are unable to sustain atmospheric concentrations of carbon dioxide of more than 450 parts per million (ppm); efforts to reduce emissions must therefore be intensified. Second, efforts to reduce local stresses (e.g. declining water quality, over-exploitation of coral reef fish stocks; see Chapter 9) must be increased as coral reefs enter a period of a rapidly shifting oceanic climate. Addressing these two aspects aggressively over the next decade represents the last chance we have of preventing the loss of coral reefs for hundreds, if not thousands, of years.

THE ENHANCED GREENHOUSE EFFECT

Radiation from the sun heats terrestrial, oceanic and atmospheric components of the Earth. About 30% of incoming energy is eventually reradiated back into space, with the remaining 70% driving a global average temperature of the Earth of around +14°C. The trapping of heat by the atmosphere (referred to as the greenhouse effect) is largely responsible for life on planet Earth. If there was no greenhouse effect, the average Earth's temperature would be −18 °C and carbon-based life forms would not be able to exist. The term 'greenhouse effect' is used because of the similarities between the trapping of heat in the Earth's atmosphere to that seen in garden greenhouses. Visible radiation from the sun passes through the glass panels of the greenhouse, warming interior surfaces that in turn reradiate long wavelength infrared radiation. But infrared wavelengths do not pass through the glass as efficiently as the incoming radiation, so heat begins to accumulate in the greenhouse. Likewise, solar energy enters the Earth's atmosphere and warms the air, rocks and other components. It too radiates infrared radiation, which becomes trapped in the atmosphere (Fig. 10.1*A*). Most (75%) of this trapping is done by the so-called greenhouse gases – principally carbon dioxide, water vapour and ozone, methane, nitrous oxide and chlorofluorocarbons (CFC). Although these components are only present in trace amounts in the atmosphere, they have a disproportionately large influence on the heat budget of the Earth.

Life would not be possible without some warming effect of the atmosphere, so it is actually incorrect to say that the greenhouse effect is a problem for life on Earth. The problem that we currently face is associated with a rapid increase in the concentration of the greenhouse gases, a phenomenon referred to as the 'enhanced greenhouse effect'. The greater the concentration of greenhouse gases, the greater the amount of heat retained by the Earth. The increase is largely due to the burning of fossil fuels and other activities such as deforestation and agriculture, all of which release CO_2 and other greenhouse gases into the atmosphere. There is now no longer any credible scientific doubt that these changes are forcing a major and relatively unprecedented change to the heat budget, climate and consequently biological systems of the Earth. The best way to familiarise yourself with the evidence is to go to scientific consensus documents such as the latest reports from the Intergovernmental Panel on Climate Change (IPCC 2007).

The instrumental records of many nations show an almost universal increase in air temperature over the

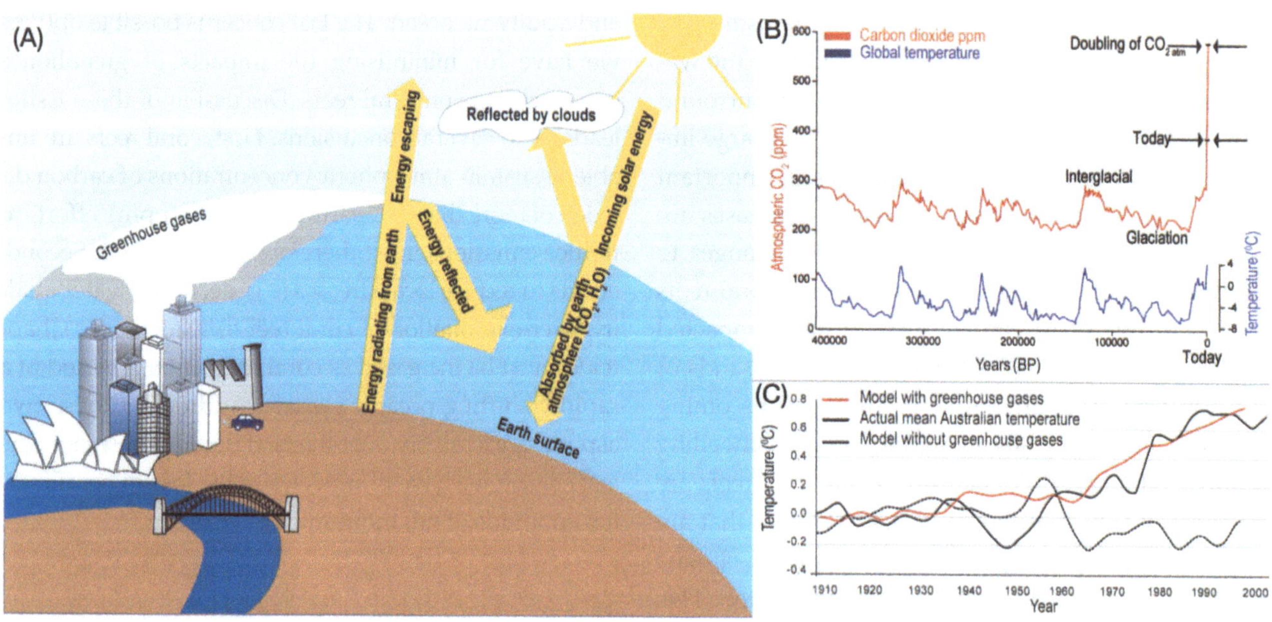

Figure 10.1 *A*, Illustration of key steps in the Enhanced Greenhouse Effect. (Figure: D. Kleine and O. Hoegh-Guldberg.) *B*, Global temperature and concentrations of carbon dioxide stretching back 400 000 years. Data derived from Vostok Ice Core (Petit *et. al.* 1999). *C*, Data illustrating projections of temperature when greenhouse gases are (upper black line) and are not (lower dotted black line) included in the calculations as compared to the mean temperature of Australia over the past 90 years. (Figure: D. Karoly and the Australian Climate Group 2006.)

BOX 10.1 THE INTERGOVERNMENTAL PANEL ON CLIMATE CHANGE (IPCC)

The United Nations triggered the formation of the Intergovernmental Panel on Climate Change (IPCC) in 1988 as a response to debate over the evidence for the enhanced greenhouse effect and its influence on the Earth's climate, biology and human societies. Since its formation, the IPCC has produced four assessment reports (1992, 1996, 2001, 2007) that represent the global consensus of over 2500 experts in matters pertaining to the greenhouse issue. There are three main working groups. The first is focused on reviewing the physical evidence and climate projections (Working Group 1: The Physical Science Basis) while the second is focused on the impacts and how human systems and society might adapt to climate change impacts (Working Group 2: Climate Change Impacts, Adaptation and Vulnerability). Working Group 3 concentrates on reviewing the ways that humans might reduce the rate at which greenhouse gases are building up in the atmosphere ('mitigation'). These working groups are formed in the years prior to each report through the nomination of the most credible and leading scientists in each area. The scale of the collaboration and consensus associated with the IPCC reporting mechanism is unparalleled in history, given that in addition to seeking the agreement of the thousands of scientists, the document is available for review by the public, governments, non-government agencies and industries. Generally, the consensus leads to a more conservative assessment of the scientific issues. Despite this, the fourth assessment report, which was released in the first half of 2007: 3, stated 'the understanding of anthropogenic warming and cooling influences on climate has improved since the Third Assessment Report (TAR), leading to very high confidence that the globally averaged net effect of human activities since 1750 has been one of warming'. The consensus concludes that there is no longer any doubt that humans have changed the climate through the burning of fossil fuels. The IPCC also reviews the projections for the future (Fig. 10.2) and has concluded that doubling the atmospheric concentrations of carbon dioxide over preindustrial values (i.e. taking it from 280 to 550 ppm) is likely to result in increased mean surface temperatures in the range of 2°C to 4.5°C over today. (*Note:* Likely is defined by the IPCC as 'something happening with a 66% probability'.) For more information on climate change and the IPCC, we recommend that you go to the IPCC web site available at http://www.ipcc.ch [Verified 19 Feb 2008].

past 100 years. An early question in the study of anthropogenic climate change was whether these increases in temperature were 'unnatural' or whether they were part of a natural cycle? The answer has come from analysis of long term records of temperatures deduced from isotopic ratios. In chemical reactions, isotopes of any chemical (e.g. ^{16}O *v.* ^{18}O, or ^{12}C *v.* ^{13}C) participate in chemical reactions in the defined isotopic ratios that are determined by temperature. In coral skeletons, the date when the chemical reaction that produced a particular growth ring of calcium carbonate can also be precisely determined, by either counting the number of growth bands back from the living surface or by dating using other isotope methods. Consequently, the isotopic ratio of oxygen in any particular growth ring can be converted into a measure of the sea temperature at a particular time. As some corals, such as *Porites*, may live for hundreds of years, with their fossil skeletons persisting for many thousands of years, we have an enormously valuable archive of past sea temperatures. Similar

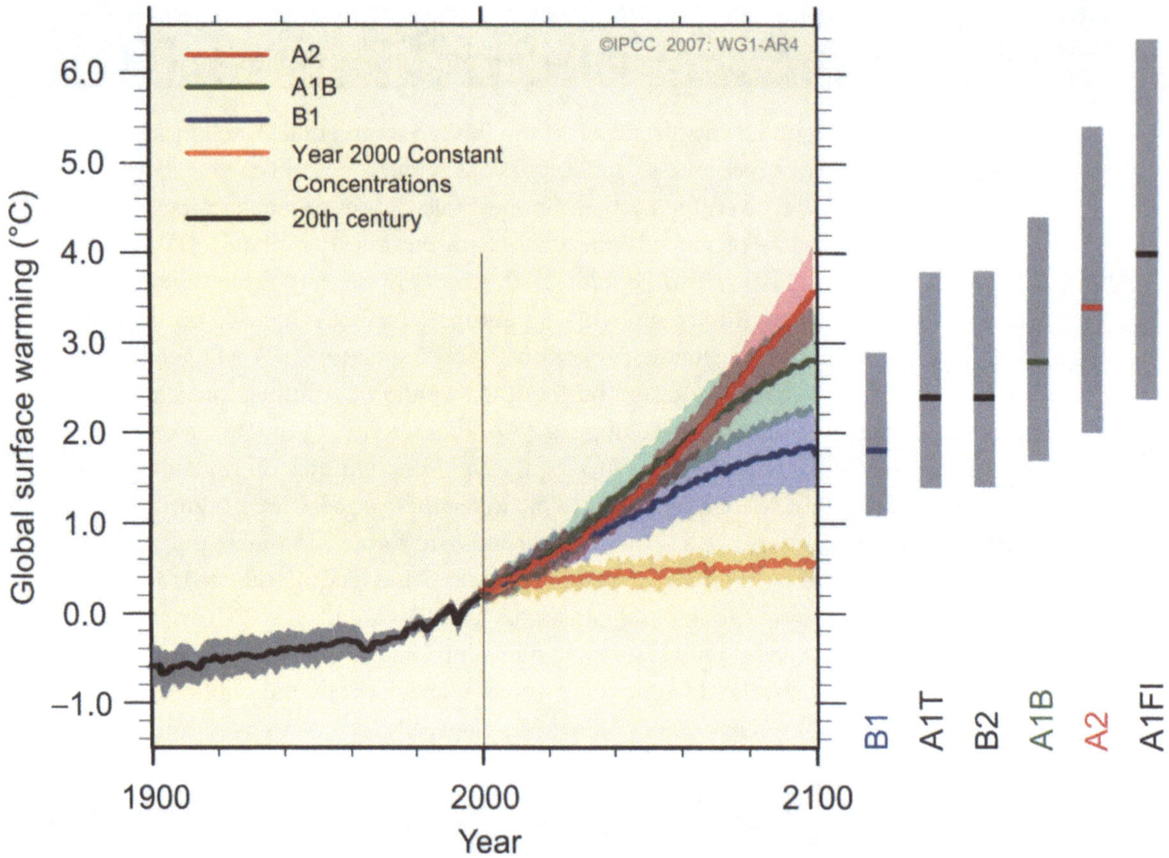

Figure 10.2 Projections summarised by the Working Group 1 and released as part of the 4th assessment report. Available at http://www.ipcc.ch [Verified 19 February 2008]. Solid lines are multi-model global averages of surface warming (relative to 1980–1999) for specific emission scenarios that involve different assumptions about technology and societal change (A2, A1B and B1). Shading denotes the plus/minus one standard deviation range of individual model annual means. The number of global climate change models run for a given time period and scenario is indicated by the coloured numbers at the bottom part of the panel. The orange line is for the experiment where concentrations were held constant at year 2000 values. The grey bars at right indicate the best estimate (solid line within each bar) and the likely range assessed for the six specific emissions scenarios. As is clear, most of the scenarios indicate an increase in global temperature of between 1.5°C and 4°C by 2100. These changes will almost certainly change the health and distribution of natural ecosystems.

techniques have been used on tree rings, lake sediments and ice cores to derive long term perspectives on how the Earth's temperature (and other key factors) has varied over thousands to tens of millions of years.

Ice cores also provide an important historic context for the recent dramatic changes in the concentration of key gases like CO_2. In some regions, precipitation has accumulated as ice that has not thawed for hundreds of thousands of years, providing an uninterrupted record of the ice deposits and their composition over time. The ice also contains small amounts of atmospheric gas trapped as bubbles in the frozen sediments. Climate scientists have drilled long cores from places such as the Vostok Station (operated by Russia) in Antarctica and analysed the bubbles. Isotopic dating has provided precise dates for when these gas concentrations occurred. Other methods allow scientists to calculate the average temperature of the planet and the total ice volume on Earth at any particular date. Dust trapped in the ice layers can give important information on the amount of volcanic activity at the time of ice deposition.

Several ice cores extend our detailed understanding of the variability of the global climate back to at least 720 000 years ago. The information generated by these cores has highlighted the unprecedented changes that humans are currently inflicting on the planet. The current concentration of carbon dioxide is 380 ppm, about 80 ppm above that of 150 years ago (Fig. 10.1*B*). Scientists quite appropriately question whether these high concentrations are part of a larger cycle that will eventually decrease as rapidly as they appeared. However, the Vostok ice core tells us that the concentration of greenhouse gases (like CO_2) has never exceeded 300 ppm in the past 720 000 years, despite repeated cycles of warm (like the one we are in currently) and cold periods (ice ages). When other data sets are taken into account, there is growing evidence that the concentration of greenhouse gases such as CO_2 may not have exceeded 300 ppm for the past 20 million years.

The immediate question for biologists is whether these seemingly small changes in CO_2 really matter. Global Circulation Models that allow us to understand the connectivity and relationships between elements of the climate system on planet Earth suggest that they do. Global Circulation Models are complex computer models that simulate the interplay between the ice, land, sea and atmosphere. They show how even such small changes to CO_2 and other components of the Earth's atmosphere could radically affect the climate of the Earth. Over 40 of these models have been built by the majority of technological nations and demonstrate a highly accurate ability to hind caste changes in global temperature (Fig 10.1C). These models all tell a similar story: recent changes in global temperature can only be explained by taking into account the recent changes in the greenhouse gas content of the atmosphere. Moreover, they tell us that projected increases in atmospheric greenhouse gases through the 21st century will drive major changes in Earth's weather systems in general. The Earth has already warmed by 0.6°C over the past 100 years, and is set to rise by between 1.4°C and 6°C by the end of this century. To gauge the significance of these projected changes, one has only to remember that the temperature difference between the modern climate and the glacial periods on the planets was only 5–7°C. The 5–7°C temperature increase from the last Ice Age to the present (over a period of 20 000–30 000 years) increased sea level (180 m) and changed the shape of the world map as continental margins were inundated and Earth's climatic zones reconfigured (see Chapters 2 and 3). Worst-case IPCC scenarios suggest that an equivalent changing temperature to that which occurred as the Earth warmed up and out of an ice age may take place in the next 100 years. Even in less extreme cases, almost every aspect of the climate will likely change, notably sea temperature, ocean currents, rainfall, and storm intensity. These in turn will shift the conditions for life on the planet, driving changes in the extent, abundance and types of ecosystems at all locations across the globe. As we will see in the next few sections, coral reefs are one of the ecosystems whose response to climate change has demonstrated that the impacts are anything but subtle, even with minor enhanced greenhouse forcing.

MASS CORAL BLEACHING AND DISEASE

Australia's tropical waters have warmed significantly over the past 150 years, with much of the increase being seen over the past 50 years. Rates of warming are similar to that observed globally for tropical/subtropical waters (+0.73°C from 1951–1990). These changes have brought corals on reefs closer to their thermal limits, with the result that warmer than average years (arising due to natural variability) now push corals beyond their upper thermal thresholds. Corals respond to stress by losing their normal brown colour ('bleaching'). Coral bleaching occurs when the symbiosis between coral and their dinoflagellate symbionts (Fig. 10.3*A*, *B*) disintegrates, with the rapid movement of the brown symbionts out of the otherwise translucent tissues of the coral (Fig. 10.3C). As a result, corals change rapidly from brown to white. Bleaching is a generic response that occurs in response to a wide array of stresses, including reduced salinity, high or low irradiance, some toxins like cyanide and many herbicides, microbial infection and high or low temperatures (see Additional reading).

Coral bleaching has been known for over 70 years from reports in which individual colonies or small patches of reefs were observed to have bleached. However, more recently 'mass' coral bleaching events have affected coral reefs over hundreds and even thousands of square kilometres. Mass coral bleaching events have only been reported since 1979. Work done during the 1980s and 1990s revealed that mass bleaching events are triggered by warmer than normal

Figure 10.3 *A*, reef building coral (*Acropora* sp.) with normal populations of dinoflagellate symbionts; *B*, the key dinoflagellate symbiont of corals, *Symbiodinium*; *C*, coral bleached due to rising environmental stress; *D*, bleached coral reef in January 2006 in Keppel I., on the southern Great Barrier Reef. (Photos: O. Hoegh-Guldberg.)

conditions and can be predicted using sea-surface temperature anomalies measured by satellites. Light is an important co-factor. Corals that are shaded tend not to bleach as severely as those under normal irradiances. Corals also differ in their susceptibility, with some corals such as *Porites* and *Favia* being more tolerant of thermal stress than *Acropora*, *Stylophora* and *Pocillopora*. Differences in sensitivity probably relate to host characteristics such as tissue thickness and pigmentation, and possibly genotype of the symbiotic dinoflagellates within coral tissues. Coral bleaching is also affected by water motion, with corals in still, warm, and sunlit conditions showing the greatest impact of thermal stress. The latter is consistent with the first observations of the association of coral bleaching with the doldrum conditions typical of El Niño years in the eastern Pacific.

Mass coral bleaching has affected almost every coral reef worldwide since the 1980s and it occurs on the GBR approximately every 3–5 years at the present time (late 20th and early 21st century). Mass coral bleaching has occurred most often in El Niño years (e.g. 1987–1988; 1997–1998) and less often in the cooler non-El Niño or La Nina years (1988–1989; 1999–2000). Bleached corals tend to recover their dinoflagellate symbiotic populations in the months following an event if the stress involved is mild and short-lived. But mortality (up to 100% of corals over large areas of coral reef) occurs following intense and long-lasting stress. This was seen in many parts of the world in 1998, in which approximately 16% of corals that were surveyed prior to the global cycle of bleaching were estimated to have died by the end of 1998. This particular figure is an average and conceals the fact that in some oceans, for example, the Western Indian Ocean, up to 46% of corals may have died.

Coral reefs in Australia have bleached repeatedly over the past 30 years. Mass bleaching events occurred in Australia in 1983, 1987, 1991, 1998, 2002 and in 2006, with large sections of the GBR bleached in each case. Mortality rates on the GBR have been relatively low compared to coral reefs elsewhere because conditions have been relatively unchanged there so far. By contrast, a very warm core of water sat above Scott Reef in the northwest waters of Australia for several months in 1998, resulting in an almost total bleaching and mortality of corals down to 30 m and 95% of reef-building corals dying in the months that followed. Recent reports indicate that recovery of these reefs has been very slow, primarily because recruitment to these remote reefs is difficult and rare. Coral disease, driven by pathogenic bacteria in addition to coral bleaching, is on the rise and may be connected to warmer than normal conditions. While coral disease affects less than 5% of the population, the incidence of diseases such as 'white syndrome' and other diseases (Fig. 10.4) are on the increase. While coral disease is currently not considered a major threat to coral reefs in Australia and many parts of the Pacific, recent experiences in the Caribbean, where coral disease decimated populations of *Acropora* corals, suggest that understanding and monitoring coral disease is important.

The strong relationship between coral bleaching and sea temperature provides an opportunity to explore how changes in sea temperature in the future might affect the incidence of mass coral bleaching. Past studies have revealed that corals in a region have particular thermal 'thresholds' for bleaching. These thresholds for triggering bleaching are reliable to the point that they predict which regions will experience coral bleaching based on their sea-surface temperatures measured from satellites. If these thresholds are compared to projections of future sea temperatures trends (produced by the Global Circulation Models, see above), it is possible to estimate how the frequency and intensity of mass coral bleaching and mortality will change over time. These changes show that an increase of 2°C over pre-industrial temperatures in the average sea temperature in tropical and subtropical Australia (expected as a result of a doubling of CO_2) will lead to annual bleaching and a major escalation in the number of mass mortality events.

The risk of mass bleaching has been examined for the GBR and revealed that the return time of severe mass coral bleaching in their models for low to moderate changes in the climate increased to the point where the ability of reefs to recover is severely compromised. Deterioration of coral populations is likely in most of the scenarios examined.

OCEAN ACIDIFICATION

The increase in atmospheric CO_2 presents a second major change for reef-building corals and other marine calcifiers. Nearly half of the CO_2 that enters the atmosphere is absorbed by the ocean, where it reacts with water to form carbonic acid (Equation 1). Carbonic acid dissociates into bicarbonate and a proton (Equation 2). The protons released by the entry of CO_2 into seawater then react with carbonate ions to form additional bicarbonate ions (Equation 3). The problem is that relatively small rises of CO_2 in the atmosphere at the pH range of seawater will cause a large decrease in the concentration of carbonate ions.

$$(1)\ CO_2 + H_2O \rightarrow H_2CO_3$$
$$(2)\ H_2CO_3 \rightarrow HCO_3^- + H^+$$
$$(3)\ CO_3^{2-} + H^+ \rightarrow HCO_3^-$$

The calcification rate of a range of marine organisms as diverse as microalgae (coccolithophores), molluscs (e.g. clams, pteropods) and corals is strongly dependent on the concentration of carbonate ions in seawater. Naturally, the concentration is highest in the warmer tropical regions due to the reduced solubility of CO_2 in warm *v.* cold water (i.e. less CO_2 dissolving into the ocean means more carbonate ions). The pH of the ocean has already decreased by 0.1 pH unit with the concentration of carbonate ions decreasing as much as 30 μmol kg^{-1}. Corals were among the first organisms identified as

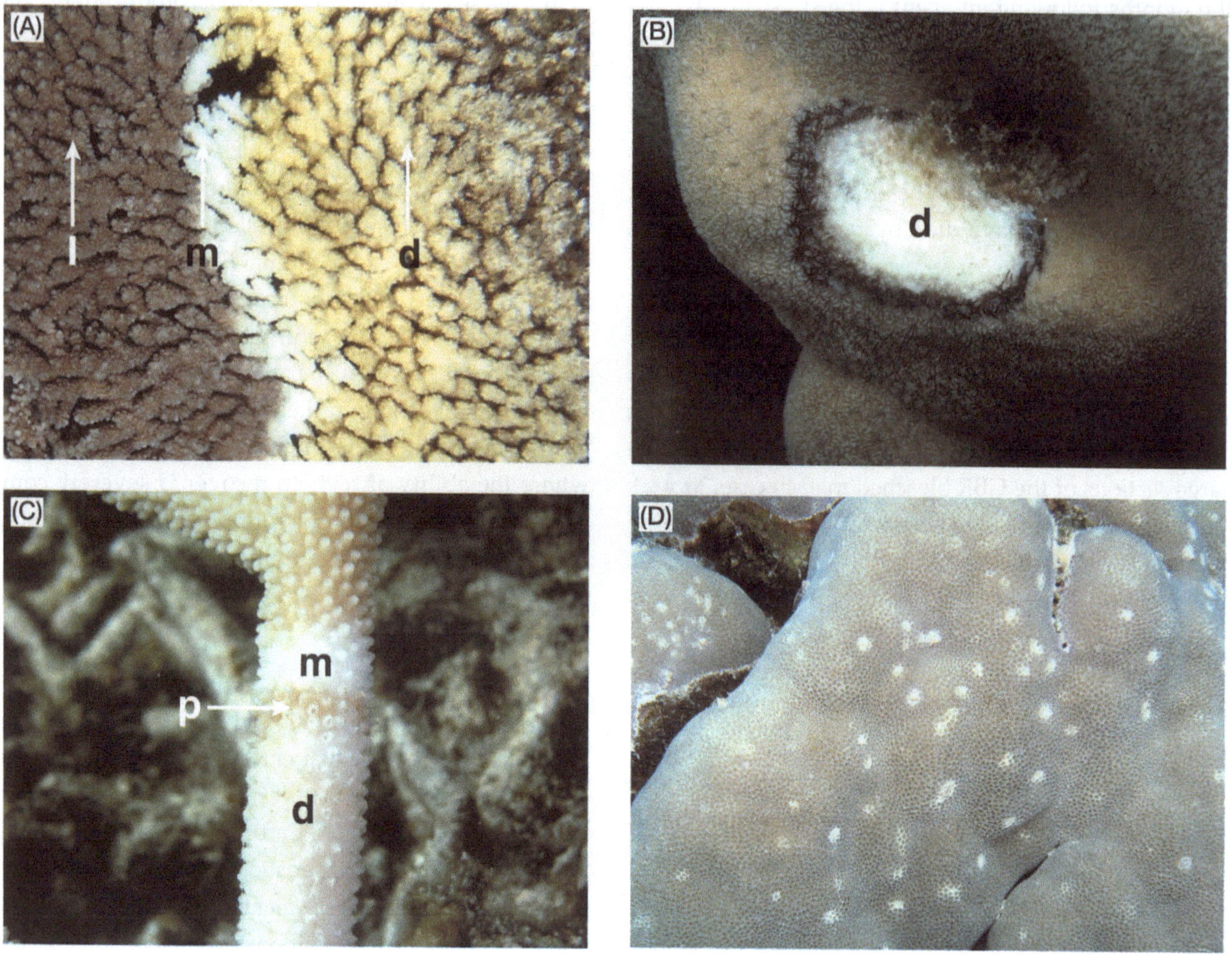

Figure 10.4 Examples of diseases reported on Great Barrier Reef corals. *A,* White Syndrome on tabulate *Acropora* (photo: G. Roff); *B,* Black Band Disease affecting *Pavona* sp. (photo: G. Roff); *C,* Brown Band Disease on a branching *Acropora* (photo: O. Hoegh-Guldberg) and *D,* White Spot syndrome on *Porites* (photo: O. Hoegh-Guldberg). i, living tissue; m, advancing margin of disease; d, dead exposed coral skeleton; p, concentrations of ciliates.

having a major problem with the rising concentration of atmospheric CO_2 and the decrease in the concentration of carbonate ions. Subsequent studies have shown that coral calcification is linearly related to the carbonate ion concentration. These studies have shown consistently that the calcification of coral reef communities effectively becomes zero at carbonate concentrations of 200 μmol kg^{-1} or less. Significantly, carbonate concentrations of 200 μmol kg^{-1} occur when atmospheric concentrations of CO_2 rise beyond 450 ppm. Given that coral reefs represent a balance between calcification and erosion (Chapter 8), it would appear that atmospheric CO_2 concentrations would need to remain well below 450–500 ppm if reef calcification (calcification minus erosion) is to remain positive against the forces of physical and biological erosion. Given that there is growing evidence that erosion (particularly bioerosion) is likely to increase under atmospheric CO_2, these thresholds become even more real (see Chapter 8).

■ OTHER GREENHOUSE DRIVEN CHANGES: SEA LEVEL AND STORM INTENSITY

The increase in the atmospheric concentration of CO_2 and other greenhouse gases impacts other parts of the environment surrounding corals. The increase in temperature, for example, is greatest in the polar regions where it affects the melting of the ice caps in the spring and summer months. The increased melting of the ice caps increases the volume of the global ocean that, in addition to the thermal expansion of the ocean, increases sea levels across the globe. Mean sea level across the planet has risen by almost 17 cm over the past 100 years and is currently rising at 1–2 mm a year. This is an order of magnitude higher than the average rate over the previous several millennia. These changes are causing concern given the enormous human population and associated infrastructure (e.g. cities, ports) located in low lying coastal areas of the world. Some low lying countries in the Pacific such as Tuvalu, Kiribati and the north coast of Java (Indonesia) are already being affected by spring tides and storm surges that reach higher than ever before and now inundate towns and human dwellings on a regular basis.

How rising sea levels will impact coral reefs is unclear at this point. On one hand, projected changes in sea level are relatively slow compared to the rates of coral growth (up to 20–30 cm per year) and hence are unlikely to present major problems for healthy coral populations. Coral reefs have also generally kept up with even the most rapid changes in the past, as discussed in Chapter 3. The problem comes, of course, with a combination of sea level rise and reduced growth as a result of warmer and more acidic oceans, which raises the prospect of coral populations that literally get left behind (drown) as sea level rises.

Changing weather patterns also have the potential to affect coral reefs through impacts on river flows and drought (and hence the amount of sediment running off coastal areas), and through the physical impact of storms. Climate change is expected to increase the intensity of storms via its influence on sea-surface temperature, a fact emphasised by recent events such as hurricanes Katrina, Wilma, Dean and the other record storms in the Gulf of Mexico in 2005. On the GBR, cyclones, which are relatively rare in the north and more common in central and southern latitudes, are important drivers of the life expectancies of corals as well as reef and island building processes. Changing cyclone strengths, wave heights and return periods along the GBR will change coastal inundation and the wave climates on reefs, land runoff regimes and coastal salinities. These changes represent potential threats, although the precise details of how they will affect coral reefs, or their overall importance, is not entirely clear at this point.

Changes to weather patterns may have other important synergies. Recent trends on the Australian mainland, for example, show a drying trend along the eastern seaboard over the past 70 years as a result of removal of vegetation, causing destabilisation of sediments in river catchment areas. Coupled with more intense storms, destabilised sediments are likely to be increasingly washed down river catchments and into coastal areas where they may have impacts on coral reefs. Given that increased sediment runoff is a major factor affecting corals in coastal regions, these changes have the potential to have similar impacts to those seen when hard-hoofed cattle were introduced into river catchments over 100 years ago along the north-east coastline of Australia. Sediment flows in this case increased by 20 fold within a few decades of cattle farmers arriving, leading to the loss of inshore coral reefs in many parts of coastal Australia.

■ GENETIC ADAPTATION TO CLIMATE CHANGE

Some of the future scenarios outlined above (e.g. responses to rising sea temperature and ocean acidity) are predicated on the assumption that genetic change within populations of organisms does not occur rapidly enough to keep up with climate change. In terms of reef-building corals, thermal thresholds have been relatively stable over several decades and have not shown any tendency to drift upwards. The very fact that bleaching and mortality are increasing is circumstantial evidence that the threshold of corals for stress is not changing rapidly enough to neutralise the effects of increasing stress from the Earth's rapidly warming seas.

Several authors have challenged the use of fixed thresholds for coral reefs, suggesting that the thermal tolerance of corals and their symbionts may evolve rapidly. Some authors have even proposed that 'reef corals bleach to survive [climate] change'. These arguments, if true, would have major impacts on how coral reefs will respond to rapid shifts in the Earth's climate. If the thermal tolerance of corals and their symbionts could keep pace with the rate of climate change, then ultimately the current frequency and intensity of coral bleaching and mortality events would remain constant.

The evidence to support these hypotheses is extremely weak (see Additional reading). Most evidence points to the evolutionary capacity of corals falling behind current and future rates of climate change. The fact that the rates of change over the past 100 years are at least 100 fold higher than the rapid shifts seen when the earth moved from ice ages to the warm interglacial periods is sobering in terms of any discussion about adaptation. Terrestrial populations are changing rapidly in response to climate change, with birds and butterfly populations experiencing extinction at lower latitudes, and establishment at higher latitudes where habitat is available. Similar trends, though less well documented, are occurring in the ocean. Fish normally found at lower, more tropical latitudes are starting to appear on reefs at higher latitudes. Responses by marine populations depend very much on the organisms involved. Reef-building by corals is limited by other factors in addition to temperature (e.g. light, concentration of carbonate ions; see Chapter 8) and hence increases in sea temperatures due to the global warming will not result in the appearance of carbonate coral reefs at higher latitudes.

A central requirement for any adaptive response to have a major impact is the stabilisation of the concentrations of greenhouse gases such as carbon dioxide and hence, of global temperature. Continuation of mid- to high-range emissions will perturb the climate for hundreds if not thousands of years, with major implications for how populations of corals and other coral reef organisms do or do not respond. For example, if the world's international agreements were to stabilise global temperatures at 2°C above present day conditions (i.e. low emission scenarios), coral populations would initially decrease as unfit genotypes disappeared from particular regions, and then would increase as fitter genotypes proliferated to establish populations under the new stable global temperature. Thermally tolerant equatorial genotypes may migrate to higher latitudes over time (probably over decades), as they track their preferred warm temperature regime. However, if greenhouse gases like CO_2 do not stabilise, there is likely be a regime of heatwaves at high latitudes too severe for even the most heat tolerant genotypes to establish. Constantly changing conditions of such a non-stabilised climate would drive corals (and the reef ecosystems) into low population densities with the prospect of massive extinction rates for corals and the thousands of coral framework-dependent species.

■ RAMIFICATIONS FOR CORAL REEF ECOSYSTEMS AND PEOPLE IN A RAPIDLY CHANGING CLIMATE

The key question for coral reef ecologists is how changes in the abundance of reef-building corals will affect those thousands of other species that are totally dependent on the coral framework for food, shelter and reproduction (Fig. 10.5). While defining the full set of relationships between corals and other reef organisms goes beyond the space allowed here, it is important to outline some of the ways that the highly interconnected coral reef ecosystem are likely to be affected by climate change.

Several studies have now demonstrated that impacts of coral bleaching and mortality on reef fish populations include local extinctions, reduced taxonomic distinctiveness and species richness, and a loss of species within key functional groups. Several studies reveal that fish diversity is directly affected by the loss of corals. Using data from Pacific and Indian Ocean studies, it is clear that fish populations appear highly sensitive to changes in coral cover, with 62% of fish species declining in abundance within three years of disturbances that resulted in greater than a 10% decrease in coral cover. Particular species appear to be more sensitive than others, with coralivorous (coral eating) species being the most sensitive. The response of other less coral dependent reef fish is not clear. Several studies have shown that the number of herbivorous fish may increase after the coral mortalities associated with mass bleaching events, primarily due to the increase in algal turfs, which are the preferred food of these fish species.

Our current understanding of how coral reef organisms other than fish are influenced by the loss of corals from reefs is limited. It is clear that other organisms are equally susceptible to the projected changes in coral cover. Obligate crab fauna that live in corals (such as *Pocillopora* spp., for example) disappear from corals that have

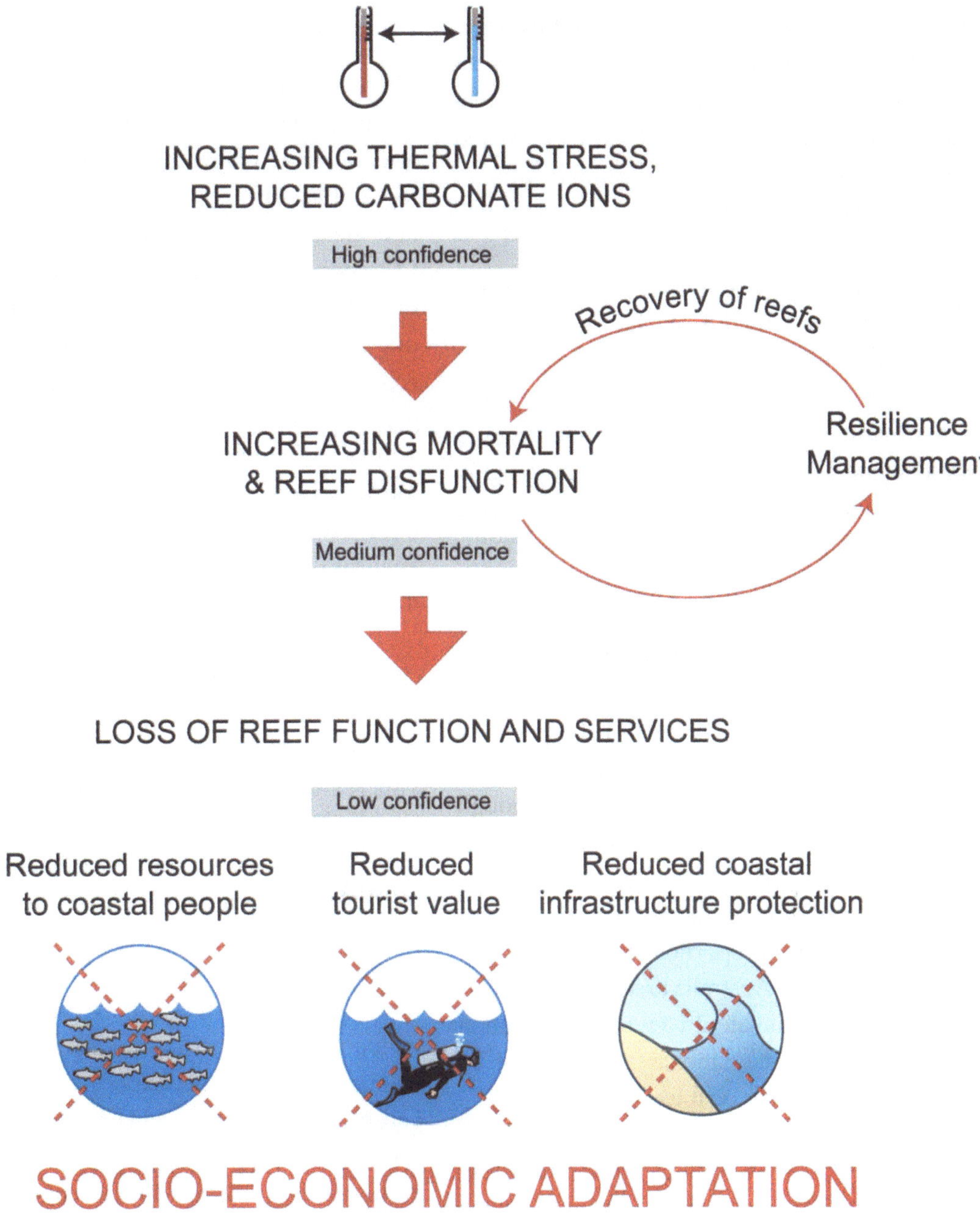

Figure 10.5 The effect of changing global conditions has ramifications on reef-building corals and on the organisms that live in association with coral reefs. The increasing mortality of corals leads to a loss of reef function and services, which leads to reduced biodiversity and effects on fish populations, among other things. These changes have ramification for the humans dependent on coral reefs through effects on tourism, resources for coastal populations and protection of coastal infrastructure from storms. Maintaining reef resilience through protection of herbivore populations is one possible way to improve the recovery of coral reefs after coral bleaching. It is important to understand that increasing uncertainty does not mean that the problem does not exist. Rather it highlights the fact that we know things will change but have little detailed information on the nature of the changes or their magnitude. (Figure: D. Kleine and O. Hoegh-Guldberg.)

bleached or die. Many other species form tight relationships with corals and hence are likely to be lost from coral reefs as coral abundance declines. There may also be an important density threshold in which some coral dwelling species require their coral colonies to be clustered closely so that they can access mates for reproduction.

The changes that are set to occur on coral reefs will also affect humans with direct and indirect dependence on them. This topic is clearly important and deserves considerably more space than a concluding paragraph (Fig. 10.5). Australia's coral reefs are currently an important economic engine for its communities and coastal economies, generating over \$6 billion per annum from GBR tourism alone. Great Barrier Reef tourism is highly dependent on its key competitive advantage of having the reputation of being the most pristine coral reef on the planet. This reputation allows the Australian tourism sector to compete with other tourist destinations in South-East Asia, the western Indian Ocean and the Caribbean that are closer to most international tourist domiciles in terms of travel time. Analysis of the potential effect of loosing that competitive edge through the deterioration of coral reefs on the Great Barrier Reef through climate change indicates that the costs could be significant. Estimated impacts vary depending on market characteristics and factors such as international trends and politics, but if one uses the proportion of the income that is due to people coming to Australia to see our Great Barrier Reef, it is clear that climate driven reef degradation could potentially reduce international tourist income by as much as \$8 billion over the next two decades.

Reefs also have major importance to people in developing countries, with over 100 million people subsisting on coral reefs as a source of food. While the details of the socio-economic impacts from climate change will vary, countries like Indonesia, Fiji and other Indo-Pacific countries could all potentially face declining coastal resources. This decline would have flow-on consequences resulting from the need to find alternative livelihoods for the many millions of people that live in coastal zones in Asia and the Indo-Pacific. Understanding the economic and social ramifications of climate change on coral reefs will consequently be critical for preparing people, industries and governments for the changes that seem almost certain to occur over the next 50 years.

RESPONDING TO THE CHALLENGE OF CLIMATE CHANGE?

There is no clearer example of how biological systems are affected by climate change than the impacts already apparent on coral reefs. The combined effects of warming and acidification have already affected coral reefs, and the effects will become devastating if atmospheric CO_2 concentrations do not stabilise below 450 ppm. Stabilisation at or below these levels of CO_2 in the atmosphere will require drastic action within political, economic and social spheres (at least a 90% reduction in greenhouse emissions by 2050). The details of the required actions are beyond the scope of this chapter. The fourth assessment report of the Intergovernmental Panel on Climate Change (IPCC 2007), particularly the analysis of working group 3, is a good place to start for those people interested in this area.

Given the scale of the projected impacts of climate change on coral reefs, there might be a temptation to conclude that preventing and managing the other impacts on coral reefs (see Chapter 12) is now irrelevant. This is a valid point of view if we do not deal with the current unrestrained growth in greenhouse gas emissions. However, if society does pursue a policy of drastically reducing emissions such that atmospheric CO_2 concentrations stabilise at 450–500 ppm, it must go hand in hand with good management of coral reefs themselves. Coral reefs will still experience increases in sea temperature and very low carbonate concentrations that will be outside any they experienced in millions of years, and bleaching events will become more commonplace. There is now substantial evidence that the ability of corals to recover from bleaching events is strongly affected by how much they are being influenced by other stresses. For example, elimination of grazing fish (from overfishing) appears to increase the recovery of corals after an ecological disturbance, like the mortality following coral bleaching, by 2–3 fold. Deteriorating water quality might act synergistically with under-grazing to further lower recovery after bleaching events. If, through improved reef management, the recovery of reefs from bleaching can be improved, then reefs will persist longer under the stresses expected over the next 50–100 years. Actively managing coral reefs in this current period of rapid global change will consequently translate as a greater abundance and diversity of coral reef organisms from which reefs will expand when the climate has finally stabilised once again.

ADDITIONAL READING

Baker, A. C. (2001). Reef corals bleach to survive change. *Nature* **411**, 765–766.

Done, T. J., Whetton, P., Jones, R., Berkelmans, R., Lough, J., Skirving, W., and Wooldridge, S. (2003). Global climate change and coral bleaching on the Great Barrier Reef. Final report to the State of Queensland, Greenhouse Taskforce through the Department of Natural Resources and Mining, Townsville.

Donner, S. D., Skirving, W. J., Little, C. M., Oppenheimer, M., and Hoegh-Guldberg, O. (2005). Global assessment of coral bleaching and required rates of adaptation under climate change. *Global Change Biology* **11**, 1–15.

Hoegh-Guldberg, O. (1999). Coral bleaching, climate change and the future of the world's coral reefs. *Marine and Freshwater Research* **50**, 839–866.

Hoegh-Guldberg, O. (2006). Low coral cover in a high-CO_2 world. *Journal of Geophysical Research* **110**, C09S06, doi:10.1029/2004JC002528. Available at: http://www.agu.org/pubs/crossref/2005/2004JC002528.shtml [Verified 21 March 2008].

Hoegh-Guldberg, H., and Hoegh-Guldberg, O. (2004). 'Biological, Economic and Social Impacts of Climate Change on the Great Barrier Reef.' World Wide Fund for Nature, Sydney.

Hoegh-Guldberg, O., Mumby, P. J., Hooten, A. J., Steneck, R. S., Greenfield, P., Gomez, E., Harvell D. R., Sale, P. F., Edwards, A. J., Caldeira, K., Knowlton, N., Eakin, C. M., Iglesias-Prieto, R., Muthiga, N., Bradbury, R. H., Dubi, A., and Hatziolos, M. E. (2007). Coral reefs under rapid climate change and ocean acidification. *Science* **284**, 118–120.

Hughes, T. P., Rodriques, M J., Bellwood, D. R., Ceccarelli, D., Hoegh-Guldberg, O., McCook, L., Moltschaniwskyj, and N., Pratchet, M. S. (2007). Regime-shifts, herbivory and the resilience of coral reefs to climate change. *Current Biology* **17**, 360–365.

Intergovernmental Panel on Climate Change Assessment Reports (1992, 1996, 2001, 2007). Available at http://www.ipcc.ch/ipccreports/assessments-reports.htm [Verified 25 March 2008].

Kleypas, J. A, Buddemeier, R. W., Archer, D., Gattuso, J.-P., Langdon, C., and Opdyke, B. N. (1999). Geochemical consequences of increased atmospheric carbon dioxide on coral reefs. *Science* **284**, 118–120.

Marshall, P., and Schuttenberg, H. (2006). 'A Reef Manager's Guide to Coral Bleaching.' Great Barrier Reef Marine Park Authority, Townsville, Australia. Available online at http://www.gbrmpa.gov.au/corp_site/info_services/publications/misc_pub/a_reef_managers_guide_to_coral_bleaching.

Parmesan, C., and Yohe, G. (2003). A globally coherent fingerprint of climate change impacts across natural systems. *Nature* **421**, 37–43.

Petit, J. R., Jouzel, J., Raynaud, D., Barkov, N. I., Barnola, J.-M., Basile, I., Benders, M., Chappellaz, J., Davis, M., Delayque, G., Delmotte, M., Kotlyakov, V. M., Legrand, M., Lipenkov, V. Y., Lorius, C., Pépin, L., Ritz, C., Saltzman, E., Stievenard, M. (1999). Climate and atmospheric history of the past 420,000 years from the Vostok ice core, Antarctica. *Nature* **399**, 429–436.

Raven, J., Caldeira, K., Elderfield, H., Hoegh-Guldberg, O., Liss, P., Riebesell, U., Shepherd, J., Turley, C., and Watson, A. (2005). Ocean acidification due to increasing atmospheric carbon dioxide. *Royal Society Special Report*, pp. 68.

Walther, G. -R., Post, E., Convey, P., Menzel, A., Parmesan, C., Beebee, T. J. C., Fromentin, J.-M., Hoegh-Guldberg, O., and Bairlein, F. (2002). Ecological responses to recent climate change. *Nature* **416**, 389–395.

Wilson, S., Graham, N. A. J., Pratchett, M., Jones, G. P., and Polunin, N. V. C. (2006). Multiple disturbances and the global degradation of coral reefs: are reef fishes at risk or resilient? *Global Change Biology* **12**, 2220–2234.

11. Terrestrial Runoff to the Great Barrier Reef and the Implications for Its Long Term Ecological Status

J. Brodie and K. Fabricius

The Great Barrier Reef (GBR) region with its two layers of protected areas (the GBR Marine Park and the GBR World Heritage Area) is a large marine ecosystem adjacent to the north-east Australian coast. The land adjacent to the GBR forms the GBR Catchment Area (GBRCA) (Fig. 11.1) from which many rivers and streams discharge into the GBR. As the GBRCA has been developed for agricultural, industrial and residential use over the last 150 years, waters discharged from GBRCA rivers have contained increasing amounts of nutrients, sediments and other pollutants.

Discharges of suspended sediments from the GBRCA to the GBR, estimated from monitoring and modelling studies, have increased by approximately five times, nitrogen exports by four times and phosphorus exports by 10 times. Some pesticides are now also being discharged to the GBR at potentially ecologically relevant concentrations. The sources of these pollutants are the principal land uses on the GBRCA: sediments and nutrients from beef grazing, primarily through increased soil erosion; sediments, nutrients and pesticides (currently mainly herbicides) from cropping (sugarcane, banana and other horticulture, cotton, and grain crops) associated with soil erosion, fertiliser use and pesticide

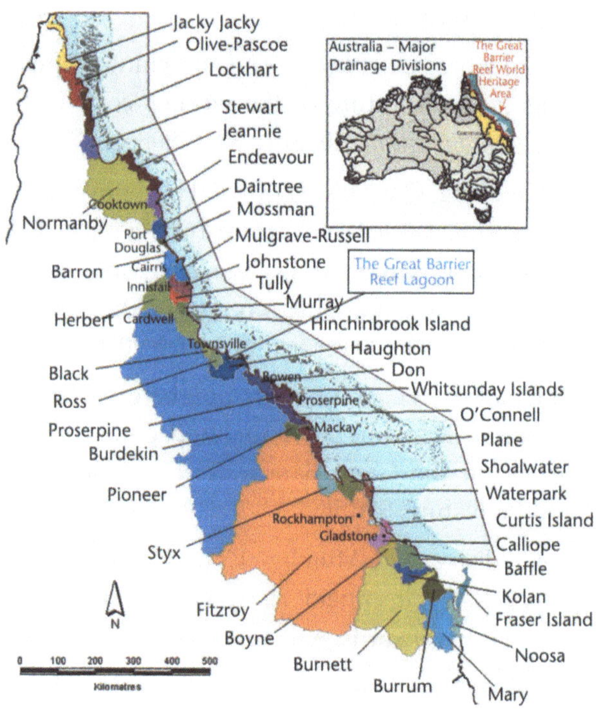

Figure 11.1 The Great Barrier Reef (GBR) and the Catchment Area (GBRCA). (Data source: Auslig Australia PL.)

use; and a diverse array of other pollutants (e.g. heavy metals, persistent organic pollutants, antibiotics, and caffeine) and nutrients from urban land use associated with sewage effluents and stormwater runoff.

Pollutants are transported through GBRCA rivers primarily in large flow events following monsoonal/cyclonic rainfall. As the rivers flood, a proportion of the pollutant load is trapped on floodplains, especially in riparian vegetation and wetlands. However, since GBRCA rivers are relatively short and fast flowing there is limited time for trapping and most rivers discharge a high proportion of their fine particulate and dissolved pollutant load to the ocean (Fig. 11.2). Most (80% south of Cooktown) lower river floodplains in the GBRCA have been drained and converted to agricultural and urban uses, further reducing their ability to trap catchment pollutants.

River discharges into the GBR can form plumes transporting pollutants hundreds of kilometres through the GBR Lagoon (Fig. 11.3). Within the plume, pollutant concentrations decline with distance from the river mouth via dilution, sedimentation and biological uptake. Most suspended sediments and particulate nutrients are initially deposited close to the river mouth (1–10 km) (Fig. 11.4) but the fine fraction then continues to be dispersed and transported via wind resuspension and longshore currents further along the coast. Dissolved nutrients and herbicides are transported in the plume, the bioavailable nutrients stimulating phytoplankton blooms (Fig. 11.3). This transport may expose large areas of the GBR lagoon and the embedded ecosystems to terrestrial pollutants. The extent of exposure of inshore ecosystems, such as coral reefs and seagrass beds, to the various land-sourced pollutants has been estimated through modelling and monitoring.

The understanding of the effects of land-sourced pollutants on coral reef, seagrass and planktonic ecosystems and specifically the GBR have slowly improved in the last 20 years. Five different types of exposure have to be distinguished when determining the effects of terrestrial runoff on coral reefs. These are: exposure to turbidity-related light limitation, enrichment with particulate organic matter (POM), dissolved inorganic nutrients, sedimentation and pesticides. The effects of enrichment with POM and pesticides on marine organisms are presently poorly understood and deserve

Figure 11.2 Russell-Mulgrave River in high discharge conditions. (Photo: A. Elliot, GBRMPA.)

Figure 11.3 Phytoplankton bloom and flood plume from the Mackay/Whitsunday region rivers, January, 2005. Satellite image from the Landsat sensor processed by Department of Natural Resources and Water, Queensland Government.

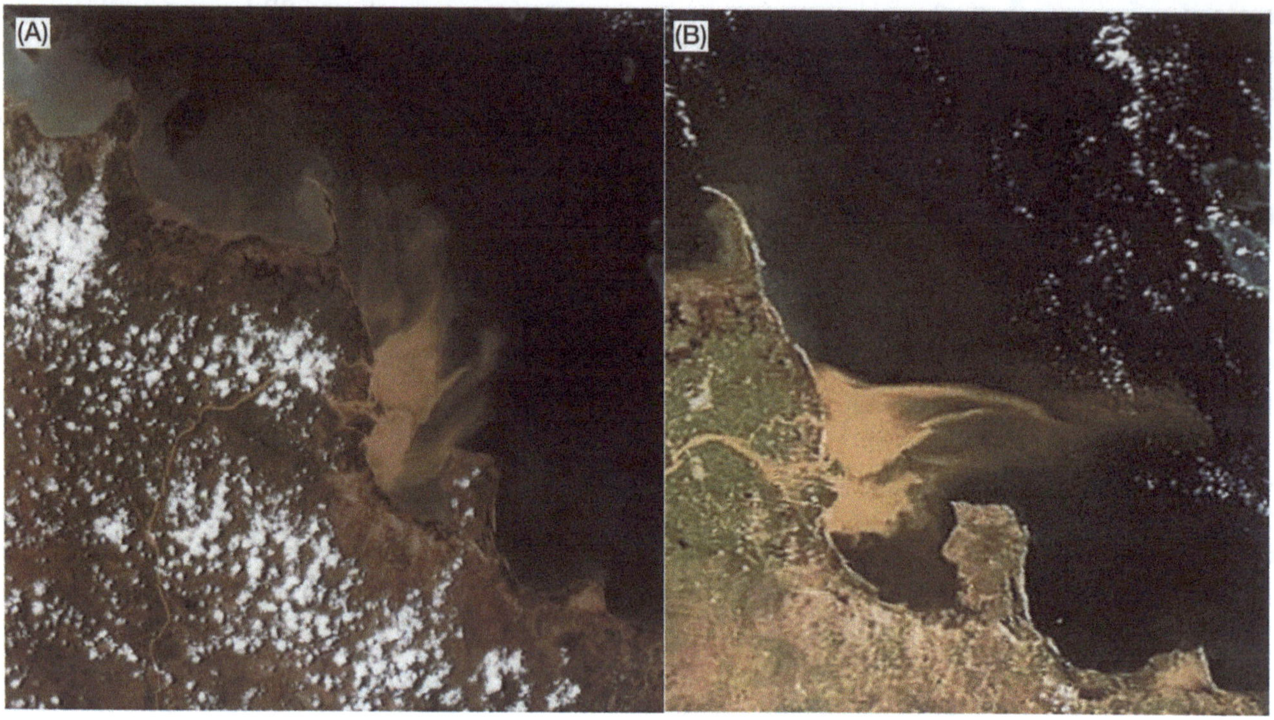

Figure 11.4 *A*, Flood plumes from the Burdekin River on 28th January, 2005 and *B*, 29th January, 2005 showing the area of intense deposition of suspended sediments near the river mouth. Satellite images from the MODIS sensor processed by Arnold Dekker and colleagues of CSIRO Land and Water, Canberra.

further research. A brief overview over these five types of water quality issues is given here.

Both suspended sediments and phytoplankton blooms can reduce light to benthic organisms via increased turbidity. Light limitation from turbidity reduces gross photosynthesis and growth at deeper depths but not in shallow water in hermatypic corals, seagrass and macroalgae. Light limitation also greatly reduces coral recruitment. The extent of trophic plasticity varies between coral species and effects of light limitation are most severe for more phototrophic species, and less influential on trophically more plastic groups. Heterotrophic species such as filter feeders may be promoted at low light.

Enrichment with POM enhances feeding rates and growth in some corals, providing a growth advantage that can partly or fully compensate for light reduction, especially in high-flow environments. However, while some corals can benefit from POM, heterotrophic filter feeders will benefit even more than corals do, hence the competitive advantage shifts from corals that can grow at extremely low food concentrations to simpler, more heterotrophic communities.

Dissolved inorganic nutrients can reduce coral calcification and increase macroalgal abundances. Excessive nutrients may shift coral communities towards algal dominated communities, although this shift is complicated by the abundance of algal grazing fish. Where grazing fish (and other invertebrate grazers) are abundant they may control algal proliferation and thus reduce the manifestation of undesirable effects of nutrient enrichment. Dissolved inorganic nutrients are generally quickly removed from the water column through biological uptake. It is therefore suspected that their main role in affecting ecosystems is that of promoting the organic enrichment of benthos, sediments and suspended particulate organic matter.

A more indirect effect of increased nutrients on coral reef ecosystems may arise through an increased probability of the formation of population outbreaks of the crown-of-thorns starfish (*Acanthaster planci*), a major coral predator. *Acanthaster planci* outbreaks have been a principal cause of coral mortality on the GBR (and throughout the Indo-Pacific coral province) over the last 40 years. Its planktonic larvae feed on large phytoplankton, and experiments suggest that the successful development of these planktonic larvae is food limited: their survivorship increases steeply with increasing availability of suitable food. Observations suggest that the location and timing of high nutrient concentrations in the many Indo-Pacific locations correlate well with the location and times of primary outbreaks of *A. planci*. In the GBR, mean summer chlorophyll concentrations in the central and southern GBR lagoon are about twice as high as in the far northern part of the GBR, and can be elevated during large flood events. If experimental findings and observations also apply to the GBR, it is likely that increased survivorship of *A. planci* larvae and subsequent adult population outbreaks may be best explained by increased larval survival facilitated by the terrestrial runoff of nutrients and sediments.

Sedimentation represents a severe disturbance for coral reefs. Sediments can smother corals and other reef organisms. Sedimentation reduces growth and survival in a wide range of coral species, although responses differ substantially between species and also between different sediment types. Coral skeletal records can incorporate trace elements that serve as proxy for sedimentation rates. These records suggest that the amount of soil washed off the land has increased about 5-fold since European settlement, and that inshore corals are exposed to (and incorporate) some of this runoff associated material. Furthermore, sediments that are enriched with nutrients are more effective than clean inorganic carbonate sediment in smothering corals, either through the formation of muddy marine snow or anoxia. Sedimentation also greatly limits recruitment and the survival of early life stages in corals: settlement rates are near-zero on sediment-covered surfaces, and sedimentation tolerance in coral recruits is at least one order of magnitude lower than for adult corals. Some of the bioeroding and space-competing groups of organisms are also sensitive to sedimentation by fine silt, and so are crustose coralline algae, with negative consequences for coral recruitment.

Herbicide residues (especially diuron, atrazine, simazine, ametryn, hexazinone and tebuthiuron) are commonly being detected in river water in high flow conditions and in river flood plumes. More sensitive monitoring tools are now deployed in the GBR Lagoon, where residues of some pesticides are also being detected, presently in quite low concentrations, during non-flood (dry season) periods. Herbicide residues (e.g. diuron) can disrupt photosynthetic activity in marine plants including mangroves, seagrasses, coral zooxanthellae and crustose coralline algae. In addition, pesticide residues of various commonly used and detected types (herbicides, insecticides, fungicides) can directly affect the physiology of coral, especially at the reproductive, larval and juvenile stages.

The type and severity of response to terrestrial runoff at any particular location depends on whether changes occurred predominantly in sedimentation, turbidity, POM, dissolved inorganic nutrients or pesticides. The severity of response also depends on the physical, hydrodynamic, spatial and biological properties of a location. Reefs that are surrounded by a shallow sea floor, reefs in poorly flushed bays or lagoons, deeper reef slopes, and frequently disturbed reefs are likely to experience changes even at low levels of pollution, in particular when populations of herbivores are low. In contrast, well-flushed shallow reef crests surrounded by deep sea floors or in areas of moderate tides are likely to have the highest level of resistance and resilience, especially when inhabited by healthy populations of herbivores that protect against overgrowth by sediment-trapping macroalgae. In most places, reduced recruitment success in corals, together with the promotion of macroalgae and *A. planci*, arguably represent the most significant direct effect of terrestrial runoff on coral reefs. In severe conditions, the overall outcome is reduced reef calcification, shallower photosynthetic compensation points (see Chapter 7), changed coral community structure, and greatly reduced species richness. Hence, reef ecosystems increasingly simplify with increasing exposure to terrestrial runoff, compromising their ability to maintain essential ecosystem functions and to recover from disturbances, including the presently increasing level of human-induced disturbances.

It can be concluded that increased pollutant discharge to the GBR resulting from agricultural and urban development on the GBRCA has degraded mangrove ecosystems in selected locations (e.g. the Mackay area) and some reef ecosystems in the inshore areas of the Wet Tropics and the Whitsundays coastal waters are also in a disturbed state, with about 50% lower coral diversity than expected, and are apparently slow to recover from disturbance. In addition, the increased frequency of outbreaks of *A. planci* has degraded inshore and offshore reefs along the entire central and southern section of the GBR. The high coral diversity and recruitment potential of inner-shelf reefs in the remote and relatively unpolluted region north of Princess Charlotte Bay is in stark contrast to the state of most inner-shelf reefs in the region of the GBR coast between Townsville and Port Douglas that has been exposed to agricultural and urban runoff (Fig. 11.5).

Management regimes to halt or reverse the decline of water quality entering into the GBR Lagoon and its ecosystems are being designed and implemented. Preliminary water quality guidelines and targets for protection of the GBR have been set. Best management practices for agricultural and urban activities have been tested for their effectiveness in improving water quality. Modelling has been used to predict the effectiveness of land management at a catchment scale on reducing pollutant loads. A management response has also been developed by the Australian and Queensland Governments through the GBR Water Quality Action Plan in 2001 followed by the GBR Reef Water Quality Protection Plan (RWQPP) in 2003. The RWQPP is now being implemented through regional natural resource management plans and the proposed regional Water Quality Improvement Plans. For the better management of fertiliser use, Nutrient Management Zones are being designated based on their estimated risk to the GBR such that management can be targeted within high source areas.

The success of the various plans will be assessed via integrated water quality monitoring programs. Monitoring programs have been designed and implemented at the various scales of the catchment to reef continuum (paddock scale, sub-catchment scale, whole of catchment scale, and GBR Lagoon scale) and the components will be integrated through a set of catchment to reef models.

Figure 11.5 McDonald's Reef. *A*, in the Princess Charlotte Bay region, in a relatively unpolluted environment, contrasts with *B*, Russell Island Reef in the Wet Tropics region exposed to agricultural and urban runoff. (Photos: K. Fabricius, AIMS.)

However, terrestrial pollution is not the only stress facing the ecosystems of the GBR, and it is the combination of multiple stresses (from global climate change, fishing and pollution principally) that will escalate threats to the long term viability of the system. Management of water quality issues alone, no matter how effective, are unlikely to prevent further degradation of the GBRWHA without concurrent management of fishing impacts and climate change.

ADDITIONAL READING

Brodie, J., Fabricius, K., De'ath, G., and Okaji, K. (2005). Are increased nutrient inputs responsible for more outbreaks of crown-of-thorns starfish? An appraisal of the evidence. *Marine Pollution Bulletin* **51**, 266–278.

Brodie, J., De'ath, G., Devlin, M., Furnas, M., and Wright, M. (in press). Spatial and temporal patterns of near-surface chlorophyll a in the Great Barrier Reef lagoon. *Marine and Freshwater Research.*

Devlin, M., and Brodie, J. (2005). Terrestrial discharge into the Great Barrier Reef Lagoon: Nutrient behaviour in coastal waters. *Marine Pollution Bulletin* **51**, 9–22.

Fabricius, K. E. (2005). Effects of terrestrial runoff on the ecology of corals and coral reefs: review and synthesis. *Marine Pollution Bulletin* **50**, 125–146.

Fabricius, K. De'ath, G. McCook, L. Turak E., and Williams. D. B. (2005). Changes in algal, coral and fish assemblages along water quality gradients on the inshore Great Barrier Reef. *Marine Pollution Bulletin* **51**, 384–398.

Furnas, M. (2003). Catchments and corals: Terrestrial runoff to the Great Barrier Reef. Australian Institute of Marine Science, Queensland.

McCulloch, M., Fallon, S., Wyndham, T., Lough, J., and Barnes, D. (2003). Coral record of increased sediment flux to the inner Great Barrier Reef since European settlement. *Nature* **421**, 727–730.

Schaffelke, B., Mellors, J., and Duke, N. (2005). Water quality in the Great Barrier Reef: responses of mangroves, seagrass and macroalgal communities. *Marine Pollution Bulletin* **51**, 279–298.

Wolanski, E., Richmond, R. H., McCook, L., and Sweatman, H. (2003). Mud, marine snow and coral reefs. *American Scientist* **91**, 44–51.

12. Planning and Managing the Great Barrier Reef Marine Park

J. Day

INTRODUCTION

The Great Barrier Reef Marine Park (GBRMPA) was established in 1975 to protect the Great Barrier Reef (GBR) and to ensure its sustainable use. The Marine Park covers 344 400 km², an area bigger than Victoria and Tasmania combined.

The GBRMP has always been a multiple-use marine park, allowing a range of uses and most reasonable activities, but with an overriding conservation objective of ecological sustainability; this means that the entire area is protected, but zoned to allow different activities in different zones and minimise impacts and conflicts.

The GBRMP extends 2300 km along the Queensland coast, and includes most of the waters from low water mark on the mainland coast, to the outer (seaward) boundary up to 280 km offshore. It is complex jurisdictionally, with both the Federal and State (Queensland) Governments involved in the management of the waters and islands within its outer boundaries. In terms of global marine protected areas, the GBRMP is:

- one of the world's largest (it certainly contains the largest network of No-Take Areas);
- probably the best known, and
- arguably the most methodically planned and comprehensively managed, particularly over such a large scale.

As the world's largest coral reef ecosystem, the GBR is also an important global resource. The GBR and its associated features directly contribute significantly to Australia's economy, with direct and indirect value added contibution estimated $5.4 billion in 2006-07. This includes $5.1 billion from the tourism industry, $153 million from recreational activity and $139 million from commercial fishing. This economic activity generates about 66 000 jobs, mostly in the tourism industry, which brings over 1.9 million visitors to the GBR each year. About 69 000 recreational vessels are registered in the area adjoining the GBR. These industries, and their flow-on activities, underpin a significant and growing proportion of Queensland's regional economy. They rely on the continued health of the GBR system for their long term economic sustainability.

Because of the iconic status of the GBR, many people believe that the entire area is a marine sanctuary or a marine national park, and therefore protected equally throughout. Many do not appreciate that the GBRMP is a multiple-use area, in which a wide range of activities and uses are allowed, including most extractive industries (but not mining nor drilling for oil), while still protecting one of the world's most diverse ecosystems. The multiple-use zoning system provides high levels of protection for specific areas, whilst allowing a variety of other uses to continue in certain zones. These include such diverse uses as shipping, dredging, aquaculture, tourism, boating, diving, military training, commercial fishing and recreational fishing.

The broad objectives of zoning in the GBRMP are set out in the *Great Barrier Reef Marine Park Act 1975* (the Act). The statutory Zoning Plan provides details on what, and where, specific activities are allowed, and which activities require a permit.

Under the Act, the Authority has power to perform any of its functions in co-operation with Queensland, with an authority of that State, or with a local governing body in that State. This is important given the adjoining Queensland marine parks, National Parks and islands.

The Australian Government's primary legislation for environmental regulation, the *Environment Protection and Biodiversity Conservation Act 1999*, also provides for such aspects as the protection of world heritage values, biodiversity conservation, and the protection of threatened and migratory species.

MANAGEMENT OF THE GREAT BARRIER REEF MARINE PARK

When the Act came into effect in 1975, the Great Barrier Reef Marine Park Authority (GBRMPA) was established as a Commonwealth Statutory Authority. The GBRMPA is the principal adviser to the Australian Government on the planning and management of the GBRMP, and is part of the Australian Government's Environment, Water Heritage and the Arts portfolio. The Goal of the GBRMPA is 'to provide for the long term protection, ecologically sustainable use, understanding and enjoyment of the GBR through the care and development of the Great Barrier Reef Marine Park'.

Field management (i.e. 'on the water') is undertaken by a number of Queensland and Australian Government agencies working under contract or other less formal arrangements or partnerships with the GBRMPA. These government agencies include the Queensland Parks and Wildlife (QPW), the Queensland Boating and Fisheries Patrol, the Queensland Water Police, Coastwatch, the Customs National Marine Unit, and the Australian Federal Police. The GBRMPA also directly participates in field management activities such as compliance, monitoring and the assessment of permits.

The Day-to-Day Management Program (DDM) is a jointly funded co-operative partnership between the Australian and Queensland Governments. It co-ordinates the day-to-day activities and field operations required for the management of the GBRMP and GBR World Heritage Area (including all the islands and intertidal waters).

In addition to the DDM arrangements, Queensland Government agencies with responsibilities for policy co-ordination, environment, local government, maritime matters, catchments, land use and fisheries are actively involved in administration and management of issues pertinent to the health and operation of the GBR. To carry out its functions effectively, the GBRMPA maintains liaison and policy co-ordination arrangements with all of these agencies, both at the operational and strategic levels. This close working partnership between Queensland and the GBRMPA has evolved over 30 years, and includes such aspects as complementary zoning and joint permits. This partnership has ensured the effective management of the complex and inter-related mix of marine, coastal and island issues, and provides for integrated management of the GBR on a whole-of-ecosystem basis.

This fundamental working relationship with the Queensland Government and its agencies is critically important for effective management of the GBR. Staff of the Authority also maintain strong partnerships with a wide range of other agencies, stakeholders, councils, traditional owners, community members and researchers with an interest in the protection, ecologically sustainable use, understanding and enjoyment of the GBR.

The first zoning plan for a small section of the GBRMP was finalised in 1981. It introduced a spectrum of zone types, ranging from a General Use Zone (the least restrictive zone, allowing most reasonable uses), through to a Preservation Zone (very small 'no-go' areas, set aside as scientific reference areas). These zone types, whilst refined, still exist today.

Within each zone type (see Fig. 12.1), certain activities are allowed 'as-of-right' (that is, no permit is required, but users must comply with any legislative requirements in force), some specified activities require a permit, and some activities are prohibited.

ACTIVITIES GUIDE

(see relevant *Zoning Plans* and *Regulations* for details)

	General Use Zone	Habitat Protection Zone	Conservation Park Zone	Buffer Zone	Scientific Research Zone[2]	Marine National Park Zone	Preservation Zone
Aquaculture	Permit	Permit	Permit[1]	✗	✗	✗	✗
Bait netting	✓	✓	✓	✗	✗	✗	✗
Boating, diving, photography	✓	✓	✓	✓	✓[2]	✓	✗
Crabbing (trapping)	✓	✓	✓[3]	✗	✗	✗	✗
Harvest fishing for aquarium fish, coral and beachworm	Permit	Permit	Permit[1]	✗	✗	✗	✗
Harvest fishing for sea cucumber, trochus, tropical rock lobster	Permit	Permit	✗	✗	✗	✗	✗
Limited collecting	✓[4]	✓[4]	✓[4]	✗	✗	✗	✗
Limited spearfishing (snorkel only)	✓	✓	✓[1]	✗	✗	✗	✗
Line fishing	✓[5]	✓[5]	✓[6]	✗	✗	✗	✗
Netting (other than bait netting)	✓	✓	✗	✗	✗	✗	✗
Research (other than limited impact research)	Permit	Permit	Permit	Permit	Permit	Permit	Permit
Shipping (other than in a designated shipping area)	✓	Permit	Permit	Permit	Permit	Permit	✗
Tourism programme	Permit	Permit	Permit	Permit	Permit	Permit	✗
Traditional use of marine resources	✓[7]	✓[7]	✓[7]	✓[7]	✓[7]	✓[7]	✗
Trawling	✓	✗	✗	✗	✗	✗	✗
Trolling	✓[5]	✓[5]	✓[5]	✓[5,8]	✗	✗	✗

PLEASE NOTE: This guide provides an introduction to Zoning in the Great Barrier Reef Marine Park. Relevant Queensland Marine Park Zoning Plans or the Queensland Environmental Protection Agency should be consulted for confirmation of use or entry requirements.

1. Restrictions apply to aquaculture, spearfishing and harvest fishing for aquarium fish, beachworm and coral in the Conservation Park Zone.
2. Except for One Tree Island Reef (SR-23-2010) and Australian Institute of Marine Science (SR-19-2008) which are closed to public access and shown as orange, all other Scientific Research Zones are shown as green with an orange outline.
3. Limited to 4 catch devices (eg. crab pots, dillies and inverted dillies) per person.
4. By hand or hand-held implement and generally no more than 5 of a species.
5. Maximum of 3 lines/rods per person with a combined total of 6 hooks per person.
6. Limited to 1 line/rod per person and 1 hook per line. Only 1 dory detached from a commercial fishing vessel.
7. Apart from traditional use of marine resources in accordance with s.211 of the *Native Title Act 1993*, an accredited Traditional Use of Marine Resources Agreement or permit is required.
8. Pelagic species only. Seasonal Closures apply to some Buffer Zones.

Detailed information is contained in the Great Barrier Reef Marine Park Zoning Plan and Regulations.

- Permits are required for most other activities not listed above.
- Commonwealth owned islands in the Great Barrier Reef Marine Park are zoned "Commonwealth Islands Zone" - shown as cream.
- All Commonwealth Islands may not be shown.
- Special Management Areas may provide additional restrictions at some locations.
- The Zoning Plan does not affect the operation of s.211 of the *Native Title Act 1993*.

ACCESS TO ALL ZONES IS PERMITTED IN AN EMERGENCY.

Figure 12.1 Activities Guide. © Commonwealth of Australia (July 2004).

BOX 12.1 GLOSSARY

bioregion. An area with habitats, communities (e.g. areas of seagrass) and physical features (e.g. sediment type, depth) that are more similar within the bioregion than those occurring in other bioregions.

No-Take Area/No-Take Zone. A zone or area where extractive activities are not permitted (see also Table 12.1).

Different sections of the GBRMP were progressively zoned, and by the late 1990s, approximately 15 800 km^2 (~4.6%) of the GBRMP were zoned as No-Take Areas (Marine National Park Zones), with a further 450 km^2 (~ 0.13% of the GBRMP) set aside for scientific baselines as 'no-go' areas (Preservation Zones). Over the last 30 years there has been a huge growth in scientific understanding of ecological aspects of the GBR, including ecosystem processes, connectivity, and the relationships between species and the physical environment.

Having good relationships with scientists helps the GBRMPA access the best available information for decision-making that is essential to high quality, scientifically-based management of the GBRMP. A list of research priorities for the management of the GBRMP has been compiled, and, whenever possible, the GBRMPA applies an adaptive management approach, using the best information available and if necessary erring on the side of caution.

Other management tools (outlined in the Conclusion) are used in conjunction with zoning to help achieve ecological protection and other management objectives. The GBRMP has always relied on this range of tools, but zoning has long been regarded as the cornerstone of effective planning and management.

WHY WAS THE GBR REZONED?

In the mid-late 1990s, there were concerns that the existing zoning did not adequately protect the range of biodiversity known to exist within the GBRMP. Furthermore, the zoning was inadequate to ensure that the entire ecosystem remained healthy, productive and resilient into the future. The location of most of the original highly protected zones reflected a historical focus on virtually only one habitat type (i.e. coral reefs), with a skewed emphasis in the more remote and 'pristine' areas.

Between May 1999 and December 2003, the GBRMPA undertook a comprehensive planning and consultative program to develop a new zoning plan for the GBRMP. A primary aim of the program was to better protect the range of biodiversity in the GBR, by increasing the extent of No-Take Areas (locally known as 'green zones'), ensuring they included 'representative' examples of all the different habitat types, otherwise known as bioregions (Box 12.1) (hence the name, the Representative Areas Program or RAP). A further aim was to maximise positive and minimise negative impacts on the users of the GBRMP. Scientific input, community involvement and agency innovation all contributed to achieving these aims.

During the rezoning program, all components of the existing Zoning Plans were open for comment and alteration. Given the previous and existing Zoning Plans had all been progressively developed over 17 years, some of the terms, management provisions, zone names and zone objectives differed slightly between various parts of the GBRMP. A new single Zoning Plan was therefore developed for the entire GBRMP enabling the planners to also address various important planning tasks, including:

- zoning 28 new coastal sections that were added to the GBRMP in 2000 and 2001;
- standardising terminology to make it consistent throughout, and
- implementation of simpler co-ordinate-based descriptions for all zone boundaries.

A number of factors contributed to the success of the new Zoning Plan:

- *Use of independent experts.* Independent experts greatly assisted in the development of a number of 'products' that were important to the planning process, and were widely available for discussion early in the planning program, in particular:
  - *GBR bioregionalisation.* A fundamental foundation for the new Zoning Plan was the map of 30 reef and 40 non-reef 'bioregions' that was developed early in the RAP.
  - A comprehensive range of biological and physical information across the GBRMP was used to define the bioregions. Staff of the GBRMPA initially collated information from numerous scientists with expert knowledge of the GBR. The most appropriate data sets were then used in classification and regression tree analyses to spatially cluster areas of similar species composition. A number of workshops, comprising reef and non-reef experts, then used all these data and analyses, plus their experience, to spatially describe the biodiversity of the GBR and develop the map of 70 bioregions.
  - It was decided to map diversity at the scale of 10s to 100s of kilometres because this was a scale over which habitats change markedly. It was also a scale at which most relevant information was available and it was a meaningful scale for subsequent planning and management. Areas of relative homogeneity were labelled 'bioregions' to facilitate communication with stakeholders. Bioregions were defined as having habitats, communities (e.g. areas of seagrass) and physical features (e.g. sediment type, depth) that are more similar within the bioregion than those occurring in other bioregions.
  - A draft version of the bioregionalisation was made available for public comment recognising that many local 'experts', including commercial and recreational fishers, coastal residents, rangers, and others have specialist knowledge about the level of variability of the GBR. This led to additional information and nine major refinements to the draft regionalisation.

  Operating principles for developing the new network. External natural science and social-economic-cultural advisory committees were used to develop 11 biophysical operational principles and four socio-economic, cultural and management feasibility operational principles. These principles clarified the planning 'rules' up front for all to see and apply before any new zones were proposed.
  - The biophysical operational principles included recommendations for minimum No-Take Areas for each bioregion and each known habitat type. Given the uncertainty about what amounts would be adequate for effective conservation, the recommendations were considered to be the minimum, in the context of global experience. The social, economic, cultural and management feasibility principles aimed to maximise complementarity of zoning with human uses and values.
  - The biophysical principles included advice on the size, shape and definition of protected zones to achieve biological objectives such as connectivity and to facilitate public understanding and compliance. The biophysical principles also provided a sound basis against which to assess the final extent of protection in No-Take Zones.
- *Integrated approaches.* A combination of expert opinion, stakeholder involvement and analytical approaches were used to identify options for possible zoning networks. The linking of science, technical support and community participation was an essential three-way dynamic in the planning process. The analytical tools applied included marine reserve design software, adapted and expanded for use in the RAP, and a suite of GIS-based spatial analysis tools. The analytical software enabled the GBRMPA to integrate a number of data layers representing biophysical, social and economic values, and enabled a number of zoning options to be generated and assessed.

COMMUNITY PARTICIPATION

The extensive public consultation program during the RAP included some 1000 formal and informal meetings and information sessions involving engagement with people in over 90 centres along and beyond the GBR coast. This included local communities, commercial and recreational fishing organisations, traditional owners, tourism operators, conservation groups, and so on. Meetings were held with representative organisations such as Sunfish, the Association of Marine Park Tourism Operators, World Wildlife Fund Australia, and all branches of the Queensland Seafood Industry Association within the GBR catchment area.

As required by the relevant legislation, there were two formal phases of community participation during the RAP; the first calling for input into the preparation of a new zoning plan, and the second providing the draft zoning plan for public comment. The resulting 31690 public submissions (10190 in the first formal phase; 21500 in the second phase), many of which included maps, were unprecedented compared to previous planning programs in the GBR. They necessitated the development of new, fast and effective processes for analysing and recording the range of information that was received by the GBRMPA.

A large number of the submissions included spatial information, including approximately 5800 maps in the second formal phase alone. All submissions, including this spatial information, were considered, coded and analysed, and the maps were digitised and/or scanned.

Many modifications were made to the Draft Zoning Plan as a result of the detailed information provided in submissions and other information received; however, in some locations there were limited options available to modify proposed No-Take Areas, particularly in inshore coastal areas and still achieve the minimum levels of protection recommended by the scientists.

The significant changes between the initial zoning, the Draft Zoning Plan and the final Zoning Plan, as accepted by Parliament, are readily seen in the 'Review of the Great Barrier Reef Marine Park Act 1975' – Review Panel Report, released in 2006. Maps 9, 10 and 11 on pages 69–71 of that report highlight the differences. Report available at http://www.environment.gov.au/coasts/publications/pubs/gbr-marine-park-act-chapters-5-6.pdf [Verified 21 March 2008]. Many of the aspects outlined in this chapter are illustrated in the Capricorn-Bunker case study shown in more detail in the same report (i.e. pp. 78–90 of the same webpage).

In accordance with the legislation, the Zoning Plan was submitted to both Houses of Federal Parliament in December 2003. Following a statutory review period, the Minister announced that the new Zoning Plan would be implemented on 1 July 2004.

In November 2004, the Queensland Government 'mirrored' the new zoning in most of the adjoining State waters (i.e. intertidal waters and some other areas deemed to be State waters), so now there is complementary zoning for virtually all the State and Federal waters within the entire GBR World Heritage Area. Table 12.1 indicates the area of each zone type in effect today (2008).

In conclusion, the approach that was undertaken in the RAP is now recognised as one of the most comprehensive global advances in the protection of marine biodiversity and marine conservation in recent decades. The final outcome, including the increase in No-Take Zones to more than 33% (over 115000 km^2) and including representative examples of every habitat, today comprises the world's largest systematic network of No-Take Zones.

The rezoning efforts alone will not ensure the future or sustainability of the GBR, however, they do provide a fundamental component for effectively conserving habitats, biological communities and ecosystem processes for future use.

Other spatial management 'tools' include statutory Plans of Management, permits, agreements with traditional owners and industry-specific accreditation. These have been found to be more effective for managing specific industries or activities such as tourism, shipping, ports, traditional use of marine resources, defence training and some fisheries regulations.

Similarly, a variety of temporal management tools are now applied in the GBRMP, including Special Management Areas, GBR-wide closures at specific fish-spawning times and a range of other fisheries regulations.

Table 12.1 Total area of zone types within the Great Barrier Reef Marine Park (as at 2008)

Zone name	Equivalent IUCN category	Area (km^2)	Area (hectares)	% of GBRMP
Preservation*	IA	710	71 000	<1
Marine National Park*	II	114 530	11 453 000	33
Scientific research*	IA	155	15 500	<1
Buffer	IV	9880	988 000	3
Conservation Park	IV	5160	516 000	2
Habitat protection	VI	97 250	9 725 000	28
General use	VI	116 530	11 653 000	34
Islands (Commonwealth[1])	*Various*	185	18 500	<1
Total		344 400	34 440 000	100

*, no take zones where extractive activities are generally not permitted.

The GBR continues to be under pressure from a wide range of human uses and natural impacts, and neither the GBR World Heritage Area nor the Marine Park are static. Use patterns and technology are constantly changing and the marine environment itself is dynamic. For example, the use of the Marine Park has escalated rapidly in the 30 years since its establishment. Despite various complexities, the integrated governance and management model that has been functioning in the GBR has proven to be effective and successful.

A range of strategies continue to ensure the GBR is protected now and into the future. These strategies, designed to maintain the health and increase the resilience of the GBR to cope with escalating pressures, include:

- maintaining compliance with the Zoning Plan and supporting Regulations;
- improving water quality through the Reef Water Quality Protection Plan;
- promoting sustainable fisheries;
- developing sound policy regarding the effects of climate change on reefs, and
- promoting sustainable tourism.

As the Marine Park moves into its fourth decade the new legislative framework along with the various other management strategies will collectively contribute significantly to the conservation of the entire GBR ecosystem. Collectively these will enhance significantly the health and resilience of the GBR, assist industry to achieve increased levels of environmental and financial sustainability, and provide benefits for all users and industries that are dependent on the GBR.

For more information about the Marine Park, visit www.gbrmpa.gov.au [Verified 22 February 2008].

Detailed information on the rezoning is on the GBRMPA website under 'Zoning' and then follow the link 'Information for Managers and Planners'.

All the zoning maps are on the website and copies are available in major centres along the GBR coast.

ADDITIONAL READING

Access Economics Pty Ltd. (2007). Measuring the economic & financial value of the Great Barrier Reef Marine Park, 2005–06, Canberra.

Bishop, M. (2004). Enforcement in the Great Barrier Reef Marine Park. In 'Proceedings of the 2nd International Tropical Ecosystems Management

Symposium (ITMEMS2)'. (Eds R. Jara, A. C. Kenchington and R. D. Magpayo.) pp. 515–520. (Department of Environment and Natural Resources: Manila.)

Day, J. C. (2002). Zoning – Lessons from the Great Barrier Reef Marine Park. *Ocean and Coastal Management* **45**, 139–156.

Day, J. C., Fernandes, L., Lewis, A., De'ath, G., Slegers, S., Barnett, B., Kerrigan, B., Breen, D., Innes, J., Oliver, J., Ward, T. J., and Lowe, D. (2003). The Representative Areas Program for protecting biodiversity in the Great Barrier Reef World Heritage Area. In 'Proceedings of the 9th International Coral Reef Symposium. Vol. 2'. (Eds M. K. Moosa, S. Soemodihardjo, A. Soegiarto, K. Romimohtarto, A. Nontji and S. Suharsono.) pp. 687–696. (Bali.)

Fernandes, L., Day, J., Lewis, A., Slegers, S., Kerrigan, B., Breen, D., Cameron, D., Jago, B., Hall, J., Lowe, D., Innes, J., Tanzer, J., Chadwick, V., Thompson, L., Gorman, K., Simmons, M., Barnett, B., Sampson, K., De'ath, G., Mapstone, B., Marsh, H., Possingham, H., Ball, I., Ward, T., Dobbs, K., Aumend, J., Slater, D., and Stapleton, K. (2005). Establishing representative No-Take Areas in the Great Barrier Reef: Large-scale implementation of theory on Marine Protected Areas. *Conservation Biology* **19**(6), 1733–1744.

GBRMPA (2005*a*). Report on the Great Barrier Reef Marine Park Zoning Plan 2003. Available at http://www.gbrmpa.gov.au/_data/assets/pdf_file/10680/gbrmpa_report_on_zoning.pdf [Verified]

GBRMPA (2005*b*). Research Needs for the Protection and Management of the Great Barrier Reef Marine Park. Available at http://www.gbrmpa.gov.au/corp_site/info_services/science/research_priorities [Verified 4 March 2008].

Kingsford, M.J. and Welch, D.J. (2007). Vulnerability of pelagic systems of the Great Barrier Reef to climate change. Chaper 18 in 'Climate Change and the Great Barrier Reef: A Vulnerability Assessment'. (Eds J.E. Johnson and P.A. Marshall). Great Barrier Reef Marine Park Authority and the Australian Greenhouse Office, Commonwealth of Australia.

13. Biodiversity

P. A. Hutchings & M. J. Kingsford

Biodiversity generally refers to species richness (i.e. number of species), but other definitions are also common. It is critical that biodiversity is defined carefully as it is sometimes considered synonymous with genetic diversity, habitat diversity, structural diversity, the diversity of functional groups (e.g. trophic groups of life history stages) or even life history traits (e.g. feeding type, growth forms, reproductive strategy, longevity). Some definitions can be a proxy for species richness. For example, species representation and abundance are known to vary among habitats. Habitat richness, therefore, can be especially relevant where the taxonomy of organisms in the various habitats is poorly known and there are concerns to protect species richness. A critical component of the zoning plan of the GBR (the Representative Areas Program, see Chapter 12) was partly based on the protection of different habitats and used these as surrogates to conserve species diversity as the biota on the GBR is poorly known apart from the corals and fish. In this chapter we focus on patterns of species richness at different spatial and temporal scales. We also note that descriptions of biodiversity and an understanding of processes influencing biodiversity (e.g. the impact of human activities) are critical for ecosystem management.

Coral reefs occur in a broad band around the equator, wherever suitable depths occur, between latitudes of 30°N and 30°S, see Chapter 7, although some coral reef development occurs further south and north where water currents ensure that for most of the year water temperatures do not drop below 18°C. For example, the gyre from the East Australian Current diverges eastwards south of the GBR and enables coral reefs to be developed on the east coast of Lord Howe Island at 31°30′S. This is the southernmost coral reef in the world. Species richness varies at scales that range from ocean basins and seas to habitats on reefs (Box 13.1 and Box 13.2). The richest reefs in the world in terms of number of species occur in the triangle that includes the northern tip of the Philippines and eastern and western tips of Indonesia, although today many of these reefs have been damaged by human activities, including dynamite fishing, overfishing and coastal developments.

An illustration of this diversity is the recent intensive survey of all the coral reef environments within an area of 295 km^2 in New Caledonia, which resulted in the collection of over 2700 species of molluscs. This is much larger than the number of species recorded from similar areas anywhere else in the world. Of these molluscs 32% were found only at one site and 20% of species were only represented by a single individual, indicating that rare species make up a considerable proportion of the fauna. In addition, many of the species were undescribed. In the entire Indo-Pacific, in the well known group of opisthobranchs (nudibranchs) it is estimated that at least 30% remain to be formally described. In other invertebrate groups much of the fauna remains even to be collected, let alone formally described.

BOX 13.1 MAJOR SOURCES OF SPATIAL AND TEMPORAL VARIATION IN SPECIES RICHNESS

Spatial variation

Among tropical seas and oceans (e.g. Caribbean *v.* Pacific)
With distance from diversity hot-spots (e.g. Indonesia, Philippines triangle)
With increasing latitude (e.g. along the Great Barrier Reef)
Cross shelf (e.g. from mainland to the shelf break of the GBR)
Among seascapes (e.g. reefs *v.* inter-reefal areas)
Among habitats within reefs
Over depth ranges

Temporal Variation

Evolutional time (speciation, extinction)
Climate change (e.g. changes species ranges)
ENSO and Pacific Decadal Oscillations events (e.g. bleaching causing local extinctions)
Seasonal change (especially at higher latitudes)
Pulse events (e.g. storms)

Spatial and temporal variation

Anthropogenic effects (fisheries, eutrophication, climate change, introductions)
Natural perturbations (cyclones, tsunami, COTS, productivity)

Many factors are responsible for the high diversity of biota found on coral reefs but in part it is due to the diversity of habitats found within coral reefs—ranging from soft sediments (both vegetated and unvegetated), various reefal habitats, and pelagic and coastal habitats that include mangroves and salt marsh. Habitats are usually characterised by particular habitat-forming organisms and variation in geomorphology. For example, *Isopora cylindrica* is a habitat-forming coral with spur and groove geomorphology, creating a shallow habitat (see Chapter 5).

For all groups of organisms associated with coral reefs, the highest number of genera occurs in the triangle of diversity delineated by Indonesia and the Philippines (the so-called 'coral reef triangle') with numbers declining both eastward and westward as well as north and south of this area. For example, while a total of 70 coral genera occur in the Indo-Pacific, most of these occur in the triangle with numbers declining across the Pacific with 10 in Hawaii and only two on the Pacific coast of America. The decline is smaller across the Indian Ocean with 50 genera present on the eastern Indian Ocean reefs. The Indo-Pacific is much richer than the Caribbean where only 20 coral genera occur with only *Acropora, Favia, Leptoseris, Madracis, Montastrea, Porites, Scolymia* and *Siderastrea* (Fig. 13.1) being shared between these two coral realms, and no species of corals occur in both realms. Similarly, as one moves north or south of this triangle, numbers of genera decline, with only 30 genera found at Lord Howe Island and 50 at Ningaloo on the West Australian coast, and numbers continuing to decline southwards down to two genera at Rottnest Island in the far south-west corner of Australia. If we now consider species of corals we find similar patterns. The genus *Acropora* represents the largest extant group of reef-building corals and over 370 species have been

BOX 13.2 HOW DO WE MEASURE BIODIVERSITY AND ANALYSE THE DATA?

Species richness is generally measured as number of species per unit area. Richness will, of course, increase with area and decisions have to be made on sample unit sizes that give estimates of diversity with a level of accuracy that suits the question being asked; particularly large areas are required for rare taxa. Species richness does not consider the proportion of each species in the assemblage, or the 'evenness' of representation. Measures of evenness are an attempt to measure dominance of a few species in a sample unit. Where one or a few species are very abundant when compared to the total number of species in the sample, then evenness is low.

Heterogeneity of an assemblage is a combination of species richness and the evenness of species representation. Heterogeneity indices (including, logarithmic series and log normal distributions as well as nonparametric measures such as Simpson's index, the Shannon-Wiener function and the Brillouin Index) have often been used to measure diversity. Krebs recommends the following when measuring biodiversity: (1) construct Whittaker plots of log abundance *v.* species rank; (2) where sampling effort varies spatially use the 'rarefaction method' for estimating the number of species expected in a random sample of individuals taken from a collection (this will give valid comparisons of richness among assemblages); (3) explore the use of log series and lognormal curves where variation in patterns are shown in the Whittaker plots; (4) use an inverse Simpson's Index of the exponential form of the Shannon-Wiener function to describe heterogeneity. Depending on your hypothesis you may want to weight common species (Simpson's) or rare species more (Shannon's) in the analysis. The latter decision is also critical in complex analyses of assemblage composition such as Multi Dimensional Scaling (MDS) (described by Clark and Warwick) and complementary tests such as analyses of similarity and dissimilarity among samples. Some studies use 'indicator species' as a measure of change in an assemblage, but the robust use of indicators needs to be tested carefully.

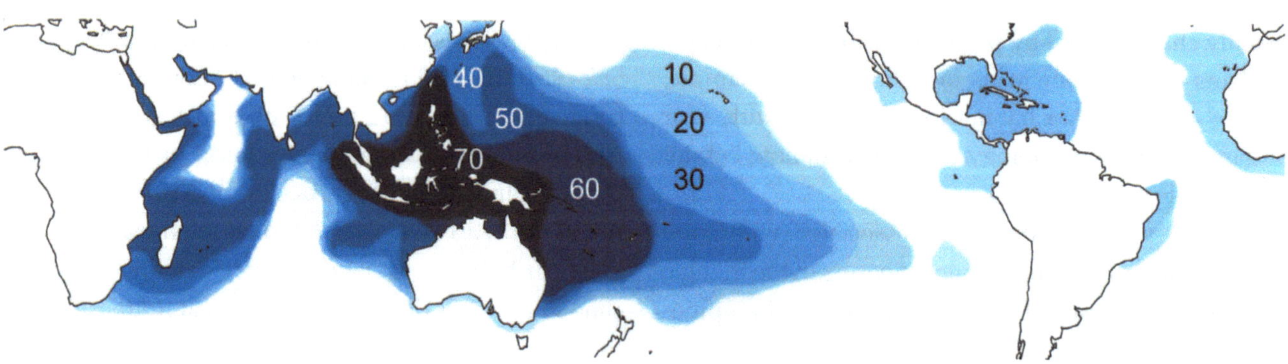

Figure 13.1 Patterns of global reef-building coral genera worldwide. The greatest generic diversity is found in the Indo-Pacific around the 'coral triangle', with numbers tapering off to the east and west as well as north and south. (Based on Veron, 2000.)

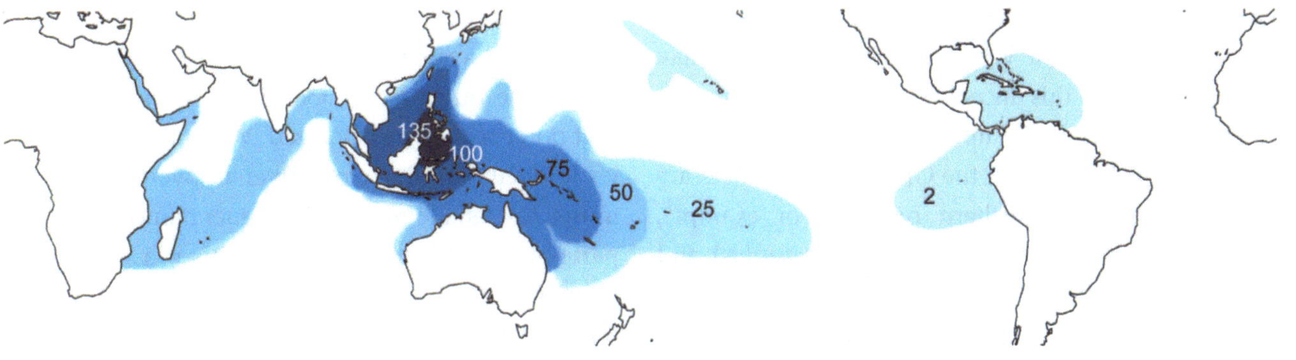

Figure 13.2 Global diversity of *Acropora* species. The greatest diversity is found in the Indo-Pacific triangle. (Based on Veron, 2000.)

described, with several major biogeographic patterns observed (Fig. 13.2). First, there is a distinct separation between Atlantic and Indo-Pacific fauna. Some species are widespread throughout the Indo-Pacific, but there is a distinctive Indian and Pacific fauna and components of this fauna extend into the Indo-Australian arc. Endemicity occurs in a number of areas including even the relatively young areas of the Red Sea and Arabian Gulf. Some of these patterns can be explained by looking at the fossil record of this group as well as the geological history of the area (see Chapter 20). Other studies are required on other invertebrate groups to see if this pattern is a universal one, but such groups need to be speciose, well known taxonomically, possess a fossil record and have detailed distribution records, which does somewhat limit the choice of taxa.

Biogeographic patterns of diversity that are similar to corals can be found among other well known groups such as fish, echinoderms, molluscs and some decapod crustaceans. For example, the distribution of species of damselfish across the south Pacific (Fig. 13.3) also clearly shows a reduction in the number of species from 110 in the coral reef triangle, slowly declining across the Pacific until members of the family are no longer found. Patterns of species representation are odd for some groups. For example, some tropical taxa have 'antitropical' distributions, in which they are found on the GBR and at high latitude reefs near Japan (e.g. Labridae, *Bodianus perdito*).

In summary, the Great Barrier Reef occurs just south of the coral reef triangle and is home to much of this Indo-Pacific fauna. Distinct regional biogeographical patterns occur within the GBR. For most groups, the northern GBR is more diverse in terms of number of species than the southern GBR. For example, 960 species of fish have been recorded for the Capricorn

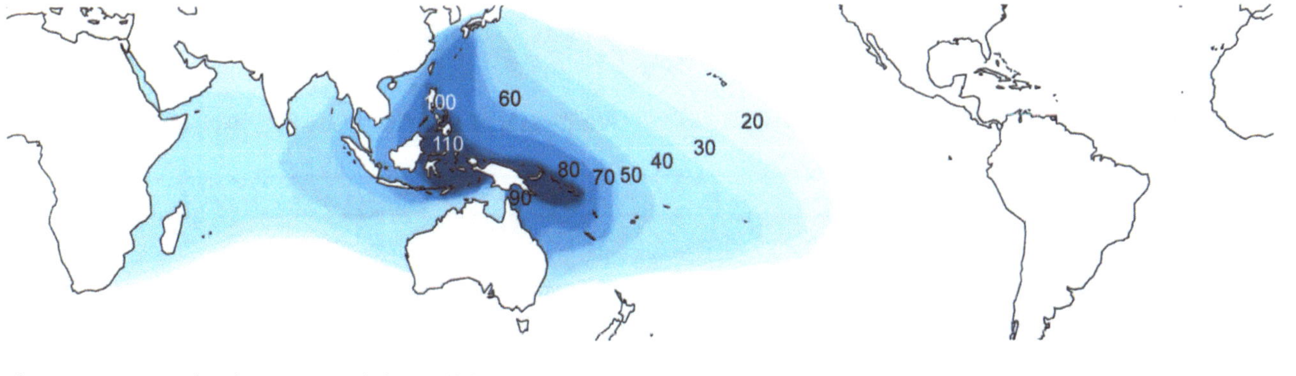

Figure 13.3 Broadscale patterns of damselfish species richness: distribution of damselfish (Pomacentridae) species across the Indo-Pacific. (Based on Allen, 1991.)

Group whereas over 1500 species have been found in the Lizard Island area. While diversity of habitats may explain some of this increase, warmer sea temperatures play a potentially important role. Certainly some species occur throughout the GBR whereas others are restricted to more northern regions. For example, the leopard coral trout (*Plectropomus leopardus*) and Chinese footballer (*Plectropomus laevis*) are found throughout the GBR whereas the square tail grouper, *Plectropomus areolatus*, is only found north of about 11°S (near Lizard Island).

In addition to latitudinal changes, the variation in diversity at spatial scales of tens to hundreds of kilometres is often greater across the shelf than with latitude. This has been well documented for fish on the GBR. Comparisons of fish faunas of inshore reefs, with those on reefs in the GBR lagoon and with those on the outer barrier reveals significant differences both in the terms of abundance and species present (Fig. 13.4). Those of the inshore reefs that are subject to strong terrigenous influences are distinctively different to those on the mid and outer continental reefs. Inshore reefs tend to be depauperate in terms of species diversity compared to offshore reefs. These differences are apparent across each of the zones of these reefal systems, for example the fish communities of the outer reef slope differ depending where that outer reef slope is located. While the initial studies to document this cross diversity were carried out in the Townsville region, subsequent studies in other parts of the GBR confirmed that this is a widespread phenomenon. The complexity of habitats within each reefal system can easily mask this cross shelf variation; it was only when the distribution of species within a reef was carefully documented against habitat that these patterns became obvious. In the case of herbivorous fish, their distributions are related to the ways in which they feed.

Inshore reefs are characterised in terms of biomass by low levels of planktivores and herbivores, whereas outer reefs are dominated by herbivorous fish. Mid shore reefs have intermediate biomasses of herbivores and the highest levels of planktivores. It is generally thought that this relates to water clarity (i.e. turbidity, Chapter 10) and food supply (e.g. availability of plankton) as well as patterns of larval dispersal (Chapter 4). Similarly, the distribution of coral species across the GBR as well as within a reefal system varies,

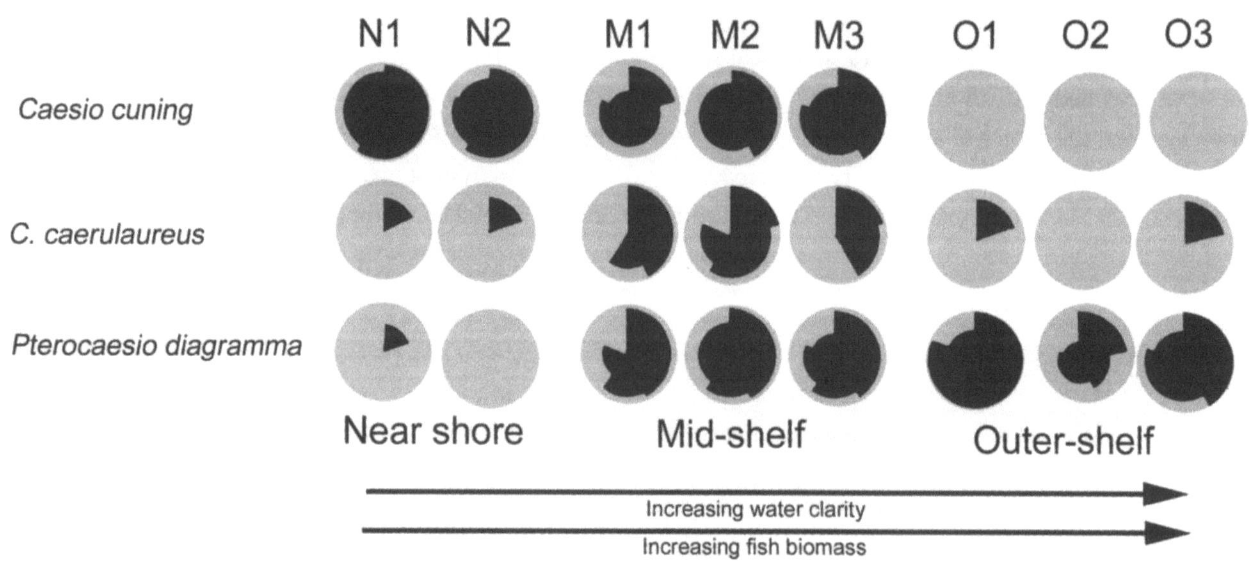

Figure 13.4 Distribution of three species of Fusiliers (Caesionidae) across the GBR shelf, with counts at nearshore (N), mid-shelf (M) and outer shelf (O) reefs. Counts were done at multiple sites at each reef and the results were combined as a circle to show spatial variation of fish in these areas. Each circle is divided into five sections that correspond to sites within reefs. The abundance of a species during each count is represented by the radius of the darkened area. The grey shading represents the extent of the circle. (Based on Williams, 1986.)

with most species having well defined habitat requirements in terms of depth, water clarity and water movement.

In the following chapters, most authors have given some idea of the diversity of each group where known. As will become clear, however, many of the specialists in the following chapters could only give partial estimates of the diversity of their groups because much of the diversity is currently undescribed. Even less is known about their distributional patterns along the GBR. This was recently highlighted by the seabed biodiversity project (Chapter 6). In general, our knowledge tends to decline with decreasing size of the organism, with the meiofauna inhabiting the soft sediments most poorly known, along with the permanent members of the plankton.

In summary, it seems likely that for most invertebrate groups the number of species is higher in the northern GBR than in the southern, and distinct distribution patterns occur across the shelf as well as within a reef depending on levels of exposure, depth and water clarity. As many of the invertebrates and many fish species recruit by pelagic larvae, larval recruitment is a critical factor in determining adult populations. Events that modify the supply of larval recruits may have significant impacts on successful recruitment many kilometres away. As discussed in earlier chapters, coral reefs are dynamic environments and anthropogenic impacts are increasingly influencing patterns of biogeography on the GBR. With climate change we may expect to see species distributions extending further south into cooler waters, providing suitable habitats exist. Key processes that need to be understood include: speciation, endemism, coexistence, extinction, the vulnerability of taxa and the habitats in which they live as well as biological and physical stressors affecting biodiversity that vary in space and time.

ADDITIONAL READING

Biodiversity measures

Clarke, K. R., and Warwick, R. M. (1994). 'Change in Marine Communities: An Approach to Statistical Analyses and Interpretation.' (Natural Environment Research Council: London.)

Clarke, K. R., and Warwick, R. M. (1999). The taxonomic distinctness measure of biodiversity: weighting of step lengths between hierarchical levels. *Marine Ecology Progress Series* **184**, 21–29.

Krebs, C. J. (1989). 'Ecological Methodology.' (Harper & Row: New York.)

Biodiversity and ecosystem management

Folke, C., Carpenter, S., Walker, B., Scheffer, M., Elmqvist, T., Gunderson, L., and Holling, C. S. (2004). Regime shifts, resilience and biodiversity in ecosystem management. *Annual Review of Ecology, Evolution and Systematics* **35**, 557–581.

Frid, C. L. J., Paramor, O. A. L., and Scott, C. L. (2006). Ecosystem-based management of fisheries: is science limiting? *ICES Journal of Marine Science* **63**, 1567–1572.

Singh, J. S. (2002). The biodiversity crisis: a multifaceted review. *Current Science* **82**(6), 638–647.

Effects of climate change

Hutchings, P. A., Ahyong, S., Byrne, M., Przeslawski, R., and Wörheide, G. (2007). Benthic invertebrates (excluding corals). GBR Ecological Vulnerability Assessment. In 'Climate Change and the Great Barrier Reef'. (Eds. J. Johnson and P. Marshall.) pp. 309–356. (Great Barrier Reef Marine Park Authority and Australian Greenhouse Office: Townsville.)

Distribution of corals across the Indo-Pacific

Veron, J. E. N. (2000). 'Corals of the World.' (Australian Institute of Marine Sciences: Townsville.)

Wallace, C. C. (1999). 'Staghorn Corals of the World.' (CSIRO Publishing: Collingwood, Victoria.)

Origin of biodiversity in the region

Wilson, M. E. J., and Rosen, B. R. (1998). Implications of paucity of corals in the Paleogene of SE Asia: plate tectonics or Centre of Origin? In 'Biogeography and Geological Evolution of SE Asia'. (Eds. R. Hall and J. D. Holloway.) pp. 165–195. (Backhuys Publishers: Leiden.)

Functional groups and indicator species

Bellwood, D. R., and Hughes, T. P. (2001). Regional-scale assembly rules and biodiversity of coral reefs. *Science* **292**, 1532–1534.

Bustos-Baez, S., and Frid, C. L. J. (2003). Using indicator species to assess the state of macrobenthic communities. *Hydrobiologia* **496**(1–3), 299–309.

Regional patterns of fish distributions and functional groups

Bellwood, D. R., Hughes, T. P., Folke, C., and Nyström, M. (2004). Confronting the coral crisis. *Nature* **429**, 827–833.

Coleman, F. C., and Willliams, S. L. (2002). Overexploitating marine ecosystem engineers: potential consequences for biodiversity. *Trends in Ecology and Evolution* **17**(1), 40–44.

Williams, D. M. (1991). Patterns and processes in the distribution of coral reef fishes. In 'The Ecology of Fishes on Coral Reefs'. (Ed. P. F. Sale.) pp. 437–474. (Academic Press: San Diego.)

See web site for an extended list, with updates: http://www.australiancoralreefsociety.org/ [Verified 22 February 2008].

14. Plankton

M. J. Kingsford, K. Heimann, C. G. Alexander & A. D. McKinnon

PLANKTON

Coral reefs of the world are surrounded by pelagic blue waters. Inter-reefal waters of the Great Barrier Reef (GBR) are mostly 10 m to 60 m deep on the continental shelf. Beyond the shelf, waters plunge to 4000 m and more. Despite the clarity of the water column a broad diversity of tiny organisms are found in the pelagic environment. Plankton are the ocean's 'wanderers' (from the Greek *'planktos'*) and are tiny organisms that drift or have limited powers of locomotion. The plants of the sea (phytoplankton, Fig. 14.1) are autotrophic and generate energy, in the form of sugars, by photosynthesis. All other planktonic organisms (bacteria, zooplankton etc., Fig. 14.2) are heterotrophs. Some plankton are 'mesotrophic', that is, both autotrophic and heterotrophic (e.g. protists, derived from the Greek 'first of all'). Phytoplankton are the basis of pelagic and benthic food chains that are now known to be complex, and involve many trophic transformations within microbial food webs. Plankton provide important links with reefal and inter-reefal environments as a source of food to suspension feeders (e.g. sponges, clams and corals), fishes and ultimately detrital feeders associated with the substratum (e.g. sea cucumbers).

Plankton that spend all of their lives in the pelagic environment are 'holoplankton' (Fig. 14.1) while others are temporary visitors. Meroplankton are larval forms of reef associated organisms such as corals, crabs, sea urchins and fishes (Fig. 14.4). At times of the year when animals are spawning, larvae can constitute a major component of total plankton; this is especially the case for corals that have 'mass spawning' (also worms, molluscs and echinoderms) and benthic algae that have mass releases of spores into the water column. Larvae range from the small and ciliated planula larvae of jellyfish and corals (about 0.5 mm across) and the transparent brachiolaria larvae of starfish, to large and essentially nektonic (organisms that can swim and move independently of currents) presettlement reef fishes. The pelagic phases of surgeonfish can be over 50 mm Standard Length (SL) at settlement and tank tests have demonstrated they may swim up to 120 km without food. The larvae of many invertebrates are often bizarre in shape and are very large compared to many other plankters. Spindly phyllosoma larvae of lobsters and sinister looking stomotopod larvae are up to 30 times the size of adult copepods and are voracious predators.

Marine snow and other structures

Historically, the pelagic environment was viewed as being occupied by a hierarchy of organisms of different sizes, from protists to jellyfishes. This perception occurred because the majority of samples were collected with nets or water bottles that disrupted or destroyed delicate structures such as are found within gelatinous zooplankton. In addition, flotsam and drifting algae

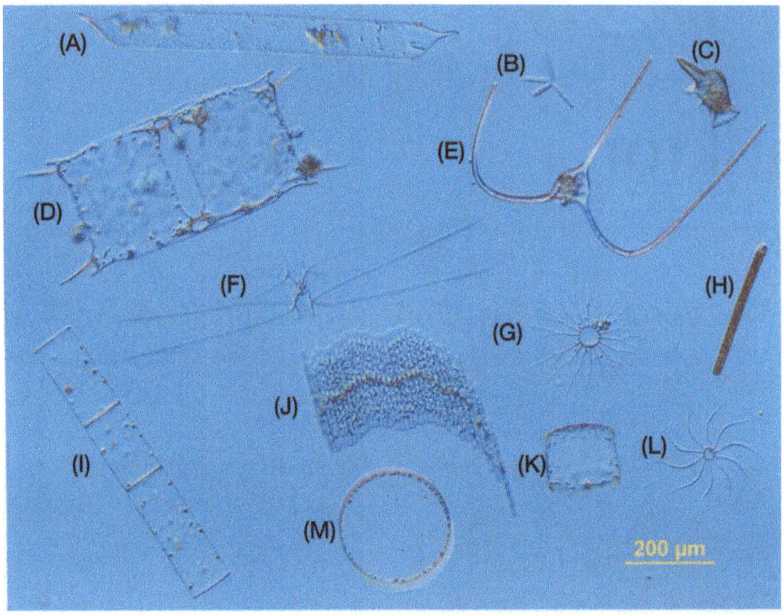

Figure 14.1 Diversity of phytoplankton. Identification from top left to right: *A*, *Rhizosolenia* sp.; *B*, *Thalassionema nitzschoides*; *C*, *Dinophysis caudata*; *D*, *Odontella* sp.; *E*, *Ceratium* sp.; *F*, *Chaetoceros* sp.; *G*, *Bacteriastrum* sp. (intercalary disc); *H*, Cyanobacterium; *I*, *Pseudoguinardia* sp.; *J*, *Bacillaria paxilifer*; *K*, centric diatom in girdle view; *L*, *Bacteriastrum* sp. (terminal disc); *M*, *Thalassiosira* sp. (Photo: K. Heimann.)

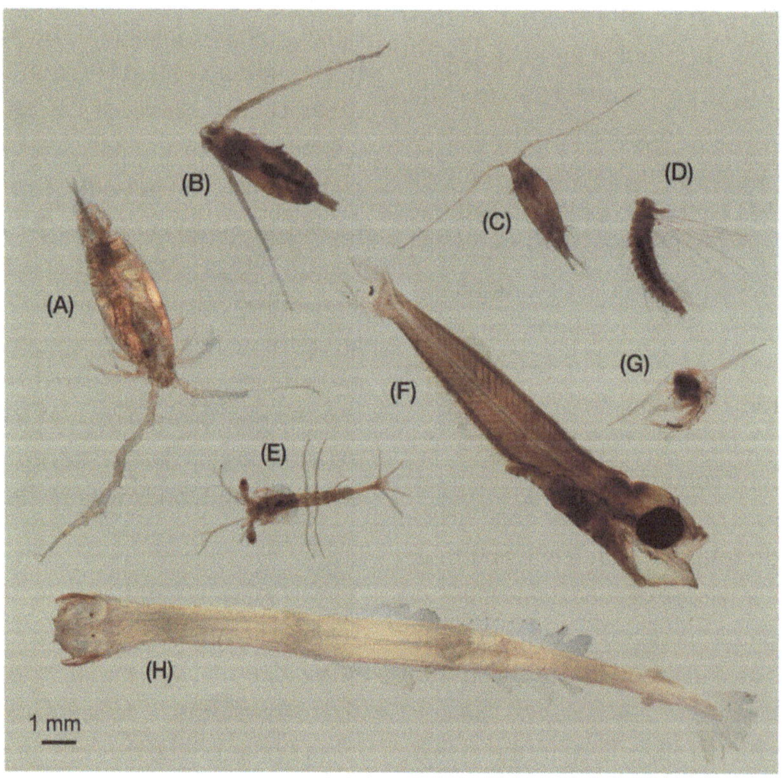

Figure 14.2 Diversity of zooplankton. *A*, large calanoid copepod (*Labidocera* sp.); *B*, calanoid copepod (*Eucalanus* sp.); *C*, calanoid *Acartia* sp.; *D*, polychaete worm larva; *E*, prawn larvae *Acetes* sp.; *F*, fish larvae; *G*, crab larval (zoel stage); *H*, arrow worm (Chaetognatha). (Photo: K. Heimann.)

BOX 14.1 HOLOPLANKTON – LIFE HISTORY OF COPEPODS

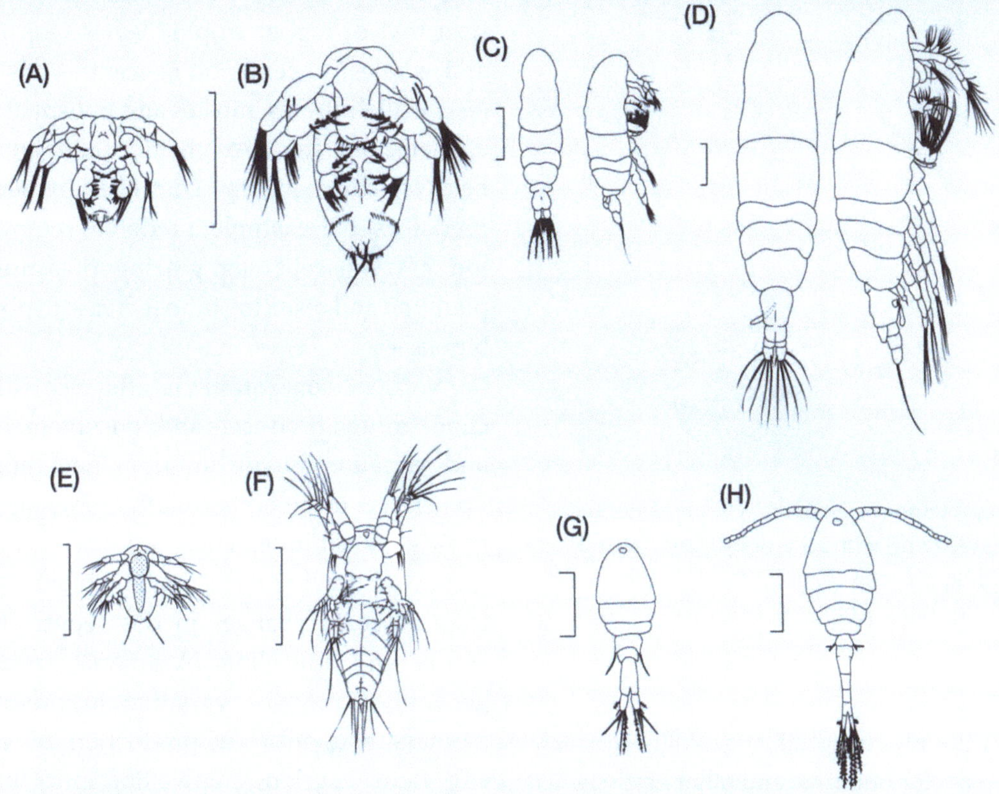

Figure 14.3 Representative developmental stages of calanoid and cyclopoid copepods. The calanoid *Acartia* sp.: *A*, nauplius stage I; *B*, nauplius stage VI; *C*, copepodite stage I, dorsal and lateral views; *D*, adult female, dorsal and lateral views, antennae omitted. The cyclopoid *Oithona* sp.: *E*, nauplius stage I; *F*, nauplius stage VI; *G*, copepodite stage I; *H*, adult female. (Scale bars = 100 μm.) Most orders of copepods brood their eggs, but the calanoids are almost all free spawners, releasing their eggs directly into the water column (some calanoids, such as *Gladioferens* and *Pseudodiaptomus*, are an exception to this rule). A small nauplius emerges from eggs of either type. There are six naupliar stages, and six copepodite stages, the last of which is an adult. Copepods show deterministic growth, and once adult no longer moult. Sexual dimorphism is apparent from the 4th copepodite stage and is most pronounced in the adult. Adults are gonochoristic (have separate sexes) and estimates of maximum age range from 80 to 160 days. (Images: Lawson, Grice, McKinnon, Trujillo-Ortiz and Uchima.)

was thrown out as 'trash' and hence disregarded. *In situ* observations have revealed that the pelagic environment has an abundance of fine structures composed of mucus, faeces and/or aggregations of phytoplankton cells. Mucus and gelatinous fragments are produced by a host of planktonic organisms that include appendicularians, heteropods, salps and jellyfishes. Mucus fragments coalesce and become a micro-assemblage of bacteria, protists, phytoplankton and other organisms that graze on them, such as copepods of the family Oncaeidae. This so-called 'marine snow' becomes heavy and slowly leaves the photic zone, providing organic matter to organisms living in deep water and ultimately sinks to the substratum as a 'snow

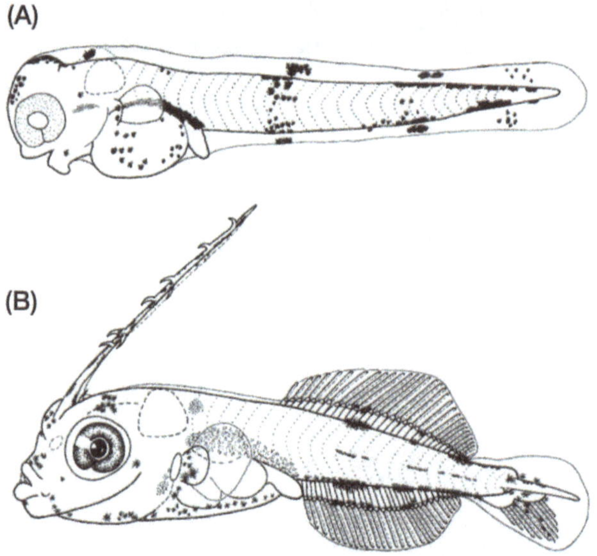

Figure 14.4 Presettlement reef fish *Oxymonacanthus* sp.: *A*, at 3 mm (total length), without a dorsal spine found in juveniles and adults on reefs; *B*, at 9 mm (TL) with an exaggerated dorsal spine. (Source: J. Leis.)

storm'. This is the raw material of geologic sedimentary successions as foraminifera and other hard-shelled plankton are deposited in the sediment. Other structures that influence the nature of marine assemblages include drifting macroalgae (e.g. *Sargassum*), rafts of cells (e.g. *Oscillatoria*), large jellyfish and flotsam. These structures influence the distribution, feeding and survival of many plankters, especially larval forms of crustacea and fishes.

PELAGIC FOOD CHAINS

Most production in pelagic systems is generated in the 'photic zone' where phytoplankton photosynthesise and produce the sugars necessary for life. About 70% of all carbon fixed by primary producers on the GBR originates from phytoplankton production and two thirds of this originates from organisms <2 μm (picoplankton) in size. Phytoplankton account for about 50% of global primary production and, therefore, have a major role in cycling atmospheric CO_2. In tropical waters the photic zone may reach a depth of about 150 m due to the clarity of the water column. Below this depth, phytoplankton respiration will exceed the energy derived from the generation of sugars (Fig. 14.5). The pelagic environment can be a 'bottom-up system' where the biomass of plankton and in turn that of higher trophic levels (e.g. fishes, squid and whales) depends on concentrations of nutrients (especially nitrates, nitrites and phosphates) and trace elements (e.g. iron). At other locations and times it can be a 'top-down system' controlled by herbivores and predators. For example, predators remove zooplankton grazers, relieving grazing pressure on phytoplankton and resulting in an increase in phytoplankton biomass.

Nutrient concentrations alter according to recycling through producers and consumers (i.e. excretion) and variation in the input of new nutrients from upwelling and riverine runoff. Production cycles vary greatly by latitude. High latitude ecosystems have great variation in production and biomass of plankton from seasonal changes in day length, temperature, storms and upwelling. In contrast, tropical systems generally have low variation in productivity and biomass, and pulses in production are event driven (e.g. floods, cyclones, upwelling intrusions). Tropical systems are generally considered high turnover, low biomass systems and the waters are generally oligotrophic (i.e. low in nutrients) and clear. Some apparent seasonality can occur in tropical waters because of increased frequency of events such as seasonal rains, which are typical of monsoonal/wet season locations. In the 'dry tropics' (e.g. the central and southern GBR) the rains are not predictable, but significant input of freshwater results during cyclone/storm events and these have a great impact on physical attributes of the pelagic environment and on planktonic assemblages and processes.

Upwelling of cold, nutrient rich, deep ocean water has a great influence on pelagic systems and it is the reason that temperate regions off the coast of Peru, the west coast of North America, and South Africa have green waters from phytoplankton growth and a wealth of consumers from copepods and krill to whales. As a result, these are the sites of some of the great fisheries of the world (e.g. Peruvian anchovy). The upwelling of nutrient rich waters into the photic zone is determined by currents, wind and topography.

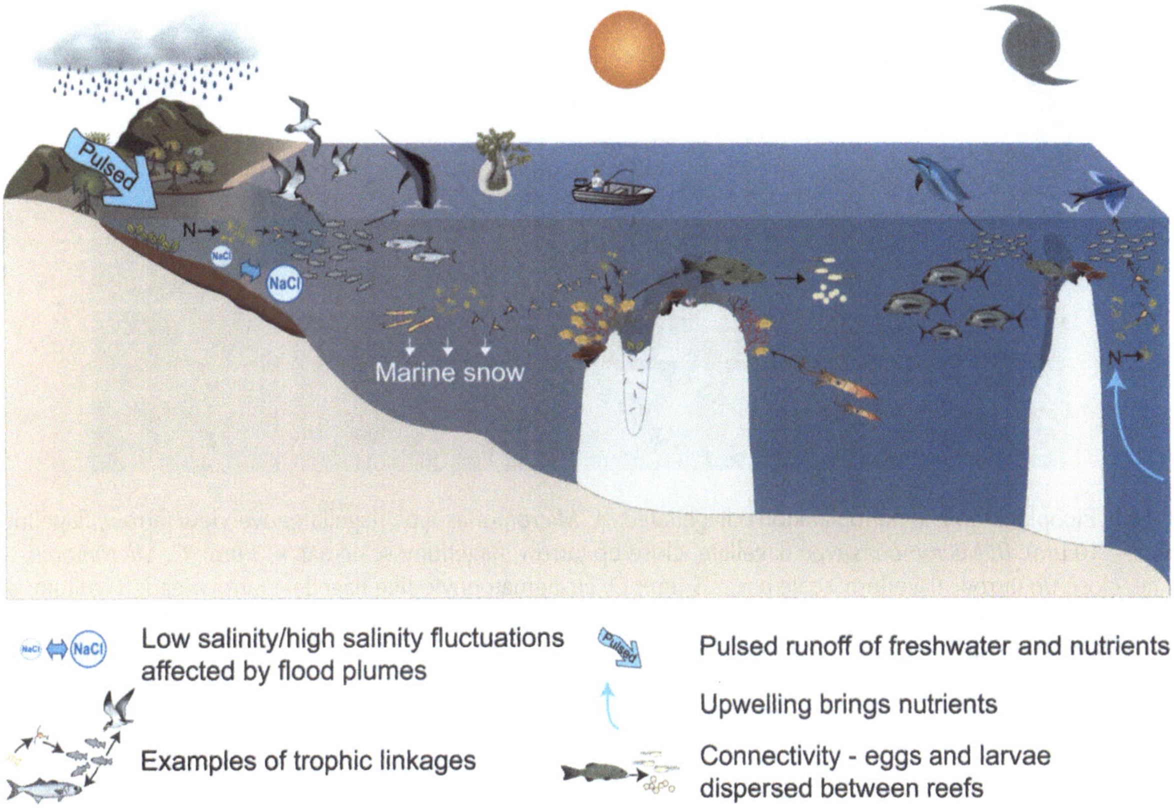

Figure 14.5 Tropical pelagic food chain showing plankton as an important source of food for pelagic nekton and reef associated species such as planktivorous invertebrates and fishes. Demersal zooplankton migrate from reefs and other substratum into the water column at night. Input of freshwater, upwelling and cyclones are factors that influence nutrient availability and growth of phytoplankton. (Image: GBRMPA.)

Tropical waters are generally considered oligotrophic and Charles Darwin considered this a great paradox given the wealth and diversity of life in tropical waters. There is regionalised and sporadic upwelling along the shelf of the GBR, but this is poorly understood. It is clearly not as great as in some temperate regions, but is of great potential importance to the tropho-dynamics of reef and inter-reefal habitats. Cool upwelled waters, and therefore high density waters, often hug the bottom rather than emerging at the surface. They immerse deep inter-reefal assemblages in nutrient rich waters. Phytoplankton thrive in these conditions and ultimately disperse throughout the water column to support a diversity of consumers. On some occasions, upwelling intrusions, probably driven by winds, can almost extend across the entire shelf of the GBR.

A SIZE HIERARCHY OF ORGANISMS

Plankton ranges in size from a few microns to metres. An understanding of planktonic assemblages, therefore, is based on a synthesis from equipment that is suited to collecting specific size fractions of plankton. Traditionally, plankton has been broken into: *femtoplankton* (<0.2 µm) (the viruses); *picoplankton* (0.2–2 µm) (including bacteria, cyanobacteria and prochlorophytes that are primary producers, heterotrophs and nitrogen fixers) (Fig. 14.6); *nanoplankton* (2–20 µm) (flagellates, diatoms and cyanobacteria that are producers, grazers, nitrogen fixers and predators); *microplankton* (20–200 µm) (diatoms, dinoflagellates and ciliates that are also producers, grazers, nitrogen fixers and predators); *Mesoplankton* (0.2–20 mm) (copepods, other zooplankton and cyanobacteria—*Trichodesmium* colonies—that are

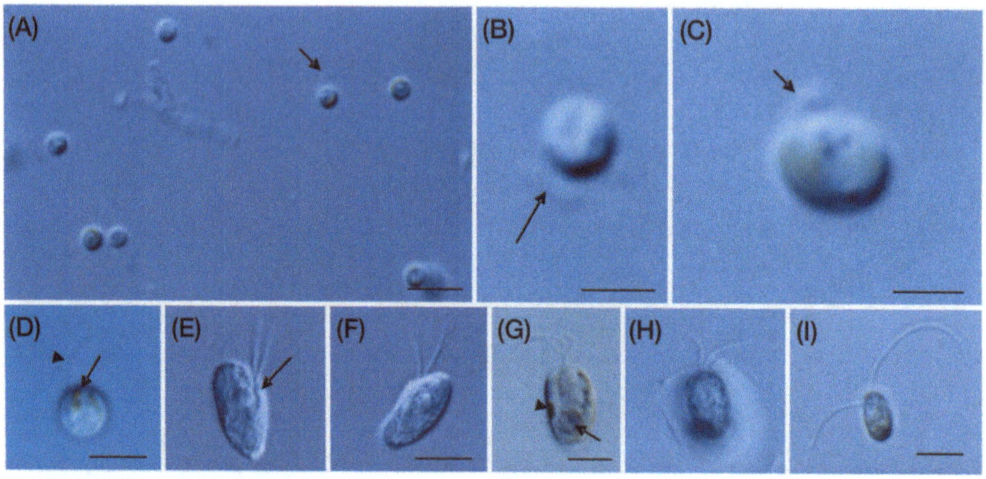

Figure 14.6 Picoplankton and nanoplankton – flagellated. *A*, *Micromonas*-type flagellate, overview (arrow, flagellum; scale bar = 10 μm); *B*, *Micromonas*-type flagellate, close-up (arrow, flagellum; scale bar = 5 μm); *C*, *Micromonas*-type flagellate, close-up (arrow, flagellum; scale bar = 5 μm); *D*, Eustigmatophyte-like flagellate (arrowhead, flagellum; arrow, eyespot; scale bar = 10 m); *E*, *F*, Cryptophyte. *Rhinomonas*-type, *E*, oblique right/lateral view (arrow, gullet); *F*, oblique left/lateral view (scale bar = 10 μm); *G*, *H*, *Tetraselmis gracilis*, *G*, ventral view (arrow, pyrenoid; arrow head, eyespot); *H*, oblique apical view showing the presence of four flagella (scale bar = 10 μm); *I*, *Nephroselmis pyriformis* (scale bar = 10 μm). (Photo: K. Heimann.)

producers, grazers, nitrogen fixers and predators) (Box 14.2); *macroplankton* (20–200 mm) (larvaceans, larval fishes and other zooplankton that are grazers and predators) and *megaplankton* (over 200 mm) (jellyfish, salps and other zooplankton that are mainly grazer and predators).

Water bottle samples (organisms <50 microns)

Our primary focus here is organisms between 2 μm and 20 μm in size: the nanoplankton. Members of the nanoplankton are very small and characterised by a high surface area to volume ratio. The surface area to volume ratio influences the amounts of materials exchanged with the environment. In general, an inverse relationship between size and reproductive capacity is observed (i.e. small cells have a much higher reproductive rate than large cells), although this relationship is not easily applied to flagellated forms of the different size classes and some diatoms can outgrow picoplankton (e.g. *Prochlorococcus*). The high reproductive capacity is linked to the ability to immediately respond to short nutrient pulses. Therefore, nanoplankton often occurs at high biomass per unit volume. For example, high biomass patches can stretch over kilometres and the resultant chlorophyll fluorescence is readily visible from space via satellites. These population/community maxima may, however, be short-lived and can change dramatically within the space of a day or two. Grazing is another factor that controls the biomass of these very small organisms, as they are readily consumed by grazers such as protozoa, pelagic tunicates, rotifers and crustaceans.

Phytoplankton of this size range may occur in colonies and as flagellated and non-flagellated single cells, which are often very delicate and generally do not preserve well. In particular, flagellated forms tend to loose their flagella and shape upon fixation with either Lugol's solution or formaldehyde. In most cases, the presence/absence and number of flagellae, as well as swimming behaviour are essential criteria for the identification to genus and species. Due to the miniscule size of these organisms, swimming behaviour and patterns of flagellation of live material must be observed at high magnifications (×400 and more) requiring special contrasting objectives to visualise these structures (i.e. differential interference contrast or phase contrast). Even if swimming behaviour can be observed in live samples, the presence and number of flagellae typically remains a mystery as these

BOX 14.2 *TRICHODESMIUM* – A SMALL BUT IMPORTANT BLUE GREEN ALGA (= CYANOBACTERIA)

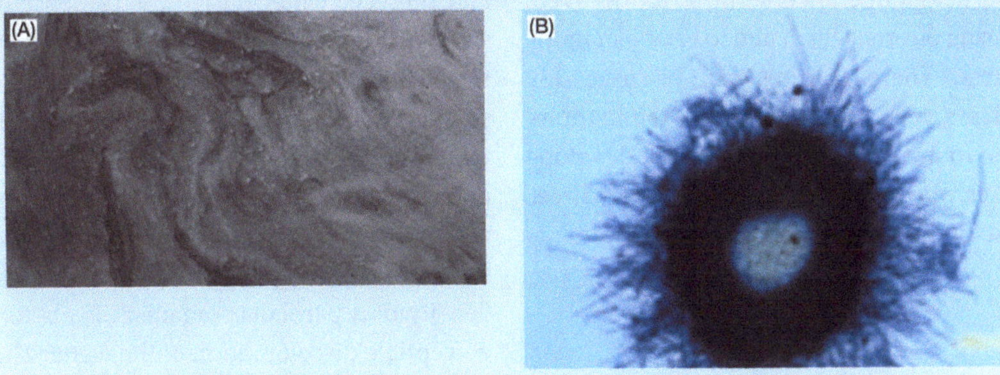

Figure 14.7 *A*, dead floating *Trichodesmium* cells in a convergence (photo: M. Kingsford); *B*, detail of *Trichodesmium* 'puff' (photo: K. Heimann).

Trichodesmium is a filamentous blue green alga that is often abundant in the water column and aggregated at the surface of tropical waters. *Trichodesmium* will form visible blooms under ideal environmental conditions (calm) and colonies (tuft colonies generated by the parallel alignment of filaments or puff colonies due to radial alignment are 0.5–3 mm in diameter, so often visible to the naked eye). Great drifts, usually of dead cells, are common on the surface of the ocean; this phenomenon has been termed 'sea sawdust' by sailors (Fig. 14.7). Concentrations of dead cells are often in the convergence zones of tidal fronts, windrows and internal waves. Although the cells within a filament are tiny, *Trichodesmium* is critical for the fixation of atmospheric nitrogen in oligotrophic tropical marine environments. Nitrogen fixation in tropical waters often exceeds 30 mg m^{-2} day^{-1}, contributing significantly to the availability of nitrogen based molecules in these nutrient-poor waters. *Trichodesmium* thrives in areas with low dissolved inorganic nitrogen levels as it out competes organisms that cannot fix atmospheric nitrogen. The abundance of *Trichodesmium* in these areas of the ocean is limited by the availability of trace elements, such as iron, molybdenium, and vanadium, essential elements of the nitrogen-fixing machinery.

organisms swim very fast, changing directions frequently diving in and out of the focal plain of the high resolution objectives. In most instances, verification by optical and electron microscopy (scanning electron microscopy for surface structures and flagellation and transmission electron microscopy for chloroplast and flagellar apparatus characteristics) is required to assign these organisms to a particular phylum/class. Another problem in analysing the species composition and abundance of nanoplankton is that unambiguous identification requires knowledge about the 'colour' of these organisms (i.e. pigments—the presence and combinations of the various chlorophylls). Typically the colour of individual cells is not readily observed due to their small size, adding to the problem of positively identifying members of the nanoplankton assemblage. Given the above impediments, it is not surprising that much less is known about the species composition and dynamics of these nanoplankton. Much of these taxonomic uncertainties can be overcome by culturing efforts, which aim to establish monoclonal cultures. These cultures can be used for pigment analyses and morphological examinations and will be the foundation for genetic probe developments, which can be employed in the field for direct qualitative (taxonomic) and quantitative (enumeration) community analysis of the nanocosm of the ocean environment.

Fine mesh samples (organisms 50–100 microns)

In comparison to our knowledge regarding community structure of nanoplankton, much more is known about the taxonomic diversity of fine net samples. The reasons for our better understanding of these organisms are two-fold: (1) many of these organisms are more robust and preserve better due to the presence of quite substantial external armour (i.e. the glass cell wall of diatoms (silica frustule) or the heavy cellulose armour of some dinoflagellates (Fig. 14.8)) and (2) the often infamous and sometimes spectacular bloom events of many members of this size class has also created much research interest. For example, blooms by dinoflagellates are commonly referred to as 'red tides', where the density of a particular organism leads to the visually perceptible discolouration of the water. These red tides can be harmful if toxin producers are involved, or harmless. For example, blooms produced by *Noctiluca scintillans* are red in colour and it is currently debated

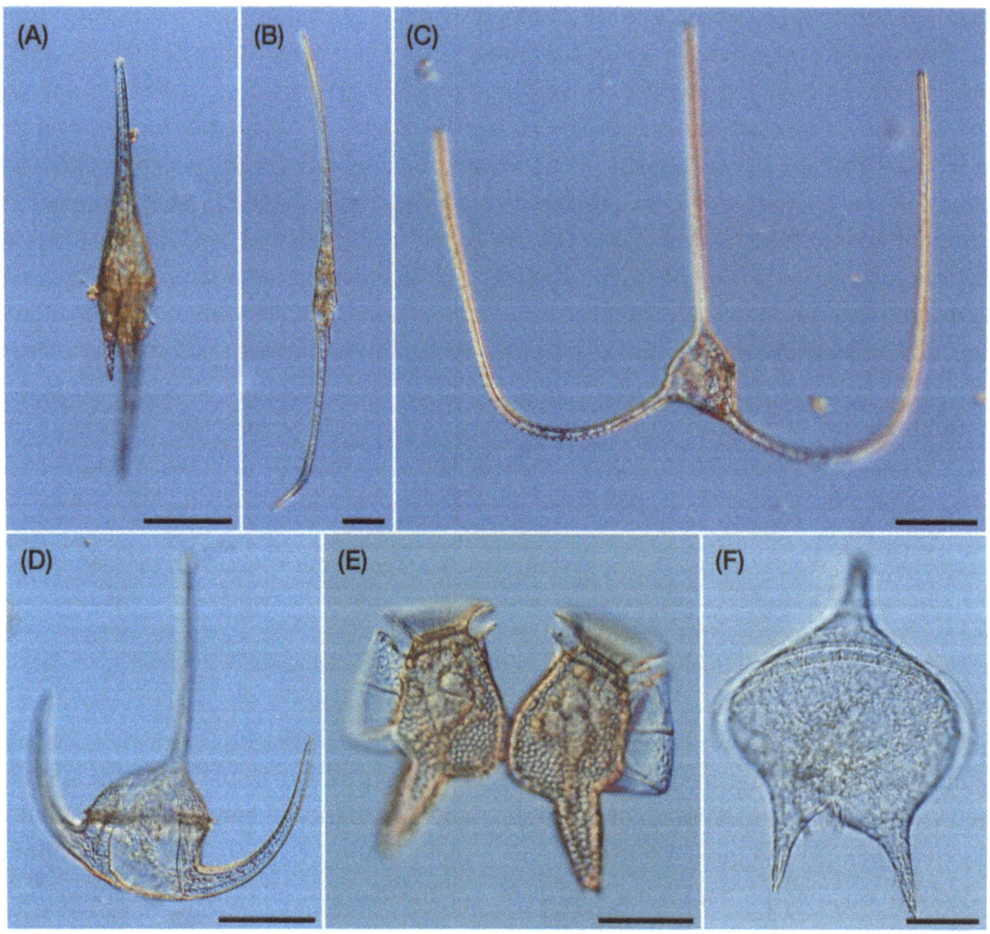

Figure 14.8 Dinoflagellates. *A*, *Ceratium incisum* (scale bar = 50 μm); *B*, *Ceratium inflatum* (scale bar = 100 μm); *C*, *Ceratium trichoceros* (scale bar = 100 μm); *D*, *Ceratium tripos* (scale bar = 50 μm); *E*, *Dinophysis caudata* (scale bar = 25 μm); *F*, *Protoperidinium oceanicum* (scale bar = 50 μm). (Photo: K. Heimann.)

whether they are harmful or harmless. In any case, these blooms are spectacular at night, as *Noctiluca scintillans* is a bioluminescent dinoflagellate (capable of producing and emitting its own light), and causes 'phosphorescence of the sea'. In contrast, many diatoms (i.e. *Pseudo-nitzschia* multiseries) and dinoflagellates (i.e. *Karenia breve* in Florida) produce potent toxins and their blooms result in large fish and marine mammal mortality. Human health can be adversely affected by consuming contaminated shellfish (shellfish filter algae and accumulate toxins in their body tissues) or via inhalation of organisms contained in ocean spray. These blooms develop when environmental conditions are perfect, that is, warm surface temperatures, stagnant and stratified water masses, low turbulence, and high light intensities are critical factors in bloom establishment and duration. An example of a bloom considered of fundamental importance in nitrogen cycling in tropical waters is produced by the cyanobacterium *Trichodesmium*. It fixes large amounts of atmospheric nitrogen and thus contributes significantly to the availability of nitrogen in these otherwise oligotrophic waters (see below).

An amazing variety of dinoflagellates, centric and pennate diatoms exist within the central GBR region in this size class of phytoplankton (Fig. 14.9). Compared to the nanoplankton, their surface to volume ratio is reduced due to their larger size and their rates of cell division are slower. Also, their nutritional strategies differ from those of nanoplankton in that these organisms can exploit nutrient patches by storing these nutrients inside their cells (e.g. phosphates). Thus, the presence and abundance of microplankton may be 'decoupled' from nutrient availability. Unlike the much smaller nanoplankton, sinking out of the photic zone is possible for larger phytoplankton. There are many means that phytoplankton use to avoid sinking. Buoyancy may be increased by the presence of gas vesicles or the storage of light oils, surface area increased by the production of elaborate cell surface extensions as observed in the centric diatoms *Bacteriastrum* and *Chaetoceros*, and mobility increased by powerful flagellation in the dinoflagellates. Susceptibility to grazing is size-dependent, with grazing pressure inversely related to cell size. The evolution of spines (setae) and the ability to form buoyant mats (*Rhizosolenia*) is a valuable anti-sinking strategy and may deter some herbivores.

Net plankton (organisms 0.1–1.00 mm)

Net plankton are often the focus of studies because they represent a critical link between microbes and larger organisms such as fish. These organisms include a broad range of zooplankton including nauplii and juveniles (e.g. copepodites) as well as adult holoplankton and many larval forms. Copepods (Subclass Copepoda) often dominate zooplankton samples (up to 80% of catches and 193 species on the GBR) and common genera in waters of the GBR include: small sized *Parvocalanus* and *Bestiolina*; medium-sized *Acartia*, *Paracalanus*, and *Temora* (~ 0.7 mm to 1.1 mm adults), and larger (~ 1.0 mm to 3.0 mm) *Candacia*, *Undinula*, *Eucalanus*, *Centropages*, *Labidocera*, and *Pontella* of the Order Calanoida; small to medium-sized *Oithona* and *Oncaea* (~ 0.7 mm to 1.1 mm adults), of the Orders Cyclopoida; and *Euterpina* and *Microsetella* of the Order Harpacticoida (Fig. 14.10). Other important members of the Phylum Crustacea that are found in net plankton include *Penilia* (Order Ctenopoda).

Common elongate plankton include *Oikopleura*, *Fritillaria* (larvaceans or appendicularians; eight species on GBR) and small arrow worms (Phylum Chaetognatha, 12 species on GBR). Larvaceans feed on picoplankton, while chaetognaths are voracious predators on meso- and megaplankton. The abundance of larval forms varies with the timing of spawning of different taxa, but common crustacean larvae include barnacle nauplii (usually too small for mesoplankton nets) and cyprids as well as crab zoea, prawns (e.g. *Acetes*), stomatopod larvae (*Squilla*) and *Lucifer*. Fish larvae can be very abundant especially after the mass release of larvae from demersal eggs around new and full moons. Coral spawn can be abundant in nets at times of mass spawnings, causing plankton nets to go pink. Other common larvae include polychaete worms and gastropod larvae. Although the cells of *Trichodesmium* are tiny masses of cells they can clog mesoplankton nets.

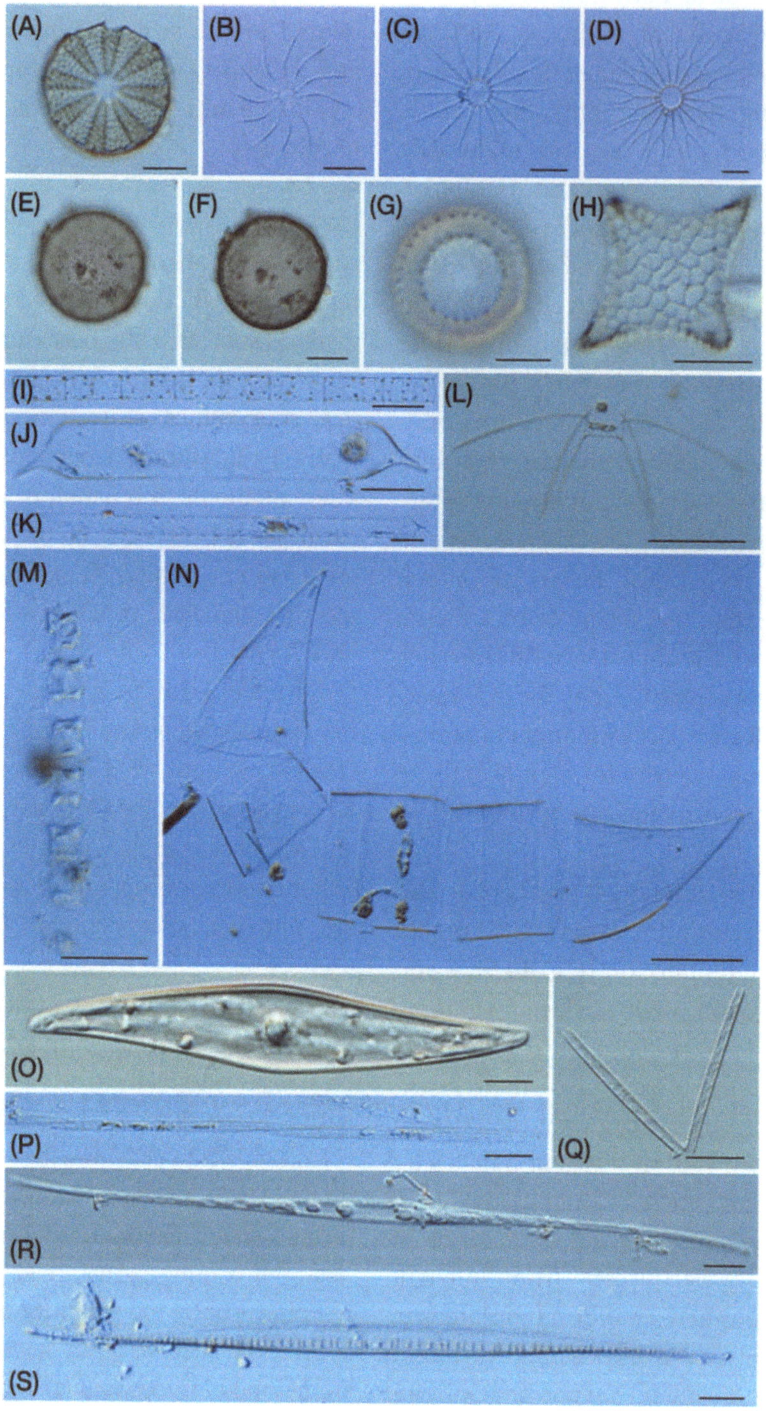

Figure 14.9 Centric diatoms. *A, Actinoptychus* sp. (scale bar = 25 μm); *B, Bacteriastrum* sp. 1, terminal valve (scale bar = 50 μm); *C, Bacteriastrum* sp. 2, terminal valve (scale bar = 50 μm); *D, Bacteriastrum* sp. 2, intercalary valve (scale bar = 50 μm); *E, F, Coscinodiscus* sp. *E*, top of valve; *F*, valve edge (scale bar = 50 μm); *G, Paralia* sp., intercalary valve (scale bar = 25 μm); *H, Triceratium* sp. (scale bar = 25 μm); *I, Pseudoguinardia* sp. (scale bar = 200 μm); *J, Proboscia* sp. (scale bar = 100 μm); *K, Rhizosolenia* sp. (scale bar = 100 μm); *L, Chaetoceros* sp. 1 (scale bar = 100 μm); *M, Chaetoceros* sp. 2 (scale bar = 100 μm); *N, Rhizosolenia robusta* (scale bar = 200 μm). Pennate diatoms. *O, Gyrosigma* sp. (scale bar = 20 μm); *P, Pseudonitzschia* sp. (scale bar = 25μm); *Q, Thalassionema nitzschoides* (scale bar = 25 μm); *R. Nitzschia* sp. (scale bar = 100 μm); *S, Nitzschia* sp. (scale bar = 50 μm). (Photo: K. Heimann.)

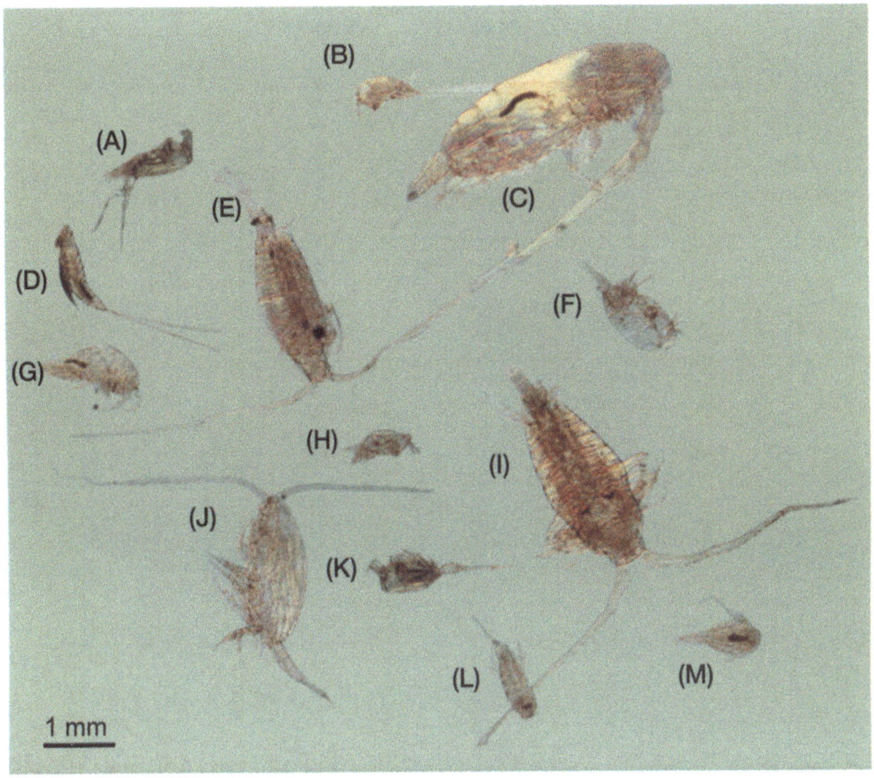

Figure 14.10 Diversity of Copepoda: cyclopoids. *A, Corycaeus* sp.; *B,* cyclopoid; *C, Calanidae; D,* harpacticoid, *Microsetella; E,* calanoid, *Centropages; F,* cyclopoid, *Oncaea* sp.; *G,* cyclopoid; *H, Corycaeus* sp.; *I,* calanoid, *Candacia; J,* calanoid; *K, Corycaeus* sp.; *L,* cyclopoid; *M, Oithona* sp. (Photo: K. Heimann.)

Large plankton (organisms 1 mm+)

Some copepods (*Pontella*), euphausiids (14 species on GBR) and many elongate zooplankters such as larvaceans/appendicularians (e.g. *Megalocercus*), chaetognaths and fish larvae are over 1 mm long (Fig. 14.11). Molluscs that include cone-shaped pteropods (15 species on GBR) and gelatinous heteropods (three species on GBR) are common. Large zooplankton are in the 'grey' zone of what is defined as 'plankton' in that many have great mobility and could almost be considered free swimming nekton. The early life history stages of fish, for example, hatch at small sizes (2 mm to 6 mm), but many settle on reefs at large sizes (e.g. 1–4 cm) and have excellent swimming and sensory abilities (i.e. olfaction, hearing and vision) that assist orientation.

Gelatinous zooplankton are among the most obvious members of the zooplankton (Fig. 14.12). The Phylum Cnidaria includes the Class Scyphozoa, represented by many large jellyfish, some of which have tentacles that extend over 3 m in length (e.g. *Cyanea, Desmonema*, 'Lions mane' jellyfish) and they can weigh over 10 kg. Others have stumpy tentacles, such as *Catostylus*, and some of these taxa are targeted by fisheries for sushi grade jellyfish (adults are up to 5 kg wet weight). Jellyfishes are close relatives of corals and some taxa (e.g. *Phyllorhiza punctata*) have symbiotic algae (i.e. zooxanthellae). One of the most common jellyfish in coastal waters is *Aurelia* (moon jelly) that is harmless and like most jellyfish can be found in great numbers. Densities of jellyfish can be so great at times that they are referred to as 'blooms'. The deadliest jellyfish belong to the Class Cubozoa and include stingers and multiple types of jellyfish that are responsible for the medical condition called Irukandji syndrome (e.g. *Carukia barnesi* and *Pseudoirukandji* spp., see Chapter 18).

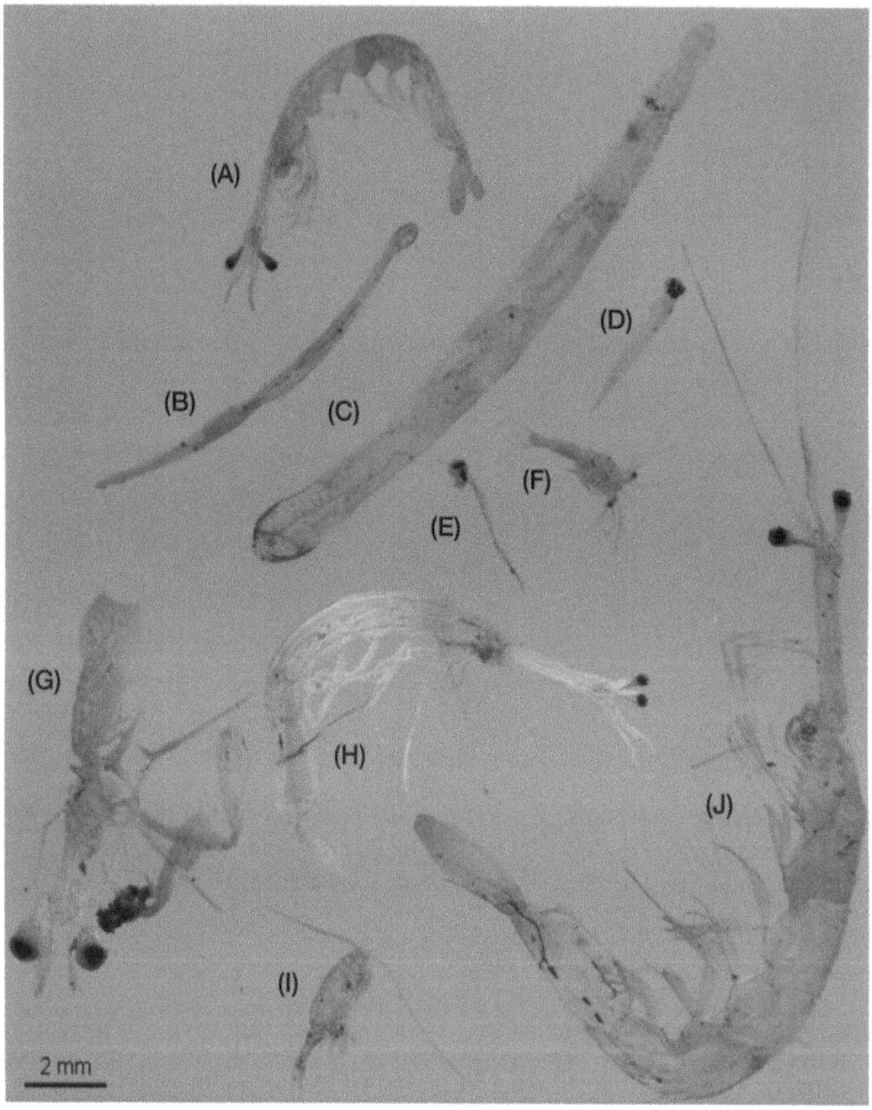

Figure 14.11 Diversity of large plankton. *A*, thalassinid, *Lucifer* sp.; *B*, chaetognath; *C*, chaetognath; *D*, appendicularian/larvacean, *Oikopleura* sp.; *E*, *Oikopleura* sp.; *F*, prawn larva, *Acetes* (zoel stage); *G*, stomatopod larva; *H*, *Lucifer* sp.; *I*, calanoid copepod; *J*, *Lucifer* sp. (Photo: K. Heimann.)

Conspicuous neustonic taxa are often species of the Class Hydrozoa. The Portuguese man-of-war or blue-bottle (*Physalia* sp.), smaller taxa that include the by-the-wind sailor (*Velella vellella*) and *Porpita* sp. float and are common as drift on beaches and are well known for their painful stings. Siphonophores (Class Siphonophora; 32 species on GBR) are voracious predators of plankton and are solitary or form long strings (colonies) that drift and swim by hydraulic pumping of seawater.

Other gelatinous zooplankters that sometimes form strings, but are unrelated to siphonophores are the salps and doliolids (Phylum Chordata, Class Thaliacea; nine species on GBR) (Fig. 14.13). Their abundance is often influenced by temperature and biologists have focused on this group as an indicator of variation in planktonic assemblages due to climate change. Individuals are often only a centimetre or so long, but they generally form strings and some colonies form cylinders that can be a few metres in length (e.g. *Pyrosoma*).

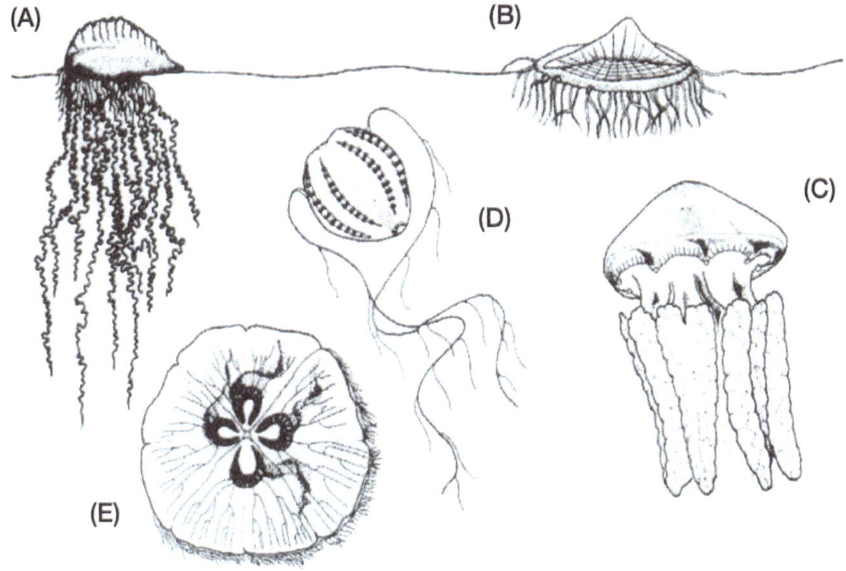

Figure 14.12 Diversity of gelatinous macroplankton. *A*, siphonophore, Portuguese man-of-war *Physalia* sp., bladder ~5 cm; *B*, siphonophore, by the wind sailor *Velella vellela*, 2.0–4.5 cm; *C*, scyphozoan jellyfish, *Catostylus mosaicus*, umbrella maximum diameter 32 cm; *D*, ctenophore *Pleurobranchia*, 1–3 cm diameter; *E*, scyphozoan jellyfish *Aurelia* sp. (Image reproduced with permission from UNSW Press.)

Figure 14.13 Gelatinous zooplankton, salp about 12 mm long. (Photo: L. Gershwin.)

Insects are periodically collected and these are not always an allochthonous (land derived) component lost from land. *Halobates*, the only truly marine insect, is a cosmopolitan pleustonic marine bug that skims the water surface feeding on surface-associated prey; mites are also occasionally found in the ocean, although common in sediments.

LINKS BETWEEN ENVIRONMENTS AND MIGRATIONS

Links with the substratum

Planktonic organisms often have strong links with the substratum. For example, most jellyfishes start life on the substratum. The fertilised eggs of jellyfishes develop into 'hairy' planula, which settle on the substratum and grow into polyps. The polyps grow and bud into other polyps. When conditions in the water column are suitable the polyps 'strobilate' (i.e. bud) and release tiny (~2 mm) 'ephyrae' into the water column. They grow into tiny jellyfish and grow rapidly to adult form. For example, the polyps of stingers are found in estuaries and the release of small jellyfish is triggered by the input of freshwater.

In temperate waters copepods often release eggs into the water column that sink to the substratum. Egg reserves on the substratum can be great and development is usually arrested (they 'aestivate') until conditions in the plankton are suitable for hatching; this is probably not true in the tropics. Some copepods spend time in the plankton and on the substratum. The harpacticoid *Euterpina*, for example, shows great variation in abundance in the plankton and this is partly related to seasonal use of the substratum.

Demersal plankton

There is great variation in tropical plankton between day and night. This is largely the result of emergence of plankters from the substratum at night. Invertebrates that include copepods, amphipods, mysids (23 species on GBR), isopods and polychaete worms migrate from the substratum after dusk. Nocturnal movements are probably for the purposes of feeding or dispersal. Demersal plankton provides a rich source of food for invertebrates and fishes and this can be of greater importance as an energy source than the import of plankton from inter-reefal waters. The feeding polyps of many corals only emerge at night and fishes that include cardinal fishes and squirrel fishes feed on large nocturnal zooplankton. Historically, the importance of demersal plankton was not understood, but is now viewed as a critical component of the food chain of reefs.

Vertical migration

Plankton of inter-reefal waters, on the continental shelf and beyond often undertake nocturnal vertical migrations. A Russian scientist, Vinogradov, referred to a 'ladder' of migrations where phytoplankton and zooplankton migrate from different daytime residence depths toward the surface. Phytoplankton may alter depth of residence by secreting and absorbing gas, or by varying their oil or salt content. Phytoplankton often migrate to deeper, more nutrient rich waters during the night and back to the sunlit surface waters at the onset of the day. This migration pattern is known as reversed diel migration. Zooplankton and larger flagellated phytoplankton migrate by swimming. In other parts of the world, copepods often swim from depths of 100 m or greater toward food rich waters at the surface at night, but there are few data on this for the GBR. Visual predators are active in the photic zone during the day and it is thought that gloomy waters at depth provide refugia for zooplankton. Some studies have demonstrated that copepods will return to deep water at night if they detect the predators through chemical or tactile senses. The timing of some nocturnal migrations varies by reproductive state.

DISTRIBUTION PATTERNS OF PLANKTON

The influence of oceanography

The movement of water will have a great influence on organisms that largely drift. However, some directional control can be obtained by varying their vertical position in water columns that have vertical stratification of current speed and direction. Currents can influence the trajectories of plankton and can influence concentration. Some eddies are called 'phase eddies' that only occur at one phase of the tide (e.g. on ebb tides behind a reef) while other eddies (and larger gyres) persist over long time periods in mainstream currents. In addition to eddies, currents can interact with abrupt topography such as reefs and seamounts to alter the supply of plankton through upwelling.

Plankton often concentrate in convergences. In tropical waters it is common to see blue green algae (Box 14.2), coral spawn, jellyfishes, spawning appendicularians and other plankters in convergence zones generated by physical phenomena that include: tidal and thermal fronts, tidal jets, the edge of eddies, windrows and internal waves. As a result, therefore, convergences are often sites of intense biological activity and nekton (e.g. planktivorous fishes, piscivores) and birds are often attracted to the abundant food in these areas.

Oceanography can also influence the distribution of plankton with depth. Although the water column on the shelf of the GBR is often well mixed (see Chapter 4), thermoclines and haloclines can stratify the water column into different water masses. For example, at the edge of the continental shelf upwelling events are common and 10 m to 20 m deep 'wedges' of cool water move along the bottom over the shelf and under lighter warmer shelf waters. Different planktonic assemblages may be found above and below the thermocline in these water masses. In addition, some plankton will concentrate in and around the thermocline. Freshwater plumes will also cause vertical stratification and variation in abundance of plankton with depth.

Interactions between plankton and oceanography are studied intensively, but the behaviour of plankton also has a role. Some plankton will meet in convergences, but will maintain their position for the purposes of reproduction (e.g. Appendicularia and jellyfishes). The larvae of prawns, crabs and fishes regulate depth at different stages of their development and this has major consequences for transport. For example west Australian lobster larvae (*Palinurus ornatis*) are transported into the Indian Ocean at the surface and migrate to deeper water later in development for transport back toward the Australian coastline.

PLANKTON AND CLIMATE CHANGE

It is well known that physical forcing, through changes in climatic conditions, has a great influence on planktonic assemblages. Phytoplankton has a major role in the recycling of global CO_2. For this reason, anything deleterious that happens to phytoplankton will have a cascade effect through the entire food chain. Pelagic ecosystems are well known for trophic cascades and it is generally assumed that nutrient supply and subsequent growth of phytoplankton will result in 'bottom-up' control of the whole food chain. Top-down effects are also possible where predators remove zooplankton grazers, relieving grazing pressure on phytoplankton and resulting in an increase in phytoplankton biomass. Changes can happen very quickly in the plankton. Long term plankton records in the northern hemisphere have demonstrated that the warming of waters has resulted in substantial changes in planktonic assemblages and there is great concern how this will affect higher trophic groups and the survival of larval forms. The impact of global change on plankton of the GBR will be greatest due to changes in conditions that include: temperature, availability of nutrients (e.g. upwelling and runoff) and change in pH (that can affect plankton with carbonate skeletons; i.e. coccolithophores). On the GBR, it is expected that changes over the next 50 years are likely to be patchy by region (e.g. central section *v.* southern section).

ADDITIONAL READING

Biological oceanography

Furnas, M. J., and Mitchell, A. W. (1996). Nutrient inputs to the central Great Barrier Reef (Australia) from subsurface intrusions of Coral Sea waters: a two dimensional displacement model. *Continental Shelf Research* **16**, 1127–1148.

Genin, A. (2004). Biophysical coupling in the formation of zooplankton and fish aggregations over abrupt topographies. *Journal of Marine Systems* **50**, 3–20.

Mann, K. H., and Lazier, J. R. N. (2003). 'Dynamics of Marine Ecosystems: Biological-physical Interactions in the Ocean.' (Blackwell: Oxford.)

Identification of plankton

Boltovskoy, D. (1999). 'South Atlantic Zooplankton.' (Backhuys Publishers: Leiden.)

Dakin, W. J., and Colefax, A. N. (1940). 'The Plankton of the Australian Coastal Waters off New South Wales. Part I.' (Australian Publishing Company Ltd: Sydney.)

Graham, L. E., and Wilcox, L. W. (2000). 'Algae.' (Prentice Hall: New Jersey.)

Leis, J. M., and Carson-Ewart, M. (2000). 'The Larvae of Indo-Pacific Coastal Fishes.' (Brill: Leiden.)

Newell, G. E., and Newell, R. C. (1977). 'Marine Plankton: A Practical Guide.' (Hutchinson: Essex.)

Todd, C. D., and Laverick, M. S. (1991). 'Coastal Marine Zooplankton: A Practical Guide for Students.' (Cambridge University Press, Cambridge.)

Ecology and behaviour of plankton

Johnson, J. M., and Marshall, P. (2007). 'Climate Change and the Great Barrier Reef.' (Great Barrier Reef Marine Park Authority: Townsville, Queensland.)

Kingsford, M. J., and Murdoch, R. (1998). Planktonic assemblages. In 'Studying Temperate Marine Environments: A Handbook for Ecologists'. (Eds. M. J. Kingsford and C. N. Battershill.) pp 227–268. (Canterbury University Press: Christchurch.)

McKinnon, A. D., Duggan, S., and De'ath, G. (2005). Mesozooplankton dynamics in inshore waters of the Great Barrier Reef. *Estuarine, Coastal and Shelf Science* **63**, 497–511.

Reynolds, C. S. (2006). 'Ecology of Phytoplankton.' (Cambridge University Press: Cambridge.)

See web site for an extended list, with updates: http://www.australiancoralreefsociety.org/ [Verified 22 February 2008].

15. Macroalgae

G. Diaz-Pulido

OVERVIEW

Generalities

Macroalgae are often referred to as seaweeds, yet they are not actually 'weeds'. Rather, macroalgae is a collective term used for large algae that are macroscopic and generally grow in the sea. They differ from other plants such as seagrasses and mangroves, in that algae lack roots, leafy shoots, flowers, and vascular tissues. In fact, algae are a polyphyletic group with diverse evolutionary histories. Whilst still used, the term 'algae' is no longer recognised as a formal taxon although it remains a useful name when referring to those protists that are photosynthetic.

The macroalgae of the Great Barrier Reef (GBR) are a very diverse and complex group of species and forms. Forms include crusts, foliose, and filamentous thalli (thallus refers to the body of an alga), ranging from simple branching structures to complex forms with highly specialised structures. The specialised structures are adaptations for light capture, reproduction, support, flotation, and/or substrate attachment. The size of coral reef macroalgae ranges from a few millimetres to plants of up to 3–4 m high (such as the brown alga *Sargassum*). However, in the more nutrient rich temperate regions of the oceans, brown algal kelps may grow to over 50 m in length (e.g. giant bull kelp in California). These organisms are not found on coral reefs.

Tropical macroalgae occupy a variety of habitats, including shallow and deep sections of coral reefs, inter-reefal areas, sandy bottoms, seagrass beds, mangrove roots, rocky intertidal areas, or even within the skeletons of healthy and dead corals, shells and limestone material (endolithic algae). Macroalgae are the major food source for a variety of herbivores, are major reef formers and create habitat for invertebrates and vertebrates of economic interest. They also play critical roles in reef degradation, when coral dominated reefs are replaced by rocky reefs covered in macroalgae.

DIVERSITY OF GREAT BARRIER REEF MACROALGAE

Taxonomic diversity and classification

Macroalgae are taxonomically classified into four different Phyla: Rhodophyta (from the Greek 'rhodon' meaning 'red rose' and 'phyton' meaning 'plant': red algae), Ochrophyta (Class Phaeophyceae, from the Greek 'phaios' meaning 'brown': brown algae), Chlorophyta (from the Greek 'chloros' meaning 'green': green algae) and Cyanobacteria (from the Greek 'cyanos' meaning 'dark blue': blue-green algae) (Table 15.1). This systematic classification is based on the composition of pigments involved in photosynthesis.

There are *c.* 630 species and varieties of macroalgae recorded for the GBR according to the Australian Marine Algal Name Index. However, this list is preliminary and it is likely that the number will increase with future field exploration. In fact, the macroalgal flora from the GBR,

Table 15.1 Higher level classification of benthic macroalgae

	Blue-green algae	Red algae	Brown algae	Green algae
Empire	Prokaryota	Eukaryota	Eukaryota	Eukaryota
Kingdom	Bacteria	Plantae	Chromista	Plantae
Subkingdom	Negibacteria	Biliphyta	Chromobiota	Viridaeplantae
Phylum	Cyanobacteria (=Cyanophyta)	Rhodophyta	Ochrophyta (=Heterokontophyta)	Chlorophyta
Class	Cyanophyceae	Florideophyceae Bangiophyceae	Phaeophyceae	Chlorophyceae

together with the northern Australian coast, is one of the lesser known floras on the Australian continent.

The Rhodophyta are the most diverse phylum for the GBR, with 323 species contained in 131 genera, with *Laurencia* (27 species), *Polysiphonia* (19) and *Ceramium* (16) the most speciose. There are 111 species of Phaeophyceae, with more than 50% of the species belonging to *Sargassum* (47 species) and *Dictyota* (11). Thirty-two genera of brown algae have been recorded for the GBR. The Chlorophyta include 195 species in 51 genera, of which *Caulerpa* (36 species), *Halimeda* (23), and *Cladophora* (19) contain the highest number of species.

Functional group diversity

Besides traditional species classification, macroalgae can also be classified based on ecological terms following a functional form group approach. This approach takes into consideration key plant attributes and ecological characteristics, such as the form of the plant, size, plant toughness, photosynthetic ability and growth, grazing resistance etc. Functional group classification is helpful in understanding the distribution of algal communities and their responses to environmental factors, since morphologically and anatomically similar algae have similar responses to environmental pressures regardless of their taxonomic affinities. The functional approach is useful particularly when identification to species level is not possible and consequently has been widely used in ecological studies on coral reefs of the GBR.

The functional group approach includes three main algal categories (Table 15.2):

- *Algal turfs.* Assemblages or multispecies associations of minute filamentous algae and the early life history stages of larger macroalgae, with high productivity, fast growth and colonisation rates. Turfs are ubiquitous and have low biomass but dominate much of the reef framework's surface. Analogous to some grasslands in terrestrial environments, turfs owe their continued existence to herbivores that graze on them, thereby preventing overgrowth by fleshy macroalgae. The term 'epilithic algal community' or EAC is often used to refer collectively to the algal assemblage that grows on the substrate; usually this refers to an assemblage dominated by filamentous algal turfs.
- *Upright macroalgae.* Large algal forms, more rigid and anatomically more complex than algal turfs, abundant in zones of low herbivory such as the intertidal, reef flats, or inshore reefs where strong wave action, heavy predation, or water quality limit grazing. They often contain chemical compounds that deter grazing from fishes.
- *Crustose algae.* Calcareous plants that grow completely adhered to the substrate forming crusts, with slow growth rates in general, and are abundant in shallow reefs with high herbivory pressure. This group includes species from the families Peyssonneliaceae and Corallinaceae.

Table 15.2 Categories and functional groups of benthic macroalgae present on the Great Barrier Reef

Algal categories		Functional groups	Examples of common genera
Algal turfs (<10 mm height)		Microalgae	*Lyngbya, Calothrix*
		Filamentous	*Cladophora, Polysiphonia*
		Juvenile stages of macroalgae	
'Upright' macroalgae (>10 mm height)	Fleshy	Foliose – Membranous	*Ulva, Anadyomene*
		Foliose – Globose	*Ventricaria, Dictyosphaeria*
		Foliose – Corticated	*Dictyota, Lobophora*
		Corticated	*Laurencia, Hypnea*
		Leathery	*Sargassum, Turbinaria*
	Calcareous	Calcareous articulated	*Halimeda, Amphiroa*
Crustose algae		Calcareous crustose	*Porolithon, Peyssonnelia*
		Non-calcareous crustose	*Ralfsia, Cutleria*

Spatial and temporal distribution

Biogeography

The algal flora of the GBR belongs to the Solanderian biogeographical province in terms of the benthic algal flora of Australia. This province is less diverse when compared to southern Australia (Flindersian province) and New South Wales (Peronian province). Endemism on the GBR is low, since most species are widely distributed in the Indo-East Pacific biogeographical region, and many GBR species are also thought to be present in the tropical Atlantic. However, recent studies using DNA markers has revealed that some of the species that look identical are actually distinct.

Spatial distribution

The distribution and abundance of macroalgae on coral reefs are determined by the resources they require (i.e. light, carbon dioxide, mineral nutrients and substrate), the effects of environmental factors (e.g. temperature, salinity and water movement), individual rates of recruitment, mortality and dispersal, and biological interactions such as competition and herbivory. Macroalgal communities of the GBR are highly variable, showing latitudinal, cross shelf and within reef variation in composition and abundance. Many GBR macroalgae are also highly seasonal.

Cross shelf distributions – offshore and inshore reefs. Offshore reefs usually have low abundance of fleshy macroalgae and high cover of algal turfs and crustose calcareous algae (CCA) compared to inshore reefs. Some fleshy macroalgae, such as the green fleshy macroalgae *Caulerpa, Chlorodesmis, Halimeda*, and the reds *Laurencia, Galaxaura* and *Liagora* are common in offshore reefs but in low abundance. Crustose calcareous algae are abundant and diverse on offshore reefs and play significant roles in reef construction. Common taxa on offshore reefs include the CCA *Porolithon onkodes*, and species of *Neogoniolithon* and *Lithophyllum*. The cross shelf distribution of the algal functional groups is predominantly affected by fish grazing and water quality (nutrient availability and sedimentation).

Inshore reefs usually have abundant and conspicuous stands of fleshy macroalgae. In particular, brown macroalgae of the Order Fucales such as *Sargassum, Hormophysa, Turbinaria* and *Cystoseira*, form dense and highly productive beds of *c.* 2 m height. Other fleshy brown macroalgae such as *Lobophora variegata, Dictyota, Colpomenia, Chnoospora* and *Padina* and the red *Asparagopsis taxiformis*, may also be abundant in shallow inshore reefs. *Lobophora variegata* can be particularly abundant in inshore reefs, especially between branches of corals and after coral disturbance such as bleaching. Crustose calcareous algae are common but are not abundant.

Within reef distribution. Algal zonation is quite clear in rocky intertidal coasts but is normally diffuse in subtidal reefs, where algal communities are distributed as a continuum along environmental gradients (e.g. depth). A number of reef zones can be recognised in a cross-section of an offshore reef from shallow to deep areas (Table 15.3):

- *Intertidal and beach rock* – diverse fleshy macroalgal communities, reduced grazing by large animals, intense solar radiation
- *Reef lagoon* – limited macroalgal growth due to sandy bottom; however, the microphytobenthos community that grows on sand can be highly productive
- *Reef flat (back-reef)* – diverse fleshy macroalgal communities, low grazing
- *Reef crest* – abundant crustose coralline algae and algal turfs, intense grazing and wave action
- *Reef front and upper reef slope* – abundance and diversity of macroalgae decreases with increasing depth, algal communities dominated by turfs and CCA, poorly developed fleshy macroalgal populations
- *Walls* – low algal cover and high coral cover, some upright calcareous macroalgae like *Halimeda* can be locally abundant

Particular microhabitats such as crevices and the territories of damselfishes play important roles in locally increasing the diversity of fleshy and turf algae.

Inter-reef areas and Halimeda beds. Macroalgae associated with seagrass beds, particularly in deep, soft-bottom areas of the northern GBR are quite abundant. In contrast to seagrasses, most algae do not attach to sand, although a number of green macroalgae have adapted to such environments by developing special anchoring features. This is the case for green algae such as *Halimeda*, *Caulerpa* and *Udotea*, commonly found intermixed with seagrasses. Macroalgae growing on leaves of seagrasses are called epiphytes and may play important roles as food for invertebrates and vertebrates in seagrass meadows. *Halimeda* mounds grow in nutrient rich upwelling water in between reefs that make up the outer barrier reef (see Box 15.1).

Table 15.3 Example of zonation of benthic algae in the Great Barrier Reef

Zones	**Crustose corallines**	**Common fleshy macroalgae**		
		Reds	***Browns***	***Greens***
Upper reef Slope and reef Front	*Porolithon* *Neogoniolithon* *Hydrolithon*	*Predaea*, *Galaxaura*	*Lobophora*	*Chlorodesmis*
Reef crest	*Porolithon* *Neogoniolithon*	*Laurencia*		*Caulerpa racemosa* *Chlorodesmis*
Reef flat	*Porolithon*	*Acanthophora* *Laurencia* *Gelidiella* *Hypnea*	*Dictyota*, *Padina* *Sargassum* *Hydroclathrus* *Chnoospora*	*Caulerpa* *Chlorodesmis* *Halimeda* *Dictyosphaeria*
Lagoon			*Hydroclathrus*	Blue-green algae *Halimeda* spp. *Caulerpa* spp.
>10 m deep and cryptic	*Lithothamnium* *Mesophyllum* *Neogoniolithon*	Turfs	*Lobophora* *Melanamansia*	*Rhipilia* *Halimeda* *Caulerpa*

BOX 15.1 INTER-REEF AREAS AND *HALIMEDA* BEDS

Seaweeds are abundant in the deep water, inter-reefal areas of the northern part of the GBR. Large mounds formed from the green calcareous alga *Halimeda* are estimated to cover up to 2000 km^2 in this region and may be up to 20 m high. These *Halimeda* meadows occur principally in the northern sections at depths between 20 m to 40 m, but there are also some in the central and southern sections of the GBR, where they have been found at depths down to 96 m. The GBR apparently contains the most extensive actively calcifying *Halimeda* beds in the world, although the real extent of such meadows is unknown. Tidal jets and localised upwelling events in the northern section of the GBR provide the nutrients needed to sustain extensive deep (30–45 m) meadows of *Halimeda*.

BOX 15.2 PRESERVATION

Macroalgae used for taxonomic purposes should be collected with the anchoring systems and preferably with reproductive structures. Macroalgae can be preserved in a solution of 4% formalin or in ethanol 70% (although decolouration may occur). However, dry herbarium specimens are easier to use and transport. White cardboard acid-free sheets are used for mounting specimens. Each herbarium sheet should contain basic information on the locality of collecting, date, depth, habitat, colour, and the collector's name. Permits are required, see Chapter 12.

Seasonality

The abundance, growth and reproduction of many GBR macroalgae are highly variable in time. Large seaweeds like *Sargassum* are strongly seasonal, with peaks in biomass and reproduction generally during the summer and lowest biomass during the winter. Extensive blooms of fleshy brown macroalgae like *Chnoospora* and *Hydroclathrus* are common on shallow reef flats during winter. Due to these strong seasonal changes, some authors argue that the seasonality of the GBR flora may be as strong as that from temperate zones.

Common genera and identification

Identification of coral reef macroalgae to the genus level is relatively easy, but the identification at the species level is more difficult and generally requires examination under a compound microscope (see Box 15.2 for preservation methods). The reproductive structures, internal tissues, cell organisation, and so on, are key features required for rigorous species identification.

Common red algae

- *Amphiroa:* heavily calcified, branches cylindrical to flattened, composed of smooth segments linked by very short non-calcareous joints. Pale pink to red, also called geniculated Corallinaceae. Common throughout the GBR (Fig. 15.1*A*).
- *Asparagopsis taxiformis:* plants soft, with creeping stems and upright fluffy or feathery tufts. Bright pink and common on inshore reefs (Fig. 15.1*B*).
- *Corallophila:* red filamentous alga with creeping axes and erect branches with pointed tips, usually found overgrowing corals (Fig. 15.1C).

Figure 15.1 Common genera and species of benthic algae from the Great Barrier Reef. *A, Amphiroa* sp.; *B, Asparagopsis taxoformis; C, Corallophila huysmansii* (=*Centroceras huysmansii*); *D, Eucheuma denticulatum; E, Hypnea pannosa; F, Jania* sp.; *G, Laurencia* cf. *intricate; H, Melanamansia glomerata* (=*Amansia glomerata*); *I, Peyssonnelia* sp.; *J, Porolithon onkodes; K,* crustose coralline algae epiphytic on *Lobophora variegate; L, Dictyota* sp.; *M, Colpomenia sinuosa; N, Hydroclathrus clathratus; O, Lobophora variegata*. (Photos: G. Diaz-Pulido.)

- *Eucheuma:* similar to some species of *Hypnea* but the thalli are tougher and rubbery. Sometimes found between branches of branched *Acropora* corals (Fig. 15.1*D*).
- *Galaxaura:* lightly calcified, dichotomously branched, branches cylindrical or flattened, smooth or hairy. Pink to red, sometimes with chalky appearance.
- *Hypnea:* branches cylindrical and generally bearing numerous relatively short spine- or tooth-like branchlets with pointed tips. Colour variable from pale brown to dark purple. Usually between branches of hard corals (Fig. 15.1*E*).
- *Jania:* heavily calcified, similar to *Amphiroa* but plants are smaller (few mm) and branches are predominantly cylindrical. Widespread in shallow and deep reefs (Fig. 15.1*F*).
- *Laurencia:* plants generally bushy, branches usually cylindrical with blunt tip branchlets. Colour variable, some with green branches and pink tips, others ranging from orange, red to pink. Common on reef flats (Fig. 15.1*G*).
- *Melanamansia:* red to reddish-brown, branches leaf-like and grouped in rosettes, with marginal teeth. Generally found on crevices and other low-light microenvironments (Fig. 15.1*H*).
- *Peyssonnelia:* encrusting, rounded to fan-shaped plants. Calcification on the lower side and fleshy surface. Colour variable, dark red, purple pink or red yellowish. Common on overhangs and cryptic microhabitats (Fig. 15.1*I*).
- *Porolithon:* encrusting, heavily calcified with chalky texture. Pink crusts of several mm thick. Common on the reef crest (Fig. 15.1*J*). Other crustose coralline algae are epiphytic (grow on the surface of other benthic algae) (Fig. 15.1*K*).
- *Polysiphonia:* filamentous, usually a few mm tall, examination under the microscope shows cylindrical polysiphonous branches with a 'banding' appearance, similar to *Sphacelaria*. Pink to red, red-brown or brown. Common and abundant on algal turfs.

Brown algae

- *Dictyota:* plants creeping or erect, flattened, strap or ribbon-like (without midrib as in *Dictyopteris*). Branching dichotomous (forked). Light brown and several species with blue-green iridescence. Common (Fig. 15.1*L*).
- *Chnoospora:* plants forming mats or cushion-like clumps, branches are slightly flattened and dichotomous. Light brown. Common on reef flats.
- *Colpomenia:* plants rounded or irregularly lobed with hollow interior. Light to pale golden brown. Common on inshore reefs (Fig. 15.1*M*).
- *Hincksia:* small (few mm) filamentous plants, fine, erect, uniseriate (one row), pale yellowish and translucent. Reproductive structures (sporangia) somewhat corn cob-like. Common on algal turfs.
- *Hormophysa:* similar to *Sargassum* but blades with internal oblong air bladders. Yellow to dark-brown. Common on inshore refs.
- *Hydroclathrus:* light to pale golden brown, net-like structure, perforated. Common on calm and sheltered waters (Fig. 15.1*N*).
- *Lobophora variegata:* creeping, rounded to fan-shaped plants, sometimes encrusting. Pale to dark brown and usually with concentric bands and radiating lines. Common to locally very abundant, particularly between branches of corals (Fig. 15.1*O*).
- *Padina:* upright, sheet-like, fan-shaped plants, with concentric bands and also whitish bands due to carbonate deposition. Similar to *Lobophora* but has characteristic inrolled outer margin. Common on reefs flats (Fig. 15.2*A*).
- *Sargassum:* plants erect, leathery, some up to a couple of metres are the tallest seaweeds in the GBR. They typically have a stipe, leaf-like fronds with a midrib (central vein), and air vesicles or floats. Very abundant on inshore reefs (Fig. 15.2*B*, C).
- *Sphacelaria:* filamentous, small (few mm) turfing plants, branch cells (observed under compound microscope) arranged in regular transverse tiers, individual cells rectangular and elongated longitudinally. Distinctive dark brown cell at the tip of each

Figure 15.2 Common genera and species of benthic algae from the Great Barrier Reef. *A, Padina* sp.; *B, Sargassum tenerrimum; C, Sargassum* spp.; *D, Turbinaria ornate; E, Caulerpa cupressoides; F, Caulerpa* sp.; *G, Caulerpa racemosa; H, Chlorodesmis fastigiata; I, Dictyosphaeria versluysii; J, Halimeda* cf. *discoidea; K, Halimeda* sp.; *L*, green band of an endolithic algal community (including *Ostreobium* spp.) within the skeleton of *Porolithon onkodes; M, Udotea* sp.; *N*, algal turf overgrowing recently dead coral; *O*, blue-green algae growing on dead coral; *P*, filamentous blue-green algae under compound microscope. (Photos: G. Diaz-Pulido.)

filament. One of the most abundant taxa of algal turfs.

- *Turbinaria:* plants erect, tough, leathery, with closely placed top-shaped branches with spiny margin, each containing an embedded air bladder. Common on reef flats (Fig. 15.2*D*).

Green algae

- *Caulerpa:* all species have a creeping stolon attached by rhizoids and erect green branches or fronds. Fronds are very variable including leaf- and feather-like, others have cylindrical to club-shaped, spherical-like branchlets. Common on sandy and reef bottoms from shallow and deep reef (Fig. 15.2*E–G*).
- *Chlorodesmis:* plants forming tight clumps or tufts of repeatedly forked filaments. Bright green and common on shallow reefs (Fig. 15.2*H*).
- *Dictyosphaeria:* plants spherical to irregularly lobed, light green, surface hard and tough composed of one

layer of angular or polygonal cells, resembling a honeycomb. Hollow to solid inside depending on the species. Common (Fig. 15.2*I*).

Halimeda: plants erect, lightly to heavily calcified, pale to dark green. Branches formed by calcified segments separated by deep constrictions. Segments can be flattened (triangular to discoid or kidney-shaped) to cylindrical (Fig. 15.2*J*, *K*).

- *Ostreobium:* microscopic green filaments, cylindrical to inflated, usually within skeletons of healthy and dead corals and other carbonate substrates. Widespread on deep and shallow reefs (Fig. 15.2*L*).
- *Udotea:* upright calcified, stalked and fan-shaped plant, anchored to the substratum by a rhizoidal mass. Grey-green (Fig. 15.2*M*).
- *Ulva:* bright green, sheet-like or membranous blades. Uncommon on reefs. Species of *Enteromorpha* (now belong to the genus *Ulva*) are small (a few mm) and have the form of a hollow tube; common on algal turfs.
- *Ventricaria:* globose plants up to several cm in diameter. Glossy dark green with bright reflective glare. Usually epiphytised by pink crustose calcareous algae. Common throughout the reef.

Reproduction

Macroalgae reproduce either asexually or sexually. Asexual reproduction involves the release of spores (propagules) or by fragmentation (pieces of plant braking off to produce new individuals). In sexual reproduction, male and female gametes are released into the water, however, there are some examples where female gametes are retained by the parent and the resulting embryo develops (at least temporarily) on the parent gametophyte.

Macroalgae have complex life cycles, of which there are three types in coral reef algae. (1) *haplontic life cycle* (meiosis of a zygote occurs immediately after karyogamy); (2) *diplontic life cycle* (the zygote divides mitotically to produce a multicellular diploid individual), and (3) *diplobiontic life cycle or alternation of generations* (the haploid and diploid phases are alternated, each phase consisting of one of two separate, free living organisms: a gametophyte, which is genetically haploid, and a sporophyte, which is genetically diploid).

ECOLOGICAL ROLES

Contribution to primary production

A large proportion of the primary production (the creation of organic matter by plants from inorganic material like CO_2 and sunlight during photosynthesis) in a coral reef comes from the contribution of benthic algae. Net primary production is variable ranging from 148–500 (g C $m^{-2} yr^{-1}$) for algal turfs, 146–1095 (g C $m^{-2} yr^{-1}$) for fleshy macroalgae, and 73–475 (g C $m^{-2} yr^{-1}$) for crustose coralline algae. Planktonic microalgae and algal symbionts of scleractinian corals also contribute to reef productivity but to a much lesser degree. The organic matter (carbon) produced by planktonic microalgae enters the reef food chain either by: (1) consumption by herbivorous fishes, crabs, sea urchins and zooplankton; (2) release of Dissolved Organic Matter (DOM) by the algae into the water column where it is consumed by bacteria that in turn may be consumed by a variety of filter feeders, or (3) export to adjacent ecosystems such as seagrass meadows, mangroves or to the sea floor by currents and tides. See further details on the energy flow through coral reefs in Chapter 7.

Nitrogen fixation and nutrient cycling

Filamentous blue-green algae like *Calothrix* living in algal turf communities (Fig. 15.2*O*) fix significant amounts of atmospheric (inorganic) nitrogen into ammonia, which is then used by the blue-green algae themselves to build organic matter. Because of the rapid growth rates of blue-green algae and intense grazing on turf communities, the organic nitrogen is then distributed throughout the reef, contributing to reef nutrition (Chapter 7). Macroalgae take up, store, and release nutrients, thereby contributing to nutrient cycling in coral reef ecosystems.

Construction

Many macroalgae make important contributions to the construction of the reef framework by depositing calcium carbonate ($CaCO_3$). Crustose calcareous algae (CCA, e.g. *Porolithon* and *Peyssonnelia*) are important framework builders and framework cementers in coral reefs, whereby they bind adjacent substrata and provide a calcified tissue barrier against erosion. This process is particularly important on the reef crest environments of the GBR. Crustose calcareous algae are also important in deeper areas at the edge of the continental platform in the southern GBR (80–120 m), where they form large algal frameworks of several metres high. Deposition of calcium carbonate within the tissues of CCA (as calcite as well as high magnesium calcite) can be up to 10.3 kg $CaCO_3\,m^{-2}\,yr^{-1}$ in some parts of the GBR (e.g. Lizard Island).

Upright calcareous algae such as *Halimeda*, *Udotea*, *Amphiroa* and *Galaxaura* contribute to the production of marine sediments that fill in the spaces between corals. The white sand of beaches and reef lagoons is largely the eroded calcium carbonate skeletons of these algae. Calcium carbonate is deposited as aragonite in *Halimeda* with an estimated production of around 2.2 kg $CaCO_3$ $m^{-2}\,yr^{-1}$. Calcification may be an adaptation to inhibit grazing (a defensive mechanism), resist wave shock, and to provide mechanical support.

Facilitation of coral settlement

Crustose coralline algae induce settlement and metamorphosis of coral larvae and a range of other invertebrates in the GBR, thus playing a critical role in reef resilience. Some evidence supports the idea that this interaction seems to be mediated by chemicals released by the alga.

Roles in reef degradation

Macroalgae play important roles in reef degradation, particularly in ecological phase shifts, where abundant reef-building corals are replaced by abundant fleshy macroalgae. Reductions in herbivory due to overfishing, and increases in nutrient inputs leading to eutrophication (e.g. sewage and fertiliser), have been suggested as causes of increased abundance of fleshy macroalgae leading to coral overgrowth and reef degradation. Coral bleaching, crown-of-thorns starfish outbreaks, extreme low tides, coral diseases, cyclones, and so on result in coral mortality, providing an environment that is rapidly colonised by diverse algal communities (Fig. 15.2*N*). Such disturbances, and particularly those due to climate change (e.g. bleaching), may lead to an overall increase in the total amount of macroalgae (see Chapters 9 and 10 for further details). Dominance by thick mats or larger, fleshy macroalgae may contribute to reef degradation by overgrowing corals, inhibiting coral recruitment, contributing to coral diseases (e.g. Fig. 15.2*P*), and thereby decreasing the aesthetic value of reefs.

Bioerosion

Endolithic algae that live within the skeletons of both healthy and dead corals as well as other calcareous substrates contribute to reef erosion and destruction. These algae are generally filamentous and microscopic but form a thin dark green band visible to the naked eye underneath the coral and crustose algal tissue (Fig. 15.2*L*). Some examples of carbonate-boring algae include the greens *Ostreobium* spp., cyanobacteria *Mastigocoleus testarum*, *Plectonema terebrans*, and *Hyella* spp. and some red algae. Endolithic algae penetrate and dissolve the calcium carbonate, weakening the reef framework and thus hasten other erosive activities. Studies at One Tree Island on the GBR have show rates of bioerosion by endolithic algae to range between 20–30 g $m^{-2}\,yr^{-1}$. For more information on bioerosion see Chapter 9.

ADDITIONAL READING

Algae of Australia Series (2007). (Eds) 'Algae of Australia: Introduction.' (Australian Biological Resources Study: Canberra/CSIRO Publishing: Melbourne.)

Australian Marine Algal Name Index. Available at http://www.anbg.gov.au/abrs/online-resources/amani/ [Verified 26 February 2008].

Borowitzka, L. J., and Larkum, A. W. D. (1986). Reef algae. *Oceanus* **29**, 49–54.

Clayton, M. N., and King, R. J. (Eds) (1990). 'Biology of Marine Plants.' (Longman Cheshire: Melbourne.)

Cribb, A. B. (Ed.) (1996). 'Seaweeds of Queensland: A Naturalist's Guide.' (The Queensland Naturalist's Club: Brisbane.)

Diaz-Pulido, G., and McCook, L. J. (2002). The fate of bleached corals: patterns and dynamics of algal recruitment. *Marine Ecology Progress Series* **232**, 115–128.

Huisman, J. M. (Ed.) (2000). 'Marine Plants of Australia.' (University of Western Australia Press: Nedlands, Western Australia.)

International Database of Information on Algae. Available at http://www.algaebase.org/ [Verified 26 February 2008].

Littler, D. S., and Littler, M. M. (Eds) (2003). 'South Pacific Reef Plants.' (Offshore Graphics: Washington.)

McCook, L. J., Jompa, J., and Diaz-Pulido, G. (2001). Competition between corals and algae on coral reefs: a review of evidence and mechanisms. *Coral Reefs* **19**, 400–417.

Price, I. R., and Scott, F. J. (Eds) (1992). 'The Turf Algal Flora of the Great Barrier Reef: Part I. Rhodophyta'. (James Cook University of North Queensland: Townsville.)

Womersley, H. B. S. (Ed.) (2003). 'The Marine Benthic Flora of Southern Australia. Rhodophyta – Part III D.' (State Herbarium of South Australia: Adelaide.)

16. Mangroves and Seagrasses

N. C. Duke & A. W. D. Larkum

OVERVIEW

Sandwiched between two of the world's iconic tropical ecosystems of coral reefs and rainforests, are two important coastal communities: mangroves and seagrasses (Fig. 16.1). While corals flourish in shallow warm seas, and rainforests cover wetter upland regions, mangroves inhabit the sheltered intertidal margins barely above mean sea level. Seagrasses occupy depths from intertidal to deeper habitats, depending on the clarity of the water column. These biota-structured ecosystems play an important role in coastal processes with highly developed linkages and connectivity between them, and to the neighbouring communities represented by coral reefs and rainforests. These relationships appear vital to the survival of each. For example, while sediment-loving mangroves depend on waters sheltered by coral reef structures, they in turn protect hypersensitive corals from excessive sediments and nutrients coming down from surrounding catchments.

GENERAL DESCRIPTION AND ROLE IN ECOSYSTEMS

Mangrove and seagrass communities are characterised by a small number of widely distributed angiosperm taxa, having evolved mostly post-Cretaceous (within the last 60–100 million years). The relatively recent evolution of these communities may explain their comparatively low species diversity today, although this is also arguably related to the harsh environmental factors that characterise these communities' habitats. The relatively rich mangrove and seagrass floras of today are testament to their adaptive and evolutionary success for living in and adjacent to the intertidal zone. These highly specialised plants flourish in soft sediments, tapping rich estuarine nutrients with their distinctively vascular root systems. Mangroves further provide significant habitat and structure where their biomass accumulation, while readily seen as contiguous with adjacent rainforests, is also analogous to that created by coral reefs.

While important details of relevant phyletic origins remain lacking, ancestral mangrove and seagrass plant taxa are known to have reinvaded marine environments in multiple episodes from the diverse selection of angiosperm lineages. Their evolution appears constrained by key functional attributes essential for survival in saline and aqueous settings where isotonic extremes, desiccation and hydrologic exposure combine as uniquely harsh constraints on organisms living in tidal zones and estuaries. To achieve this, mangroves and seagrasses share a number of ecophysiological traits and have evolved mechanisms to cope with life at the land-sea interface, for example, salt tolerance, translocation of gases to aerate their roots, and reproductive strategies. Both plant communities perform important ecosystem services such as sediment stabilisation, nutrient processing, shoreline protection, and provide habitat and

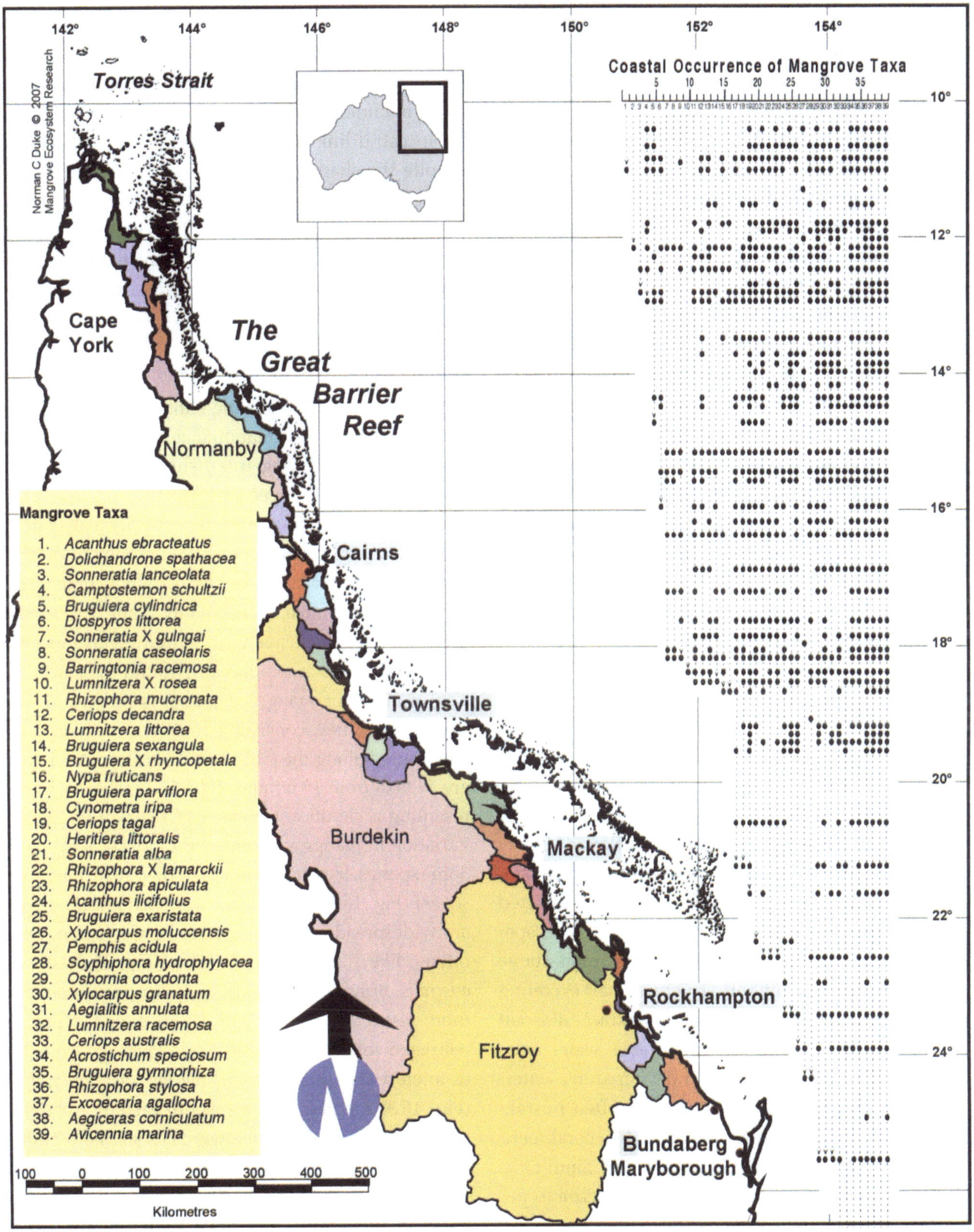

Figure 16.1 Distribution of mangrove species along the coastline of GBR catchments in Queensland.

nursery grounds. While the species diversity of mangroves and seagrasses is relatively low compared with adjacent communities like tropical rainforests and coral reefs, the diversity of organisms that reside in, or utilise, mangroves and seagrasses is high. Many fish and shellfish spend all or part of their life cycle in mangroves and/or seagrasses, including important commercial, recreational and artisanal fisheries.

The land-sea interface is a dynamic environment, where subtle natural changes in climate, sea level, sediment and nutrient inputs have dramatic consequences in the distribution and health of mangroves and seagrasses. Local human disturbances include eutrophication, dredging/filling, overfishing and sedimentation. The combined pressures of human disturbances and global climate change have led to mangroves and seagrasses becoming 'endangered communities'. Small scale restoration projects have demonstrated the extreme difficulty of scaling up to effective, large scale restoration. Urgent protective measures need to be implemented to avoid further loss of mangroves and seagrasses and the resulting environmental degradation of coastal ecosystems of the GBR coast.

MANGROVES

Overview

Mangroves are a diverse group of predominantly tropical trees and shrubs growing in the upper half of the intertidal zone of coastal areas worldwide. They are well known for their morphological and physiological adaptations for life coping with salt, saturated soils and regular tidal inundation, notably with specialised attributes like: exposed breathing roots above ground, extra stem support structures, salt-excreting leaves, low water potentials and high intracellular salt concentrations to maintain favourable water relations in saline environments, and viviparous water-dispersed propagules. Mangroves are often mistakenly thought of as a single entity. But, like coral reefs, mangroves are as functionally diverse and complex as the range of species, variants and morphotypes present at particular locations. Also, like coral reefs, they provide essential structure and habitat for a host of marine and intertidal species, including residents among their dense forests and complex roots, and as visitors with each flooding tide (Box 16.1, Fig. 16.2). Mangroves are also analogous in function to tropical rainforests, providing comparable canopy habitat for birds, mammals and insects. This overlap is reinforced by ancestral links between these plant habitats. But, despite the shared features, mangroves include specialist attributes and dedicated resident biota found nowhere else. Examples include specialist mangrove forms of the robin, mistletoe bird, mistletoes, grapsid crabs, molluscs, herbivorous insects, and numerous floral visitors.

Taxonomy and functional morphology

Mangroves are not a genetic entity, but an ecological system. Mangrove vegetation includes a range of functional forms, including trees, shrubs, a palm, and a ground fern. These generally exceed 0.5 metres in height, and normally grow above mean sea level in the intertidal zone of marine coastal environments and estuarine margins. Mangrove plants do not come from a single genetic source, and the only plant families that are comprised exclusively of mangrove taxa are Avicenniaceae and Sonneratiaceae. Around the world, the total number of mangrove plants is around 72 taxa from 21 families, consisting mostly of angiosperms. Distributed along the GBR World Heritage Area, there are 39 taxa from 19 families (Fig. 16.1, Table 16.1), representing a significant portion of the worlds' genetic variation in mangrove plants, and most of Australia's. Some species, like *Avicennia marina* (Fig. 16.3), *Rhizophora stylosa* (Fig. 16.4) and *Bruguiera gymnorhiza* (Fig. 16.5), are widespread in the Indo-West Pacific region, while others, like *Aegiceras corniculatum* (Fig. 16.6), *Ceriops australis*, *Bruguiera exaristata*, and *Diospyros littorea*, are more restricted to the Australasian region. None are restricted solely to the GBR World Heritage Area. It is of interest also that *Avicennia marina* var. *eucalyptifolia* (Fig. 16.3) merges with the south-eastern Australian variety, *A.* var. *australasica* towards its southern boundary, south of the Tropic of Capricorn.

The widespread group of species mentioned above are also ubiquitous mangrove representatives commonly found throughout the GBRWHA. They are commonly located on many mainland islands, sand cays

BOX 16.1 FRUITING MANGROVES AND GRAZING GREEN TURTLES

Green turtles seek out maturing mangrove propagules to include in their diets. Several observations come as evidence and clues of this previously unknown behaviour. It has been noted for some time that turtles frequent mangrove-lined waterways at high tide. They are also known for occasionally, as tides recede, becoming caught up in tree limbs and stranded on exposed mudflats. Firm evidence of feeding, however, has only recently been described. In 2000, Limpus and Limpus reported purposeful cropping of mature propagules of *Avicennia marina* in Shoalwater Bay. This has been further supported by observations of *A. marina* propagules within mature turtle guts interspersed with seagrass. The feeding behaviour to be interpreted from these observations is that some turtles feed on seagrass at low tide, and, for a short time at high tide, they take in the nutritious mangrove propagules. This behaviour is dependant on periods of higher tides and the distinct seasonality of propagule availability. However, the extent to which turtles seek out this occasional food bonanza is seemingly shown by more recent observations in Great Sandy Straits of some turtles purposely stranding themselves on mudflats near mangroves (risking predation and ignoring low tide feeding) to gather more of this previously unrecognised food source.

Figure 16.2 Green turtle (*Chelonia mydas*) cropping grey mangrove (*Avicennia marina*) propagules from foreshore trees at high tide. (Image: Fran Davies 2006, original artwork commissioned by N. C. Duke based on observations of Limpus and Limpus 2000.)

and reef atolls, never falling below 60% presence and often above 80% and include: *Avicennia marina, Aegiceras corniculatum, Bruguiera gymnorhiza, Excoecaria agallocha* (Fig. 16.7), *Osbornia octodonta* (Fig. 16.8) and *Rhizophora stylosa*. Only *R. stylosa* and *A. marina* are considered major dominants. All species of this group are generally described as less climatically sensitive. Their wide ecological amplitude makes them tolerant of a wide range of salinities from seaward exposure to periodic pulses of freshwater. Each species has further individual traits, for example: *R. stylosa* are confined to the leading edge of the mangrove zone, often fronting the sea and associated with moderate but not extreme salinities; *E. agallocha* are mostly associated with dry margins; while *O. octodonta* is characteristically common on sandy substrates.

Role of mangroves and the filters of the coast

Mangroves and tidal wetlands are essential to the sustainability of highly productive natural coastal environments. Mangroves have many well-acknowledged roles in coastal connectivity, supporting enhancements in biodiversity and biomass not possible otherwise. At another level, commercial advantages decry the

Table 16.1 The 39 mangroves of GBR coastal catchments. Those highlighted in bold are commonly found on islands within the GBRWHA. See the key to differentiate taxa.

Family	Taxa
Acanthaceae	*Acanthus ebracteatus*
	Acanthus ilicifolius
Arecaceae	*Nypa fruticans*
Avicenniaceae	***Avicennia marina*** – two varieties including: northern, *A.* var. *eucalyptifolia*; and southern, *A.* var. *australasica*
Bignoniaceae	*Dolichandrone spathacea*
Bombacaceae	*Camptostemon schultzii*
Caesalpiniaceae	*Cynometra iripa*
Combretaceae	*Lumnitzera littorea*
	Lumnitzera racemosa
	Lumnitzera X rosea
Ebenaceae	*Diospyros littorea*
Euphorbiaceae	***Excoecaria agallocha***
Lecythidaceae	*Barringtonia racemosa*
Lythraceae	*Pemphis acidula*
Meliaceae	*Xylocarpus granatum*
	Xylocarpus moluccensis
Myrsinaceae	***Aegiceras corniculatum***
Myrtaceae	***Osbornia octodonta***
Plumbaginaceae	*Aegialitis annulata*
Pteridaceae	*Acrostichum speciosum*
Sterculiaceae	*Heritiera littoralis*
Sonneratiaceae	*Sonneratia alba*
	Sonneratia caseolaris
	Sonneratia X gulngai
	Sonneratia lanceolata
Rhizophoraceae	*Bruguiera cylindrica*
	Bruguiera exaristata
	Bruguiera gymnorhiza
	Bruguiera parviflora
	Bruguiera X rynchopetala
	Bruguiera sexangula
	Ceriops australis
	Ceriops decandra
	Ceriops tagal
	Rhizophora apiculata
	Rhizophora X lamarckii
	Rhizophora mucronata
	Rhizophora stylosa
Rubiaceae	*Scyphiphora hydrophylacea*

Figure 16.3 *A*, Pubescent rounded fruits and lanceolate leaves of the northern grey mangrove, *Avicennia marina* var. *eucalyptifolia*. *B*, Small tree of northern grey mangrove, *Avicennia marina* var. *eucalyptifolia*, growing on a rocky shoreline of Snapper I. near Daintree River. (Photos: N. Duke.)

Figure 16.4 A distinctly rooted tree of long-style stilt mangrove, *Rhizophora stylosa*, on the rocky foreshores of Snapper I. near the Daintree River. (Photo: N. Duke.)

importance of mangroves, where around 75% of the total seafood landed in Queensland comes from mangrove estuarine related species. These messages clearly indicate that healthy estuarine and nearshore marine ecosystems are biologically and commercially important, and, these natural systems are intimately related, connected and dependent. So, where one is impacted, the effect will be felt more widely than might otherwise be expected. This is the case whether these ecosystems are viewed as sources of primary production with complex trophic linkages, as nursery and breeding sites, or as physical shelter and buffers from episodic severe flows and large waves.

These ecosystems and their linkages are seriously threatened. The GBR of Australia, for example, is threatened by another insidious factor in addition to global warming and coral bleaching. This immense and unique natural wonder is seriously threatened by smothering plumes of mud that greatly exceed levels of prior natural flooding. This corresponds with a century of large scale land clearing and conversion of coastal forested wetlands of GBR catchments into agricultural, port, urban and industrial developments. Coastal rivers have become little more than drains transporting eroded mud to settle in lower estuaries, in coastal shallows, and on inshore reefs. Mangrove-lined estuaries have offered some respite and have dampened this effect. In recent years, however, this final bastion of coastal sediment filtration has also begun to succumb to the increasing and unrelenting pressures of human populations that are expanding across coastal and estuarine regions of the GBR catchments.

Figure 16.5 *A*, Knobbly knee roots of large-leafed orange mangrove, *Bruguiera gymnorhiza*, in Daintree River estuary. *B*, Showy red flowers and buds of large-leafed orange mangrove, *Bruguiera gymnorhiza* (note also the characteristic insect bite indentations around leaf margins). (Photos: N. Duke.)

Figure 16.6 Horn-like fruit capsules and rounded leaves of river mangrove, *Aegiceras corniculatum*. (Photo: N. Duke.)

Figure 16.7 Unusually exposed twisty roots of milky mangrove, *Excoecaria agallocha*, at Flying Fish Point at the mouth of the Johnstone River. (Photo: N. Duke.)

Figure 16.8 *A*, Shrubs of myrtle mangrove, *Osbornia octodonta*, growing on an exposed sandy foreshore north of Agnes Waters. *B*, A shrubby hedge of myrtle mangrove, *Osbornia octodonta*, growing at the mouth of Coopers Creek north of the Daintree. (Photos: N. Duke.)

KEY TO THE 39 MANGROVES OF THE GBR CATCHMENT COAST

1. Palm or ground fern, both trunkless 2
   – Shrub or tree 3
2. Palm to 10 m .... *Nypa fruticans*
   – Ground fern to 1 m .... *Acrostichum speciosum*
3. Exuding white sap (oozes from broken leaf or cut bark) .... *Excoecaria agallocha* (Fig. 16.7)
   – No exuding sap 4
4. Compound leaves 5
   – Simple leaves 7

5. Flower larger than 5 cm, seed pod long curved .... *Dolichandrone spathacea*
   – Flower less than 1 cm, seed pod rounded, compact 6
6. Seed pod rugose (bivalved, pubescent, irregular), leaflet tip imarginate, buttresses or pneumatophores absent .... *Cynometra iripa*
   – Seed pod smooth (obscurely 4-valved, coriaceous, globate), leaflet tip apiculate, with conical pneumatophores or plank buttresses .... *Xylocarpus* (2 taxa)
7. Simple opposite leaves 8
   – Simple alternate leaves 16
8. Leaf undersurface pubescent (silvery grey appearance) .... *Avicennia marina* (2 varieties) (Fig. 16.3)
   – Leaf undersurface not pubescent 9
9. Flower zygomorphic, leaf often spiny (small shrub to 2 m) .... *Acanthus* (2 taxa)
   – Flower symmetric, leaf never spiny (shrub or tree) 10
10. Prop roots and/or buttresses, fruit germinates on tree, viviparous (long, >6 cm, green) 11
    – Prop roots or buttresses absent, fruit with seeds, not viviparous (rounded, <6 cm, often not green) 13
11. Fruit inverted pear-shaped, calyx lobes less than six (flat-expanded tube) 12
    – Fruit within calyx, calyx lobes 8–15 (turbinate tube) .... *Bruguiera* (6 taxa)
12. Calyx lobes five (thick sinuous buttresses) .... *Ceriops* (3 taxa)
    – Calyx lobes four (sturdy prop roots) .... *Rhizophora* (4 taxa)
13. Stipules persistent, pod deeply 8-ribbed (barrel-like) .... *Scyphiphora hydrophylacea*
    – Stipules absent, pod or fruit not deeply ribbed 14
14. Fruit to 4 cm (globular), tree medium to large (leaves large >5 cm), pneumatophores present .... *Sonneratia* (4 taxa)
    – Fruit less than 1 cm (calyx pod), shrub to small tree (leaves small <2 cm), pneumatophores absent 15
15. Leaf smooth (glabrous) with oil glands (aromatic when crushed), flower petals absent .... *Osbornia octodonta* (Fig. 16.8)
    – Leaf pubescent without oil glands, flower petals conspicuous .... *Pemphis acidula*
16. Leaf undersurface pubescent (hairs or fine scales, silvery appearance) 17
    – Leaf undersurface not pubescent (glabrous) 18
17. Leaf flat, less than 12 cm, seed pod ovate, scaly (less than 1 cm) .... *Camptostemon schultzii*
    – Leaf floppy, greater than 12 cm, seed pod keeled, smooth (greater than 5 cm) .... *Heritiera littoralis*
18. Leaf petiole envelopes stem (leaving annular scars) .... *Aegialitis annulata*
    – Leaf petiole does not envelop stem 19

19. Leaf tip imarginate, bark rough, pneumatophores wiry................*Lumnitzera* (3 taxa)
   – Leaf tip not imarginate, bark finely fissured, pneumatophores absent..................20
20. Flower with more than eight stamens, flowers on long raceme (>20 cm), fruit pod 4-cornered (leaf >20 cm).. *Barringtonia racemosa*
   – Flower with less than eight stamens, flowers not on raceme, fruit pods not 4-cornered (leaf <10 cm) 21
21. Flower bud on simple umbel, fruit horn-like .... *Aegiceras corniculatum* (Fig. 16.6)
   – Flower bud on obscure peduncle, fruits ovoid................................ *Diospyros littorea*

Avicennia marina varieties (2)

1. Bark smooth, greenish, leaves lanceolate, calyces incompletely pubescent ....var. *eucalyptifolia*
   – Bark rough, grey, leaves ovate-elliptic, calyces entirely pubescent............................ ....var. *australasica*

Genus *Xylocarpus* (2)

1. Buttresses small, pneumatophores conical, bark grey, vertical fissured, flaky rough, mature fruit 8–12 cm diameter (deciduous in early spring)... ....*Xylocarpus moluccensis*
   – Buttresses plank-like (sinuous), pneumatophores absent, bark pale orangy, flaky patches, smooth, mature fruit 10–25 cm diameter.................... *Xylocarpus granatum*

Genus *Acanthus* (2)

1. Inflorescence open, flower white or deep purple, bract single, bracteoles absent, stem without axillary spines .... *Acanthus ebracteatus*
   – Inflorescence clustered, flower mauve streaked with white, bract single, bracteoles two, stem with axillary spines .... *Acanthus ilicifolius*

Genus *Bruguiera* (6)

1. Flowers multiple (2–5), small; petal spine exceeds lobes...2
   – Flowers solitary, large; petal spine shorter than lobes or absent 3
2. Fruit calyx ribbed, lobes adpressed (1/4–1/5 length of calyx) ....*Bruguiera parviflora*
   – Fruit calyx smooth, lobes reflexed (1/2 length of calyx)............ *Bruguiera cylindrica*
3. Petal spine absent or minute ....*Bruguiera exaristata*
   – Petal spine shorter than lobes 4
4. Petal bristles absent or minute...*Bruguiera sexangula*
   – Petal bristles 1–3..5

5. Petal bristles 1–2, less than 2 mm .... *Bruguiera* X *rhynchopetala*
   – Petal bristles three, greater than 2 mm .... *Bruguiera gymnorhiza* (Fig. 16.5)

Genus *Ceriops* (3)

1. Peduncle length greater than width, petal tip with three lobes, fruit calyx tube flat expanded, mature hypocotyl to 35 cm long...2
   – Peduncle length equal to width, petal tip fringe-like, fruit calyx tube cup-shaped, mature hypocotyl to 27 cm long..*Ceriops decandra*
2. Mature hypocotyl to 35 cm long, deeply ribbed .... *Ceriops tagal*
   – Mature hypocotyl 9–15 cm long, smooth with ribs absent or minor .... *Ceriops australis*

Genus *Rhizophora* (4)

1. Leaf undersurface without small corky spots, peduncle length less than petiole length, stamens 9–15, calyx lobes thick and angular..2
   – Leaf undersurface with small corky spots, peduncle length greater than petiole length, stamens 6–8, calyx lobes thin and rounded..3
2. Bracts corky brown, style length 0.5–1.3 mm, inflorescence branches 1(–2), stamens (9–) 11–12 (–15) .... *Rhizophora apiculata*
   – Bracts smooth green, style length 1.5–2.7 mm, inflorescence branches 1–2, stamens (7–) 8–11 (–15)..*Rhizophora* X *lamarckii*
3. Style length greater than 2.5 mm, inflorescence branches (0–) 2–4 (–6), stamens (6–) 8 ....*Rhizophora stylosa* (Fig. 16.4)
   – Style length less than 2.5 mm, inflorescence branches 0–1(–2), stamens (7–8) ....*Rhizophora mucronata*

Genus *Sonneratia* (4)

1. Calyx smooth and shiny, fruit calyx cup-shaped, fruit diameter <4 cm, less than or equal to hypanthium..2
   – Calyx leathery, fruit calyx flat-expanded, fruit diameter > 4 cm, mostly 0.5 cm or more, greater than hypanthium...3
2. Petals white if present; stamens white; calyx shiny, fruit rounded dull....................... ....*Sonneratia alba*
   – Petals red, stamens red or white; calyx satiny luster or dull; fruit indented about style base, satiny .... *Sonneratia* X *gulngai*
3. Petals red, stamens red, leaf rounded, calyx grooved often deeply, surface satiny coriaceous, mature flower bud medial constriction........................ *Sonneratia caseolaris*
   – Petals red, stamens white, leaf lanceolate or linear, calyx tube rounded, surface dull smooth, mature flower bud with no medial constriction.................................. .... *Sonneratia lanceolata*

SEAGRASSES

Australia is fortunate in having nearly half the 60 species of the world's seagrasses represented, 14 of which occur on Australian coral reefs. However, in reef waters in Australia, seagrasses may be inconspicuous because they are grazed heavily by invertebrates, fish, turtles and dugongs (Box 16.2, Fig. 16.9). In protected areas they form dense, lush meadows, which harbour a diversity of animal and algal life.

BOX 16.2 SEAGRASS BEDS AND MACROGRAZERS

The larger species of seagrass in the GBR, *Zostera muelleri, Cymodocea serrulata, C. rotundata, Thalassia hemprichii, Enhalus acoroides,* and *Thalassodendron ciliatum* form dense beds near the coast and in sheltered bays offshore on continental islands. These beds are usually formed from several species of seagrass and present a wide diversity of habitats for infauna and inflora. Many epiphytes, epizoans and an extensive microflora occur on the seagrass fronds. Many animals and algae grow on or in the sediment between the seagrass plants. Organic substances and oxygen released into the sediments from the roots of seagrasses support a specialised habitat of microflora, protists and invertebrate grazers around the roots.

It is therefore not surprising that macrograzers are found in or near seagrass beds. The dugong (*Dugong dugon*) has received the most attention in Australia, because this marine mammal is almost entirely dependent on seagrass beds, and the seagrasses themselves, for their food. Marine turtles also feed on seagrass beds, although this has been much better documented in the Caribbean that in Australian waters.

Figure 16.9 A dugong feeding on a shallow bed of *Halodule uninervis*. (Photo: Marine Themes: the marine wildlife stock library, www.marinethemes.com.)

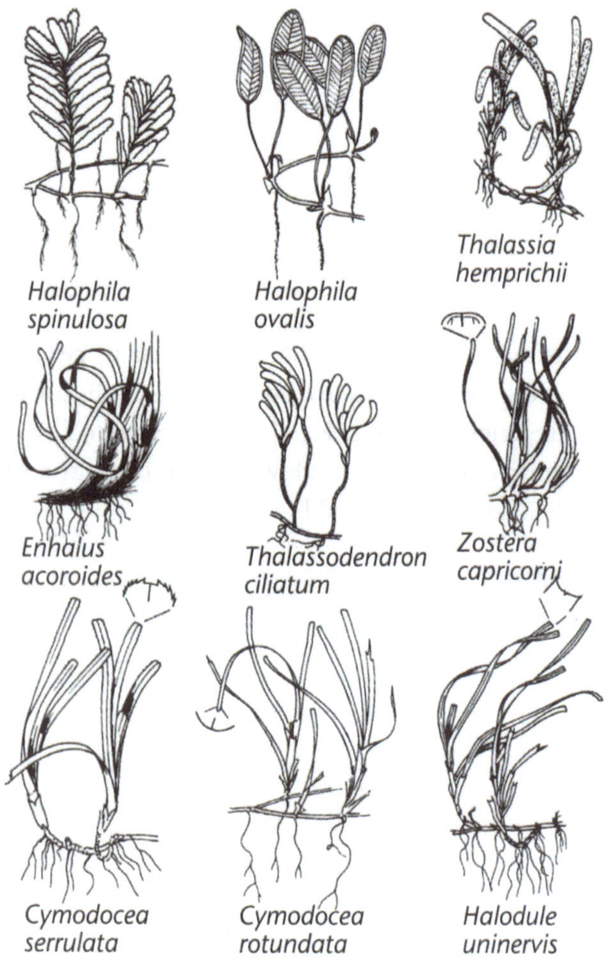

Figure 16.10 Seagrass species.

Seagrasses are not a monophyletic group, being found in five families of monocotyledonous (grass-like) plants. However, they all have a creeping stem (rhizome) and underwater flowering and pollination and are the only vascular plants that are aquatic and inhabit the sea.

Seagrasses grow mainly on soft substrates (silt and sand) in shallow, sheltered marine or estuarine situations. There are two exceptions to this generalisation: (1) *Thalassodendron ciliatum*, which grows directly on hard substrates (such as dead coral), as it has specialised roots that allow it to become firmly attached to these substrates; (2) many species of *Halophila* that can grow at considerable depths, especially *H. capricorni*, *H. decipiens*, *H. spinulosa* and *H. tricostata*, which can grow down to 60 m. Thus, these latter species of *Halophila* are often found growing on deep sediments outside coral reefs or in inter-reefal environments on the GBR.

In the main, seagrasses on coral reefs are found in lagoons or in shallow inter-reefal areas, where they support large populations of herbivores. They therefore form a very important source of primary production. In the GBR Zone these same species of seagrass are often found growing on shallow, sheltered sediments in coastal situations, where there are no corals.

Only three species of seagrass, all in the genus *Halophila*, are present in the Bunker-Capricorn Groups at the southern end of the GBR (*H. capricorni*, *H. decipiens* and *H. spinulosa*). Further north, seagrasses are both more numerous in terms of species (all the species in the Key (14) may be represented in a fairly small area) and they are more conspicuous. Further information is available in Additional reading.

KEY TO THE SEAGRASSES OF THE GREAT BARRIER REEF

1. Leaves ligulate (i.e. a tongue-like structure at junction of the leaf blade and sheath, which attaches to the rhizome)....................8
   – Leaves without ligules....................2
2. Leaves strap-shaped, neither compound nor petiolate (stalked)....................7
   – Leaves compound or petiolate (with a stalk)....................*Halophila* 3 (6 taxa)

3. Erect, lateral shoots bearing a number of compound leaves 4
 – Erect lateral shoots bearing one pair of simple leaves 5

4. 10–20 pairs of opposite leaflets (serrate margins) on an erect shoot. .... *Halophila spinulosa* (Fig. 16.10)
 – 2–3 leaflets (serrate margin, no cross-veins), 6–18 nodes per shoot .... *Halophila tricostata*

5. Leaf margin finely serrulate, hairs on leaf surface 6
 – Leaf margin entire, no hairs present 7

6. Leaf hairs throughout and on both sides, cross veins 9–14 .... *Halophila decipiens*
 – Leaf hairs only at base on abaxial side, cross veins 6–8 .... *Halophila capricorni* (Fig. 16.10)

7. Leaf blade paddle shaped, <5 mm long, <10 cross-veins .... *Halophila minor*
 – Leaf blade paddle shaped, >5 mm long, >10 cross-veins .... *Halophila ovalis* (Fig. 16.10)

8. Rhizome of two forms, scale-bearing and leaf-bearing, short shoots. Root hairs abundant on roots .... *Thalassia hemprichii* (Fig. 16.10)
 – Rhizome of only one form, covered with long, black bristles and bearing leaves all over. Root hairs inconspicuous .... *Enhalus acoroides* (Fig. 16.10)

9. Leaf blade terete (round) .... *Syringodium isoetifolium*
 – Leaf blade flat 10

10. Erect stem long bearing clusters of leaves at the end. Leaf apex rounded and coarsely toothed. Rhizome woody. .... *Thalassodendron ciliatum* (Fig. 16.10)
 – Erect stem short. Rhizome monopodial and not woody 11

11. Leaf blade 2 mm wide or less with three lateral veins running down each leaf .... *Halodule* 12
 – Leaf blade more than 3 mm wide and with nine or more lateral veins .... *Cymodocea* 13

12. Tip of leaf with three teeth, the middle one being blunt .... *Halodule uninervis* (Fig. 16.10)
 – Tip of leaf rounded; three teeth, but side teeth poorly developed .... *Halodule pinifolia*

13. Tip of leaf rounded to blunt with dense teeth, leaf 4–9 mm wide .... *Cymodocea serrulata* (Fig. 16.10)
 – Tip of leaf blunt (or faintly heart-shaped), leaf 2–4 mm wide. Short, erect shoot with leaf scars right around shoot .... *Cymodocea rotundata* (Fig. 16.10)

ADDITIONAL READING

den Hartog, C. (1970). The sea-grasses of the world. *Verhandl. der Koninklijke Nederlandse Akademie van Wetenschappen, Afd. Natuurkunde* **59**(1), 1–275.

Duke, N. C. (2006). 'Australia's Mangroves. The Authoritative Guide to Australia's Mangrove Plants.' (The University of Queensland and Norman C. Duke: Brisbane.)

Duke, N. C., Meynecke, J.-O., Dittmann, S., Ellison, A. M., Anger, K., Berger, U., Cannicci, S., Diele, K., Ewel, K. C., Field, C. D., Koedam, N., Lee, S. Y., Marchand, C., Nordhaus, I., and Dahdouh-Guebas, F. (2007). A world without mangroves? *Science* **317**, 41–42.

Jupiter, S. D., Potts, D. C., Phinn, S. R., and Duke, N. C. (2007). Natural and anthropogenic changes to mangrove distributions in the Pioneer River Estuary (Queensland, Australia). *Wetlands Ecology and Management* **15**, 51–62.

Larkum A. W. D., Orth, R., and Duarte, C. A. (2006). 'Seagrasses: Biology, Ecology and Conservation.' (Springer Verlag: Berlin.)

Lanyon, J. (1986). Seagrasses of the Great Barrier Reef. Special Publication Series No. 3. Great Barrier Reef Marine Park Authority, Townsville.

Limpus, C. J., and Limpus, D. J. (2000). Mangroves in the diet of *Chelonia mydas* in Queensland, Australia. *Marine Turtle Newsletter* **89**, 13–15.

Mumby, P. J., Edwards, A. J., Arias-Gonzalez, J. E., Lindeman, K. C., Blackwell, P. G., Gall, A., Gorczynska, M. I., Harborne, A. R., Pescod, C. L., Renken, H., Wabnitz, C. C. C., and Llewellyn, G. (2004). Mangroves enhance the biomass of coral reef fish communities in the Caribbean. *Nature* **427**, 533–536.

Tomlinson, P. B. (1994). 'The Botany of Mangroves.' 2nd edn. (Cambridge University Press: Cambridge.)

Waycott, M., McMahon, K., Mellors, J., Calldine, A., and Kleine, D. (2004) 'A Guide to Tropical Seagrasses of the Indo-West Pacific.' (James Cook University: Townsville.)

17. Sponges

J. N. A. Hooper

OVERVIEW

The Phylum Porifera is the most primitive of the multicellular animals, with a most ancient geological history. Porifera appeared early in the history of life on Earth, established in the Proterozoic, with the major Class Demospongiae first appearing in the Precambrian Ediacaran age (about 750 million years (My) ago). By the Middle Cambrian (about 500 My) demosponges were thriving, and today represent about 85% of all living species. During the massive radiation of life forms in the Ordovician (490–460 My), and for the next 100 million years or so, sponges formed extensive barrier and fringing reefs around the ancient continents and were the primary reef builders in these ancient oceans. Sponges declined as the primary reef-builders at the end of the Devonian (350 My), and today they are not significant reef-builders in shallow waters, unable to compete with the faster growing zooxanthella-bearing corals, although several hard-bodied (lithistid and hypercalcified) reef-building groups of sponges persist in modern day reefs. Although sponges are still a major component of modern-day coral reef ecosystems they are often overlooked by the curious naturalist because they are frequently hidden among the more prominent corals on the reef, or live in the less frequently visited deeper waters surrounding reefs. Nevertheless, in terms of their species diversity, they outnumber both the hard and soft corals combined. In some habitats they provide pivotal ecological services, such as significant contributions to coral reef primary productivity, filtration of the waste products and toxins from other reef organisms, and major recycling of calcium carbonate back into the reef system through bioerosion.

DIVERSITY

Sponges are predominantly marine, living from the intertidal to the abyssal zone, with a small number also found in freshwater habitats. Worldwide there are approximately 8500 species described in the literature that are considered to be 'valid', with about twice this number estimated for all oceans. Approximately 1500 species have been described among the Australian fauna so far, although an escalated collection effort over the past couple of decades, mainly from tropical and subtropical waters and spurred on by drug discovery from marine organisms, has discovered a fauna three times this size. This work leads to an estimated Australian fauna of at least 5000 species. So far only about 400 species are described in the literature for all Queensland waters, including the coast, the GBR and Coral Sea island territories. Extensive surveys over the past two decades, however, reveal that more than 2500 sponge species actually live here, with the majority remaining undescribed (new species).

Amongst the better known sponge faunas on the GBR are, not surprisingly, those from the most frequently visited localities in the vicinity of the major marine research stations, Heron and One Tree Islands in the Capricorn-Bunker Group off Gladstone (387 species), the vicinity of Orpheus and Palm Islands off Townsville (107 species), Low Isles off Port Douglas (134 species), and Lizard and the Direction Islands NE of Cairns (212 species known so far). We also know of other species-rich 'hotspots' in more remote areas on the GBR that are less frequently visited (and hence not necessarily biased by collection effort), including the Swain Reefs (304 species) at the southern end of the GBR, the Ribbon Reefs (204 species), and the Howick-Turtle Island group (210 species known so far) in the northern region of the GBR. In between these 'hotspots' is a variable mosaic of diversity and species richness, with the central region of the GBR generally less rich than either the northern or southern sectors. These observations also have some genetic support from phylogeographic analysis of rDNA ITS sequences of a widely distributed calcareous sponge, *Leucetta chagosensis* (see Fig. 17.4C), which shows clear genetic divergence between northern and southern GBR populations, both of which are genetically more closely related to Indonesian populations than they are to each other, suggesting weak connectivity between northern and southern regions and indicative of significant exogenous larval recruitment and colonisation from regions outside the GBR.

Although we now know the GBR contains a highly diverse sponge fauna, and that sometimes sponges occur in dense local populations ('sponge gardens'), there still remains a significant challenge to place these faunas into an international context through the processes of rigorous taxonomy. Only by this strategy can we accurately determine which of these species are unique/endemic to the GBR (or to a particular reef system within the GBR), and which are truly widespread and distributed over large (international) spatial scales. Like many marine invertebrate taxa, a number of sponge species have long been perceived to be widely distributed, ranging from the Red Sea to the central western Pacific islands (reported as 5–15% of regional faunas). This notion of cosmopolitanism is now gradually diminishing with increased application of molecular techniques at population levels, with the outcome being that many so-called widely distributed morphospecies may consist of several sibling species with high genetic diversity that is not, or only barely, manifested at the morphological level across their wide geographic ranges. This problem is exacerbated by the high plasticity of growth form renowned among Porifera, challenging even the most experienced taxonomists to differentiate 'regional variants' of widespread morphospecies. Estimates of sponge diversity, therefore, based on morphospecies, may be grossly underestimated.

DISTRIBUTION AND ABUNDANCE

Sponges may live in all types of coral reef habitats, but in reality they exhibit very patchy distributions, such that in one particular area they may form the dominant structural benthos, whereas in another adjacent area they may be practically absent. This is not unusual for marine invertebrates, where at small (local) spatial scales (i.e. encompassing different habitats within a single reef, up to groups of adjacent reefs tens of kilometres apart), spatial heterogeneity is common (in terms of both species diversity and abundance/biomass), and has been widely reported for sponges across all ocean basins. Many factors may significantly influence local sponge distributions. Terrestrial influences, such as freshwater input, turbidity, sedimentation, light penetration, nutrient levels, food particle size availability and so forth have been found to explain differences between sponge faunas in the lagoon, closer to the land, and those living on the outer reefs. Geomorphological differences between reefs may also markedly influence the composition and distribution of their resident sponge faunas, including factors such as microhabitat availability, the nature and quality of the substrate (coralline *v.* non-coralline, soft *v.* hard), aspect of the seabed, exposure to waves and currents, depth and other factors. In fact, adjacent reef systems (only tens of kilometres apart) have been reported to have as little as 15% similarity in their species compositions, with the presence or absence of particular niches (such as caves, a reef-flat, a lagoon, spurs and grooves) showing strong correlation with the presence or absence of particular species. Other factors that influence patchy sponge distributions include small scale random events, such as patterns and timing of arrival and survival of larvae and asexual propagules, effects of severe storm events on fragmentation and dispersal, and the history of (and changes to) current patterns and other barriers to larval/propagule dispersal. From our current understanding, sponges appear to have very limited sexual reproductive dispersal capabilities (an absence of any pelagic larval stage), with short larval lives (with a reported maximum of 72 hrs in the water column before settlement). Oviparous species, like the ubiquitous *Xestospongia testudinaria* (see Fig. 17.7*G*) that broadcast eggs and sperm into the water, last only several days at most. Conversely, at smaller spatial scales at least, it is thought (also now with some molecular support) that clonal dispersal and larval recruitment are predominant, and small scale endemism appears to be common among sponges, possibly through genetic isolation of remnant populations of once widespread species.

At larger spatial scales (i.e. from biogeographic provinces to ocean basins), factors such as historical changes to physical barriers and current patterns, climate change impacts, presence or absence of carbonate platforms have had large scale influences on present day sponge faunas. Unlike some other marine invertebrate phyla there are no apparent latitudinal gradients of sponge species richness from temperate to tropical waters (both have patchy mosaics of very rich faunas, on both sides of the Australian continent). The sponge fauna composition changes substantially, however, along the east coast of Australia, with subtropical-tropical faunal transition zones (or species turnover points) occurring in the vicinity of the Tweed River, Hervey Bay-Fraser Island, the Mackay-Townsville region, Cape Flattery north of Cooktown, and on the eastern side of Cape York.

REEF HABITATS

Once thought of as predominantly niche generalists, opportunistically scattered over reefs wherever larvae settled or propagules landed, sponges are increasingly recognised as being predominantly niche specialists with marked habitat preferences. Distinct species assemblages characterise particular habitats, although there are a number of ubiquitous species found vicariously throughout the reef. Some of the more prominent sponge habitats occurring on the GBR are described here.

Reef flats and rock pools

Of all the habitats, the reef flats and shallow parts of lagoons contain probably the most significant of the coral reef sponge faunas, at least in terms of their visibility and provision of ecological services. These are the phototrophic (or autotrophic) species, belonging primarily to two orders of Demospongiae, the Dictyoceratida and Haplosclerida, that derive most of their nutrition from the photosynthetic products of their resident symbiotic cyanobacteria, and in the process contribute significantly to overall net primary productivity of entire reef systems. Phototrophic species (which derive most of their energy from sunlight through photosynthetic symbionts, see Chapter 7) include representatives from several orders, with the common species on the GBR being *Phyllospongia papyracea*, *Carteriospongia foliascens*, *Strepsichordaia lendenfeldi*, *Haliclona cymaeformis*, *Cymbastela coralliophila* and *Lamellodysidea herbacea* (Fig. 17.1*A–H*). Some of these species can be found in very large populations, especially in the clear waters of outer reefs. These phototrophic sponges are also unusual in having (probably truly) widespread geographic distributions, appearing to be very similar on both sides of the continent, with some also common in the Indo-Malay archipelago and south-western Pacific islands, although no molecular study of these widespread populations has yet been attempted to test their alleged conspecificity.

Another highly diverse fauna on the reef flat and in the lagoon shallows are the coral rubble, under-rubble and boulder sponges, living in crowded, encrusting and sciaphilic communities. These sponges range from thin crusts of no more than several millimetres in thickness, competing with each other for space and other resources using an arsenal of chemicals (Box 17.1 and Fig. 17.2), to massive slimy sponges that bind the coral rubble together and form the paving substrate. Examples of these include: *Myrmekioderma granulata*, *Clathria aceratoobtusa*, *Leucetta microraphis*, *Aplysinella rhax*, *Neopetrosia exigua*, *Gelliodes fibulatus* and *Hyrtios erecta* (Fig. 17.3*A–H*, Fig. 17.4*A*).

Figure 17.1 Phototrophic sponge species living on coral reef flat and in coral rubble in shallow waters: *A*, *Collospongia auris*; *B*, *Lendenfeldia plicata*; *C*, *Cymbastela coralliophila*; *D*, *Carteriospongia foliascens*; *E*, *Strepsichordaia lendenfeldi*; *F*, *Phyllospongia papyracea*; *G*, *Haliclona cymaeformis*; *H*, *Lamellodysidea herbacea*. (Photos: J. Hooper.)

BOX 17.1 CHEMICAL DEFENCE

Sponges are among the most toxic of animals, with toxins derived from one or a combination of the sponge's own metabolic products, sequestered chemicals excreted by other reef organisms and filtered by sponges, and symbioses with a variety of microbial infloras. These toxins are thought to have a variety of biological functionality including repelling predators, out-competing other sessile reef organisms for space, controlling parasites and microbes, chemical recognition between host and resident in-faunas, and bioerosion of the substrate. Many of these chemicals have also demonstrated toxicity against human pathogens, cancers, parasites and other therapeutic properties of interest to the pharmaceutical industry, and over the past few decades the majority of new chemical structures and new classes of chemical compounds have been isolated from sponges, including species from the GBR.

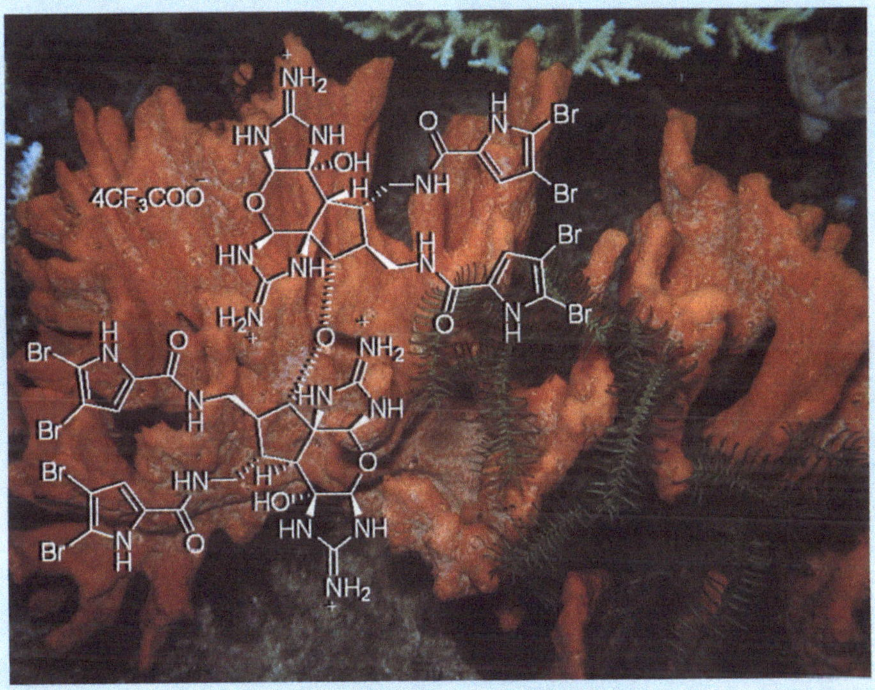

Figure 17.2 'Flabellazole A', a new P2X7 antagonist, isolated from *Stylissa flabellata* from the GBR. P2X7 antagonists may provide new treatment for inflammatory diseases. (Photo: J. Hooper; molecular structure T. Carroll, Natural Products Discovery Griffith University.)

Reef-builders, caves and crevices

In prehistoric reefs many hard-bodied sponges, like 'lithistids', 'stromatoporoids', 'chaetetids' and 'sphinctozoans' (all now considered grades of construction in body plan, and not taxonomic clades), were the major structural components of reefs. In modern day reefs hard-bodied sponges have only a minor structural role, although they are still considered to be of some importance in accreting coral skeletons in deeper waters where light is limiting and coral growth is rudimentary. Nevertheless, living representatives of these 'reef-building sponges' are still found on coral reefs and other deeper reefs, with their ancient body plans persisting over many millions of years. Some of these sponges (such as *Astrosclera willeyana*) have a solid calcitic skeleton (contributing to reef-building, analogous

Figure 17.3 Sponges of the reef flat and shallow lagoon, including the cryptic, coral rubble, and under-boulders communities: *A*, *Myrmekioderma granulate*; *B*, *Leucetta microraphis*; *C*, nudibranchs feeding on *Clathria* (*Microciona*) *aceratoobtusa* (photo: B. Rudman); *D*, *Stelletta* sp.; *E*, *Haliclona nematifera*; *F*, *Aplysinella rhax*; *G*, *Chelonaplysilla* sp.; *H*, *Gelliodes fibulatus*. (Photos: J. Hooper, except where noted.)

Figure 17.4 *A–B,* sponges of the reef flat and shallow lagoon, including the cryptic, coral rubble, and under-boulder communities: *A, Hyrtios erecta; B, Suberea ianthelliformis. C–H,* reef-builders and cave faunas: *C, Leucetta chagosensis; D, Levinella prolifera; E, Ulosa spongia; F, Soleniscus radovani; G, Astrosclera willeyana; H, Petrosia (Strongylophora) strongylata*. (Photos: J. Hooper.)

to the skeletons of modern hermatypic corals), as well as discrete siliceous spicules within the soft tissues, virtually identical to those seen in their fossil ancestors from the Lower Cretaceous (160 My) and Triassic (250 My) respectively. Others have only a solid skeleton composed of aragonitic crystals (a form of calcium carbonate) with no free spicules (e.g. *Vaceletia crypta*, with a continuous fossil record from the Middle Triassic, 245 My, to present). Others have a solid skeleton of either linked or rigid calcareous spicules (e.g. *Plectroninia hindei*, with a body plan known from the Mid Miocene, 23 My) or a rigid basal mass of calcite, and some (the so-called 'lithistids') have only siliceous skeletons composed of special spicules called desmas, ranging from entirely fused and forming a rock hard skeleton to loosely articulated rendering the body more flexible (e.g. *Theonella swinhoei*, with ancestors recorded from the Tertiary, 65 My). These so-called 'living fossil' sponges are usually found in shaded or dark habitats, such as in crevices, deep caves or under coral rubble, rarely in full light, and due to their hard coralline texture may be confused with corals by the novice. Similarly, caves and dark crevices are also home to a variety of soft-bodied demosponges, sometimes lacking pigments (e.g. *Petrosia (Stongyl-ophora) strongylata*), sometimes colourful (e.g. *Ulosa spongia*), and especially the multitude of frequently brightly coloured calcareous sponges living on the reef (e.g. *Leucetta chagosensis*, *Levinella prolifera* and *Soleniscus radovani*) (Fig. 17.4*C–H*).

Reef slope

Between the reef crest and the base of coral reefs, across the shelf of the GBR, occur large, predominantly heterotrophic sponges that feed on food particles and waste products filtering down from the coral reef above. On some reefs, particularly those closer to the coast, these faunas number several hundred species, which in some instances occur as large populations. Their morphologies are as diverse as the conditions they live under, such as flexible whips, fingers and fans adapted for coping with high currents (*Ianthella basta*, *Axos flabelliformis*, *Clathria (Thalysias) cervicornis*); soft tubes, vases and other shapes that predominate in silty, turbid water where inhalant and exhalant pores are located on different surfaces to prevent smothering (e.g. *Echinochalina (Protophlitaspongia) isaaci* and *Fascaplysinopsis reticulata*), and a number of ubiquitous, amorphous, bulbous, massive, spherical (and other shaped) forms that appear nearly anywhere they can settle and survive (e.g. *Cinachyrella schulzei*, *Stylissa massa*, *Stylissa carteri*, *Acanthella cavernosa*, *Amphimedon terpenensis*, *Pericharax heterorhaphis*, and *Agelas axifera*) (Figs 17.5*A–H*, 17.6*A–H*).

Deeper lagoon and inter-reef

Sponges living on the seabed in between the reefs are generally very different to those found on the coral reefs, living in high current, low light, turbid waters where they burrow into soft sandy and muddy sediments (e.g. *Oceanapia renieroides* and *Disyringia dissimilis*), in seagrass and *Halimeda* beds (e.g. *Oceanapia sagittaria*), or attached to hard objects on the sea floor (e.g. *Xestospongia testudinaria*, *Melophlus sarassinorum* and *Liosina paradoxa*) (Fig. 17.7*A*, *C*, *G*, *H*). Growth forms include elongate species with root-like tufts or bulbs for anchoring in soft sediments and long tubes to prevent smothering, flexible fans and whips in high current areas, and massive barrels and volcanoes attached to rock and coral outcrops. This fauna comprises a significant proportion of the 'benthos', providing important habitat for other marine species, ranging from aggregating fishes to numerous crustacean in-faunas. In a recent major survey across the length and breadth of the GBR, conducted under the auspices of the Cooperative Research Centre for the GBR (CRC Reef), approximately 1300 morphospecies of sponges were discovered from the inter-reef region, which is particularly susceptible to human impacts such as trawling (see also Chapter 6). Whereas a number of species living on the coral reef flats may be found on both sides of the Australian continent, these deeper water GBR lagoonal and inter-reef species appear to differ significantly in composition from the fauna found at similar latitudes on the west coast of the continent.

Reef bioeroders (sponges as 'parasites')

A special group of sponges are responsible for significant carbonate recycling on the reef, collectively termed excavating ('boring' or bioeroding) sponges (see also Chapter 8). They are responsible for extensive damage

Figure 17.5 Reef slope faunas: *A, Axos flabelliformis; B, Clathria (Thalysias) cervicornis; C, Ianthella basta; D, Echinochalina (Protophlitaspongia) isaaci; E, Fascaplysinopsis reticulata; F, Cinachyrella schulzei; G, Stylissa massa; H, Amphimedon terpenensis*. (Photos: J. Hooper.)

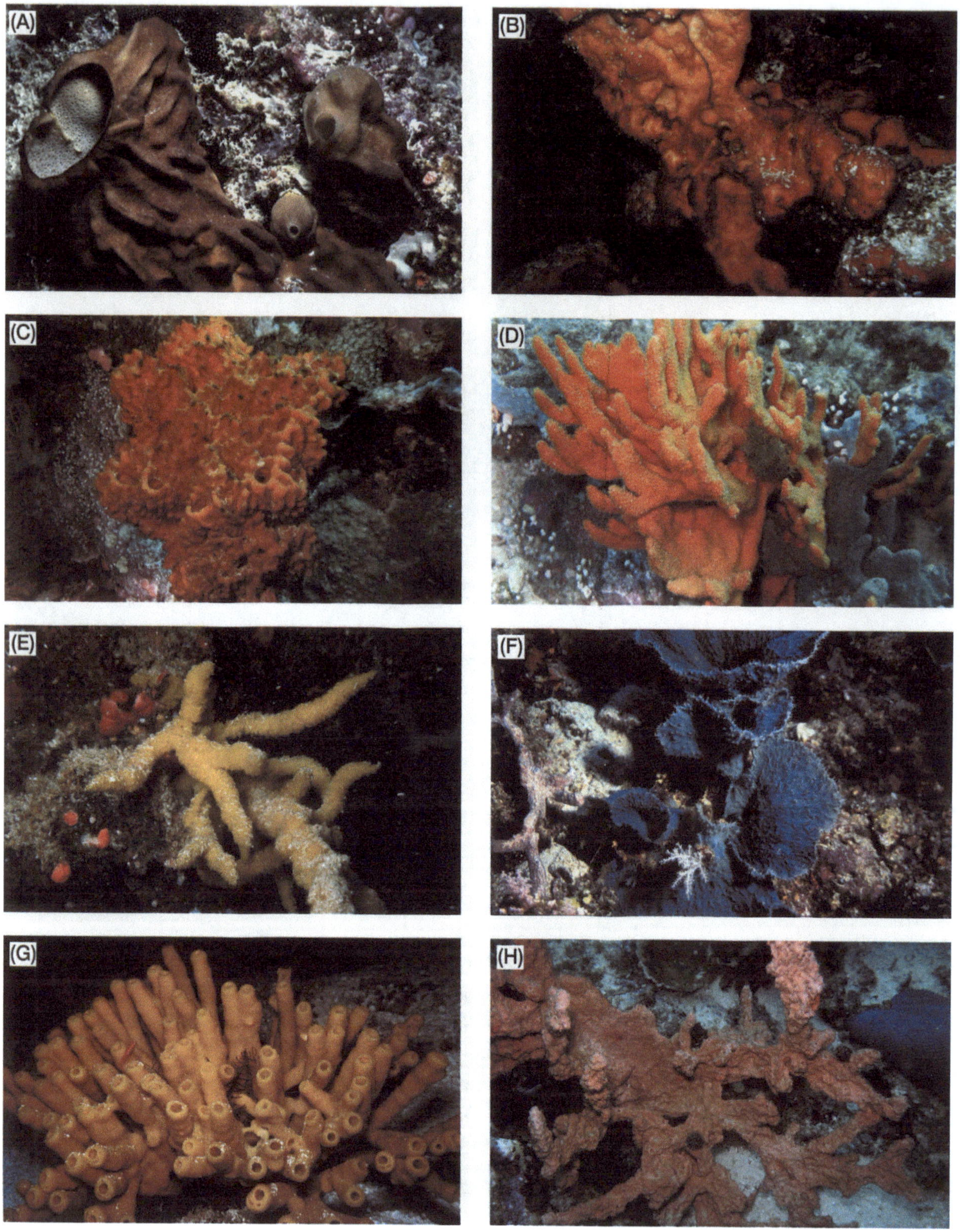

Figure 17.6 Reef slope faunas: *A, Pericharax heterorhaphis; B, Agelas axifera; C, Acanthella cavernosa; D, Reniochalina* sp. (photo: Chris Ireland); *E, Phycopsis fusiformis; F, Callyspongia aerazusa; G, Pipestela candelabra; H, Stylissa carteri*. (Photos: *A–C, E–H*, J. Hooper.)

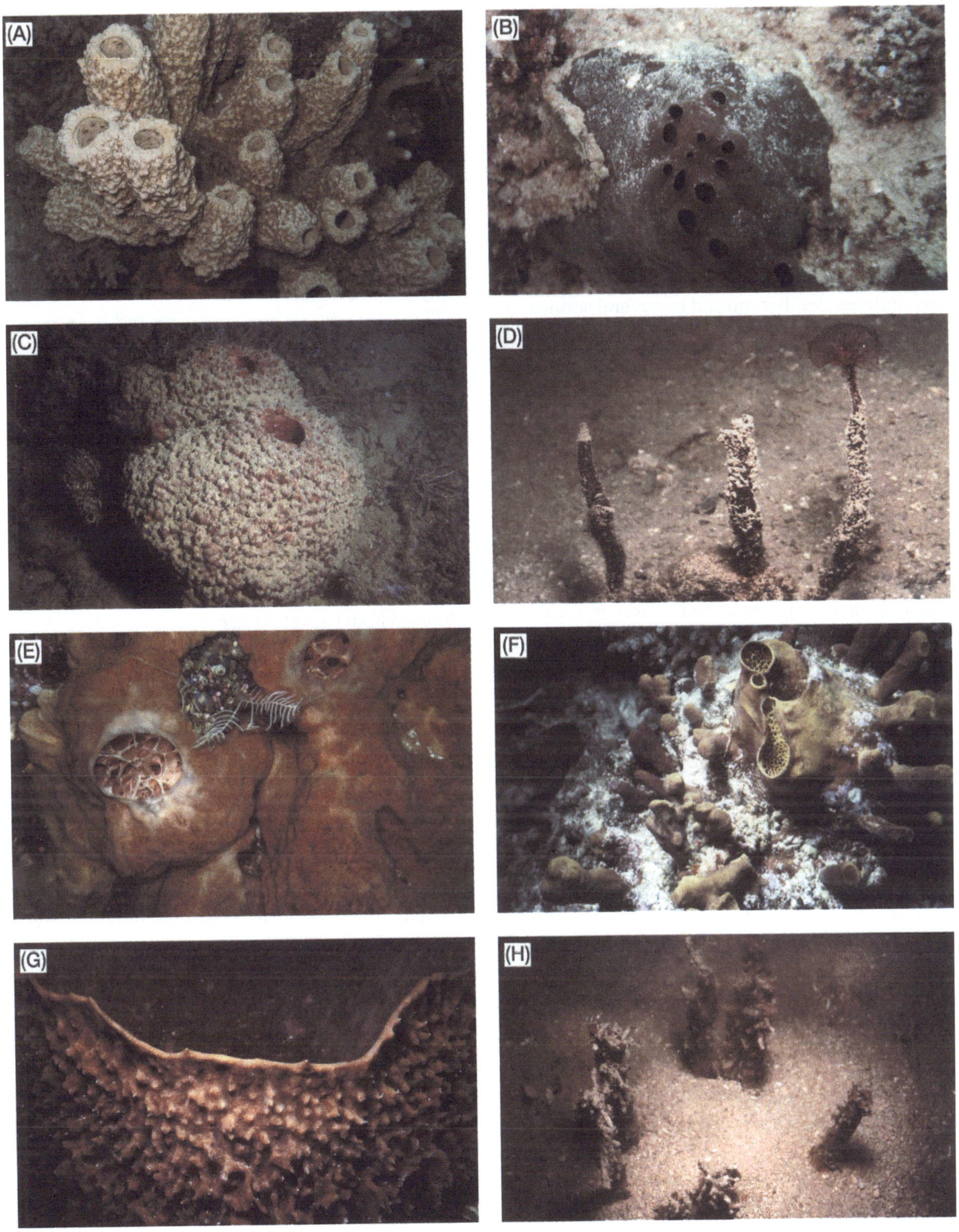

Figure 17.7 Deeper lagoon and inter-reef faunas, reef bioeroders and soft sediment faunas: *A, Liosina paradoxa*; *B, Spheciospongia vagabunda*; *C, Melophlus sarissinorum*; *D, Oceanapia sagittaria*; *E, Cliona* sp.; *F, Coelocarteria singaporensis*; *G, Xestospongia testudinaria*; *H, Oceanapia renieroides*. (Photos: J. Hooper.)

to hard and soft corals, as well as shellfish and other molluscs, and are among the most destructive internal bioeroding organisms of coral reefs both in terms of effects (such as weakening coral platforms and producing dead coral rubble). The rates of destruction by these organisms range up to 15 kg m^{-2} per year. Much of the damage caused to corals during storms has been attributed to weakening of basal structures by bioerosion (see also Chapter 8). An excavating mode of existence has been independently acquired by several sponge orders. Some of these (e.g. *Terpios*) simply overgrow coral at rapid rates, periodically resulting in extensive tracts of coral bleaching and the destruction of large tracts of coral. Others burrow into dead coral, eventually occupying the entire original coral head, with breathing tubes (fistules) protruding (e.g. *Coelocarteria singaporensis* and *Aka* sp.). The most significant of these are the Hadromerida ('clionaids') belonging to the families Clionaidae, Alectonidae and Spirastrellidae (e.g. *Cliona* sp., *Spheciospongia vagabunda* and *Cliona montiformis*) (Fig. 17.7*B*, *E*, *F*). Clionaids excavate chambers within the coral skeleton using a cellular process undertaken by special etching cells secreting acid phosphatase and lysosomal enzymes that dissolve organic matter and produce limestone chips that are physically liberated into the sea water via the sponge exhalant canal. Etching initially produces a cavity with sponge papillae protruding outside the coral (alpha stage), after which the external papillae fuse to produce a continuous sponge crust covering the coral (beta stage), eventually becoming massive and consuming the entire coral (gamma stage). Although clionaid sponges have the ability to invade living coral tissue and to survive direct contact with coral polyps, their ecological success may be largely due to their ability to undermine and erode the coral skeletal base, thus avoiding contact with the coral polyp defensive mucus and nematocysts.

Sponge body plans and classification

The Phylum Porifera is defined by their unique possession of chambers lined by a single layer of flagellated cells (choanocytes or collar cells) that actively beat to produce a unidirectional water current through the body, connected to the external water column by a system of differentiated inhalant and exhalant canals with

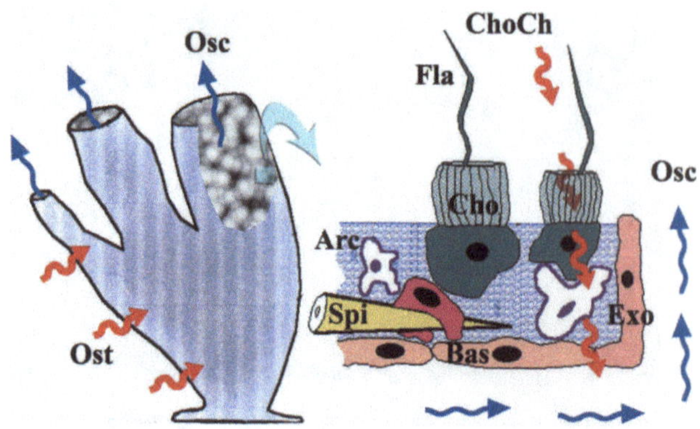

Figure 17.8 Diagrammatic sponge morphology: Arc, totipotent phagocytotic cells (archaeocytes); Bas, basipinacocytes lining internal aquiferous system; Cho, choanocytres or collar cells; ChoCh, choanocyte chamber (lined by choanocyte cells); Exo, exopinacocytes (lining exterior surfaces); Fla, flagellum on choanocytes; Ost, inhalant pores (ostia); Osc, exhalant pores (oscula); Spi, spicules (siliceous or calcitic depending on class). Red arrows (inhalant water current with food particles etc.); blue arrows (exhalant water current with waste products). (Modified from UCMP Berkeley.)

external pores (ostia and oscula, respectively), together forming a highly efficient aquiferous system that maintains basic metabolism and contributes significantly to reef filtration (Fig. 17.8). Sponges have a cellular grade of construction without true tissues, with their highly mobile populations of cells capable of differentiating into other cell types (totipotency), thus conferring a plasticity to growth form. The outer and inner layers of the sponge individuals are formed by special cells (exopinacocytes and basipinacocytes) that lack a basement membrane (except in some members of one group, the homoscleromorphs). The middle layer (or mesohyl) is variable among the orders of sponges but always includes motile cells and usually some skeletal material. Sponge skeletons are essentially divided into the ectosome ('skin') and choanosome (body containing the choanocyte chambers). Adult sponges are generally sessile, attached to the seabed or other substrate for most of their lives (although some are capable of slow movement), and most have motile larvae that swim or crawl away from their parent. Body plans range from simple (asconoid and syconoid, found in a few calcarean sponges) through to complex (leuconoid, occurring in most sponges), produced by varying degrees of infolding of the body wall and complexity of water canals throughout the sponge. Adults are asymmetrical or radially symmetrical, and have evolved an amazing range of growth forms best described as highly irregular and sometimes completely plastic, frequently altered by prevailing external conditions (currents, turbidity, salinity etc.). Sponges also have evolved an amazing array of colours, some linked to dietary carotenoid proteins and others with a photoprotection functionality.

The current classification of the Porifera is based primarily on features of the organic (collagen fibres and filaments) and inorganic skeletons (discrete and/or fused spicules composed of calcium carbonate or silicon dioxide), with some species also having a hypercalcified basal skeleton of solid limestone. The taxonomic scheme is primarily morphologically-based, and as complex as the diversity of sponges—the study of sponge taxonomy is not for the faint-hearted. Applying taxonomic principles to sponges is made even more difficult by the occurrence of frequent character losses, modifications and apparently convergent features reappearing within the classification. No attempt is made here to provide more than a very basic summary, with a list of further reading provided. There are three distinct classes of living sponges (plus a fourth extinct one): Calcarea, having calcitic spicules with three or four rays; Hexactinellida, with discrete and/or fused siliceous spicules, the larger ones three or six rayed; and Demospongiae, with siliceous spicules in many (but not all) species, and/or a fibrous skeleton, and spicules with one, two or four rays divided into megasclere and microsclere categories. Only Calcarea and Demospongiae have so far been recorded from the GBR, although Hexactinellida live in deeper waters on the continental slope and shelf adjacent to the GBR. An overview of the phylum, including a taxonomic revision and identification keys for approximately 25 orders, 127 families and 700 genera, has recently been undertaken but species-level identifications remain appallingly difficult, with few easily accessible taxonomic publications that would be useful to a non-specialist audience. Further useful reading is listed below, including general reading on sponge biology, sponge cell biology, a web checklist of the published Australian sponge fauna (including Queensland species) with keys to genera, a web list of all published sponge species worldwide, and sponge higher classification. The recent escalation of the molecular study of sponges will certainly have a major impact on our current ideas of the phylogeny and classification of Porifera, and to this end a Sponge Barcoding Project (based on a systematic use of molecular tools) has commenced and is also available on the web.

REPRODUCTION AND LIFE HISTORY

Sponges utilise a number of reproductive strategies based around their characteristic cellular totipotency. Asexual reproduction involves the production of propagules such as buds and fragments containing a sufficient number of cells from which complete sponges can develop. A few euryhaline species of the genus *Mycale* produce gemmule-like bodies, but true gemmules are restricted to freshwater sponges of the Order Haplosclerida. Most groups have considerable means of asexual propagation, such as fragmentation from storm events, which is thought to be an important mechanism for sponge recruitment, and all have extensive regenerative powers that appear to be vital for sustaining local populations. Sponges have sexes that are separate, or sequencially hermaphroditic, producing eggs and sperm at different times. Although there are no gonads or reproductive ducts, sexual reproduction involves the production of gametes by the choanocytes and totipotent archaeocytes, with fertilisation often (but not always) internal. Individuals release sperm externally via the exhalant current, whereas their oocytes reside in the incurrent aquiferous system to minimise self fertilisation. Sperm are engulfed by choanocytes, which become amoeboid, travelling to and transferring them to the oocytes. Cleavage leads to a solid steroblastula or hollow coeloblastula, with internally brooded, viviparous embryonic development in many cases, and larvae leaving the parent for dispersal. Other sponges are oviparous, with females shedding their eggs externally as zygotes or early embryo stages, rarely as unfertilised oocytes, although the details of embryology still remain unknown for most species. Other forms of development have also been recorded, such as elimination of the free-swimming larval stage and embryos brooded in the maternal sponge before being expelled as young adults. Eight different larval types are known but few of these have been adequately investigated. Most embryos develop into free-swimming (lecithotrophic) or demersal crawling larvae, ciliated to a greater or lesser extent, 0.05–5.00 mm long, with a brief planktonic phase, short longevity (maximum of 72 hrs recorded), and, unlike most marine invertebrates, have no planktotrophic stage.

Release of propagules (gametes, zygotes or embryos) is asynchronous in viviparous species but highly synchronous in oviparous sponges, triggered by factors such as temperature and lunar cycle. A prominent member of the GBR sponge community, *Xestospongia testudinaria,* is oviparous and broadcasts eggs in spawning events that were synchronised among populations of the same species, with timing found to be correlated with the lunar cycle. Molecular studies of individuals in local sponge populations show that most have high levels of genetic variability, not high genetic relatedness as would be expected if asexual recruitment was predominant, with some evidence that both asexual and sexual propagules are important for population structure, whereby sponge fragments that disperse and reattach may contain incubated sexual propagules.

Larval settlement and metamorphosis is thought to be influenced by a variety of environmental stimuli (such as light, gravity, physical and chemical features of the substrate), with the former best studied to date. There are examples of both photonegative and photopositive responses among the phylum. Larval competence (the threshold and duration of larval maturity required for settlement) is not thought to be as important for sponges as for many invertebrates since the high cellular totipotency allows fragmented larvae to attach unselectively. Growth rates, regenerative abilities after damage, and longevity is still poorly understood, but what little is known to date demonstrates that these vary considerably across groups of sponges and the habitats they occupy. Using various direct (C_{14}) and indirect measurements (e.g. growth rate extrapolation indices), some species are known to reproduce and die in less than one year (such as some soft-bodied *Haliclona* spp.), or are highly seasonal in their growth, biomass and ultimately survival (*Chondrilla australiensis*); some species live for many decades (*Aplysilla* sp.), to over 400–500 years (*Astrosclera willeyana*) and it is claimed that sponges belonging to the hexactinellid family Rosellidae living in Antarctica are among the oldest living animals on the planet, with individuals estimated at 1515 years old.

FEEDING

Sponges filter sea water to eat, exchange gases and excrete waste products. Filtration is an active process involving choanocytes lining chambers. Each choanocyte has a central flagellum that actively beats to create a water current, surrounded by a collar of cilia that traps food particles such as plankton and bacteria, as well as detritus. A water current containing food enters the sponge through an osculum and is initially filtered through a series of sieve-like pores (diminishing in size), finally ending up at the collar cells. Food particles are actively carried across the cell wall, engulfed by archaeocytes, and are transferred throughout the mesohyl to other cells. Filtered water leaves the sponge via the exhalant canal system. Unlike most multicellular animals, digestion and excretion of waste products occurs within cells, not within any common body cavity. There may be 7000–18 000 choanocyte chambers per cubic millimetre of sponge, and each chamber may pump approximately 1200 times its own volume of water per day. Thus, a sponge is capable of pumping around 10 times its body volume each hour, making them the most efficient vacuum cleaners of the sea. Some sponges, particularly those growing on coral reef flats, also have a unique symbiosis with cyanobacteria, providing the sponge with nutrients derived from photosynthesis to supplement those obtained by the sponge from normal filter feeding activities (phototrophy or autotrophy). These extra nutrients greatly augment sponge growth rates and competitive ability in coral reef systems. There are also often huge populations of bacteria and Archaea living within sponge cells and/or within the sponge mesohyl (hence the term 'sponge hotels'), with

BOX 17.2 SPONGE FARMING

Sponges have been used for cosmetic, bath, or industrial applications since early Grecian times (one early reference is in Homer's *The Iliad*). However, modern supply, which is predominantly from wild harvest fisheries, is unable to meet a well-established global demand. This shortfall presents an opportunity for sponge production through in-sea aquaculture. Sponges grow easily from cuttings, on lines or mesh panels suspended in the water column. Research has shown that commercially viable sponge farming can be achieved within a sustainable environmental footprint using basic infrastructure, and this opportunity is currently being explored by remote coastal Aboriginal communities in the GBR region and elsewhere in the south Pacific.

There is further opportunity to expand the species targeted for sponge farming to include those that elicit promising bioactive compounds. To date, many of these have not progressed in drug development due to the lack of reliable supply. Bulk wild collection is not suitable because the compounds are often produced in low yields, and their structural complexity makes them initially difficult to synthesise. Access to large quantities of sponges is necessary, and development of sponge aquaculture has provided a production option. While this opportunity is potentially quite lucrative, it is also high risk and at best transient for any one chemical entity. While aquaculture has already played an important role in supply for drug development research and proof of concept, history shows that industry prefers synthesis for global market supply, often based on a variation and simplification of the natural molecule. (Libby Evans-Illidge, Carsten Wolff & Alan Duckworth, AIMS.)

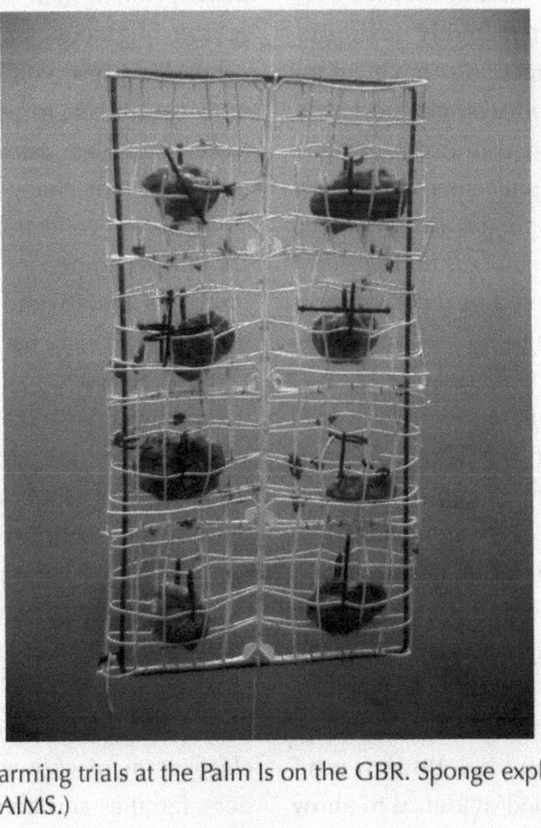

Figure 17.9 Sponge farming trials at the Palm Is on the GBR. Sponge explants being grown in mesh panels. (Photo: AIMS.)

which the sponge cells interact at various levels (from predation to commensalism). In addition to nutrients acquired through filter feeding activities sponges may ingest a myriad of toxic chemicals excreted by other plants and animals from the coral reefs above, which they modify (sequester) and reuse for their own purposes. The combination of chemicals produced by normal sponge metabolism, those sequestered from the seawater, and those produced by or in combination with the resident microbial populations makes sponges among the most toxic of all life forms, and hence of great interest to the pharmaceutical industry (Box 17.1 and Fig. 17.2).

PREDATION AND DEFENCE

Sponges are most unappetising by human standards due to a combination of high toxicity and a generally low ratio of soft tissue to mineral skeleton. The ability to digest and modify the waste products and chemicals produced by other organisms which live in, on or near sponges may at least partly account for their diverse, frequently novel and often highly toxic biochemistry. Nevertheless, sponges do have many recorded predators such as molluscs, echinoderms, fishes and turtles. On the GBR and in north-western Australia, sea cucumbers (*Synaptula* species) are frequently seen congregating on sponges (in particular branching and lamellate *Haliclona*, *Axos* and *Ianthella* species), feeding on the mucus exudate. Nudibranchs are also active feeders on sponge mucus and collagen, and sometimes these molluscs are quite specific as to the species or genus of sponge upon which they prey. Nudibranchs also ingest the sponge's toxic chemicals, concentrating, modifying and reusing (sequestering) them for their own chemical protection. The predators are often the same colour as the sponge, having ingested the sponge's characteristically brightly coloured carotenoid pigments. Other documented predators of sponges on the GBR include green and hawksbill turtles, many species of grazing fishes and asteroid, ophiuroid and holothurian echinoderms. There is now some good evidence to show that sponges also use their extensive arsenal of chemicals as both offensive and defensive weapons, such as repelling predators, deterring parasites and competing for space, and that the concentration, toxicity and/or secondary modification of particular compounds may vary seasonally and in response to predation intensity.

PRESERVATION AND IDENTIFICATION

An information sheet on collecting, preserving, histological preparation and lists of characters used for identification of sponges can be found on the web (see Additional reading). Collection of sponges within the Marine Park requires a permit and should also be undertaken with care (both for the sponge and the collector), owing to the often fragile nature of many specimens that disintegrate upon collection, and the sharp spicules and toxic mucus/chemicals that may injure the collector. Underwater and/or on-deck photographs of living specimens are highly recommended given that body shape and colouration may change dramatically following preservation. Freezing specimens prior to their preservation may be useful to fix soluble pigments in colourful species. Sponges are generally preserved separately in 70–80% ethanol, with care taken to prevent leaching of pigments between samples, particularly those with aerophobic pigments that change from yellow to blue and may stain entire collections. Formaldehyde preservative should be avoided for most sponges although it is used briefly as a fixative for calcareans, and air dried specimens are virtually useless for taxonomic identification. Other specialised fixatives include gluteraldehyde for detailed cellular ultrastructure studies (do not freeze), and 100% ethanol, DMSO or laboratory grade silica gel for storage of DNA samples. Routine histology is required for identification, and includes nitric acid or chlorine bleach digestion of the silica and calcitic spicule skeletons, respectively. Thin sections of the whole skeleton can be hand or microtome cut to include details of the ectosome and choanosome. Staining sections for different cellular elements and scanning electron microscopy are now widely used. Keys to orders, families and genera (see Additional reading) are largely based on features of the inorganic and organic skeletons.

ADDITIONAL READING

ABIF-Fauna. (2004). Australian Faunal Directory. Available at: http://www.deh.gov.au/biodiversity/abrs/online-resources/fauna/afd/group.html [Verified 3 March 2008].

Berquist, P. R. (1978). 'Sponges.' (Hutchinson:London.)

CRC Reef. (2006). Reef Futures. Great Barrier Reef Seabed Biodiversity Project. Available at http://www.reeffutures.org/topics/biodiversity/seafloor_maps.cfm [Verified 3 March 2008].

Hooper, J. N. A. (2000). Sponguide: Guide to sponge collection and identification. Available at http://www.qm.qld.gov.au/organisation/sections/SessileMarineInvertebrates/index.asp [Verified 3 March 2008].

Hooper, J. N. A., and Soest, R. W. M. Van (2002). 'Systema Porifera: A Guide to the Classification of Sponges. Vols 1 and 2.' (Kluwer Academic/Plenum Publishers: New York.) [For keys to families, genera and species.]

Simpson, T. L. (1984). 'The Cell Biology of Sponges.' (Springer-Verlag: New York.)

Soest, R. W. M. Van (Ed.) (2006). World Porifera Database. Available at http://www.vliz.be/vmdcdata/porifera/ [Verified 3 March 2008].

Wörheide, G. (Ed.) (2006). The Sponge Barcoding Project. Available at http://www.spongebarcoding.org/ [Verified 3 March 2008].

18. Pelagic Cnidaria and Ctenophora

L. Gershwin & M. J. Kingsford

The jellyfishes are conspicuous, but poorly understood, members of the Great Barrier Reef (GBR) fauna. Much attention has been given over the past 50 years to two forms in particular, the so-called 'box jellyfishes' and 'Irukandjis', both of which are highly dangerous and are known to kill humans. However, the dangerous species comprise only a small fraction of the jellyfishes that make their home in the GBR region.

Jellyfishes in general are among the most intriguing of animals, because they tantalise the child in all of us with their strange shapes, often bright colours, and sometimes flashing lights, and yet, even the milder stinging varieties seemingly represent the forbidden signal: DO NOT TOUCH! (Box 18.1.)

While appearing to be simple creatures consisting of no more than mucus and stinging cells, jellyfish have great potential to alter ecosystems through predation and uptake and excretion of nutrients. The medusa stage is the most obvious and often the most voracious part of the life cycle; however, the polyp stage represents the 'seed bank' of the species, making most jellyfish species incredibly difficult, if not impossible, to eradicate.

Several well-studied cases exist of jellyfish species being introduced to exotic environments and

BOX 18.1 DID YOU KNOW. . . .

- Box jellyfish are found primarily close to the coast, but Irukandji jellyfish are found on reefs and islands as well as along the coast.
- There are well over 100 species of jellyfishes that make their home in the GBR region, and all but about 10 are regarded as harmless.
- At least 10 species of jellyfish are believed to cause Irukandji syndrome, ranging from as far north as the United Kingdom and as far south as Victoria.
- All jellyfishes are animals with no bones, no brain, no heart, and no lungs or gills.
- Box jellyfish and Irukandjis have well developed eyes with lenses, retinas, and corneas, just like our eyes, and we know experimentally that they can see, even without a brain!

causing great harm to the ecosystem in the process (e.g. *Mnemiopsis* in the Black Sea and *Phyllorhiza* in the Gulf of Mexico). In other cases, serious impacts on the local biota have been caused by extreme concentrations of native species (e.g. blooms of *Chrysaora* in the Gulf of Alaska and *Aurelia* in California). Jellyfish can also be a food source for many species, such as fishes and turtles, and they are heavily fished in some parts of the world (e.g. China).

TAXONOMIC OVERVIEW

Jellyfishes of the GBR fall into two phyla, the Cnidaria, or medusae, and the Ctenophora, or comb jellies. Most zoology textbooks give a good account of the classification and morphological differences between the two phyla. Other gelatinous zooplankton include salps, or pelagic tunicates in the Phylum Urochordata; these are not dealt with in this chapter (see Chapter 14).

Cnidaria is the more familiar of the two phyla, comprising numerous quite different forms, including:

- *Hydroid jellyfishes* (Class Hydrozoa, Figs 18.1, 18.2). Hydroids are characterised by having small, clear bodies with a shelf-like velum, gonads on the radial canals or manubrium; and they lack true oral arms (e.g. *Obelia, Aequorea, Craspedacusta* and *Physalia*).
- *True jellyfishes* (Class Scyphozoa, Figs 18.3, 18.4). These are characterised by having usually large, fleshy, colourful bodies, a greatly reduced velum, and gonads in internal pockets; the Rhizostomeae (blubbers) have eight oral arms and they lack marginal tentacles, whereas the Semeostomeae (sea nettles, moon jellies and their kin) have four oral arms and well developed marginal tentacles (e.g. *Chrysaora* and *Aurelia*). *Cassiopea* spp. are benthic rhizostomes that spend most of their time upside down, that is, lying with the umbrella down and oral arms up, gently pulsating water over their algal symbionts.
- *Box jellyfishes* (Class Cubozoa, Figs 18.5, 18.7). These are characterised by a box-shaped body, with eyes in cavities within the lower edge of the bell wall, and a gelatinous pedalium that emerges from each corner; the Carybdeida (Irukandjis and jumbles) have a single tentacle on each unbranched pedalium, whereas the Chirodropida (box jellies) have branched pedalia, with each branch leading to a tentacle (e.g. *Carybdea, Chironex* and *Chiropsalmus*).
- *Benthic jellyfishes* (Class Staurozoa). These less familiar jellyfishes are characterised by a stalked body, flared out into an 8-rayed flower-like form (e.g. *Haliclystus, Manania and Lipkea*)

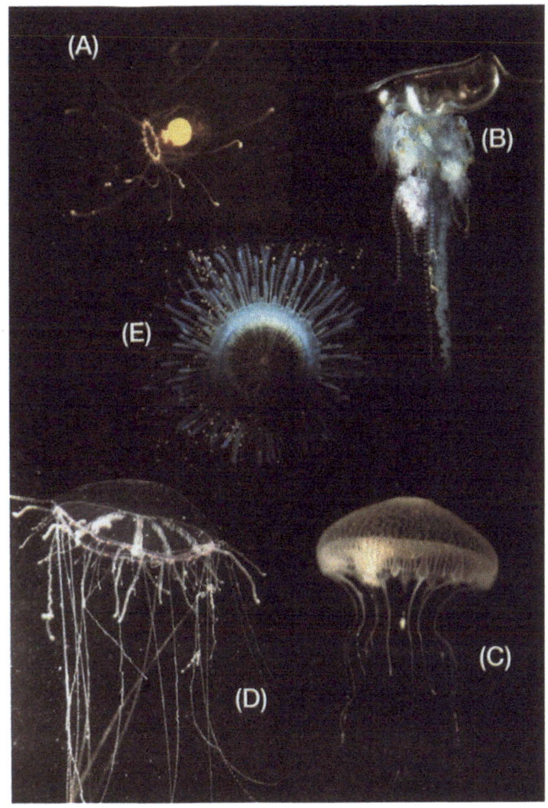

Figure 18.1 Some hydrozoans from the GBR. Clockwise from upper left: *A, Turritopsis lata*; *B, Physalia utriculus*; *C, Zygocanna* sp.; *D, Olindias* sp.; *E, Porpita porpita*. (Photos: L. Gershwin.)

The other jellyfish Phylum, Ctenophora (Figs 18.8, 18.9), includes sea gooseberries, sea walnuts, and Venus's girdles, none of which are harmful to humans. Ctenophores are divided into two classes: Nuda, which comprises only the genera *Beroe* and *Neis*, both found in Australia, which lack tentacles, and Tentaculata, which comprises most of the pelagic and all of the benthic forms, characterised by having two tentacles,

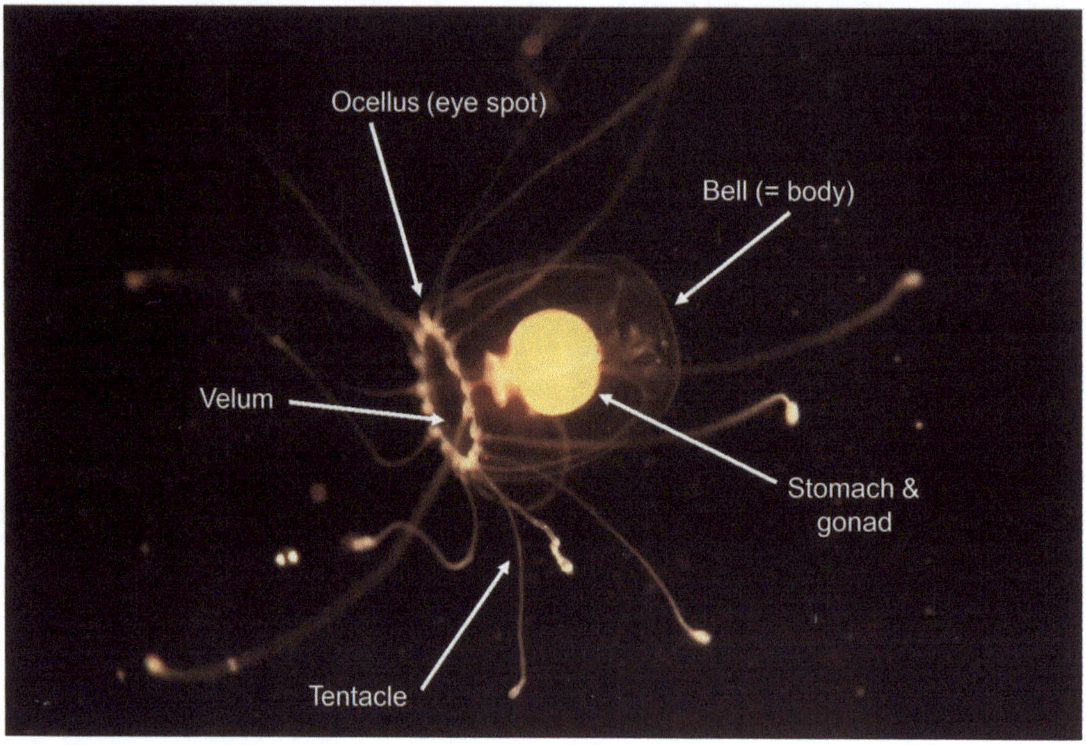

Figure 18.2 Features of hydrozoan jellyfish. (Photo: L. Gershwin.)

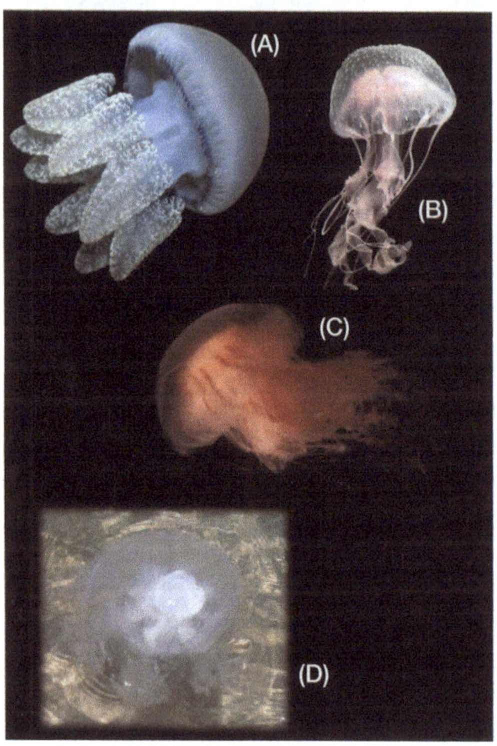

Figure 18.3 Some scyphozoans from the GBR. Clockwise from upper left: *A*, *Catostylus* sp.; *B*, *Pelagia* sp.; *C*, *Cyanea* sp.; *D*, *Netrostoma nuda*. (Photos: L. Gershwin.)

which may be secondarily lost. Some of the more familiar forms on the GBR include:

- Order Cydippida (*Pleurobrachia, Euplokamis*), the members of which look like small grapes with two tentacles (Fig. 18.8*A*);
- Order Lobata (*Bolinopsis, Ocyropsis*), which are about the size and shape of an egg with two large lobes (Fig. 18.8*C, D*);
- Order Cestida (*Cestum, Velamen*), which look like ribbons, and
- Order Platyctena (*Coeloplana, Ctenoplana*), which look like a creeping, gliding, or swimming flatworm with two tentacles.

Jellyfishes are an ancient group, found in the fossil record before the Cambrian explosion; credible examples of fossil jellyfishes from the famous Ediacaran fauna date back about 585 million years, and appear relatively unchanged through the eons. Many interpretable fossils exist from the Cambrian and scattered more recent periods. Even the box jellyfishes, which are considered 'the pinnacle of development' among

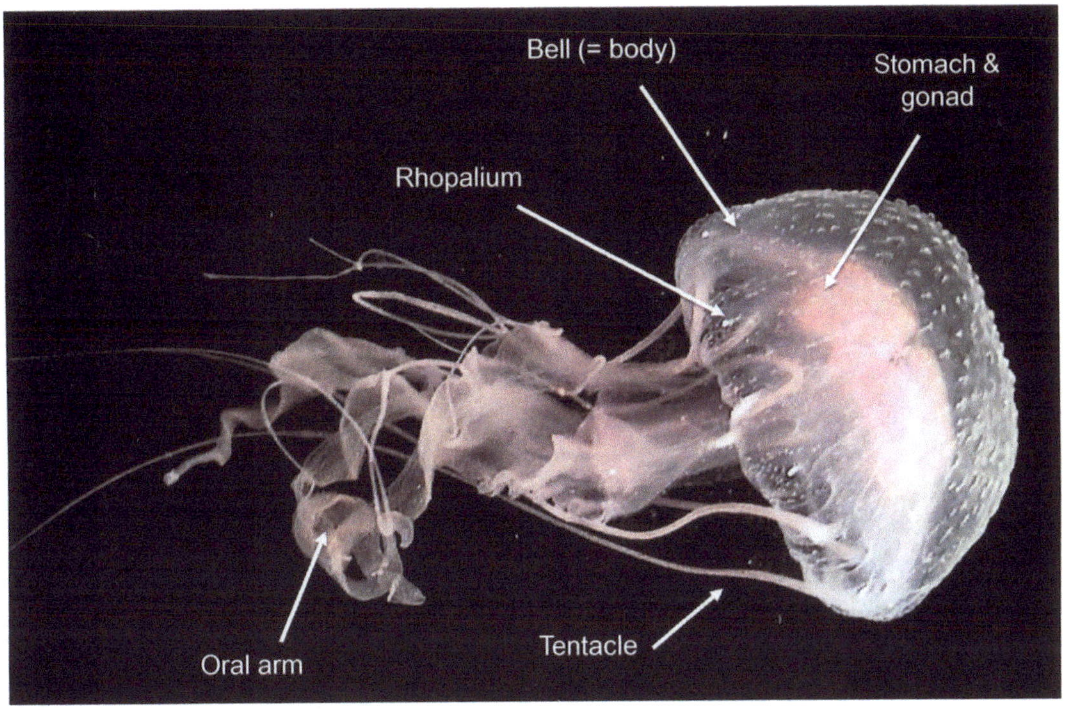

Figure 18.4 Features of scyphozoan jellyfish. (Photo: L. Gershwin.)

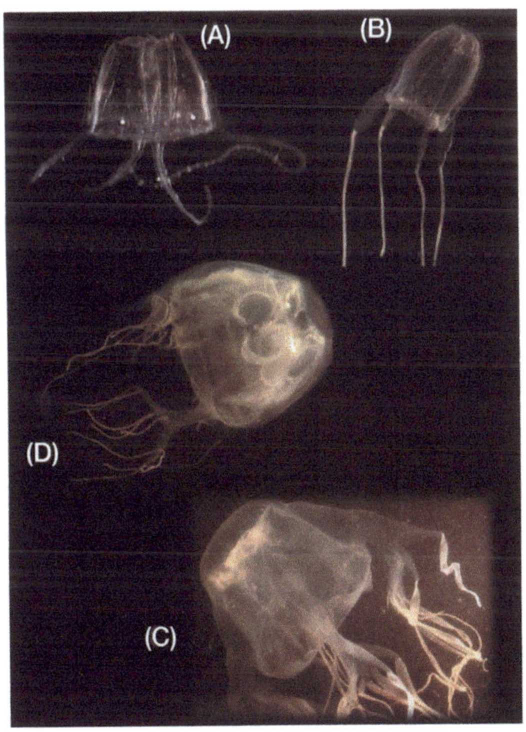

Figure 18.5 Some cubozoans from the GBR. Clockwise from upper left: *A, Carukia barnesi; B, Malo kingi; C, Chironex fleckeri; D, Chiropsella bronzie.* (Photos: L. Gershwin.)

jellyfishes, were fully developed by the Pennsylvanian (about 300 Mya).

Currently, well over a hundred jellyfish species are recorded from the GBR region. Many of these species, however, are still awaiting formal classification. Most are tiny and inconspicuous hydromedusae, presenting no serious medical threat, giving a 'sea lice' sting at the very most. However, they may at times occur in such dense aggregations that visibility may become severely diminished and even minor stings become annoying if in large number. Furthermore, Irukandji jellyfishes are often found aggregating with dense swarms of salps (pelagic tunicates, see Chapter 14) and hydromedusae, so it is advisable to take extra safety precautions or not enter the water on days when there are dense gelatinous zooplankton blooms.

BIOLOGY AND ECOLOGY

The cnidarian jellyfishes typically have a complex (metagenic) life cycle wherein the medusa (sexual jellyfish stage) alternates with a sessile polyp (asexual hydra stage, Figs 18.10, 18.11). In most cases, the jellyfish are either male or female, and spawn freely into the water

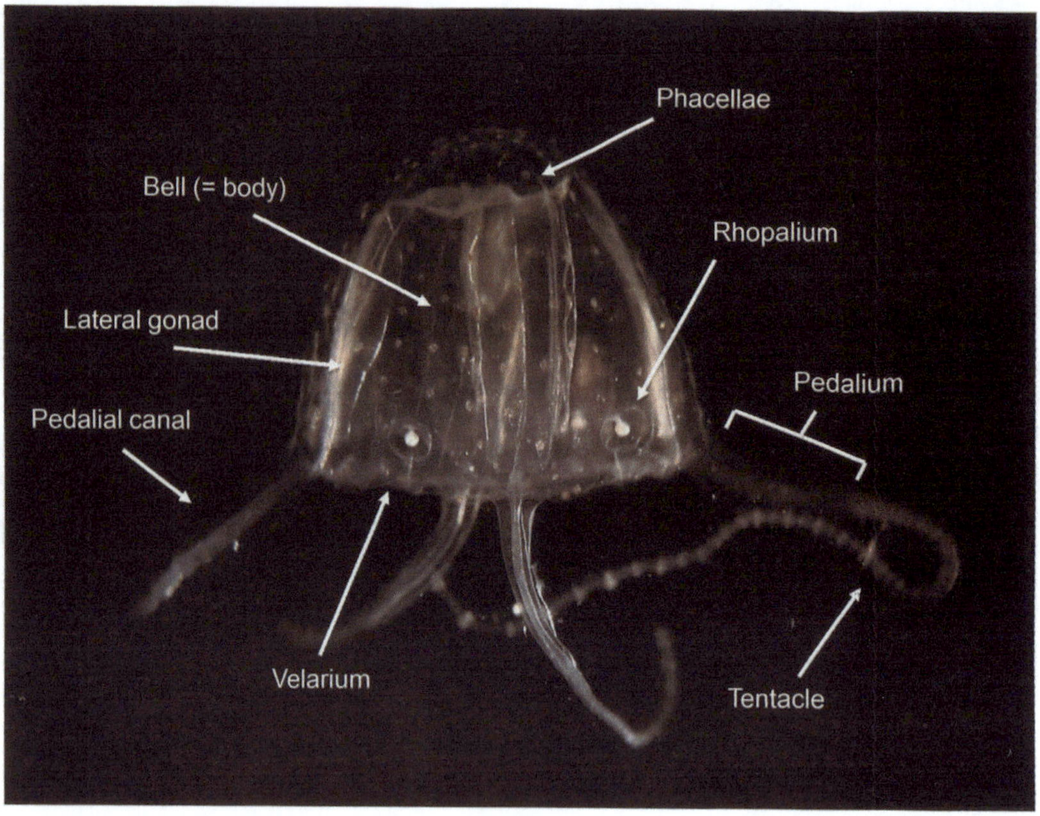

Figure 18.6 Features of cubozoan jellyfish. (Photo: L. Gershwin.)

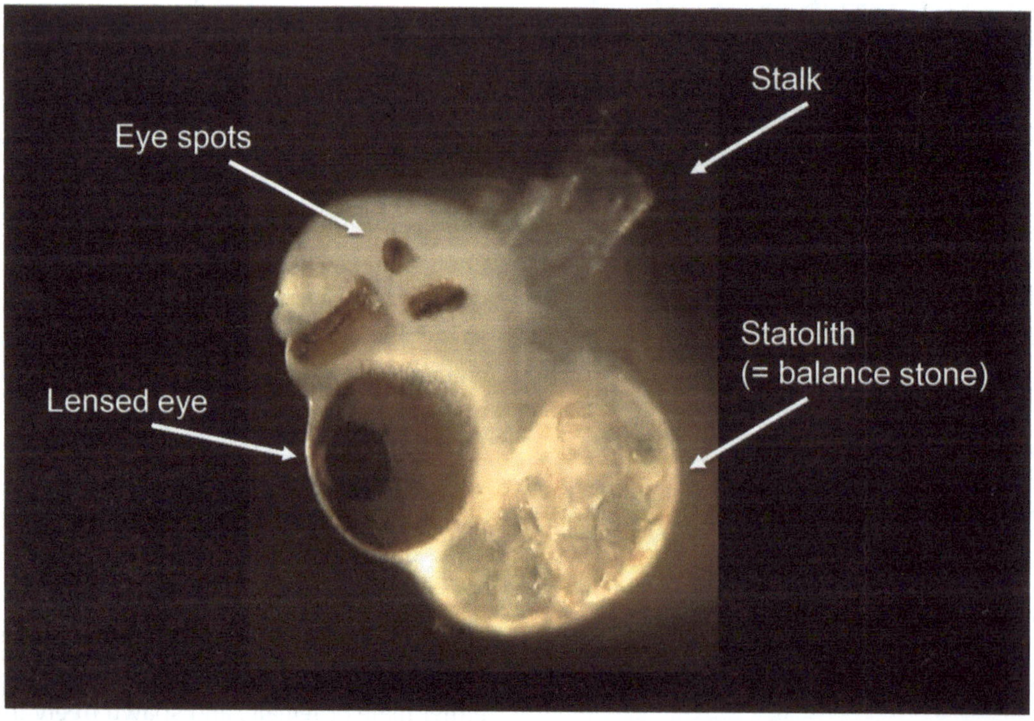

Figure 18.7 Features of the cubozoan rhopalium. (Photo: L. Gershwin.)

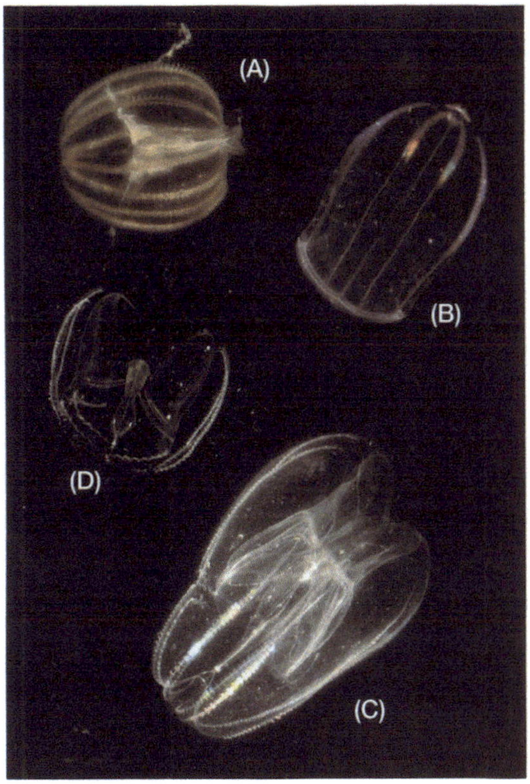

Figure 18.8 Some ctenophores from the GBR. Clockwise from upper left: *A, Pleurobrachia* sp.; *B, Beroe* sp.; *C, Bolinopsis* sp.; *D, Ocyropsis* sp. (Photos: L. Gershwin.)

column, though fertilisation is sometimes internal. The fertilised embryo grows into a 'planula larva' stage, a tiny ciliated teardrop-shaped creature, which seeks a suitable place to settle and transform into a polyp. The polyp produces many clones, through asexual reproduction, and sometimes environmentally resistant cysts (podocysts). When the conditions are right, the benthic forms undergo a metamorphosis to produce many tiny juvenile jellyfish. The baby jellyfish are voracious feeders on a diversity of plankton and generally grow very fast. For most of the GBR species, life cycles are largely assumed through knowledge of closely related species from other regions (Figs 18.10, 18.11).

The life cycle of most ctenophores remains unknown, but most are thought to be simultaneous hermaphrodites, bearing both sperm and eggs at the same time. Ctenophores lack a polyp stage, and instead develop directly from the fertilised egg into a larva. Ctenophores often occur in dense aggregations, which may facilitate external fertilisation. Most ctenophores are carnivorous, and like their cnidarian cousins, may decimate available food sources.

Because jellyfishes are pelagic, they occur throughout the GBR region, though some have restricted

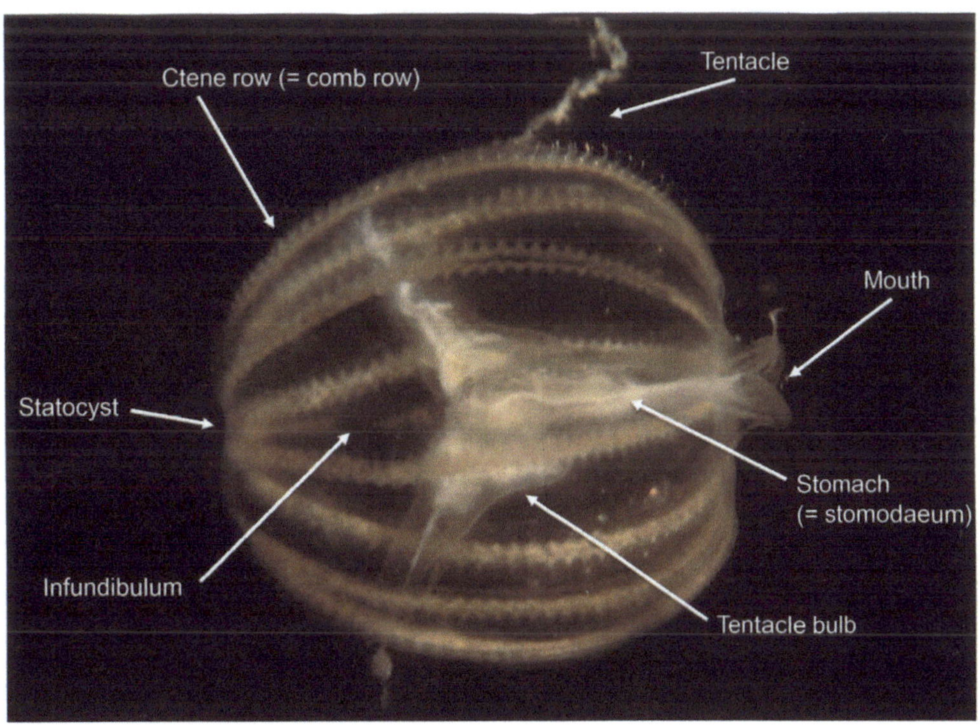

Figure 18.9 Features of ctenophores. (Photo: L. Gershwin.)

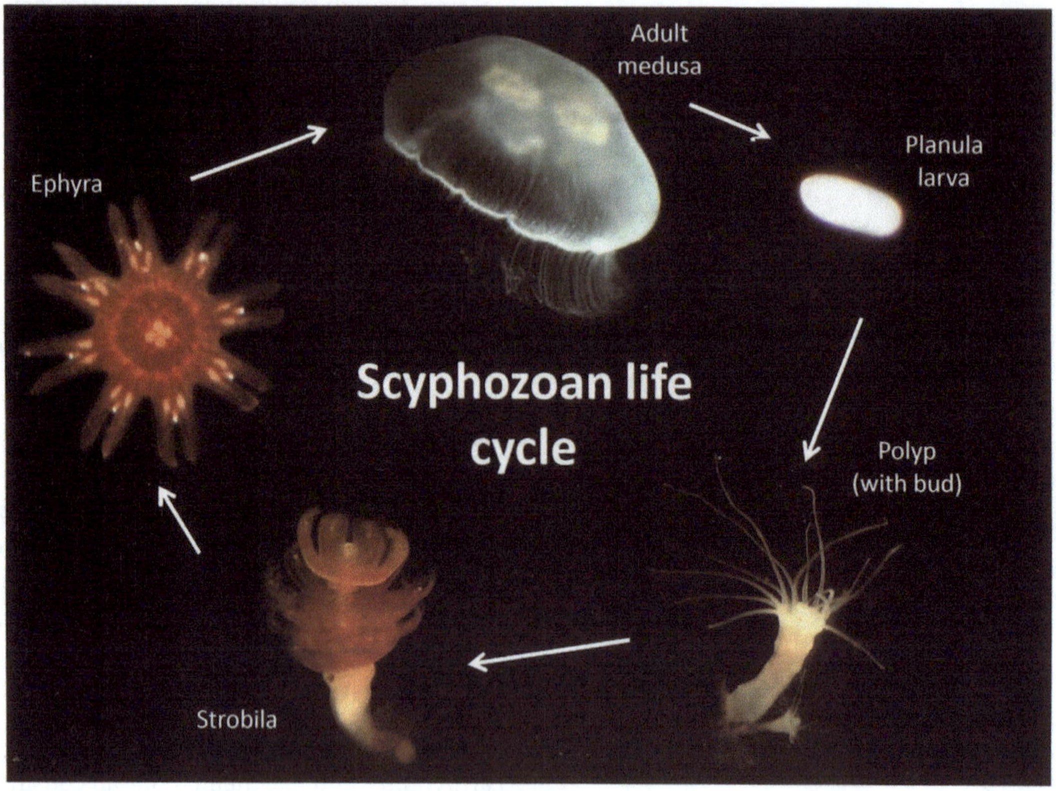

Figure 18.10 Life history of a scyphozoan jellyfish, *Aurelia* sp. (stages from multiple taxa shown). (Photo: L. Gershwin.)

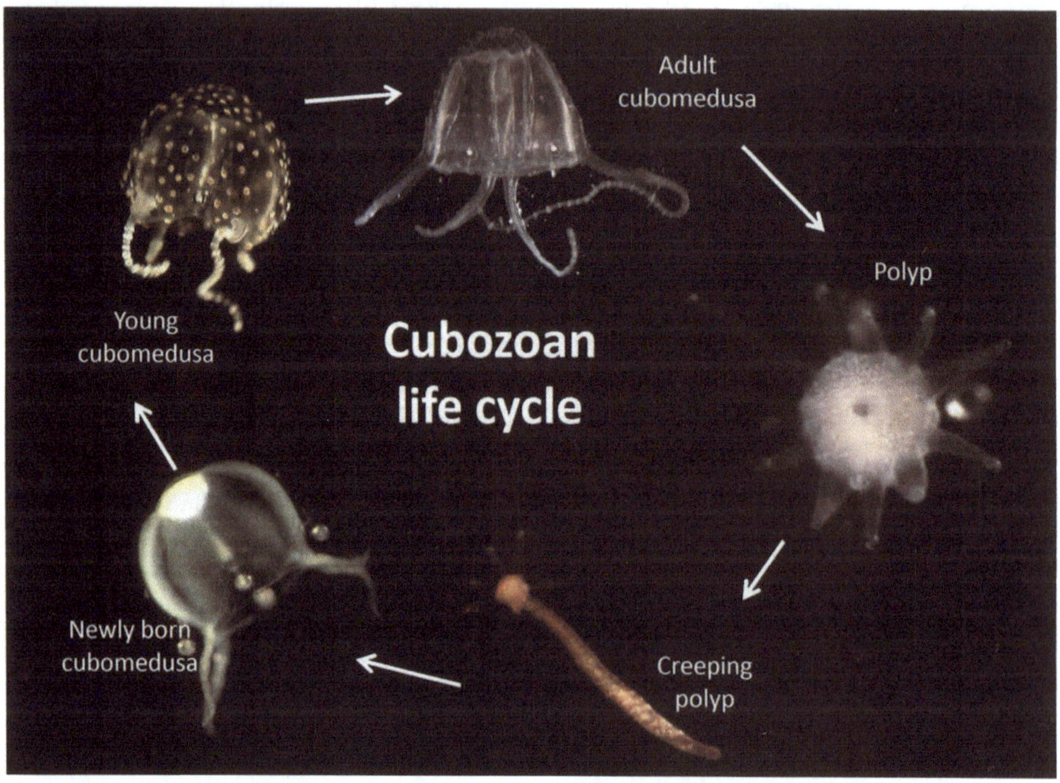

Figure 18.11 Life history of a cubozoan jellyfish (stages from multiple taxa shown). (Photo: L. Gershwin.)

BOX 18.2 IRUKANDJI SYNDROME

About 50 stings each year in the Great Barrier Reef region send people to hospital with a bizarre constellation of symptoms known as 'Irukandji syndrome'. At least 10 species of jellyfish have been linked with the syndrome. The initial sting is typically mild, and is followed after a 5–40 minute delay by any or all of the following: severe lower back pain, nausea and vomiting, cramps, spasms, muscular restlessness, difficulty breathing, profuse sweating, or a feeling of impending doom. Irukandji syndrome is not usually fatal, but it can be; more frequently it resolves entirely or, rarely, results in permanent heart damage.

The best prevention against Irukandji stings is a full-body lycra suit (or neoprene wet suit), which provides good protection for about 80% of the body's skin surface. It is still possible to be stung on the hands, feet, or face, but this rarely happens compared to the larger body surfaces. For those wishing additional protection, gloves, booties, and hood are recommended.

Irukandjis are more prevalent on days with dense 'jelly button' or 'sea lice' blooms, especially if salps are present (see Chapter 14). These sorts of days only occur a few times a year, but it is wise to avoid the water if possible.

Treatment priorities for known or suspected Irukandji stings:

1. If the person is already feeling systemic symptoms, call '000' for an ambulance, then douse the sting well with vinegar. Rest and reassure the patient.
2. If systemic symptoms have not yet onset, douse the sting site well with vinegar and monitor the patient for 30–40 minutes to see if symptoms develop; if so, call '000' for an ambulance.

Treatment priorities for box jellyfish stings:

1. Call for help	dial '000' for ambulance or get a lifeguard
2. Treat the victim	provide emergency care (CPR if necessary)
3. Treat the sting	douse sting area with vinegar
4. Seek medical aid	transport to hospital

distributions. Box jellyfish, for example, are generally restricted to coastal and near-shore island localities. In contrast, stings from Irukandji jellyfishes have been reported from nearly every coastal and island/reef locality where recreational activities take place throughout the GBR (Box. 18.2). Curiously, the moon jellyfish (*Aurelia* spp.) is one of the commonest species globally, but is only sporadically found in the GBR region, where it may occur in dense but short-lived blooms. However, the majority of jellyfish species have only been reported from one reef or island but this is probably more due a paucity of studies than a reflection of their true abundance. In general, the dangerous species have received the most attention but some are yet to be formally classified.

A hot topic of modern jellyfish research is bloom dynamics, or the cycles and triggers of jellyfish blooms. Most jellyfish are capable of exploiting perturbed ecosystems, whether through pollution (especially eutrophication), overfishing, species introductions, or climate change. They have changed little in design in over 600 million years, suggesting that they are tolerant of changes in pelagic assemblages and environmental change. Water quality conditions that threaten the survival of some organisms may facilitate the development of jellyfish blooms. Numerous examples exist around the world where jellyfish have essentially 'flipped' the local ecosystem, becoming the dominant predator. In this capacity, jellyfish can have direct and indirect effects on other species. For example, they may eat the larvae of other species as well as the food of these larvae (that, in turn, influences larval survival).

PREDATION AND DEFENSE

Most medusae and ctenophores will eat a diversity of plankton and some nekton. They are typically carnivorous, preferring zooplankton prey rather than phytoplankton. Their own kind can be also be on the menu. *Cyanea* spp. (lions mane) eat *Aurelia* and other medusae, while the ctenophore *Beroe* preys on other ctenophores such as *Pleurobrachia.* Some species of 'blubbers' (Order Rhizostomeae) have symbiotic algae (zooxanthellae) in their tissues, similar to those of corals. These symbiotic jellyfishes are typically found in warm shallow waters, where they 'farm' their algal symbionts; the jellyfish may obtain a significant portion of its nutritional requirements from its algae.

Some jellyfish have definite feeding preferences. For example, one species of box jellyfish from the GBR region has been demonstrated to change its dietary habits as it grows, starting out preferring prawns, and switching to fish later in life.

While the huge appetites of jellyfish make them fearsome predators, their soft bodies also make them an easy meal. Turtles, crabs, many fishes, such as sunfish and triggerfish, and seabirds actively hunt jellyfishes.

All cnidarian jellyfishes have venomous stinging organelles (nematocysts, Fig. 18.12), which they use for defence and prey capture. The ctenophores lack nematocysts, and instead, they have strange organelles called colloblasts, which lack venom but ensnare prey with adhesive. A nematocyst might be thought of as a tiny harpoon with poison, while a colloblast is more like a rope covered in honey.

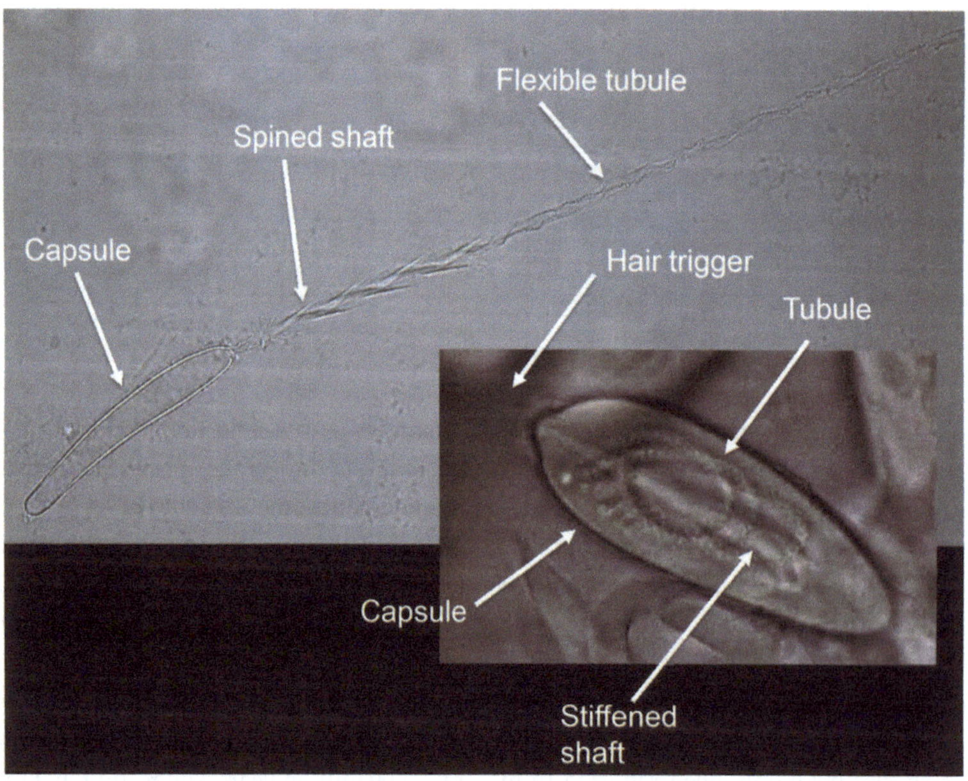

Figure 18.12 The morphology of nematocysts (stinging cells). (Photo: L. Gershwin.)

Nematocysts, the weapons of jellyfish, are basically a capsule with a spring-loaded thread coiled up inside, and an external hair-trigger at one end. The thread has rows of spines near the base, and usually a relatively smooth shaft along the rest of its length. When the hair trigger is activated, the thread is discharged inside-out (i.e. spines first), with tremendous force, about 40 000 times the force of gravity. Nematocysts can easily penetrate crab carapaces, fish scales, and human skin, so delivering their subcutaneous dose of venom. Nematocysts discharge following a combination of mechanical and chemical stimulation.

COLLECTING AND PRESERVATION

In general, jellyfishes, especially in the GBR region, should not be handled as their stings can be painful or even life-threatening. If it is necessary to handle them, gloves should be worn and safety precautions should be strictly adhered to at all times. Jellyfish can be challenging to photograph, with the diver having to manage lighting, background, currents, and typically uncooperative subjects; however, truly beautiful images can be taken by a skilled and patient photographer. It is important to ensure that the correct permits are in place before collecting or handling any species in the GBR region. Specific collecting and preserving techniques for different groups are available on the internet at www.medusozoa.com/curating.html [Verified 13 February 2008].

Jellyfishes do not preserve well in most standard preservatives, and must be preserved in dilute (2–4%) formaldehyde to help retain their original form. Alcohol may be needed to preserve hard parts such statoliths in cubozoans.

IDENTIFICATION

Some of the more conspicuous jellyfishes are easily identifiable, even at a glance:

- The cubozoan box jellyfish, *Chironex fleckeri* (Fig. 18.5*D*), is large, more or less transparent, and has up to 15 fettuccini-like tentacles hanging off each corner; a close relative, *Chiropsella bronzie* (Fig. 18.5C), has spaghetti-like tentacles and rarely more than seven or nine per corner.
- Irukandjis (Cubozoa, Fig. 18.5*A*, *B*) are also transparent, and may be as small as 1 cm or as large as 18 cm in height, but always have box-shaped bodies with only four tentacles, one extending from each corner.
- Scyphozoan blubber jellyfishes, Order Rhizostomeae, *Catostylus* (Fig. 18.3*A*), *Mastigias*, *Netrostoma* (Fig. 18.3*D*), and so on, are large and heavy, and often brightly coloured; they have eight frond-like structures hanging down from the centre, and never have tentacles hanging from the bell margin.
- Scyphozoan hair jellies, *Cyanea* spp. (Fig. 18.3C), have a large flat bell, with a mop of hundreds of fine tentacles clustered underneath.
- The scyphozoan mauve stinger, *Pelagia* (Fig. 18.3*B*), has a small dome-shaped pink-purple body, with eight fine marginal tentacles and four ribbon-like pleated oral arms extending from the centre of the underside of the body.
- Hydromedusae (Class Hydrozoa) have small transparent bodies, typically with numerous tentacles around the margin; often they are not seen but are felt as 'sea lice'. Members of the Order Leptomedusae (Fig. 18.1C) have flat bodies with gonads on the radial canals, hollow marginal tentacles and external statocysts; Order Anthomedusae (Fig. 18.1*A*) have tall bodies with gonads on the stomach, and with hollow or solid marginal tentacles and no statocysts; Order Narcomedusae have stiff tentacles emerging from the upper surface of the bell, and with the gonads in pouches; Order Trachymedusae have deep bodies with gonads on the radial canals, and with solid marginal tentacles and free statocysts; and Order Limnomedusae (Fig. 18.1*D*) have flat bodies with gonads on the radial canals or stomach, and have hollow tentacles and enclosed statocysts.
- Blue bottles (Subclass Siphonophora), *Physalia* spp. (Fig. 18.1*B*), have a floating bubble structure atop a cluster of blue and clear tentacles and filaments, with one or more bright blue main fishing tentacles trailing out to a distance of about a metre.
- Ctenophores all have eight bands of cilia along the body. *Pleurobrachia*-type comb jellies (Order

BOX 18.3 COMB JELLIES: ALIEN INVADERS

Comb jellies are well known as invaders. For example, *Mnemiopsis leidyi*, a lobate ctenophore in the Class Tentaculata were inadvertently carried in ships' ballast water from the Chesapeake to the Black Sea in the early 1980's. Without natural predators, by 1990 they had bloomed so much that surveys estimated that *Mnemiopsis* now comprised one billion tons of biomass in the Black Sea. This one species single-handedly caused the collapse of an ecosystem and the fisheries that went with it. This case has been well studied, and it is believed that the jellyfish ate the larvae of other species including sardines, as well as the food sources that the larvae would eat; this combination of direct and indirect effects prevented other species from regaining viable populations.

Elsewhere in the world, other ctenophore and medusa species have been considered pests, either through exotic species introductions, or through local perturbations that allowed the jellyfishes to alter existing ecosystems; perturbations include overfishing, pollution, and changes in water temperature. Jellyfish can bloom rapidly when conditions are ideal or go for long periods of time without food if necessary. Their potential to quickly change pelagic assemblages, therefore, is great.

Cydippida, Fig. 18.8*A*) have a small spherical body with two tentacles. *Bolinopsis*-type comb jellies (Order Lobata, Fig. 18.8C) have an egg-shaped body with two large lobes. *Beroe*-type comb jellies (Order Beroida, Fig. 18.8*B*) have a flattened palm-of-the-hand-shaped body without tentacles. (Box 18.3.)

ADDITIONAL READING

Arai, M. N. (1997). 'A Functional Biology of Scyphozoa.' (Chapman & Hall: London.)

Brusca, R. C., and Brusca, G. J. B. (2002). 'Invertebrates.' 2nd edn. (Sinauer Associates: Sunderland, MA.)

Gershwin, L. (2005). 'Taxonomy and Phylogeny of Australian Cubozoa.' (School of Marine Biology and Aquaculture: James Cook University, Townsville.)

Harbison, G. R. (1985). On the classification and evolution of the Ctenophora. In 'The Origins and Relationships of Lower Invertebrates'. (Eds S. C. Morris, J. D. George, R. Gibson and H. M. Plattpp.) pp. 78–100. (Oxford University Press: Oxford.)

Kingsford, M. J., Pitt, K. A., and Gillanders, B. M. (2000). Management of jellyfish fisheries, with special reference to the order Rhizostomeae. *Oceanography and Marine Biology: An Annual Review* **38**, 85–156.

Kramp, P. L. (1961). Synopsis of the medusae of the world. *Journal of the Marine Biological Association of the United Kingdom* **40**, 1–469.

Mayer, A. G. (1910). 'Medusae of the World. Vols 1 and 2, the Hydromedusae; Vol. 3, the Scyphomedusae.' (Carnegie Institution: Washington, DC.)

Totton, A. K. (1965). 'A Synopsis of the Siphonophora.' (British Museum of Natural History, London: London.)

Williamson, J. A. Fenner, P. J. Burnett, J. W., and Rifkin, J. F. (Eds) (1996). 'Venomous and Poisonous Marine Animals: A Medical and Biological Handbook.' (NSW University Press: Sydney.)

See web site for an extended list, with updates, available at http://www.australiancoralreefsociety.org/ [Verified 21 March 2008].

19. Hexacorals 1: Sea Anemones and Anemone-like Animals (Actiniaria, Zoanthidea, Corallimorpharia, Ceriantharia and Antipatharia)

C. C. Wallace

OVERVIEW

Because they have no skeleton and do not participate in the building of reefs, sea anemones and their kin (Class Anthozoa; Orders Actiniaria, Zoanthidea, Corallimorpharia Ceriantharia and Antipatharia) are not as intensely studied as their relatives, the hard corals (Order Scleractinia in the same class). Nevertheless, they are prominent, and in some cases well-known, members of coral reef communities. The largest sea anemones, reaching almost a metre in diameter, are hosts to anemone fish, crustaceans, and various invertebrates that live among the anemones' tentacles in fascinating symbiotic relationships. Other anemones play guest rather than host and gain mobility by attaching to molluscs, crabs or shells housing hermit crabs. Zoanthids and corallimorphs can be abundant occupiers of space, sometimes when this is made available by bleaching or other damage to corals, and they too may have symbiotic associations with many other reef organisms, thus securing space in the reef column, protection from predators and/or access to food and energy sources.

Like hard and soft corals, the members of these groups have life cycles involving a short larval stage and a long polyp phase. The adult occurs as a single polyp in Actiniaria (sea anemones) and Ceriantharia (tube anemones) and usually, but not invariably, as a colony or clone of interconnected polyps in Corallimorpharia (jewel anemones) and Zoanthidea (zoanthids). Antipatharia (black corals) are exclusively colonial. The major external characteristics of the polyp are a column, an oral disc bearing the mouth, siphonoglyph(s) (flagellated furrows that direct water currents into the gastrovascular cavity) and tentacles, and (in solitary forms) the pedal disc or bulb allowing attachment or anchorage. In colonial forms polyps are joined by basal stolons or common coenenchyme. Main internal characters include the actinopharynx leading from the mouth and siphonoglyphs, and the paired mesenteries that may be complete or incomplete (reaching or not reaching the actinopharynx at their upper limit) and on which the gonads develop. (See Fig. 19.1 for examples of these characters in various orders.)

Animals from these orders contribute greatly to the taxonomic and ecological diversity of the GBR and to the dominant role of the Anthozoa in the coral reef environment. They are mostly carnivorous, feeding on tiny planktonic or benthic animals or particulate matter, and sometimes utilising symbiotic dinoflagellates in a

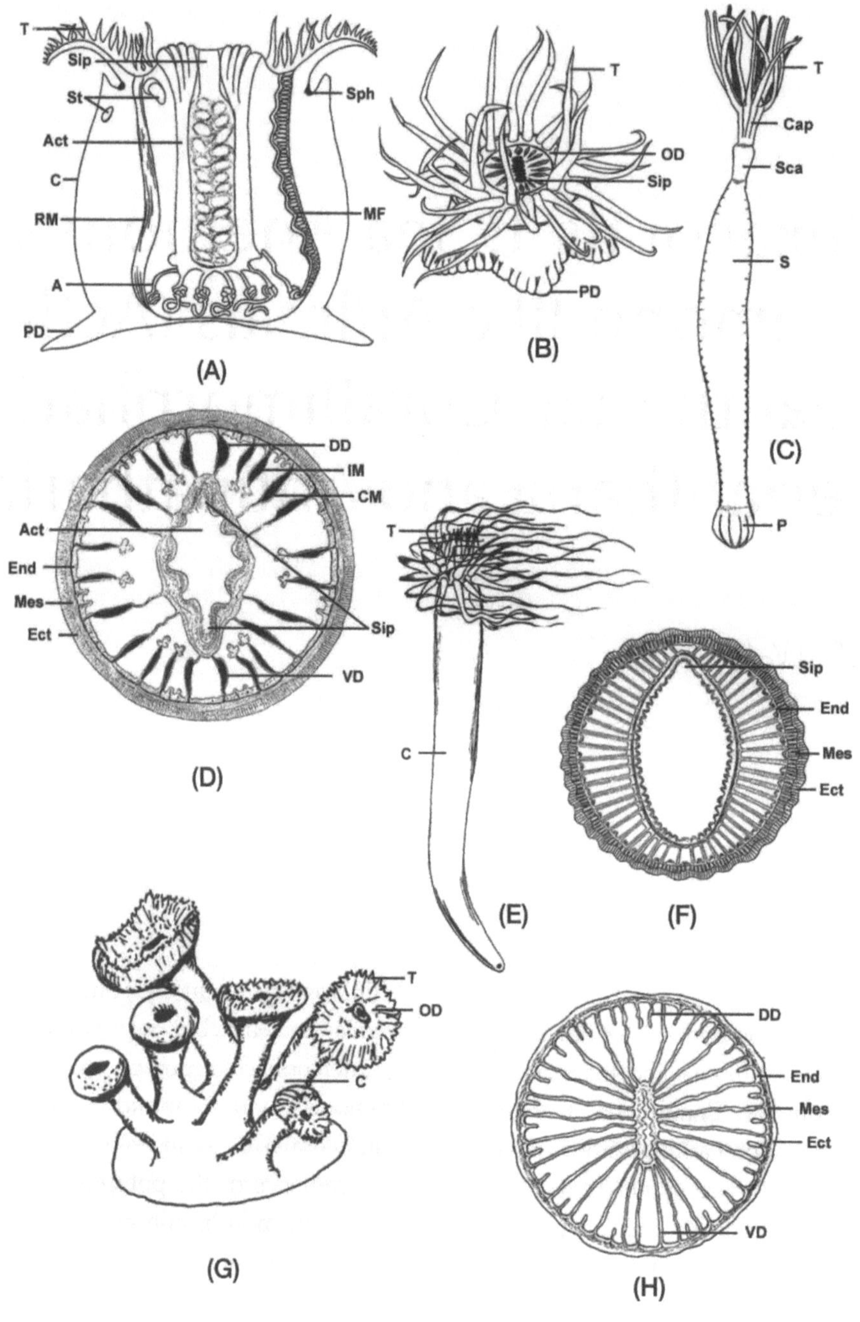

Figure 19.1 Anatomical features. *A–D*, Actiniaria: *A*, diagrammatic sea anemone internal features; *B*, external features (*Anthopleura handi*); *C*, external features, Edwardsiidae; *D*, diagramatic TS through actinopharynx region. *E–F*, Ceriantharia: *E*, external features; *F*, TS through actinopharynx region. *G–H*, Zoanthidea: *G*, external features; *H*, TS through actinopharynx region. A, acontia; Act, actinopharynx; Cap, capitulum; CM, complete mesentery; DD, dorsal directives; Ect, ectoderm; IM, incomplete mesentery; St, stomata; MF, mesenterial filament; Mes, mesoglea; OD, oral disc; PD, pedal disc; P, physa; RM, retractor muscle; S, scapus; Sca, scapulus; Sip, siphonoglyph; Sph, sphincter; T, tentacles; VD, ventral directive. (Images redrawn from various sources by W. Napier.)

similar manner to the hard corals. Their attractive radially symmetrical form and dramatic colouration, combined with relatively simple environmental requirements, at least in some cases, make this group very popular with the aquarium trade. Because of this, some are well known to hobbyists and their food and habitat requirements and even breeding and dividing patterns are recorded on numerous web pages, although scarcely documented in the scientific literature.

Whereas the classification and identification of Scleractinia and Alcyonaria are primarily based on skeletal elements (solid skeleton in hard corals and elaborate spinules in soft corals), other features of the polyp, both macro- and microscopic, must be used for these predominantly skeleton-free orders. Valid identification of smaller specimens often requires verifying details of external and internal anatomy under a dissecting microscope as well as histological preparation of thin sections and tissue squashes for high power microscopic study. Fortunately for the field observer, many coral reef species have large size, bright colouration, unique habitat preferences or typical associates that allow for field identification with reasonable certainty.

■ ORDER ACTINIARIA (SEA ANEMONES)

Throughout the world, sea anemones occur in most marine environments, from rocky and soft intertidal to deep abyssal substrata, and in many sizes from more than a metre in diameter down to a millimetre or two. Large sea anemones can be seen in both tropical reefs and deep Antarctic waters, but the families represented differ, with many of the deep sea species having strong gelatinous column walls or heavy external coatings and very large mouths while tropical species tend to have soft bodies, smaller mouths and strong defence mechanisms. Not surprisingly, many of the tiny sea anemones remain undescribed or unrecorded for Australia and their presence in ecosystems may go unnoticed. Some sea anemones harbour symbiotic dinoflagellates and small-bodied species are sometimes used as indicators for laboratory study of heat tolerance and other characteristics in these organisms.

Sea anemones occupy many habitats and microhabitats on coral reefs. Some favour elevated sites on the reef surface, some extend from recesses within the framework and others burrow within the sandy floor. Sometimes closely related species have obvious differences in habitat preference, for example the two large anemonefish hosts *Heteractis magnifica* and *H. crispa* favour exposed high points and protected crevices respectively.

The 'swimming anemone' *Boloceroides mcmurrichi* lives unattached or very lightly attached, and is observed sporadically on reefs, sometimes in 'swarms' of large numbers of individuals. Many other anemones have the capability of changing location slowly by actions of the pedal disc. Most anemones are highly retractile and expandable, capable of greatly varying the column and tentacle length, closing the oral disc through the actions of a strong marginal sphincter, or even of disappearing from sight completely by withdrawing into the substratum.

While all Cnidaria have nematocysts (stinging capsules) in their ectodermal tissues, sea anemones have a great variety of these and some coral reef inhabitants are extremely dangerous because of the toxic properties of their nematocysts. Anyone wishing to explore the taxonomy of anemones must master the various categories of these cellular features, as the types of nematocysts present are used in family determinations, and genera and species can have typical dimensions and proportional composition of these nematocysts in different body regions (see Fig. 19.2 for some common

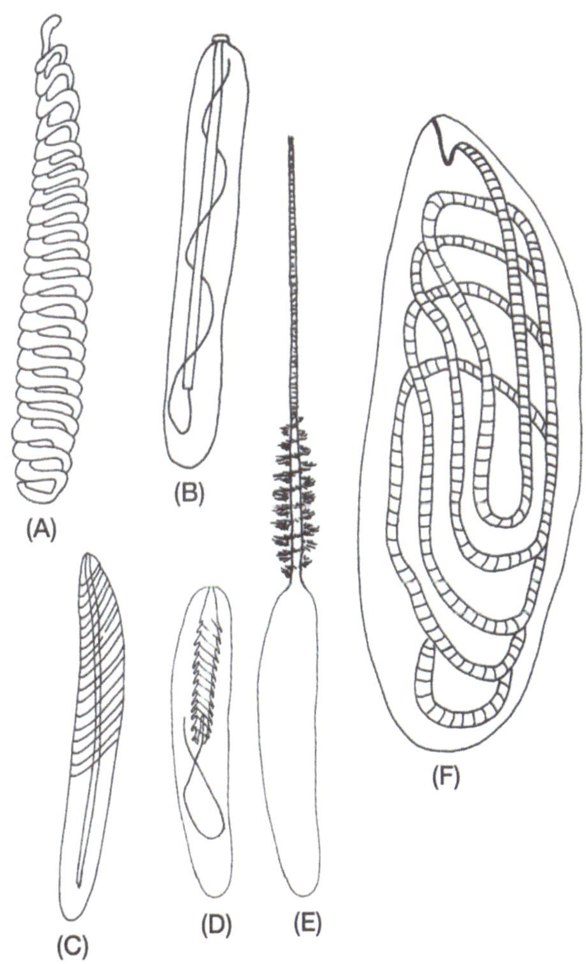

Figure 19.2 *A*, spirocyst and some nematocyst types in Actiniaria; *B*, basitrich; *C*, microbasic b-mastigophore; *D–E*, microbasic p-mastigophore (*E* released); *F*, holotrich. (Images redrawn from various sources by W. Napier.)

types). Much remains to be resolved in the classification of sea anemones, and relationships between and among genera and families are not always clear. This is an active research area and new developments can be seen as updates in the 'Hexacorallia of the World' database (see Additional reading).

External anatomy

The sea anemone column may have up to three recognisable regions (from the base: scapus, scapulus and capitulum: see Fig. 19.1) and above the column is a flat or domed oral disc bearing the mouth and tentacles in two to many cycles. The mouth is oval or slit-like and has one, two or more siphonoglyphs. The margin of the oral disc is often folded into a groove or fosse. Tentacles in Actiniaria are mostly hollow and simple but may be branched, sometimes elaborately so, or carry structures containing numerous nematocysts. Other projections with various names such as verrucae (adhesive projections from the column), acroragi (projections bearing nematocysts) and marginal spherules (projections without nematocysts) may occur on the margin, in the fosse and on the column. A large group of families (known as 'acontiate' anemones) have nematocyst-bearing structures called acontia (fine hair-like projections from the base of the mesenteries), which can be extruded through various parts of the body, and sometimes through permanent holes in the column called cinclides. All these external features are involved in the identification process, but internal features such as the structure of the marginal sphincter between the top of the column and the oral disc, the number and extent of mesenteries and the position of gonads on them are all used in the complete description and identification of a species.

Internal anatomy

The mouth leads to an actinopharynx, below which is the gastrovascular cavity. The body cavity is divided by paired mesenteries, some complete (reaching the actinopharynx) and others incomplete, in typical arrangements. Gonads (oocytes and testes) develop from gastrodermal interstitial cells that ripen within the mesenteries. The arrangement of gonads is used in the diagnostic process, with either hermaphroditic (both sexes occurring in each polyp) and gonochoric (sexes separate) states occurring in different species. As might be expected in soft-bodied animals without skeletons, anemones are equipped with a variety of muscles, which serve to maintain body shape, control tentacles and maintain the position of the animal on or in the substratum. Often the presence/absence, position, size and shape of the muscles (especially of the marginal sphincter muscle responsible for closing off the oral disc) are diagnostic.

Life history and ecology

All sea anemones are capable of sexual reproduction and many are also capable of asexual reproduction by division of the polyp and separation of the two resulting polyps. The bulb-tentacle anemone, *Antacmaea quadricolor*, and the magnificent sea anemone, *Heteractis magnifica*, are well known for forming huge clones of individuals, occupying many square metres, by this method. Other species, most notably the aquarium pest *Aiptasia pulchella*, produce numerous cloned individuals by a process known as pedal laceration, and the swimming anemone, *Boloceroides mcmurrichi*, is able to reproduce by shedding tentacles, which then bud a complete new anemone. Sexual reproduction involves the development of eggs and sperm in separate gonads, followed by release of the gametes into the water column for external fertilisation (in hermaphroditic and possibly some gonochoric species) or fertilisation of eggs in situ by sperm released from other individuals and brooding of larvae inside or close to the female parent (in gonochoric species).

Diversity

Of about 150 species of anemones known from Australian waters (including Antarctica), around 40 are known so far from the GBR (Figs 19.3–19.7, Box 19.1). This is likely to be a gross underestimate, particularly in regard to smaller and cryptic species and sand-dwelling species, especially those from the family Edwardsiidae, which is under review worldwide at present. As with corals, many of the anemone species found on the GBR have broad ranges through the Indo-Pacific.

Figure 19.3 Some sea anemones of the Great Barrier Reef. *A*, **Heteractis magnifica*; *B*, **Heteractis crispa* (with *Amphiprion clarkii*); *C*, **Heteractis aurora*; *D*, **Heteractis malu*; *E*, **Stichodactyla mertensii*; *F*, **Cryptodendrum adhaesivum*; *G*, **Entacmaea quadricolor*; *H*, *+Telmatactis* sp.; *I*, *Thallassianthus* sp.; *J*, *+Megalactis griffithsi*; *K*, *Actinodendron glomeratum*; *L*, sand-burrowing anemone, Edwardsiidae. *, anemonefish host; +, dangerous because of stinging potential. (Photos: *A*, *E*, *F*, *G*, A. Crowther; *B*, *C*, *D*, *J*, *L*, P. Muir; *H*, W. Napier; *I*, *K*, R. Steene.)

Figure 19.4 Merten's sea anemone *Stichodactyla mertensii* with the clownfish *Amphiprion clarkii*. (Photo: P. Muir).

Figure 19.5 Clown anenomefish in *Entacmaea quadricolor* in ReefHQ aquarium, Townsville. (Photo: GBRMPA.)

Figure 19.6 *Callianthus polypus* is often seen in association with hermit crabs, and can be seen when the crab is on the move at night, feeding on the reef surface. (Photo: P. Harrison.)

Figure 19.7 The 'boxer crab' (*Lybia* sp.) carries juveniles of the sea anemone *Triactis producta* in its claws, presumably for protection from predators. Presumably, also, this is a dispersal mechanism for the anemone. (Photo: R. Steene.)

Preservation and study

For taxonomic research, sea anemones should ideally be examined both alive in the field and laboratory, to note external morphological features using a microscope for small specimens. For preservation and histology, the animal should be relaxed before being fixed in 4% formaldehyde and then 75% alcohol. The most popular relaxant is magnesium chloride, added in very small amounts to a concentration of about 1.8% over a period of one to three hours (depending on the size and response

BOX 19.1 SEA ANEMONES AS HOSTS AND GUESTS

Hosts

One of the best known and loved associations on coral reefs is that between large tropical sea anemones and the small, brightly coloured fish known as 'clownfish' or 'anemonefish'. Fish and sea anemones live together in a close association known as 'symbiosis', where both species benefit from the arrangement and concessions are made by the partners to make it possible for the two to coexist. Ten species of sea anemone are known to play host to anemonefishes. These come from the families Stychodactylidae (species of *Heteranthus* and *Stichodactyla*), Actiniidae (*Entacmaea quadricolor* and *Macrodactyla doreensis*) and Thalassianthidae (*Cryptodendrum adhaesivum*) (see Fig. 19.3 for some of these species). Like most sea anemones, these hosts have a well-developed defensive system based on nematocysts within the ectoderm of the tentacles. The fish swim around among the tentacles despite this hazardous feature and benefit from the protection from larger predators that would be harmed by the tentacles. The anemones themselves have also been found to benefit by protection from predators due to the defensive activities of the anemonefishes and a number of other ecological advantages have been documented.

At least 28 fish species are involved in anemone associations throughout the Indo-Pacific, all from the family Pomacentridae and most from the genus *Amphiprion* (see Figs 19.3*B*, 19.4). Some anemonefish are quite host specific, but most can occur in many hosts. A hierarchy exists among the fish species themselves, which means that certain species are much more likely than others to successfully colonise and monopolise the available hosts. Other factors too, such as habitat and geography, play a role in the availability and likelihood of settlement in a host of a particular fish species. The fish species differ in the way they approach new hosts, but all have some mechanism for developing a tolerance to the sting of the anemone via a mucous surface coating developed following gradual contact with the tentacles.

Guests

There are also numerous associations of sea anemones with other host organisms, where again the anemone's stinging capabilities provide some protection for the host and the host gives the anemone access to resources. One thing that is difficult for anemones (at least after the larval phase) is moving around and some of them avoid this problem by living in association with motile animals such as slow-moving molluscs or mollusc shells inhabited by hermit crabs.

of the animal), until the animal is unresponsive to touch. Permits are required to collect, see Chapter 12.

■ ORDER ZOANTHIDEA ('ZOANTHIDS')

These are mostly colonial forms (except for species of the large tropical free living zoanthid genus *Sphenopus*), in which the polyps lack a pedal disc and are joined by coenenchyme, which often contains a thick mesoglea (Fig. 19.8*A–F*). In one group of tropical zoanthids, including *Palythoa* and *Parapalythoa*, this coenenchyme incorporates sand and other material within a mesogleal matrix, so that the columns of the polyps are not visible. The number of valid zoanthid species known to be present on the GBR is around 13, but revisions and new

Figure 19.8 *A–F*, Zoanthidea; *G–J*, Corallimorpharia; *K*, Ceriantharia and *L*, Antipatharia: *A*, *Protopalythoa* sp.; *B*, *Palythoa heliodiscus*; *C*, *Zoanthus* cf *vietnamensis*; *D*, *Palythoa tuberculosa* on reef flat; *E*, *Palythoa tuberculosa* on giant clam; *F*, *Acrozoanthus australiae* (note open polyps below water line); *G*, *Discosoma* sp.; *H*, *Discosoma* sp.; *I*, *Ricordia* sp.; *J*, *Amplexidiscus fenestrafer*; *K*, *Cerianthus* sp.; *l*, Black coral *Antipathes* sp. (Photos: *A, B, I, J, K, L*, P. Muir; *C, F, G, H*, A. Crowther; *D, E*, GBRMPA.)

studies on zoanthids from the Indo-Pacific (including genetic studies) may increase this number, and clarify the identity of species representatives of many genera. Identification of genera is often possible from external features but even this needs to be confirmed by internal observation in many cases, for example, the genus *Zoanthus* has a divided marginal sphincter. Confirmation of some identifications requires histological sectioning. Much progress is being made in zoanthid taxonomy and natural history, and many new publications on this group can be expected over the coming years.

Anatomy

The column wall is without special external structures as seen in Actiniaria. The oral disc has unbranched tentacles, which occur in an exococelic and an endocoelic cycle (these cycles often appearing as a single marginal ring). The mouth is usually slit-like and has a single syphonoglyph, which is ventral. The septal arrangements are unique in living Anthozoa and new septa form only to either side of the ventral septal pair. In general, each pair of septa has one large septum (macroseptum, extending to the actinopharynx) and a small septum (microseptum), except for the dorsal directives, which are both microsepta, and the ventral directives, which are both macrosepta. This is called the 'brachynemous' arrangement (as in Fig. 19.1*H*), but a few zoanthids also have another complete pair of macrosepta on each side (the 'macronemous' arrangement).

Life history and ecology

Colonies of zoanthids may grow attached to the substratum by a broad colony base (e.g. in *Palythoa* species) and many are epizoic, often using specific animal species, worm tubes (e.g. *Acrozoanthus australis*), sponges, gorgonians and numerous other organisms as an attachment surface. Many species host symbiotic dinoflagellates (zooxanthellae) in the ectodermal tissues. Sexual reproduction probably occurs in all species and follows similar patterns to those of corals and sea anemones. At least one species has been seen to spawn during the coral mass spawning event on the GBR. Zoanthids also form clones by division of colonies and large areas of reef flat may contain representatives of only a few genotypes because of this.

The intertidal reef flat often appears to be a preferred habitat for zoanthids, probably due to the rapid growth and cloning ability of the genera *Palythoa* and *Protopalythoa* that occur here and are particularly obvious a few months after damage to corals and other organisms, for example by events such as cyclones. Species of *Protopalythoa* and *Zoanthus* occur in calmer sublittoral and back-reef areas. Zoanthids are very toxic, their tissues containing a toxin known as palytoxine. For this reason they are not as popular as corallimorphs for aquarium culture.

■ ORDER CORALLIMORPHARIA ('JEWEL ANEMONES' OR 'CORALLIMORPHS')

Usually grouped with sea anemones, this relatively small order of around 40 named species is sometimes regarded as 'skeleton-free corals', although the close relationship with Scleractinia is questioned by a recent genetic study. They occur as clones of polyps connected by stolons (Fig. 19.8*G–J*). The polyps have a broad, thin oral disc and short, featureless column. The mouth protrudes from the centre of the oral disc. There are a variable number of short tentacles, usually occurring throughout the oral disc, but sometimes in rows and sometimes there are specialised tentacles (bearing nematocyst batteries in rounded tips). The nematocyst composition (cnidom) is less complex than in Actiniaria, but some corallimorph nematocysts are very large. Corallimorphs are often very brightly coloured and a single species may occur in many colour morphs. There are probably around 20 species on the GBR, although further study may discover more.

Corallimorphs may be non-zooxanthellate (family Corallimorphidae) or zooxanthellate (Discosomatidae and Ricordaeidae). Many are formidable predators. The large discosomatid *Amplexidiscus fenestrafer* has been observed to feed by enveloping small fish in its oral disc. Other discosomatids may be able to form monocultures by defending territory using nematocysts stored in the tips of the marginal tentacles. They are mostly subtidal and many avoid bright light. They are generally very tolerant of environmental conditions and are popular as aquarium subjects for this reason as well as their attractive appearance. Histological sectioning of these disc-shaped animals is very difficult, and external features are regarded as very important for identification.

■ ORDER CERIANTHARIA ('SAND ANEMONES' OR 'CERIANTHIDS')

As the common name suggests, these animals occur within soft sediments with only the extended tentacles and a little of the long, soft column visible (Fig. 19.8*K*). The mouth is usually hidden by a crown of short tentacles and this is surrounded by much longer tentacles that may extend in the direction of water flow. The individual produces and occupies a flexible tube that may be up to a metre long. They are identified by dissection and identification of internal features.

Cerianthids occur from the intertidal to the deep sea, and they are quite frequently encountered on reef lagoons and the off-reef sea floor. Most feed at night, retracting (at least partially) into their tubes by day. Associates from a number of phyla, including lophophorates, may occur in the tube.

■ ORDER ANTIPATHARIA ('BLACK CORALS')

These are colonial anthozoans in which the polyps occur on the outside of an axial skeleton of hard, black, proteinaceous material that bears tiny thorn-like projections (Fig. 19.8*I*). The polyps are non-retractile and bear only six tentacles. Two main colonial structures occur: single whip-like forms that may be coiled but do not branch (genus *Cirripathes*), and bushy forms, in which a thick central trunk divides into increasingly smaller branches and eventually very thin branchlets (genus *Antipathes*). While *Cirripathes* are common on the GBR, the large tree-like forms mostly occur out of scuba diving range.

There is a long history of exploitation of black coral skeleton for jewellery. Colonies large enough to have formed the desired dense, shiny black skeleton at their bases are believed to be very old and usually occur far deeper than 20 m: these are sometimes found on Australian beaches after heavy weather. All international trade in black coral, even in the form of the final product, is subject to regulation under the Convention for International Trade in Endangered Species (CITES).

ADDITIONAL READING

Burnett, W. J., Benzie, J. A., Beardmore, J. A., and Ryland, J. S. (1997). Zoanthids (Anthozoa, Hexacorallia) from the Great Barrier Reef and Torres Strait, Australia: systematics, evolution and a key to species. *Coral Reefs* **16**, 55–67.

Carlgren, O. (1949). A survey of the Ptychodactiaria, Corallimorpharia and Actiniaria. *Kungliga Svenska Vetenskaps-Akademiens Handlingar* **1**, 1–121.

Erhardt, H., and Knoop, D. (2005). 'Corals Indo-Pacific Field Guide.' (IKAN: Undterwasserarchiv, Frankfurt.) [Has very good section on Actiniaria and kin.]

Fautin, D. G. (2006). Hexacorallians of the World. Available at http://geoportal.kgs.ku.edu/hexacoral/anemone2/index.cfm [Verified 21 March 2008].

Fautin, D. G., and Allen, G. R. (1997). Anemone Fishes and Their Host Sea Anemones. (Western Australian Museum: Perth.)

Fosså, S. A., and Nielsen, A. J. (1998). 'The Modern Coral Reef Aquarium. Vol. 2.' (Birgit Schmettkamp Verlag: Bornheim.)

Napier, W. R., and Wallace, C. C. (no date). Available at http://www.environment.gov.au/biodiversity/abrs/online-resources/species-bank/records-cnidaria.html [Verified 21 March 2008]. [Species Bank entry for *Acrozoanthus australiae*].

Opresko, D. M. (1972). Redescriptions and reevaluations of the Antipatharians described by I. F. De Portales. *Bulletin of Marine Science* **22**, 951–1017.

Saville-Kent, W. (1893). The Great Barrier Reef of Australia: Its Products and Potentialities. (W. H. Allen & Co.: Waterloo Place, London.)

Scott, A., and Harrison, P. L. (2005). Synchronous spawning of host sea anemones. *Coral Reefs* **24**, 208.

Wallace, C. C., and Richards, Z. Available at: http://www.environment.gov.au/biodiversity/abrs/online-resources/species-bank/records-cnidaria.html [Verified 21 March 2008]. [Species Bank entries for *Boloceroides mcmurrichi*, *Cryptodenron adhaesivum*, *Entacmaea quadricolor*, *Heteractis magnifica*, *Heterodactyla hemprichii*, *Macrodactyla doreensis*, *Stichodactyla haddoni* and *Triactis producta*.]

20. Hexacorals 2: Reef-building or Hard Corals (Scleractinia)

C. C. Wallace

OVERVIEW

The hard corals (Class Anthozoa, Order Scleractinia) are among the most intensely studied marine invertebrates in the world today, due to their essential role in the formation and maintenance of coral reefs, and the threats to this role presented by human impact and global climate change. Through the process of laying down a skeleton of calcium carbonate and interaction with the symbiotic dinoflagellates that enhance this process, corals have built the present fabric of the GBR over some 6000 to 8000 years and previous reef formations over many millennia, as the carbonate structures of the Queensland continental shelf and Coral Sea testify. Like trees in the forest, living corals also provide the three dimensional structure in and around which the other creatures of the reef spend their lives. Their health is vital to the well-being of innumerable organisms and ultimately to the continued existence of the GBR itself.

Hard corals, like other members of the Class Anthozoa, are coelenterate animals that have a life cycle involving a short larval stage and a long polyp phase. The polyp has the ability to lay down an external skeleton by the activity of specialised cells located within the external tissue layer (the 'calicoblastic ectoderm') at its base. Although about half the members of the Order Scleractinia throughout the world occur as single polyps, the majority of those on coral reefs occur as colonies of interconnected polyps produced by growth and division from an original single polyp. The presence of symbiotic dinoflagellates (*Symbiodyndium* or 'xoozanthellae') in the tissues of these colonies enhances the ability of the polyps to produce skeleton, and these colonial, rapid-growing corals are known as the 'hermatypic' 'zooxanthellate', or 'reef-building' corals. As explained in Chapter 8, hermatypic corals have a dependency on the conditions suitable to their symbionts and the carbonate accretion process, and are thus limited mostly to the tropical and subtropical regions of the world and mostly to depths of less than 40 m. It is important to also note that azooxanthellate corals (those without zooxanthellae) occur in great diversity right around Australia (more than 250 species recorded to date), especially at depths greater than 40 m.

ANATOMY

In corals, the basic polyp is simple in structure (Fig. 20.1). The column is cylindrical and without adornment, the tentacles hollow and (with few exceptions) simple, although one or two specialised larger tentacles may be developed. The tentacle number is based on a primary cycle of six, with cycles being

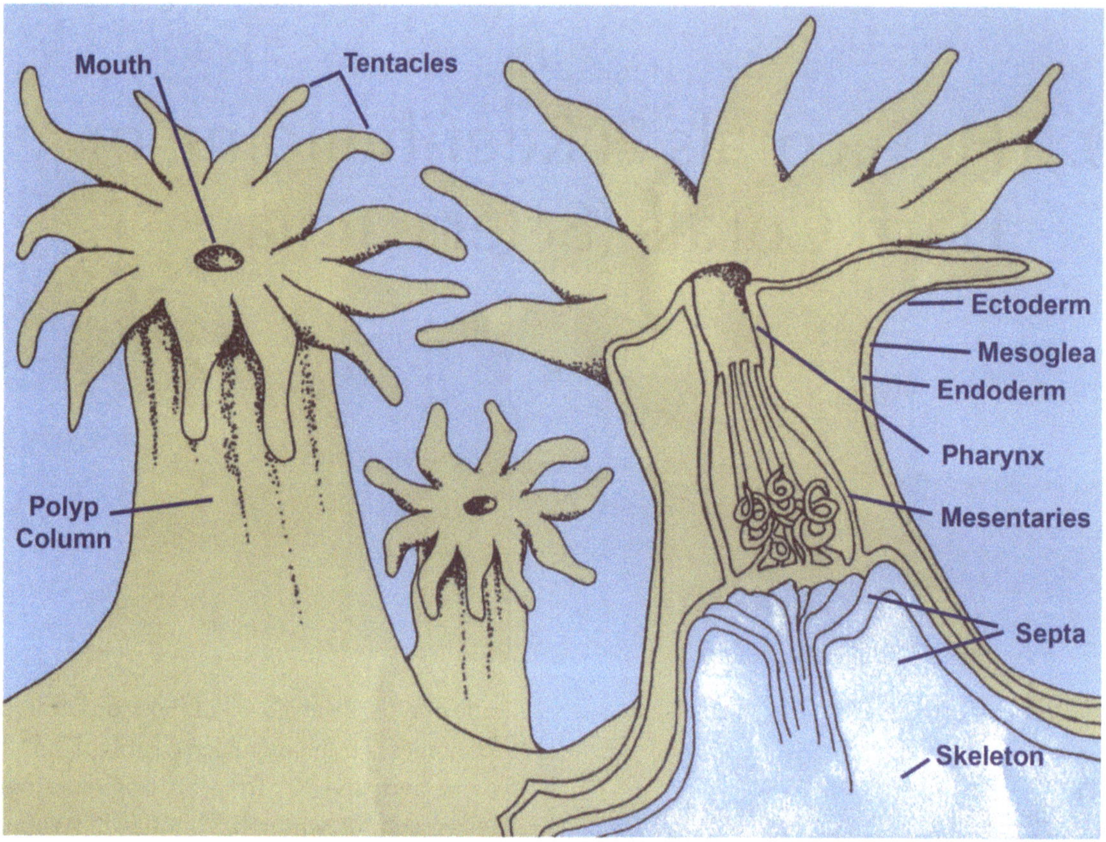

Figure 20.1 Relationship of tissue to polyp in an idealised coral colony. Entire polyp on left, cut-away polyp, exposing skeleton, on the right. (Adapted from Wallace and Aw 2000.)

added between the preceding cycles. The oral disc bears a round or oval mouth without syphonoglyphs (the flagellated furrows found in anemones and zoanthids), and this leads to a pharynx and the paired mesenteries, which may be complete or incomplete (reaching or not reaching the pharynx at their upper limit). In the process of replicating polyps to form a colony, division may occur either within the polyp (as evidenced by the presence of dividing corallites on the surface of the coral) or external to the polyp. When division is incomplete, corallites may exist in continuous connection (as in the brain corals) but little is known about the actual form of polyps when wall structures are combined. In general, the coral colony can be thought of as the sum of its polyps, with processes such as feeding and gonad formation being performed in individual polyps and contributing to the nutrition and sexual propagation of the colony.

CLASSIFICATION AND IDENTIFICATION

The classificatory scheme for corals was developed by reference to the elements of the skeleton. Understanding this requires knowledge of the parts of the skeleton underlying various polyp features and the way these vary from family to family (Fig. 20.2). Paradoxically, the exoskeleton of a coral appears to an observer to be inside the polyp, but actually it is underneath. The remarkable process of skeletal production leads to a structure that supports almost the entire three-dimensional form of the polyp as well as the connecting tissue and continuation of the body cavity (coenenchyme) between polyps. This creates a colony architecture that allows species to be identified both in life and after death, without reference (in most cases) to the tissue structure of the polyps themselves. After death, the skeleton persists, although it may be subject to erosion by other reef organisms, and as a

Figure 20.2 Structures and terms used in the identification of reef corals. *A–F*, terms relating to overall colony shape and polyp arrangements in corals; *G–L*, terms relating to forms of corallite arrangement in corals. *A*, massive; *B*, encrusting; *C*, plating; *D*, solitary free-living; *E*, branching arborescent; *F*, branching tabular; *G*, plocoid; *H*, ceriod; *I*, meandroid; *J*, phaceloid; *K*, hydnophoroid; *L*, plocoid with coenosteum features. (Photos: P. Muir.)

consequence, corals have perhaps the best fossil record of all animals. Much can be learnt about the history, geography, and future of living corals by reference to fossils.

A revolution in higher level classification of corals is being led by molecular biology (genetics), which is also being used to study boundaries and relationships between species, population level genetic diversity and even the identification of individuals and species. Differences in skeletal morphology, often microscopic and previously missed in the skeletal classification scheme, are being found to validate many of the distinctions found by molecular analyses and these will lead to some name changes in the classificatory scheme over coming years.

Because of the major role of the skeleton in the identification of corals, a vast terminology has developed to cover the various skeletal components and some of these parallel the tissue components. The skeleton formed by the colony or part of the colony is the corallum and that formed by the polyp is a corallite; this has dividing elements called septa (singular septum) that support the mesenteries and often a central element, the columella, which is also part of the support system of the polyp. Septa may extend outside the polyp as costae (singular costa). Interconnecting areas of skeleton are referred to as the coenosteum. All of these elements may be ornamented in various ways and even the basic process by which they are laid down affects their appearance and forms the basis of the classification of families. Supporting structures below and between the polyps are dissepiments and these represent layers of skeleton laid down as the polyps and connecting tissues move upwards in the colony as they (and the colony) grow. A set of terms has also been developed to describe the layout of corallites within the corallum and these are used so often in descriptions of corals that it is very useful to master the basics of them.

More than 450 species of corals have been recorded from the waters of the GBR, eastern coastal Australia, and continental and Coral Sea islands. This chapter concentrates on the field appearance of corals and their identification in situ on the reef. When accurate identification to species level is required, especially for those corals with microscopic polyps, further microscopic identification may be necessary. Guidance for this can now be found in numerous publications, and coral collections for consultation are held in universities and museums; coral identification workshops and coral identification services can be found at some of these.

LIFE HISTORY AND ECOLOGY

All corals have the ability to reproduce sexually, by producing male and female gametes for fertilisation and subsequent development of large numbers of ciliated larvae known as planula larvae (or planulae), which settle on the reef surface, transform into individual polyps and begin to lay down skeleton. The coral colony is then formed by division of this founder polyp. Individual polyps in a colony may all be male or female (the gonochoric or dioecious condition) or both sexes can be developed within each polyp (hermaphroditic or monoecious condition). There are two very different modes of production of larvae, usually characteristic of a genus or family. Sperm may be released by the polyps of one colony to find their way into the body cavity of other colonies, where eggs are fertilised and develop in or on the polyps (Fig. 20.3*A*). This is the brooding condition, with corals having this life cycle being referred to as 'brooders'; until the 1980s, this was thought to be the only type of life cycle in corals (see Box 20.1). In an alternative life cycle, shown by corals referred to as 'broadcast spawners', eggs and sperm are released into the water column for external fertilisation and development (Fig. 20.3*B*). Since gametes from same colonies do not usually self-fertilise, this second mode requires many colonies of the same species to release gametes at the same time (see Box 20.1).

The larvae of brooding corals are released fully developed and ready to settle on the reef, whereas embryogenesis and development of larvae by broadcast-spawning corals requires a developmental period of up to ten days before the planula is ready to settle. Dispersal of larvae outside the home reef is thus far more likely in broadcasters than brooders. Some corals are also subject to asexual reproduction by breakage and redistribution of branches, formation of polyp balls, and other mechanisms, and this may contribute considerably to population size and biomass.

Feeding and nutrition with a coral colony generally involves two sources: (1) acquisition of food by individual polyps, digestion in the polyp's gut and distribution of digestate through the colony via the

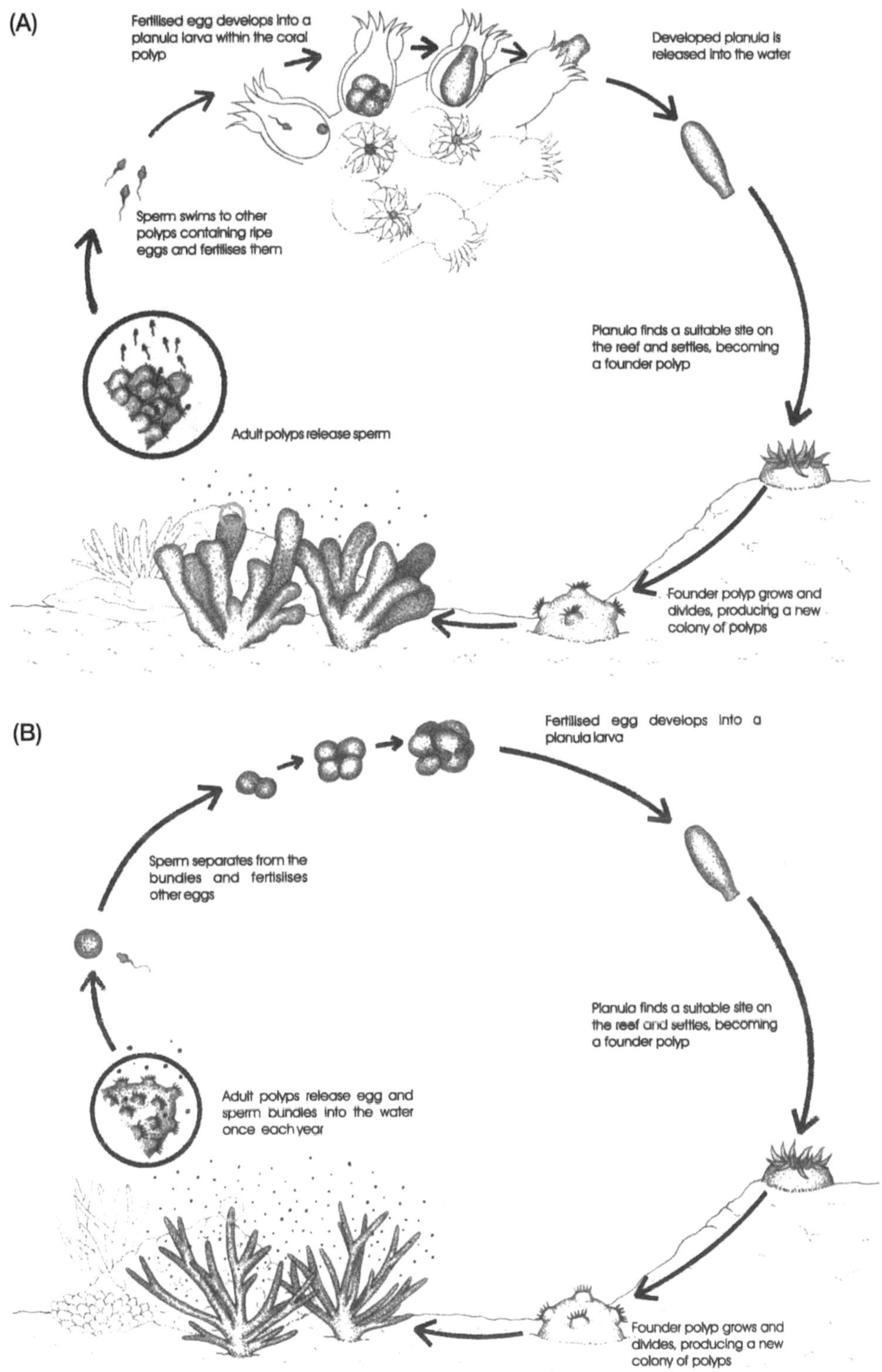

Figure 20.3 Reproductive cycles in corals. *A*, cycle in brooding species; *B*, cycle in hermaphroditic broadcast-spawning species. Other variations such as broadcast spawning in gonochoric species, occur in some corals. (Image adapted from Wallace and Aw 2000.)

BOX 20.1 THE CORAL MASS SPAWNING PHENOMENON

Corals from the majority of species on the GBR develop eggs and sperm once a year and release these for external fertilisation in multispecies spawning events. These occur at night during the week following the full moon in October to February, with November being the major spawning month. The discovery of this extraordinary phenomenon in Australia in the 1980s was one of the major advances in coral reef science of the 20th century. Mass spawning, as it has been dubbed, presents a paradox: on the one hand it ensures ample reproductive material for the cross-fertilisation required to ensure development of the next generation of each species, at the same time, saturating the food supply of predators for a very short period; on the other, it also provides opportunities for mis-matches, in which sperm could fertilise eggs of the wrong species to form hybrids or non-viable embryos. The discovery of mass spawning led to revision of ecological and evolutionary models and predictions and provided unprecedented opportunities for research, management planning and experimentation using gametes. Its impact has been worldwide and the phenomenon has now been reported from most countries bearing coral reefs.

Figure 20.4 Coral colonies spawning during the mass coral spawning event on the Great Barrier Reef. *A*, *Acropora tenuis* spawning egg/sperm bundles for external fertilization (note gametes on the water surface); *B*, female *Fungia* species spawning eggs (these will be externally fertilised by sperm released by a male); *C*, *Galaxia fascicularis* spawning large egg/sperm bundles for external fertilisation. (Photos: GBR.)

coelenteric connections between polyps; (2) photosynthetic activity of the symbiotic dinoflagellates that live in the polyp's tissues (see Chapter 8). Coral polyps are basically carnivorous, with their food source being to some extent determined by polyp, mouth and tentacle size. Food sources include planktonic organisms, incidental detritus and mucous with contained micro-organisms.

The ecology of reef corals is a broad subject that could not be dealt with adequately in this section, but is alluded to in the descriptions of major families below. The combined processes of sexual and asexual reproduction, dispersal, growth and response to physical and chemical conditions on different parts of the reef, play a role in zoning of corals on the reef, such that particular species, genera and even families, can be seen to predominately occur on, and sometimes dominate, particular parts of the reef, such as the inner reef flat, reef crest, mid slope or lagoon floor. Interaction and competition with other corals and occupiers of benthic space, impact of predators, and the impact of unpredictable events, and even coral diseases, also play a role in the presence of particular corals in any particular reef location, as well as in the size and shape of coral colonies.

■ MAJOR CORAL FAMILIES AND THEIR SIGNIFICANCE IN CORAL REEFS

While all corals contribute in some way to the ecology of the reef, certain families and/or genera dominate by making a major contribution to the reef framework, coral cover, species diversity, age structure, viability, coral/organism interactions and many other aspects of the overall operation of the reef. These groups tend to be the focus of the most research and to provide much of the information relevant to understanding and managing coral reefs in the face of a multiplicity of demands and threats to their persistence as healthy ecosystems. Presented below is a discussion of the six major families represented on the GBR. See Box 20.2 for a listing of all families and genera.

(1) Family Acroporidae (Fig. 20.5*A–F*)

This family includes the two most diverse genera of living zooxanthellate corals, *Acropora* and *Montipora* and also the genus *Isopora*, with just a handful of species. These genera often dominate the outer reef flat and the reef front, down to around 15 m, after which more species from the less prominent families can be found. *Acropora* (known as the staghorn coral genus), has a unique branching mode in which one type of corallite (the axial) forms the centre of the branch and buds off many more of a second kind (the radials) at its tip as it extends. This has led to numerous colony shape possibilities. Additionally, growth of the colony may be symmetrical and determinate (slowing down as a determined shape is reached) or asymmetrical and indeterminate, the colony growing to fill available space on the reef. As many as 40 species of this genus may occur in 500 m^2.

Acropora has over 120 species worldwide and over 70 on the GBR. Probably because of its great abundance, *Acropora* suffers the most from storm and cyclone damage, coral diseases, grazing by predators such as the crown-of-thorns starfish *Acanthaster planci* and aggregations of *Drupella* gastropods, and coral bleaching due to elevated sea-surface temperatures. *Acropora, Isopora* and *Montipora* are usually abundant among recruiting corals in experimental and natural situations and some species of *Acropora* also recruit asexually by growth of fragments after storm damage. *Isopora* is unique in the family in that its species are brooders, whereas all other Acroporidae are broadcast spawners.

(2) Family Faviidae (includes many of the 'brain corals') (Fig. 20.5*G–L*)

In this family the genera are massive, encrusting, or, less commonly, branching, and corallites have numerous septa, well-developed columellae and walls formed by simple fan systems of calcium carbonate crystals (trabeculae). Faviids, the largest family in terms of the number of genera, contribute to the reef by their densely constructed skeletons and their wide-ranging environmental tolerances. Genera are distinguished by the type of polyp division (extra- or intra- tentacular) and the arrangement of corallites (either separated or in contact, or combined in valleys to form the meandroid form that leads to the name 'brain corals'). Species distinctions within genera are based on dimensions of the corallites, features of the septa, costae, coenosteum and columella and number, shape, ornamentation and elevation of the septa above the corallite.

Faviids can be seen in most reef habitats, except perhaps very sandy locations. They are common on the intertidal reef flat and submerged shallow shoals, and also on fringing reefs along the coastline and continental islands. They can be the dominant coral on fringing reefs, such as the reefs of Moreton Bay at the port of Brisbane and on the exposed reef flats on the western side of Magnetic Island, off Townsville. Faviids form a major component of the coral mass spawning events in eastern

BOX 20.2 FAMILIES AND GENERA OF SCLERACTINIA (HARD CORALS) ON THE GREAT BARRIER REEF

Families 1 to 6 are dominant in the ecology of the GBR. Genera and families in bold are suggested as a first learning tool for the beginner. Genera in bold here are illustrated in this chapter. Asterisked genera are azooxanthellate.

Table 20.1

(1) Family Acroporidae
Acropora
Isopora
Montipora
Astreopora
Anacropora

(2) Family Faviidae
Favia
Favites
Goniastrea
Platygyra
Montastrea
Leptastrea
Leptoria
Oulophyllia
Plesiastrea
Echinopora
Caulastrea
Cyphastrea
Diploastrea
Moseleya
Barabattoia

(3) Family Poritidae
Porites
Goniopora
Alveopora

(4) Family Pocilloporidae
Pocillopora
Stylophora
Seriatopora

(5) Family Fungiidae
Fungia
Cycloseris
Heliofungia
Ctenactis
Herpolitha
Halomitra
Podobacia

(6) Family Mussidae
Lobophyllia
Symphyllia
Acanthastrea
Micromussa
Blastomussa
Cynarina
Scolymia

(7) Family Astrocoeniidae
Stylocoeniella
Palauastrea

(8) Family Siderastreidae
Siderastrea
Psammocora
Coscinaria

(9) Family Agaricidae
Pavona
Leptoseris
Gardenoseris
Pachyseris

(10) Family Pectiniidae
Pectinia
Echinophyllia
Oxypora
Mycedium

(11) Family Euphyllidae
Euphyllia
Catalophyllia
Physogyra
Plerogyra

(12) Family Oculinidae
Galaxea

(13) F. Dendrophylliidae
Turbinaria
Duncanopsammia
Tubastrea
*Heteropsammia (asterisked)

(14) Family Merulinidae
Merulina
Hydnophora
Scapophyllia

(15) F. Trachyphyllidae
Trachyphyllia

(16) F. Caryophyllidae
*Heterocyathus (asterisked)

Australia, and exhibit a number of different strategies in the way they release their gametes; for example, gametes may be negatively rather than positively buoyant in some species. The genera *Goniastrea* and *Leptastrea* appear to be among the most tolerant of all corals to heat stress and exposure to silty conditions. On reef flats, *Platygyra* and *Goniastrea* commonly develop into 'microatolls', with living coral surface confined to the vertical sides of the colony, surrounding a dead top, which sometimes gets colonised by algae or other corals. This morphology comes about when the top of the colony gets regularly exposed to the air and/or sunlight on low tide.

Figure 20.5 Examples of genera and species from the major coral families Acroporidae (*A–F*) and Faviidae (*G–L*). Species: *A, Acropora echinata; B, Acropora muricata; C, Acropora hyacinthus; D, Isopora palifera; E, Montipora incrasata; F, Astreopora gracilis; G, Favia maritima; H, Favites halicora; I, Goniastrea aspera; J, Platygyra sinensis; K, Echinopora gemmifera; L, Diploastrea heliopora*. (Photos: P. Muir.)

(3) Family Poritidae ('golf ball' corals, 'bommie' corals) (Fig. 20.6*A–C*)

Colonies in this family are massive, encrusting, plating or branching. In all cases, corallite walls are formed by upward growth of fine trabecular elements. This family contains one of the most remarkable zooxanthellate genera, *Porites*. Although species of *Porites* can exist in branching and plating forms, it is the massive species that provide examples of the longest-lived and largest solid coral forms. Growth in these massive colonies is in usually in increments of 1–2 centimetres per year, with two growth seasons being recorded by a 'dark' and a 'light' (dense and less dense) band. By counting and measuring these bands, it is possible to calculate the age of the colony and witness the history of environmental conditions that the colony has encountered. Colonies of over 700 years old have been identified and these sometimes form the basis of the huge 'bommies' found within lagoons and just outside the reef slope on reefs of the GBR. Not surprisingly, species of *Porites* form the subject of numerous studies of coral growth, vulnerability to environmental conditions, effects of coastal runoff and many other research topics.

Some situations favour the development of *Porites*-dominated coral assemblages, for example, calm, shallow offshore shoals or fringing reefs. *Goniopora* may also be present in these habitats. This genus is most notable for its long-columned polyps that are extended during the day. It is a popular aquarium subject and is apparently tolerant of high turbidity and low-light situations. Another genus, *Alveopora*, looks superficially like *Goniastrea* because its polyps are also extended during the day, but genetic studies indicate it may belong in the Acroporidae.

(4) Family Pocilloporidae ('brown stem corals') (Fig. 20.6*D–G*)

This is a family of genera with relatively small colonies, a branching mode of growth, simple, small corallites with reduced walls, only two septal cycles and always separated by coenosteum. All genera are brooders, and in some species of *Pocillopora*, planulae can be produced asexually by budding. *Pocillopora* is regarded as a 'weedy' genus, its species recruiting soon after catastrophic events and living short lives, with relatively regular monthly release of larvae throughout the year. Early observations of the breeding patterns of genera in this family led to the misconception that all corals release brooded larvae throughout the year, making larvae available for settlement on the reef at all times. This misconception influenced ecological thinking about reef corals until the 1980s, when mass broadcast spawning in corals was discovered (see Box 20.1). The best-known species are *Stylophora pistillata* and *Pocillopora damicornis*, which have been used for much published experimentation on the effects of water chemistry, light, temperature and other factors on corals.

(5) Family Fungiidae ('mushroom corals') (Fig. 20.6*H–K*)

Most species in this family occur as individual polyps that grow very large instead of multiplying, and that remain free living (unattached to the reef) during adult life. These corals contain zooxanthellae, are mostly regarded as non reef-building, or 'ahermatypic', because they generally do not contribute much to the reef structure (though there are notable exceptions when they and their dead predecessors occur in large mounds of thousands of individuals). Sexual reproduction often involves gonochoric gamete development (individuals of separate sexes), followed by simultaneous release of eggs or sperm by many individuals of the same species. Polyps settle on the reef to form an attached juvenile coral that may bud to form a 'stack' of cloned individuals: as these grow, they break off the stack and lie free on the reef floor. Generally found in the subtidal parts of reefs, fungiids may occur in great abundance and diversity.

(6) Family Mussidae ('spiky brain corals') (Fig. 20.7*A–C*)

Genera in this family have massive or encrusting colonies, with pronounced teeth on the septa and corallite walls formed by complex trabecular growth, which leads to very sturdy corallites, including some of the largest corallites to be seen in colonial corals. Polyp tissues are very fleshy and sometimes very colourful. These corals contribute great bulk to the reef structure.

Figure 20.6 Examples of genera and species from the major coral families: Poritidae (*A–C*), Pocilloporidae (*D–G*); Fungiidae (*H–K*) and minor family Oculinidae (*L*). Species: *A*, *Porites lutea*; *B*, *Porites cylindrical*; *C*, *Goniopora fruticosa*; *D*, *Pocillopora damicornis*; *E*, *Pocillopora eydouxi*; *F*, *Stylophora pistillata*; *G*, *Seriatopora hystrix*; *H*, *Fungia valida*; *I*, *Heliofungia actiniformis*; *J*, *Ctenactis echinata*; *K*, *Herpolitha webberi*; *L*, *Galaxea astreata*. (Photos: P. Muir except *G*, GBRMPA.)

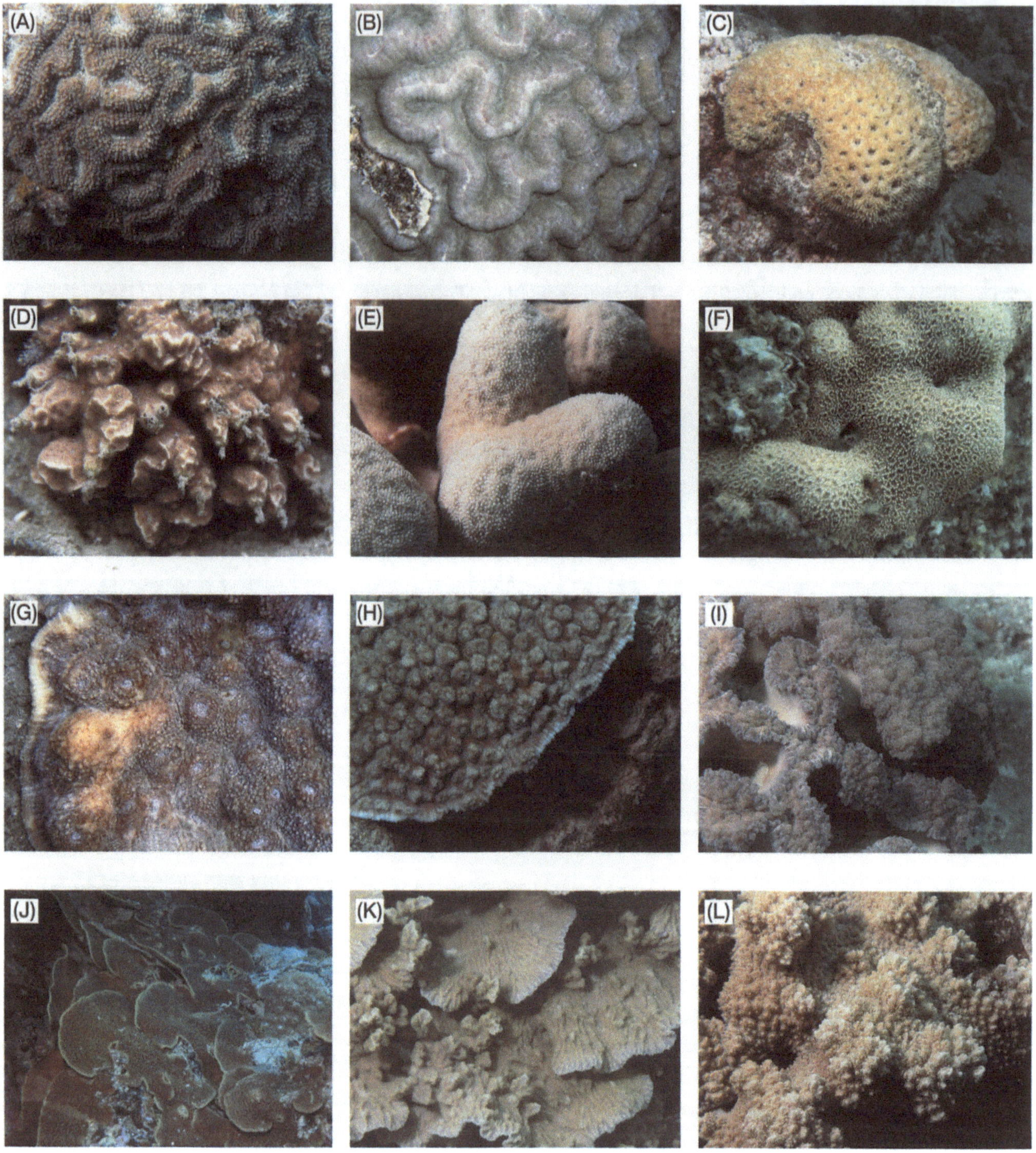

Figure 20.7 Examples of genera and species from the major coral family Mussidae (*A–C*) and minor families Siderastreidae (*D, E*); Agaricidae (*F*); Pectiniidae (*G, H*); Euphyllidae (*I*); Dendrophyllidae (*J*) and Merulinidae (*K, L*). Species: *A, Lobophyllia hemprichi; B, Symphyllia recta; C, Acanthastrea echinata; D, Psammocora haimeana; E, Coscinarea exesa; F, Pavona maldivensis; G, Echinophyllia aspera; H, Mycedium elephantotus; I, Euphyllia ancora; J, Turbinaria mesenterina; K, Merulina ampliata; L, Hydnophora exesa*. (Photos: P. Muir.)

One genus, *Acanthastrea*, is common in high latitude reefs, such as Moreton Bay.

Some of the minor coral families are shown in Fig. 20.6*L* and Fig. 20.7*D–I*.

PRESERVATION

It is important to remember that taking corals from the wild is illegal without a permit from the relevant organisation, usually the Great Barrier Reef Marine Park Authority, Moreton Bay Marine Park Authority or the state Environmental Protection Agencies. Corals are prepared for laboratory identification by bleaching with household bleach and rinsing in fresh water, then drying them bone dry in the sun. For genetic studies, a very small piece of tissue is taken and placed in a specified medium, most frequently full strength ethanol. For studies of tissue structures (e.g. gonads) and for histological sections, a sample should be fixed in 4% formaldehyde for 48 hours and then placed in 75% ethanol.

ADDITIONAL READING

Cairns, S. D. (2004). The azooxanthellate Scleractinia (Coelenterata: Anthozoa) of Australia. *Records of the Australian Museum* **56**, 259–329.

Deas, W., and Deas, J. (2005). 'Coral Reefs: Nature's Wonders.' (Western Australian Museum: Perth.)

Fautin, D. G. (2006). Hexacorallians of the World. http://geoportal.kgs.ku.edu/hexacoral/anemone2/index.cfm [Verified 21 March 2008].

Harrison, P. L., and Booth, D. J. (2007). Coral reefs: naturally dynamic and increasingly disturbed ecosystems. In 'Marine Ecology'. Chapter 13. (Eds S. D. Connell and B. M. Gillanders.) pp. 316–377. (Oxford University Press: Melbourne.)

Harrison, P. L., and Wallace, C. C. (1990). Reproduction, larval dispersal and settlement of scleractinian corals. In 'Ecosystems of the World: Coral Reefs'. (Ed. Z. Dubinsky.) pp. 133–208. (Elsevier: Amsterdam.)

Veron, J. E. N. (1987). 'Corals of Australia and the Indo-Pacific.' (Angus and Robertson: North Ryde.)

Veron, J. E. N. (2000). 'Corals of the World. Vols 1–3.' (Australian Institute of Marine Science: Melbourne.)

Wallace, C., and Aw, M. (2000). 'Staghorn Corals: A 'Getting to Know You and Identification Guide.' (OceanEnvironment: Singapore.)

Wallace, C. C. (1999). 'Staghorn Corals of the World: A Revision of the Coral Genus *Acropora* (Scleractinia; Astrocoeniina; Acroporidae) Worldwide, with Emphasis on Morphology, Phylogeny and Biogeography.' (CSIRO Publishing: Melbourne.)

Wallace, C. C., Lovell, E. R., and Alderslade, P. N. (1984). Corals: animal, vegetable or mineral? In 'Reader's Digest Book of the Great Barrier Reef'. (Ed. F. Talbot.) pp. 95–131. (Reader's Digest: Sydney.)

Wells, J. W. (1956). Scleractinia. In 'Treatise on Invertebrate Paleontology'. (Ed. R. C. Moore.) pp. F328–F444. (Geological Society of Australia and U. Kansas Press.)

21. Octocorals

P. Alderslade and K. Fabricius

TAXONOMIC OVERVIEW

Octocorals are sessile animals with a mobile larval phase that are only found in marine systems. Octocorallia are a subclass within the Class Anthozoa. The distinguishing characteristic of this subclass is that their polyps always bear eight tentacles (hence octo-coral), which are fringed by one or more rows of pinnules along both edges (Fig. 21.1). The popular term 'soft coral' points to the fact that most octocorals, in contrast to the related hard corals, have no massive solid skeleton. Instead, their colonies are supported by tiny calcareous granules called sclerites, which in most cases are separately embedded in the tissue and further supported either by a hydroskeleton or a proteinaceous or calcareous axis, or, in a few cases, by sclerite fusion into a solid structure.

The terminology used in the literature to refer to octocorals can initially be confusing. The term 'soft coral' is commonly used to refer only to those octocorals that have no massive skeleton or internal axis, but sometimes it also includes the sea fans, and occasionally it is used to refer to all Octocorallia. The term 'gorgonian' is most conveniently used when referring to octocorals (other than sea pens and blue coral) that arise from the substrate with the support of an internal axis. The term 'sea fan' is often used synonymously with the term gorgonian, however, some people use the term sea fan exclusively for gorgonians with a fan-shaped morphology.

Figure 21.1 Species of the Subclass Octocorallia are characterised by bearing polyps with eight tentacles, which in most cases are fringed by one or several rows of pinnules on both edges. (Photo: K. Fabricius.)

Taxonomically, there is no distinct morphological dividing line between soft corals and gorgonians, although the terms are useful to retain. Earlier separations into different groups have now been abandoned, because continua of intermediate forms exist. They are therefore now all included under the one scientific category—the Order Alcyonacea. Recent phylogenetic research suggests that further significant changes to the current family level classification are likely to occur in the next decade.

Octocorals are a diverse group of reef inhabiting organisms on Indo-Pacific coral reefs. There are three scientific orders in the Subclass Octocorallia: first, the Order Alcyonacea, which contains all species commonly known as soft corals and sea fans; second, the Order Helioporacea, (blue coral), and third, the Order Pennatulacea (sea pens). By far the greatest majority of octocorals found on the Great Barrier Reef (GBR) belong to the Order Alcyonacea (the true 'soft corals' and 'gorgonians'). Within this order, about 100 genera in 23 families have been described from shallow waters of the Indo-Pacific to date. In comparison, the Pennatulacea (sea pens) and Helioporacea (blue coral) play a minor role in the octocoral fauna of shallow coral reefs. In sea pens, nine genera in five families are presently known from shallow waters of the tropical and subtropical Indo-Pacific, where they tend to live in soft bottom habitats often only emerging at night. The blue coral is represented by a single species in the Indo-Pacific. Relatively few detailed octocoral studies exist (e.g. from the GBR and other parts of Australia, New Guinea, New Caledonia, Micronesia, Japan, south-eastern Africa and the Red Sea), and there are still major gaps in the understanding of octocoral biogeography. Furthermore, the number of shallow water Indo-Pacific octocoral species is unknown, as many species still await taxonomic description and many genera are in urgent need for revision. The only systematic taxonomic inventory of Indo-Pacific coral reefs is from Palau in Micronesia, where 150 species have been recorded. On the GBR, the species number might be similar, but verification of such an estimate will remain impossible until taxonomic research advances.

Biology and ecology

Octocorals of the GBR include a diverse range of species with widely contrasting biological properties and ecological requirements. While many species appear to be quite long-lived and slow-growing, some octocoral species (especially among the family Xeniidae) are fast colonisers with a short life expectancy. Some large *Sinularia* colonies (family Alcyoniidae) are probably hundreds of years old, and some of the large gorgonian colonies may also be many decades old. Life expectancy and growth rates of most soft corals and gorgonians are largely unknown, but many of them appear to be quite slow-growing, extending by only one to a few centimetres per year. Octocoral colonies also commonly shrink, for example, when torn by storm waves or damaged by moving rubble or predation; as in other modular organisms there is therefore only a weak relationship between size and age.

Dispersal strategies also vary greatly among species, and include asexual propagation and sexual reproduction. Rapid colonisation of small patches of substratum often includes the asexual generation of daughter colonies at the terminal ends of stolons ('runners') formed by the mother colony. Other forms of asexual propagation involve budding of miniature colonies that fall off the mother colony and settle nearby, and fragmentation, when a middle part of a colony dies yet the edges survive and reorganise into complete colonies. Larger patches of bare substratum tend to be colonised by the settlement of pelagic larvae, the product of sexual reproduction. Some octocoral species are gonochoric, that is, males and females are separate colonies, while other species are hermaphroditic, with all mature colonies producing both male and female gametes. Two modes of sexual reproduction exist: first, 'broadcasting' species release their male and female gametes into the water column where they are fertilised; the developing pelagic larvae are dispersed by currents over many kilometres until ready to settle some days later. Second, 'brooding' species have their eggs fertilised within the mother colony; in these cases the resulting larvae develop on the colony surface until they are ready to detach and settle near the mother colony some days later.

The food of octocorals consists of small suspended plankton particles filtered from the water column; they rely on water currents to carry the particles towards their polyps. Actively swimming zooplankton is not ingested as the stinging cells are only weakly developed and unable to paralyse large zooplankton items. Much of the food is therefore derived from phytoplankton, minute detrital particles, flagellates and very small zooplankton. More than half of the warm shallow-water Indo-Pacific octocorals also contain endosymbiotic dinoflagellate algae (often called zooxanthellae) in their tissue, which fix carbon through photosynthesis and hence supply energy to their host. The coral in return provides nutrients and shelter to the algae. This finely tuned symbiosis between animal and alga depends on the availability of light for photosynthesis. Symbiotic taxa (also called zooxanthellate or phototrophic taxa) include many representatives within the 'true' soft corals (the Alcyoniina group), especially many of the genera within the abundant families Nephtheidae, Alcyoniidae and Xeniidae, but also members of most of the other large octocoral groupings. In contrast, most sea fans, and also several of the soft corals (e.g. *Dendronephthya*) do not contain dinoflagellate endosymbionts. These asymbiotic (azooxanthellate or heterotrophic) taxa are suspension feeders that strongly depend on currents to transport food particles towards the polyps; they are mostly found in high-flow environments. Heterotrophic taxa are easily visually distinguished from their phototrophic relatives by their bright yellow, orange, red, pink, purple or snow-white colouration.

As sessile organisms without a protective skeleton, octocorals would appear to be vulnerable to predation. However, with the exception of a few snails (e.g. the cowry shell *Ovula ovum*), fish (e.g. a few species of butterfly fish that selectively feed on coral and octocoral polyps) and the odd or accidental grazing by an echinoderm such as *Diadema* sea urchins, remarkably few organisms are able to feed on octocorals, and overall feeding pressure appears low. Many species are protected against predation, fouling by algae or overgrowth by neighbouring organisms through feeding-deterrent, toxic or allelopathic secondary metabolites. Many of these substances have been investigated for their bioactivity, and some may one day become pharmaceutically relevant. Octocoral colonies, although not contributing to reef growth, nevertheless provide shelter to a range of other reef-inhabiting organisms. For example, some species of brittle star (Ophiuroidae), feather star (Crinoidea), shrimps, ctenophores and fish (gobies and pygmy sea horses) are exclusively found living on the surface of specific octocoral colonies. Most of these associates use the octocoral colony exclusively as a perch or for shelter; however, a few of these associates appear to also feed on the mucus of the octocorals.

After hard corals, octocorals are the second-most common group of macrobenthic animals on the GBR. Mean octocoral cover of the GBR regions ranges from 3% to 35% on outer reef slopes, but cover can be as high as 70% in current-swept yet wave-protected environments such as channels between reefs or islands, and near zero on wave-exposed macro-algal dominated turbid and silty inshore reef crests.

Octocorals are highly diverse, not only taxonomically, but also ecologically, including species with widely contrasting ecological niches and life history strategies. The GBR is situated on a wide continental shelf, and due to this geomorphologic setting contains a remarkable range of marine habitats. Habitats for octocorals in the GBR include wave-beaten outer barrier reef walls with steep drop-offs and oceanic water clarity, more protected midshelf reefs with sheltered lagoons and current-flushed flanks, extensive inter-reefal areas with soft bottom environments and outcrops of hard substratum, seagrass meadows and *Halimeda* mounds, and inshore coral reefs within the reach of terrestrial influences. Each of these habitat types houses a distinct octocoral assemblage. For example, outer-shelf reefs contain diverse octocoral communities that are often characterised by high abundances of members of the family Xeniidae. Midshelf reefs have the highest species richness of all reefs on the GBR. Deep-water reef slopes and inter-reefal habitats are inhabited by azooxanthellate taxa such as many gorgonians and *Dendronephthya*, as well as ubiquitous and tolerant taxa such as *Sinularia* and *Sarcophyton*. Inshore reefs with fluctuating salinity and water clarity contain some species that are rarely found in clear-water habitats (e.g. *Sinularia flexibilis*, and several species of *Briareum* and *Solenocaulon*), as well as a subset of those genera found in the clear-water habitats. Many of the Xeniidae and Nephtheidae are missing on turbid inshore reefs.

Ecological surveys have shown that the taxonomic richness of octocorals is strongly related to water clarity and amounts of sediments deposited, and that richness and the abundance of specific taxa are therefore more suitable indicators of change in environmental conditions than are hard coral cover, octocoral cover and hard coral richness, which are poorly explained by these variables. Changes in taxonomic richness and community composition in octocorals have therefore been suggested to be suitable as indicators of past and recent disturbance by poor water quality on the GBR and other reef environments.

Species richness at a given site is affected by three factors. First, the biogeographic location and colonisation history of a region determines the regional species pool present. On the GBR, the species richness in octocorals strongly attenuates with increasing latitude: many more genera and species occur in the tropical far northern part than on the southern end of the GBR. Second, environmental conditions determine what cross-section of the local species pool occurs at that locality. In octocorals, abundances of particular taxa are strongly determined by the physical environment, especially turbidity, light availability and water currents. Third, at any point in time local and regional species richness also depend on disturbance history, specifically the nature and intensity of the disturbance, and the time since past disturbances have removed colonies. For octocorals, disturbances include storms with high wave energy (dislodging or damaging colonies), episodes of high water temperatures (causing coral bleaching), chronically reduced water clarity (reducing photosynthesis) and sedimentation (smothering colonies or hampering larval settlement). After a disturbance, the speed and efficiency of recolonisation of a taxon will determine whether the taxon will again be present or not: fast colonisers continuously reestablish if propagules from surviving colonies (locally or further upstream) are available, whereas slow-colonising or slow-growing taxa are unable to quickly return to their previous abundance. Similarly, chronic disturbance such as water pollution reduces diversity, because only persistent species can survive and flourish. In order to understand regional and local biodiversity patterns, biogeographic settings, environmental requirements of taxa, and consequences of disturbances need to be investigated simultaneously.

A brief guide to major octocoral genera of the Great Barrier Reef

The identification of octocorals is based primarily on colony form, the nature, location and arrangement of the sclerites and the nature of the central axis if one is present. (See Box 21.1 for collection and preservation methods.) Octocorals are modular animals that, except in rare occurrences, are constructed of a number of polyps united in a common tissue mass called the coenenchyme. The growth form of a colony may be stolonate, membranous, encrusting, fleshy, erect, and either whip-like or branching (Fig. 21.2). In some species, a central axis is present (Fig. 21.3). If such an axis contains sclerites, the coenenchyme is divided into an outer cortex, which includes the polyps, and an inner axial medulla. The medulla may be continuous or segmented, but always contains densely grouped sclerites that can be free or may be fused to various degrees and combined with gorgonin, a horn-like material. Most gorgonians, however, have an axis that does not include sclerites. In this case, the outer layer with the polyps is just called the coenenchyme, and the inner layer is simply called the axis that may be continuous or segmented, and is often made of gorgonin with or without the inclusion of various amounts of fibrous calcium carbonate.

Two types of polyps are found in octocorals. The first type, autozooids, are generally responsible for feeding and reproduction, and are present in all species. Autozooids always have eight tentacles that, except in a few cases, bear pinnules along each edge. The second type of polyp, called a siphonozooid, is primarily for water circulation, and is found in representatives of several families. Siphonozooids are very small, and have rudimentary or no tentacles. Taxa that have both polyp types are referred to as dimorphic (e.g. the *Lobophytum*).

Calcareous sclerites and axis structures are other important features used to identify octocorals. Sclerites are present in most species, ranging in size from 0.02 mm to >10 mm. Depending on their shape, they

BOX 21.1 TECHNIQUES USED TO COLLECT, PRESERVE AND INVESTIGATE OCTOCORAL COLONIES

When taxonomic identification is attempted, a photograph showing the entire colony and its growth form is advantageous, and close-ups of details of the surface and polyp structures are useful for reference. Although with practice it is possible to identify a number of octocorals to genus level underwater or from photographs, it is rare that they can be identified to species level because a compound microscope is needed to investigate sclerites in detail. Most soft corals have different sclerites in the upper polyp-bearing surface of the colony (lobes, branches), the interior of the polyp region, the surface and interior of the base, and within the polyps. The arrangement of the sclerites within the polyps and polyp tentacles can also be informative as it can vary widely between species. For a full diagnosis, until a satisfactory level of field confidence is achieved, it is therefore necessary to collect a sample with all of the main colony regions present. Note that a sampling permit is required for collection on the GBR, and in general, sampling should be minimised wherever possible. If a sampling permit exists and sampling is considered essential, with small colonies the whole specimen may have to be collected. With larger specimens this is neither practical nor necessary. In broad, thickly encrusting colonies it is sufficient to remove a segment shaped like a pie-slice, which includes upper lobes (branches or ridges, and polyps), the surface layer of the colony side down to the base, and the attached interior portions. In gorgonians, one branch is sometimes sufficient to obtain surface, interior, and polyp material for sclerite examination, but for species determination more material may have to be collected, as sclerites can vary between branches and from colony base to upper (younger) colony parts. The branching pattern is an extremely important diagnostic feature that may be captured by a good photograph to minimise sampling. Growth form is especially important in the family Ellisellidae because a number of genera only differ in this character.

Long term storage of small to medium sized octocoral specimens is best in 70% ethanol (ethyl alcohol) in fresh water, in which colonies will last indefinitely. It is not recommended to initially fix octocorals in dilute formalin, as the calcareous octocoral sclerites are eventually eroded by the acids derived from the oxidation of formaldehyde to formic acid (even if buffered), which will make identification impossible and complicate DNA analyses. If dilute formalin is initially used to reduce shrinkage, the storage alcohol needs to be regularly changed until all traces of formaldehyde are removed. Numerous specimens reliably tagged, or individually placed in plastic bags with holes, can be kept together in a single large container suitable for alcohol storage. Larger gorgonian branches may have to be air-dried unless a container of sufficient size can be found, but air-dried samples are more difficult to work with, they are more prone to abrasion and breakage, and susceptible to mould and insect attack. A compromise then is to keep a portion of a large specimen in alcohol.

To examine sclerites they must be freed from the coenenchyme in which they are embedded by the use of concentrated bleach (sodium hypochlorite), which dissolves the organic tissues and leaves the sclerites untouched. Unless the sclerites are large, a sample <2 mm² is sufficient for examination. The sample is cut from the relevant part of the colony with a scalpel under a dissecting microscope and placed into 1–2 drops of bleach on a microscope slide. Once the bubbles have ceased, the sclerites are spread out by stirring, a cover-slip is applied and the sample is investigated under a compound microscope. Preparations that have dried can be rehydrated with 1–2 drops of water.

Permits are required, see Chapter 12.

have common names such as spindles, rods, capstans, platelets, thorn scales, double heads, rooted-leaves, 8-radiates, double stars and needles, and their details will differ from one species to another (see figures accompanying the descriptions of the genera).

The brief descriptions of some of the more commonly encountered genera presented here provide a first glimpse at the diversity of features, shapes, colours and forms found in octocorals of the GBR. Each genus is introduced by a brief text, together with a plate showing a representative underwater photograph, and the forms of sclerites typically found in that genus. In some cases, these brief descriptions will be insufficient to facilitate the reliable identification of a genus, as the variability between species within genera can be high (a single photograph often does not capture the range of shapes and forms found within a genus). To prevent the premature naming of such genera, we list the names of similar looking genera in the text, even if they are not shown in this chapter. It is advised to refer to a more comprehensive field guide and taxonomic literature before an identification is attempted.

Order Pennatulacea

The sea pens, or Pennatulacea (Fig. 21.4), are a diverse but relatively unknown group of octocorals. Nine genera in five families are represented in shallow waters of

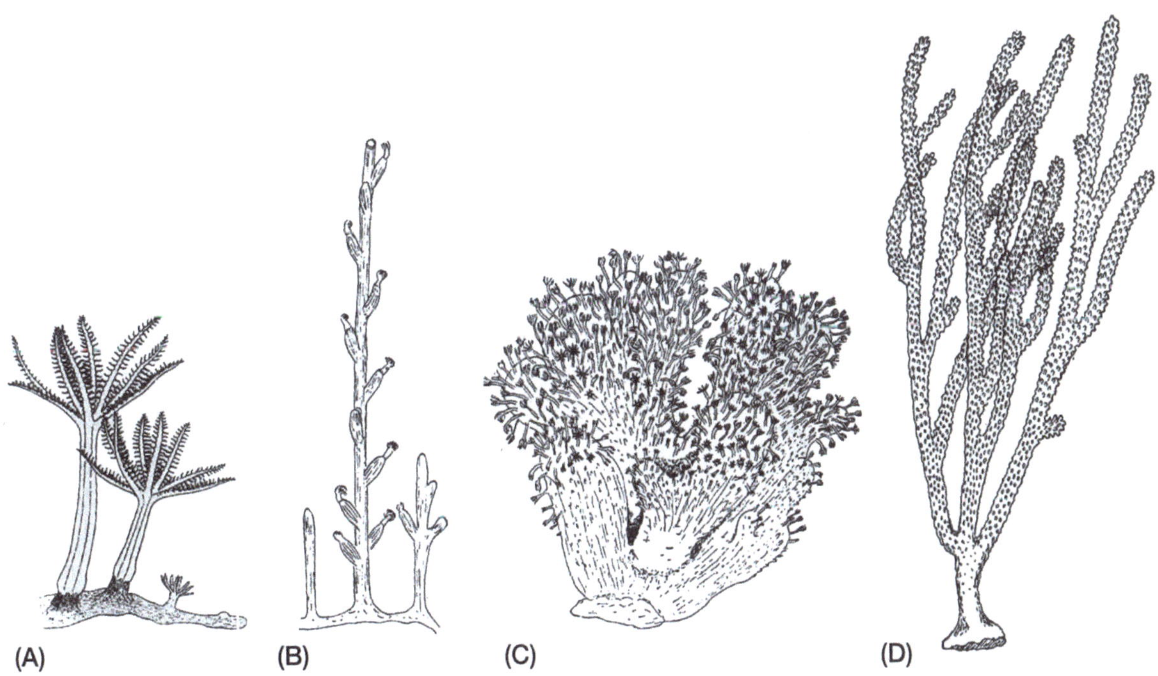

Figure 21.2 The most commonly encountered colony growth forms. *A*, single polyps connected by stolons as found in *Clavularia*; *B*, tall axial polyps that bud off lateral polyps and are united at their base with stolons; *C*, prototype 'soft coral' where polyps are embedded in a thick fleshy tissue mass called coenenchyme; *D*, prototype 'gorgonian' with upright growth form and polyps arranged around a central axis. (Modified from Hyman 1940.)

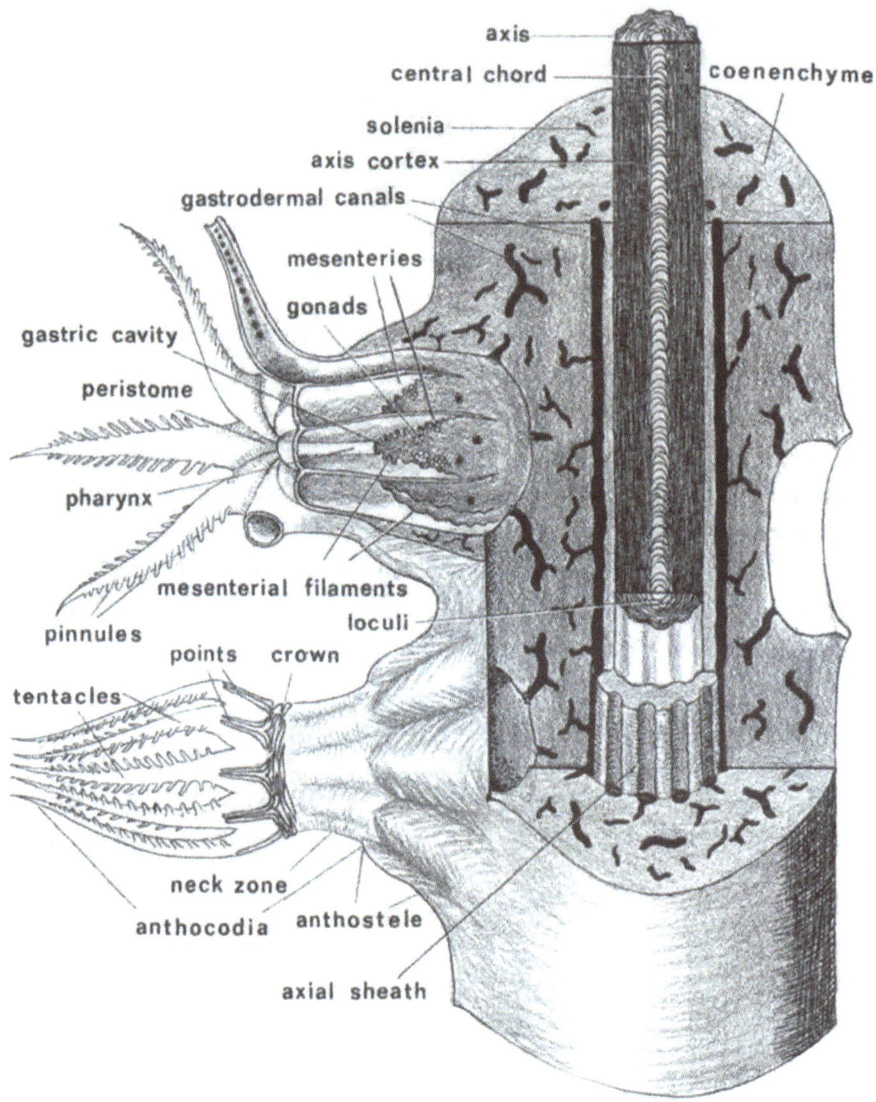

Figure 21.3 Anatomical details of octocoral polyps, here arranged along an axis of a gorgonian colony belonging to the family Plexauridae. (Modified from Bayer 1981.)

the tropical and subtropical Indo-Pacific. Because they inhabit soft bottom habitats and often are completely contracted during daylight and only emerging at night, they are only infrequently encountered by divers and snorkellers, and will not be covered in detail in this chapter. Pennatulacea are characterised by their large central primary or axial polyp ('oozooid'), which is usually supported by a proteinaceous or scleritic axis. Half of the oozooid forms the colony 'foot' or penduncle that digs into sand or mud, anchoring the colony in soft substratum. The rachis reaches into the water column when expanded, and bears the autozooids and siphonozooids. In some shallow-water species, the emergent part looks just like a feather (hence the name sea pen), others are round and columnar.

Order Helioporacea

Family Helioporidae *Heliopora* (Fig. 21.5). This order and family is represented by a single species on the GBR, *Heliopora coerulea. Heliopora coerulea* has a calcified branching skeleton with a brownish-blue, very smooth surface and blue core. Colonies can reach several metres in diameter. Polyps when expanded are <3 mm in diameter, and very fine. Its distribution is restricted to shallow-water areas in which the water temperature remains above 22°C all year round. Similar genera: none.

Figure 21.4 Pennatulacea, or sea pens, generally inhabit soft bottom habitats and are often hidden in the sand during the day. *A, Cavernularia* sp.; *B, Virgularia* sp. (Photos: K. Fabricius.)

Figure 21.5 The blue coral, *Heliopora coerulea* (family Helioporidae). *A*, whole colony; *B*, cross-section of a broken branch and close-up of the colony surface. (Photos: K. Fabricius.)

Family Clavulariidae. Genera in this family consist of cylindrical or bluntly conical polyps usually joined only at their bases by reticulating stolons, which may coalesce into thin membranous expansions. In some genera, tall cylindrical polyps develop long secondary polyps that resemble branches. In a few instances, polyps are also connected by extra, transverse, bar-like stolons above the basement layer. Nearly all of the species in this family have sclerites, and it is not unusual

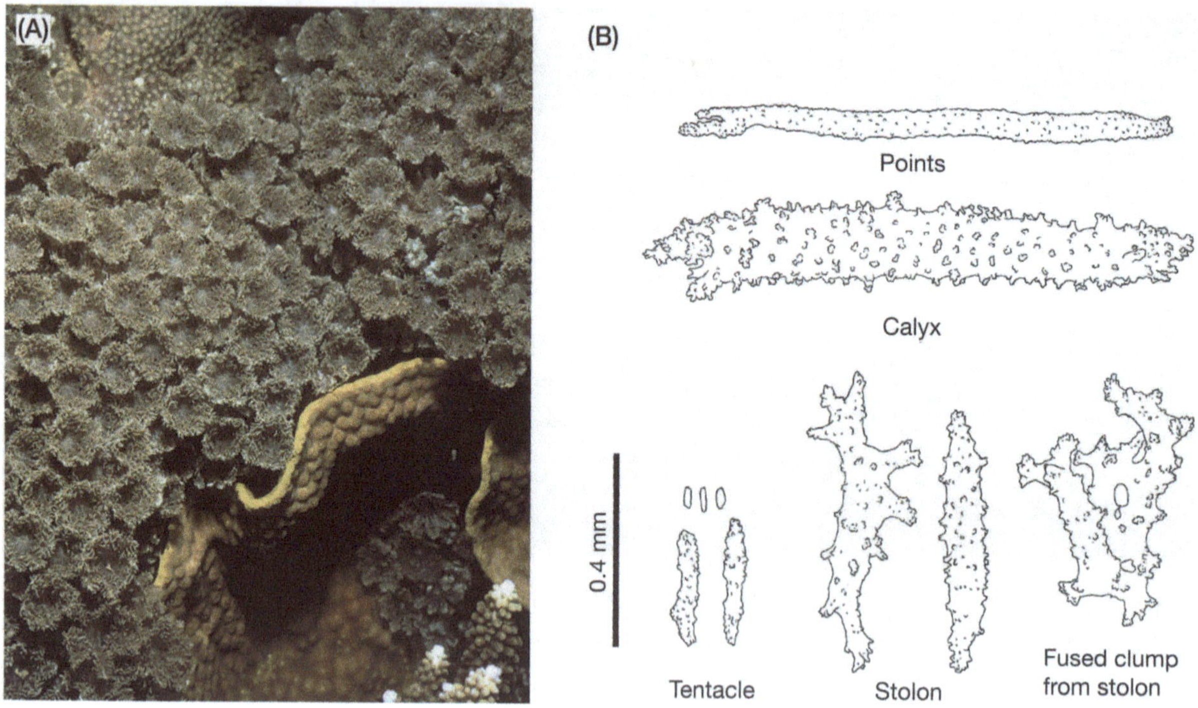

Figure 21.6 *Clavularia* sp. (family Clavulariidae). *A*, colony overgrowing a hard coral; *B*, some sclerites representative of this genus. (Photo: K. Fabricius; illustration: P. Alderslade.)

for these to be fused into clumps or tubes. Sclerites include smooth branched rods, and prickly or tuberculate 6-radiates, spindles and platelets.

Clavularia (Fig. 21.6). Individual polyps are large, 10–40 mm tall, united by ribbon-like stolons. The polyp head is completely retractile into the lower part of the body, which is stiffened by spindle-shaped sclerites to form a calyx. Sclerites of the stolons are warty spindles, often fused, while those of the tentacles are short rods and small platelets. Colour: brown, cream or greenish. On the GBR, *Clavularia* is restricted to latitudes north of 20°S. Similar genera: *Anthelia* and *Sansibia* (family Xeniidae).

Family Tubiporidae (Fig. 21.7). *Tubipora* (Organ-pipe coral) is the only genus in this family. Colonies are hemispherical and massive to thick and encrusting, consisting of a large, solid skeleton of red, hard calcareous tubes connected at regular intervals by horizontal, stolonic platforms. Each tube is formed and occupied by a single polyp, which is connected to the other polyps in the colony by canals inside the horizontal plates. Stolons and non-extendable parts of the polyps are covered in a thin, soft tissue layer.

Tentacles are cream or white in colour and do not always bear visible pinnules. The red skeletal tubes are formed from fused sclerites and the tentacles contain minute platelets. Uncommon but widely distributed, colonies are sometimes >20 cm in diameter. Similar genera: none.

Family Alcyoniidae. Members of this family often dominate inshore octocoral communities. Growth forms are often massive or encrusting, with some colonies measuring several metres across. Most colonies have a bare basal section (the stalk or trunk), and an upper, polyp-bearing part that may be flat, undulating, or divided into lobes, ridges or short branches. Monomorphic and dimorphic forms are represented; for example, siphonozooid polyps are present among the autozooids in some genera. The interior of a colony may be compressible and jelly-like if the sclerite content is low, and rigid and solid if it is high. Sclerites

Figure 21.7 *Tubipora* sp. (Organ-pipe coral, family Tubiporidae). *A*, close-up of polyps; *B*, a broken colony showing the skeletal structure; *C*, some sclerites found in the polyp tentacles. (Photos: *A*, *B*, K. Fabricius; *C*, P. Alderslade.)

include tuberculate or prickly spindles, clubs, 6– or 8–radiates, ovals and dumb-bells.

Sinularia (Fig. 21.8). Members of this genus form encrusting colonies, with the upper surface formed into knobs, ridges, simple or branched lobes. Very diverse in shape, encrusting forms can grow several metres in diameter. The polyps are retractile and monomorphic. The interior sclerites are large warty spindles; surface sclerites are mostly clubs. Colour: brown, cream, yellow or green. *Sinularia* is a very species-rich genus that is widely distributed and found even at high latitudes and dark depths (despite the presence of zooxanthellae). Similar genera: *Lobophytum*, *Lemnalia* and *Paralemnalia*.

Sarcophyton (Fig. 21.9). Colonies have a distinct stalk and a rounded capitulum with a straight or undulating margin. Members of this genus can grow up to a metre in diameter. Polyps are abundant and dimorphic. The autozooids are retractile. The interior sclerites are spindle-shaped; surface ones include many clubs. Colour: brownish, cream, yellow or green. The genus is very widely distributed. Similar genus: *Lobophytum*.

Lobophytum (Fig. 21.10). Colonies are encrusting, often large. The upper surface is lobate, digitate or with ridges, and covered in abundant dimorphic polyps. Autozooids are retractile. Sclerites of the interior are spindle to barrel-shaped with tubercles often in girdles.

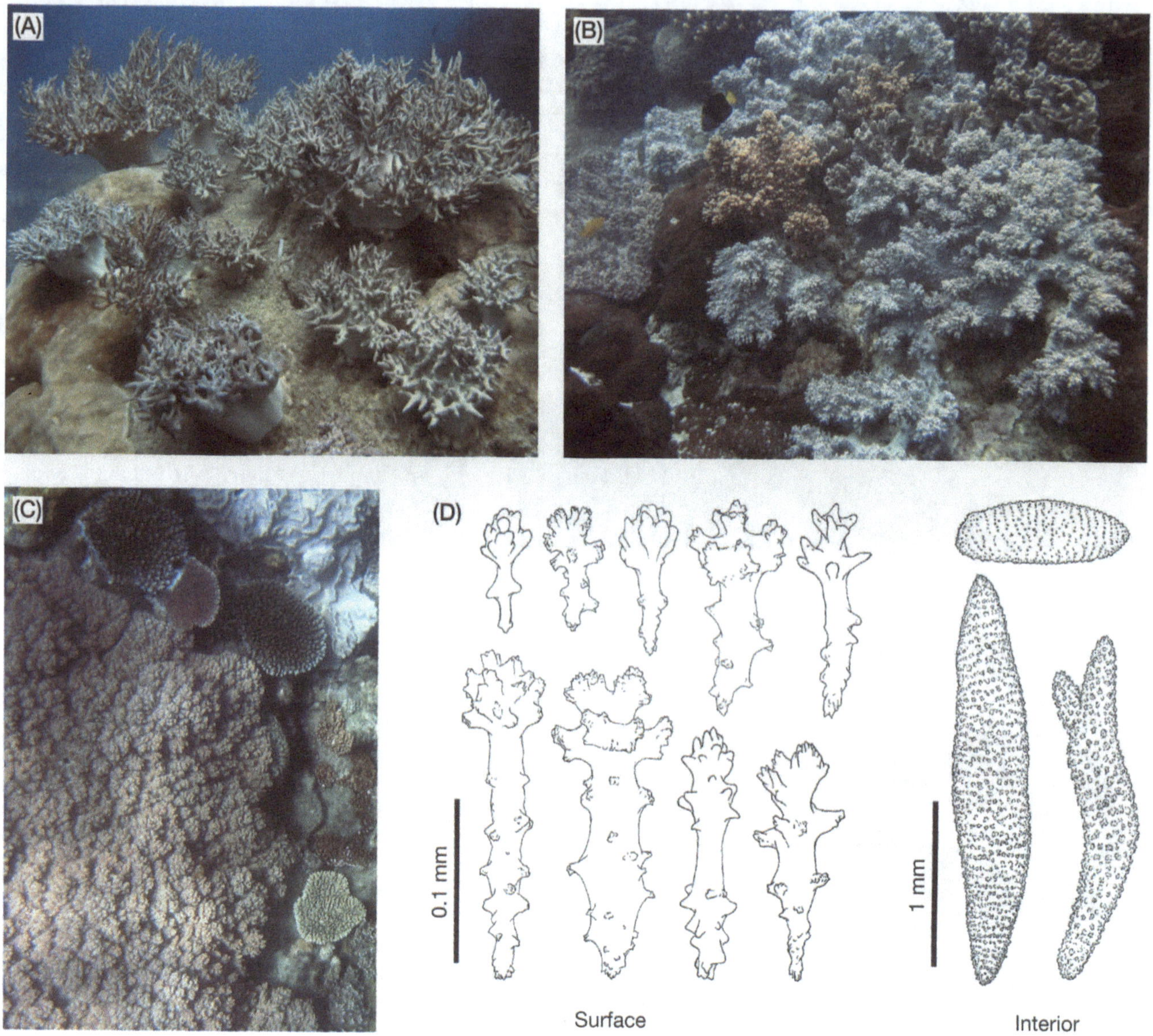

Figure 21.8 *Sinularia* spp. (family Alcyoniidae). *A–C*, some of the many and diverse growth forms found in this species-rich genus; *D*, some sclerites representative of this genus. (Photos: K. Fabricius; illustration: P. Alderslade.)

Surface sclerites are generally poorly formed clubs. Colour: brown, cream, yellow or green. Members of this genus are widely distributed, and abundant in high light environments. Similar genus: *Sarcophyton*.

Family Nephtheidae. A wide range of genera with a wide range of ecological characteristics are grouped in this family, some probably erroneously. Most have bushy, globe-shaped or arborescent growth forms, while a few are massive, or consist of finger-like lobes united by a common base. Many genera contain highly coloured species that are azooxanthellate. In most cases, the polyps, singly or in small clusters, are more or less restricted to the upper and outer twigs or branches. In a few cases, polyps grow directly on main branches. A small number of broad primary polyp canals extend longitudinally through the stem of arborescent colonies, subdividing into groups that extend up into the distal lobes and branches. The canal

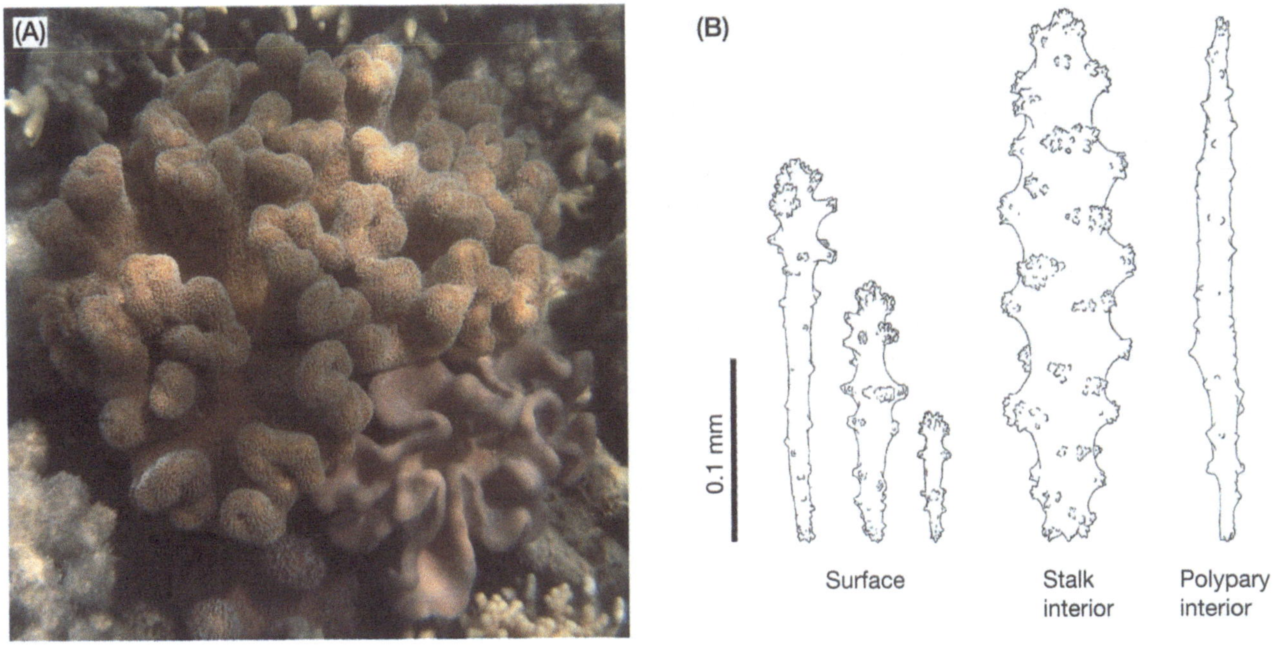

Figure 21.9 *Sarcophyton* sp. (family Alcyoniidae). *A*, colonies with expanded and retracted polyps; *B*, some sclerites representative of this genus. (Photo: K. Fabricius; illustration: P. Alderslade.)

walls are generally thin, with few sclerites, permitting colonies to easily inflate with water, dramatically increasing their size ('hydroskeleton'). Although some genera are soft and floppy, others have a rough or distinctly prickly feel due to long, protective sclerites projecting beyond the polyp head. In the stem and branches, the sandpaper-like texture can be attributed to numerous, strongly sculptured, spiny sclerites in the surface layer. Sclerite forms include prickly needles, leafy clubs, irregular shaped spiky forms, and tuberculate and thorny spindles, often extensively ornamented along one side. Members of this family are most often found in clear-water habitats.

Nephthea (Fig. 21.11). Colonies are arborescent. Non-retractile polyps, supported by a bundle of sclerites, are arranged in lobes. Sclerites are irregular, spindle or caterpillar-like. Colour: generally brown to green. *Nephthea* are found mostly in clear waters, often growing in clusters of several to many colonies. Similar genera: *Stereonephthya*, *Chromonephthya*, *Lemnalia* and *Paralemnalia*.

Dendronephthya (Fig. 21.12). Colonies are arborescent with the terminal twigs forming rounded, umbellate or tree-like groupings. Non-retractile polyps are arranged in small groups on the end of the twigs, protected by large supporting sclerites. Colonies are brightly coloured, azooxanthellate. The sclerites are similar to *Nephthea*. *Dendronephthya* are found in high currents, generally growing as individual colonies. Similar genera: *Scleronephthya*, *Chromonephthya*, *Nephthea* and *Stereonephthya*.

Lemnalia (Fig. 21.13). Colonies are arborescent, formed from bare stalks and branches and generally thin twigs. Nonretractile polyps are isolated on the twigs. Colour: cream to brownish. Interior sclerites are long, thin needles. Surface forms include capstans, crescents and brackets. *Lemnalia* are found in clear waters, growing as individual colonies, small clusters or large assemblages. Similar genera: *Nephthea* and *Paralemnalia*.

Capnella (Fig. 21.14). Colonies are small, arborescent to lobed. The lobes are crowded with incurved polyps that are covered in club-like sclerites. Interior sclerites are often globular, with the surface ones leafy to spiky capstans. Colour: grey or beige. *Capnella* are found in a wide range of habitats, as individual colonies or in small groups of colonies.

Figure 21.10 *Lobophytum* sp. (family Alcyoniidae). *A–C*, some of the growth forms found in this genus; *D*, some sclerites representative of this genus. (Photos: K. Fabricius; illustration: P. Alderslade.)

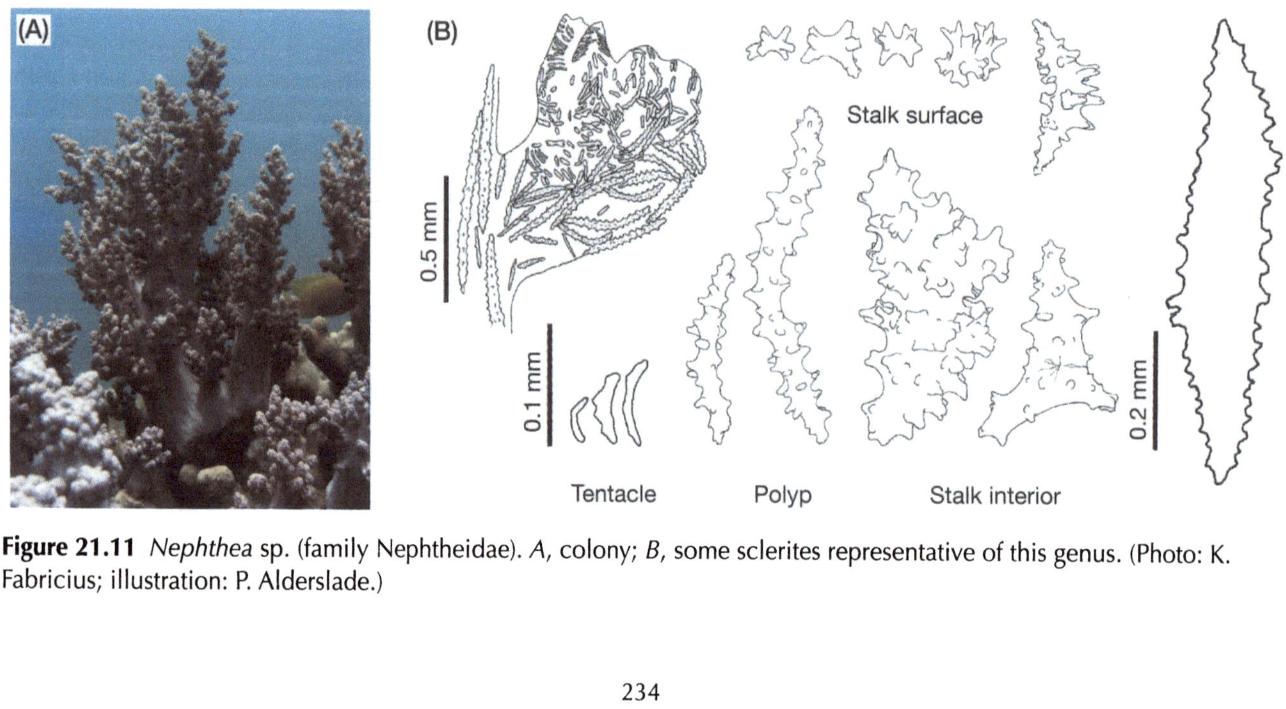

Figure 21.11 *Nephthea* sp. (family Nephtheidae). *A*, colony; *B*, some sclerites representative of this genus. (Photo: K. Fabricius; illustration: P. Alderslade.)

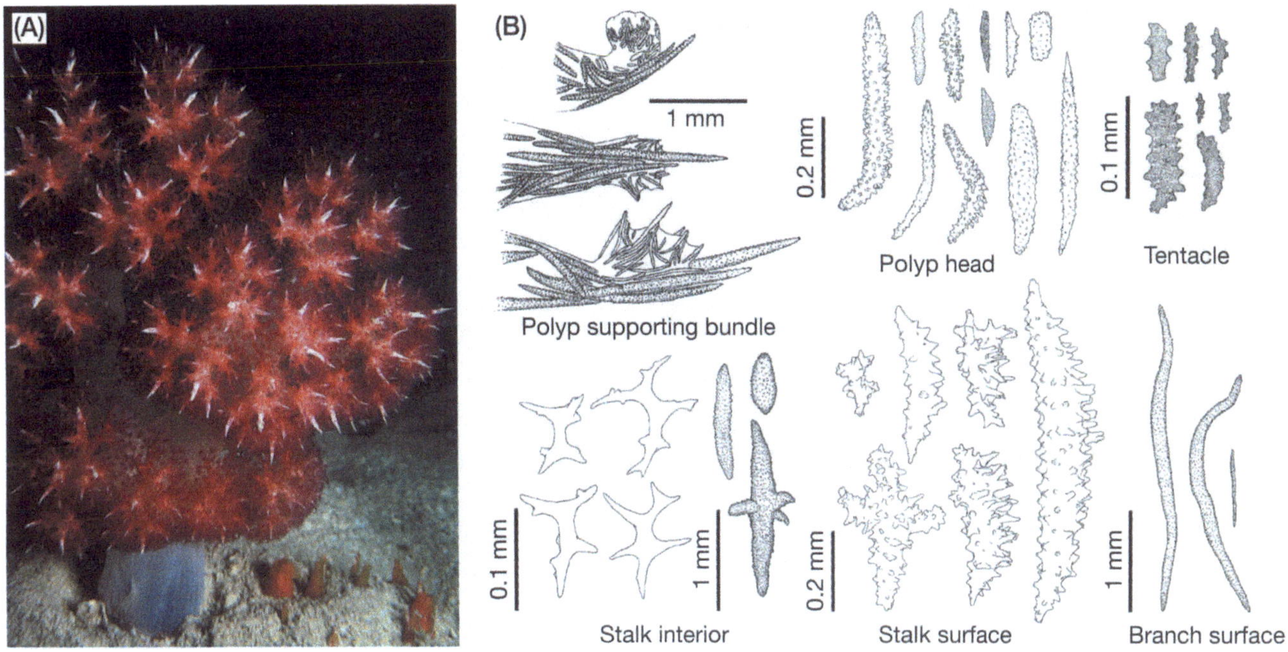

Figure 21.12 *Dendronephthya* sp. (family Nephtheidae). *A*, colony; *B*, some sclerites representative of this genus. (Photo: K. Fabricius; illustration: P. Alderslade.)

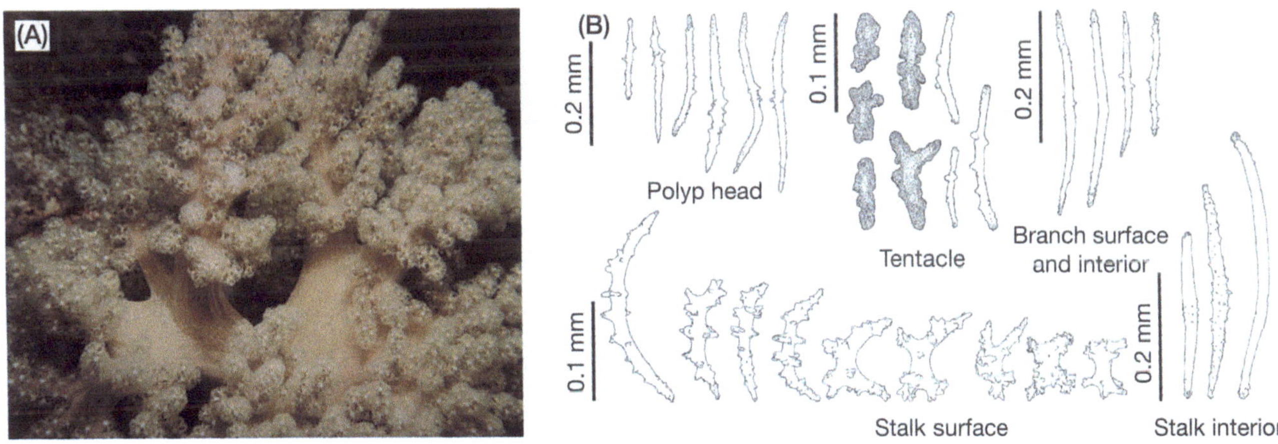

Figure 21.13 *Lemnalia* sp. (family Nephtheidae). *A*, colony; *B*, some sclerites representative of this genus. (Photo: K. Fabricius; illustration: P. Alderslade.)

Family Xeniidae. Members of this family often dominate offshore octocoral communities. Xeniid genera are mostly small and soft, and often quite slippery to touch. Colony growth forms are lobate, or thin and membranous with small retractile or tall polyps, or short, cylindrical stalks terminating with a domed polyp-bearing region. All species are zooxanthellate, often with opalescent pastel colours (white, pink, light iridescent blues and greens); some inshore species are dark brown. Some species have pulsating polyps, where the autozooids continually open and close the tentacle basket. Members of one genus have siphonozooids at least during some periods of their lives. Not all of the species in this family have sclerites.

Figure 21.14 *Capnella* sp. (family Nephtheidae). *A–B*, colonies; *C*, some sclerites representative of this genus. (Photos: K. Fabricius; illustration: P. Alderslade.)

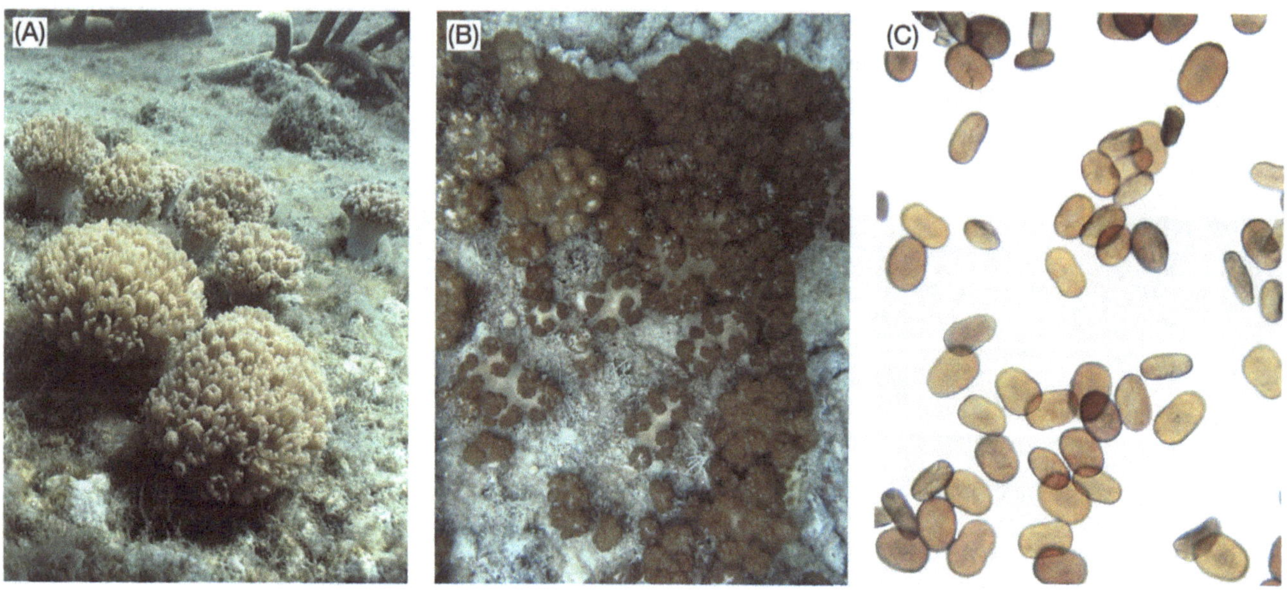

Figure 21.15 *Xenia* sp. (family Xeniidae). *A–B*, colonies; *C*, the small platelets representative of this genus. (Photos: K. Fabricius.)

Where sclerites are present, they are nearly always minute platelets or corpuscle-like forms, with almost smooth surfaces, often appearing opalescent. Only one genus has sclerites in the form of minute rods with a coarse crystalline surface. A microscopic feature of the family, observed only in histological preparations, is that only the dorsal pair of mesenteries retain their filaments in the adult polyps; the filaments on the other six mesenteries are absent or rudimentary.

Xenia (Fig. 21.15). Members of the genus *Xenia* have sparsely branched colonies with the nonretractile polyps confined to the domed branch ends. Individual colonies tend to be small (often <5 cm in diameter), but often larger super-colonies (colony clusters) are formed. Sclerites are minute platelets. Colour: white, grey to blue, yellow to brown. *Xenia* are abundant in clear waters; rarely found in inshore waters where they are dark brown. Similar genera: *Heteroxenia* and *Asterospicularia*.

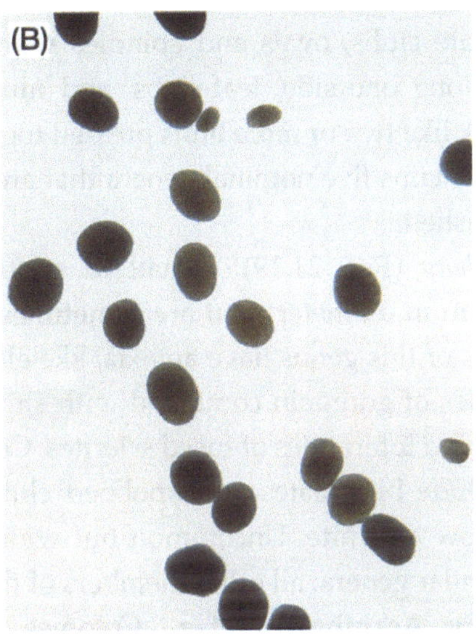

Figure 21.16 *Efflatounaria* sp. (family Xeniidae). *A*, colonies; *B*, the small sclerites representative of this genus. (Photos: *A*, K. Fabricius; *B*, P. Alderslade.)

Efflatounaria (Fig. 21.16). Species in this genus have highly contractile polyps that can appear to have retracted. Daughter colonies are produced from 'runners' (stolonic outgrowths of terminal polyps). Sclerites are minute platelets or irregularly subspherical bodies. Colour: white, pink, yellow or blue. They are abundant in clear waters. Similar genus: *Cespitularia*.

Family Briareidae. *Briareum* is the only genus in the family. On the GBR, different species of *Briareum* grow as thin sheets encrusting rock, or dead or live substrate. Most species resemble encrusting or lobular soft corals, but the division of the coenenchyme into a cortex and a medulla places them with other gorgonians. The polyp-bearing surface of the colony may be virtually smooth, or it may have calyces that can be up to 10 mm tall. The basal layer, or medulla, which is attached to the substratum is deep magenta due to the colour of the sclerites. The upper layer, or cortex, may be magenta or almost white depending upon the amount of coloured sclerites it contains. Except for the most basal layer of the medulla, the sclerites are all spindles, sometimes branched, with low or tall, spiny tubercles arranged in relatively distinct girdles. The most basal layer generally includes multiple branched, reticulate and fused forms with very tall, complex tubercles.

Briareum (Fig. 21.17). On the GBR colonies form encrusting sheets, which in some species may form filo-pastry style colonies through successive layering. The polyps are fully retractile. Calyces are tall to exceedingly small. The sclerites are spindle-shaped and multiradiate, sometimes fused in clumps. Commonly the surface layer is beige-brown and the basal layer magenta, but in some species the surface is also magenta, while polyps may be iridescent green or yellow. Members of *Briareum* can be abundant on inshore reefs where they can grow to sheets >1 m in diameter. Similar genus: *Rhytisma*.

Family Subergorgiidae. In this family the medulla is relatively consolidated: sclerites are branched, interlocked, partially fused, and also embedded in a tough matrix of gorgonin. There is a ring of boundary canals directly outside the medulla below the cortex, but there are virtually no canals running through the medulla. Colonies are either arborescent with free branches, or as net-like fans. Axial sclerites are usually smooth and sinuous, relatively long, and form a network. Sclerites of the cortex are predominantly tuberculate spindles and ovals, and small, irregularly shaped dumb-bell-like sclerites often referred to as double-heads or double-wheels. Using just external features, species in this family are easily confused with those in families with no sclerites in the axis.

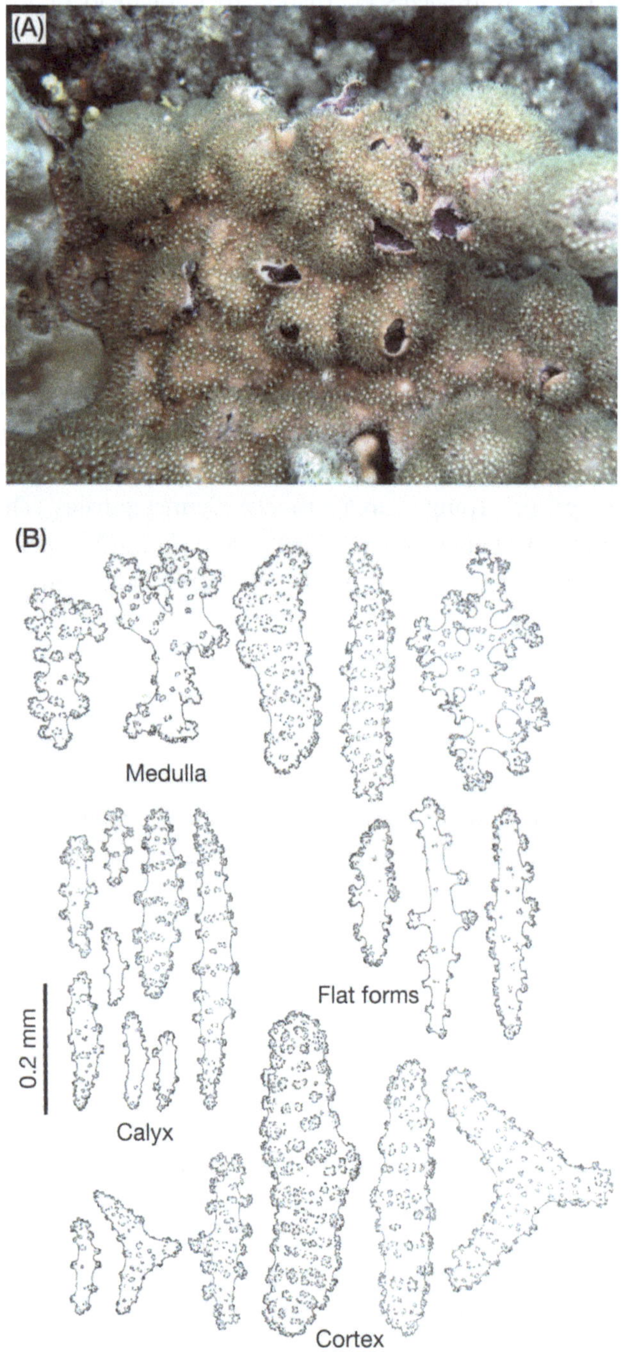

Figure 21.17 *Briareum* sp. (family Briareidae). *A*, colony; *B*, some of the sclerites representative of this genus. (Photo: K. Fabricius; illustration: P. Alderslade.)

Annella (Fig. 21.18). Members of this genus are large reticulate fans up to 2 m in diameter. The axial skeleton is comprised of gorgonin and partially fused, smooth, sinuous sclerites. Sclerites of the cortex are double discs and spindles. Colour: brown, orange, red or yellow. Uncommon. Similar genera: *Echinogorgia* and *Verrucella*.

Family Melithaeidae. The Melithaeidae consists of fan-shaped or tangled bushy colonies in which the axial medulla is segmented, comprised of a series of short, rounded nodes alternating with longer, narrower internodes. The swollen axial nodes on the stem and main branches are generally conspicuous, while those at the points of branching within the fan may not be so obvious. With few exceptions, branching occurs at the spongy gorgonin nodes. Colonies are quite fragile, breaking at node-internode joints. The sclerites are usually coloured, and besides the axial rods they include tuberculate clubs, ovals and spindles with ridges or spines along one side, leaf-clubs, and multirotulates (that look like two or more buns pressed together). The family contains five nominal genera that are not easily distinguished.

Melithaea (Fig. 21.19). Small to relatively large fans (>1 m in diameter) that are sometimes reticulate. Members of this genus have an axial skeleton of alternate nodes of gorgonin combined with small rod-like sclerites and internodes of fused sclerites. Cortical sclerites include birotulates and knobbed clubs. Colour: red, yellow or white. Uncommon but widely distributed. Similar genera: all other members of the family.

Family Acanthogorgiidae. Colonies are richly branched, fan-like, net-like, bushy, or untidy and tangled. The most noticeable features are the very conspicuous nonretractile polyps, which are completely covered with straight and curved spindle-shaped sclerites. These spindles are commonly arranged in eight double rows. Some species have a conspicuous crown of sharp spines around the top of the polyp. Colonies have a black, purely horny hard axis, and a wide, hollow, soft, cross-chambered central core. In some species, the coenenchyme is so thin that the black axis can be seen through it, which can be a clue to their identity; but in others it is quite thick, full of sclerites and opaque. Sclerites are mostly tuberculate spindles, but tripods and capstans also occur. In other genera, however, the coenenchyme is thick, and the polyps can look like the calyces found in the family Plexauridae inside which a retractable polyp can be found.

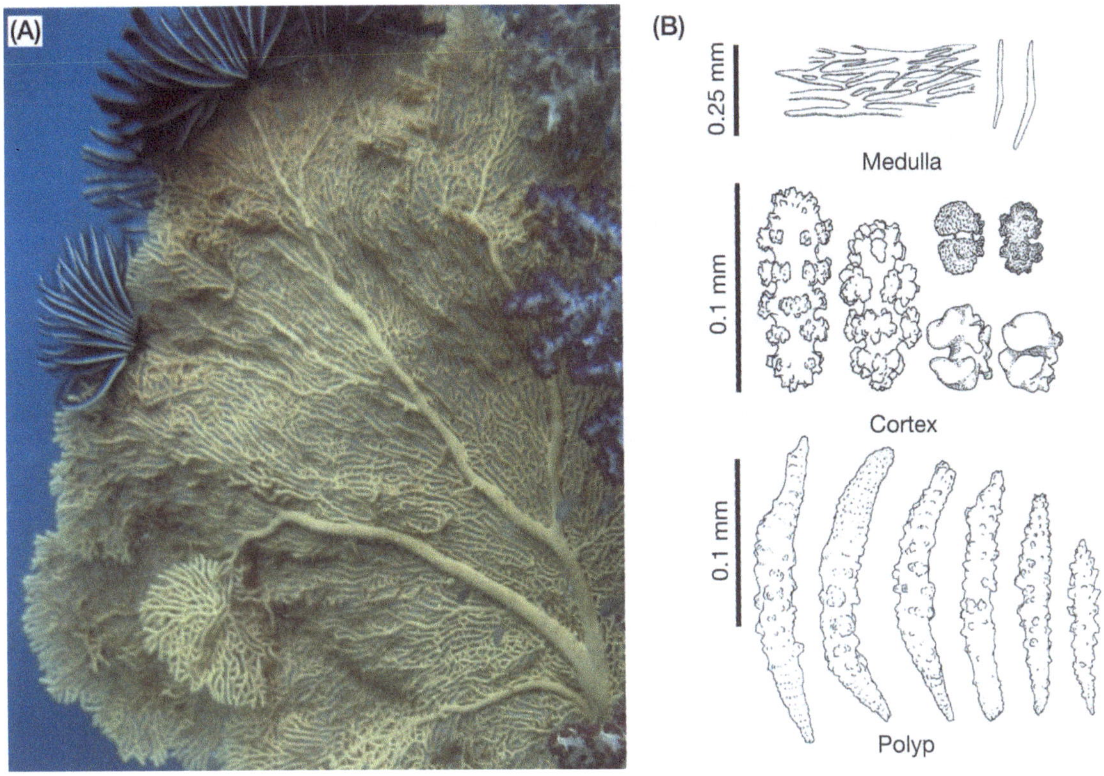

Figure 21.18 *Annella* sp. (family Subergorgiidae). *A*, colony; *B*, some of the sclerites representative of this genus. (Photo: K. Fabricius; illustration: P. Alderslade.)

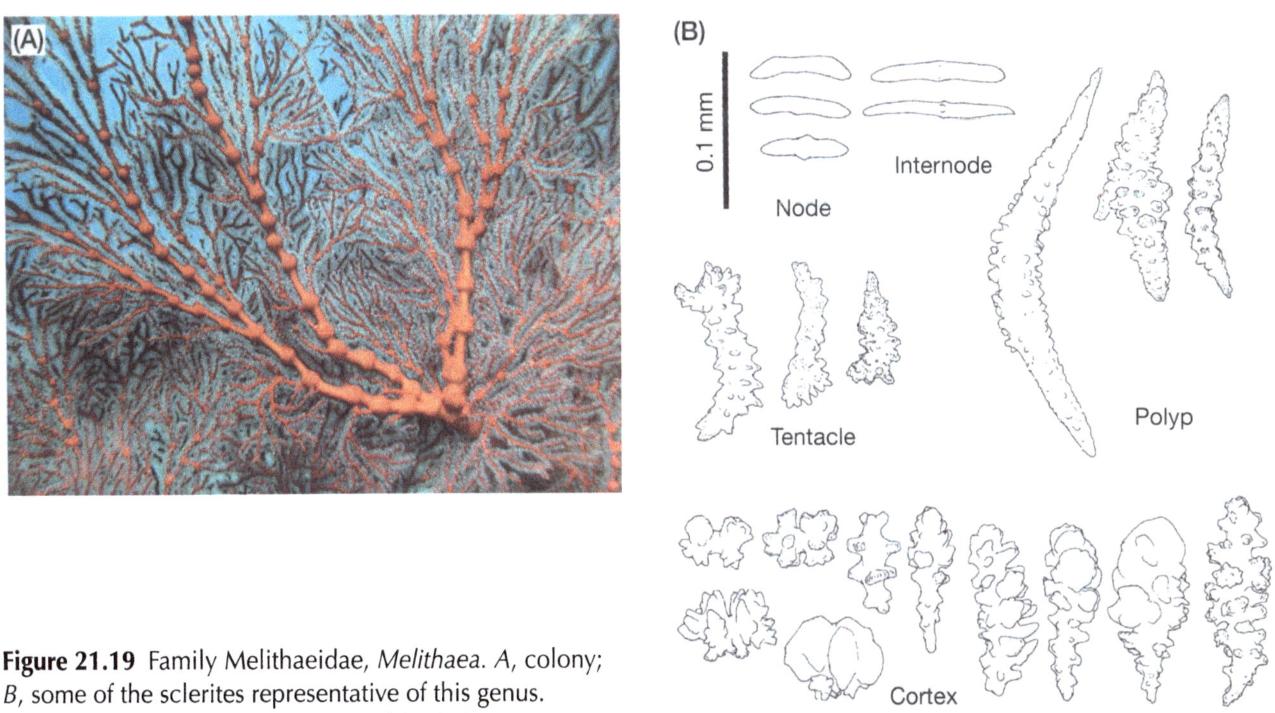

Figure 21.19 Family Melithaeidae, *Melithaea*. *A*, colony; *B*, some of the sclerites representative of this genus. (Photo: K. Fabricius; illustration: P. Alderslade.)

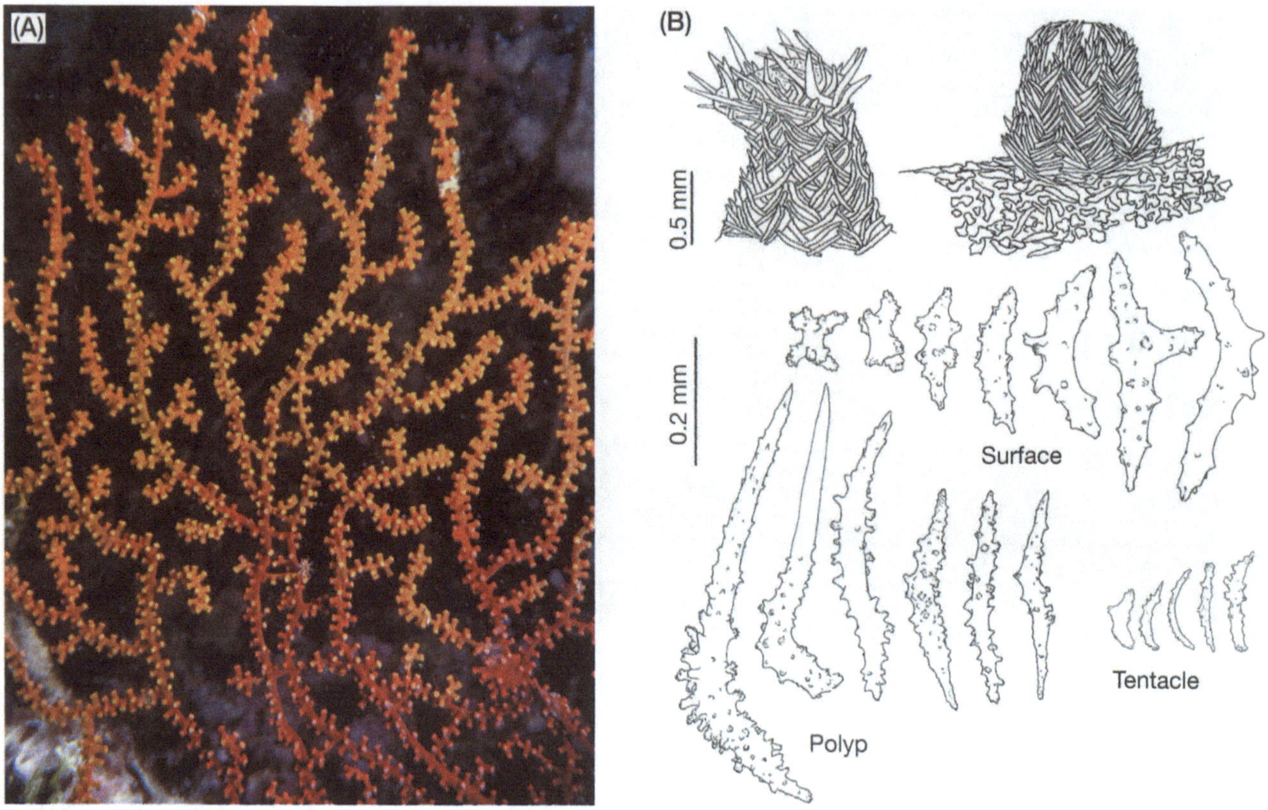

Figure 21.20 *Acanthogorgia* sp. (family Acanthogorgiidae). *A*, colony; *B*, polyp drawings and some of the sclerites representative of this genus. (Photo: K. Fabricius; illustration: P. Alderslade.)

Acanthogorgia (Fig. 21.20). Members of this genus usually have small fans with partial or full reticulation, but are sometimes bushy. The axis is usually black and visible through the thin coenenchyme. Nonretractile polyps are often very tall, and covered in eight double rows of thin, bent spindles. The distal sclerites often form a spiny crown around the polyp head. Coenenchymal sclerites are small spindles and may include thorn-stars and capstans. Colour: bright, often multicoloured. Uncommon. Similar genera: all other members of the family.

Family Plexauridae. Colonies in this family are bushy or fan-shaped (often net-like), and are sparsely to richly branched. They have a black to brown horny axis with a wide, hollow, soft, cross-chambered central core, and numerous spaces filled with fibrous calcite. The polyps are retractile, often within prominent calyces, and are usually armed with large sclerites in a collaret and points arrangement. Sclerites are often quite large, longer than 0.3 mm, and some as long as 5 mm. They are tuberculate, sometimes thorny, and the tubercles are rarely arranged in regular whorls. Sclerites called 'thorn-scales' often occur in the walls of calyces; these usually have a basal, spreading, root like structure, and large, distal spines or blades that often protrude beyond the rim of the calyx. The sclerites in the coenenchyme come in a very wide range of forms.

Euplexaura (Fig. 21.21). The genus comprises planar colonies, not richly branched, with branchlets usually arising at right angles and then bending upwards. Sclerites of the thick coenenchyme are plump spindles and spheroids. Colour: white, grey or brown. Uncommon.

Family Gorgoniidae. Colony growth forms in this family are tree-shaped, bushy, pinnate or fan-shaped (sometimes net-like), and are sparsely to richly branched. They have a black to brown, horny axis, but in contrast to the previous two genera, the hollow, soft,

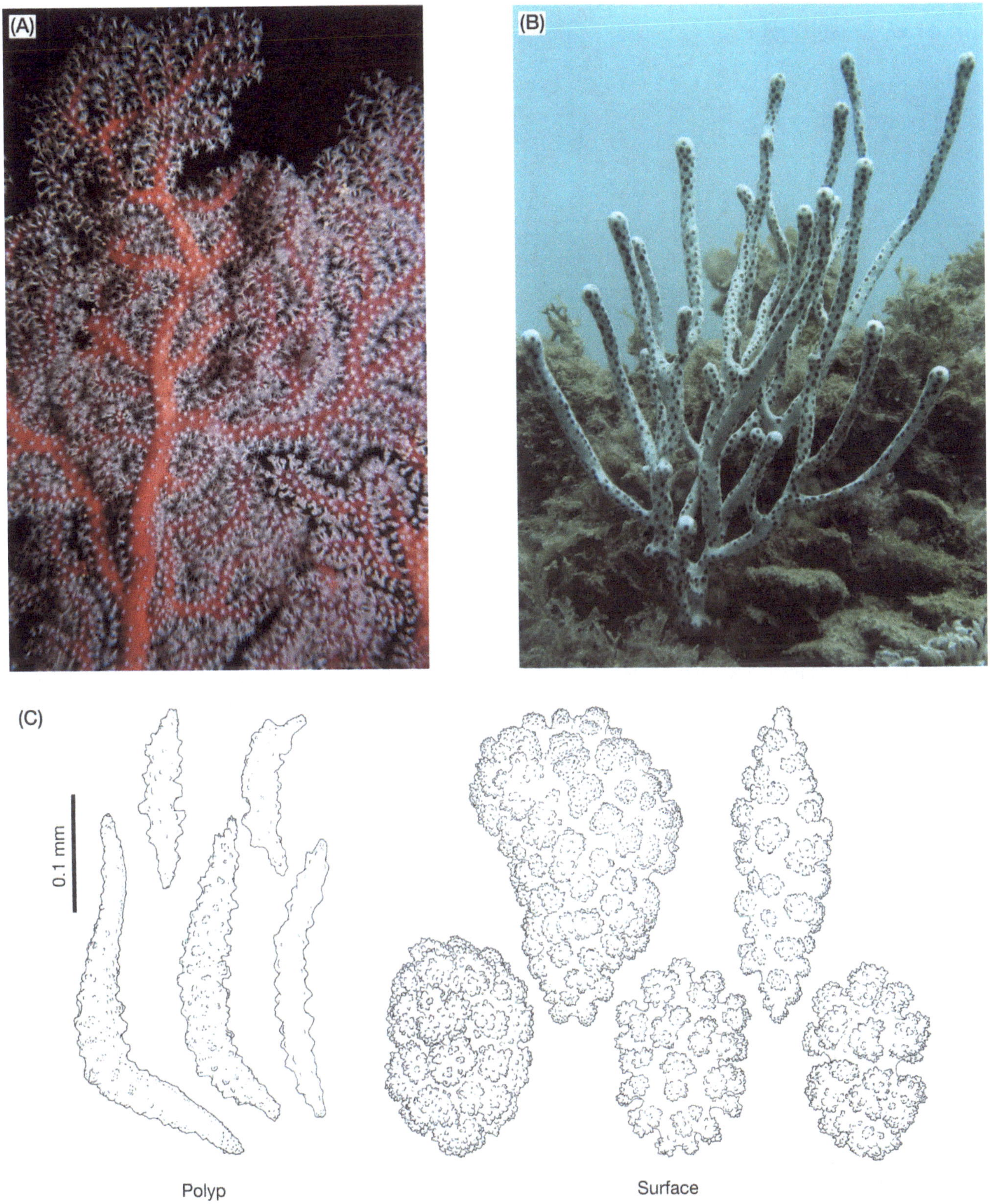

Figure 21.21 *Euplexaura* sp. (family Plexauridae). *A–B*, colonies; *C*, some of the sclerites representative of this genus. (Photos: K. Fabricius; illustration: P. Alderslade.)

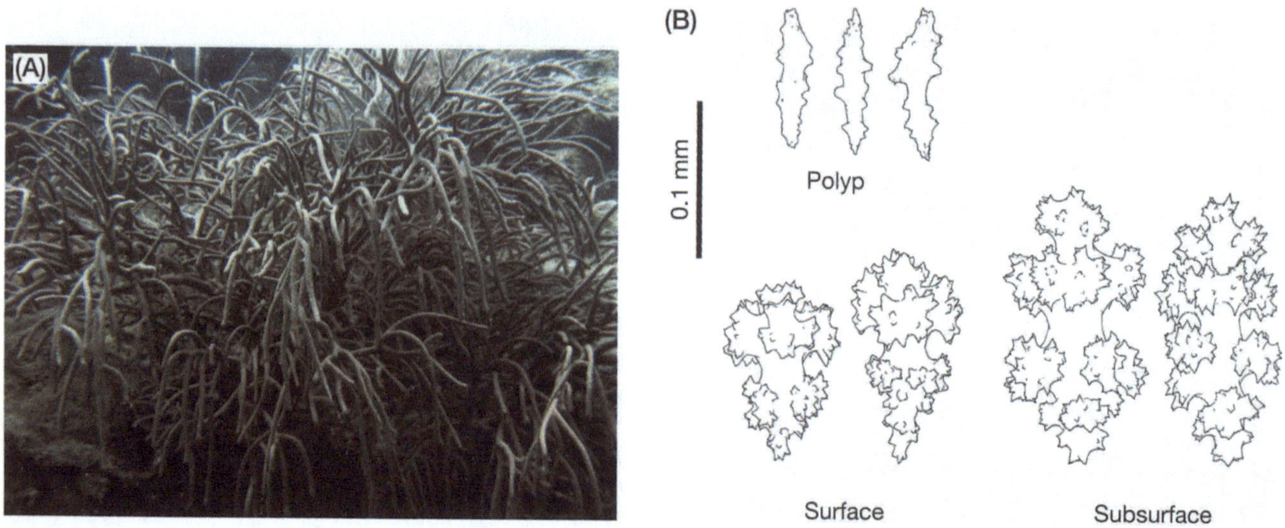

Figure 21.22 *Rumphella* sp. (family Gorgoniidae). *A*, colony; *B*, some of the sclerites representative of this genus. (Photo: K. Fabricius; illustration: P. Alderslade.)

cross-chambered central core is usually narrow, and the axial cortex surrounding the core is very dense, with little or no loculation. Non-scleritic calcareous material may sometimes be present in the axis. Sclerites are generally small, usually less than 0.3 mm in length. Polyps are always retractile, sometimes into low calyx-like mounds. The polyps may have no sclerites at all, but if present they are generally small, flattened rodlets with scalloped edges. The sclerites in the rest of the colony are nearly always spindles with the tubercles arranged in whorls. Some species have curved, asymmetrically developed spindles.

Rumphella (Fig. 21.22). Colonies are bushy, often with a large, calcareous holdfast. Polyps are very small and retractile into the thick coenenchyme. Calyces are absent. Sclerites are symmetrical clubs and spindles with the tubercles in girdles. Colour: brown to greenish-grey. Uncommon, but widely distributed.

Family Ellisellidae. Colonies have a strongly calcified, continuous axis, and can be unbranched, loosely branched, or form broad, flat fans that may be net-like. It is very important to note the way a colony branches, because genera with the same sorts of sclerites can be distinguished based on growth form. Polyps are highly contractile, but not retractile. When contracted they may fold over and lie against the branch surface, or just close up to form a small mound, which may look like a calyx. The major feature of this family is the characteristic form of the tiny sclerites, which are shaped like clubs, double heads and spindles with distinctly separate, papillate tubercles. There are also capstans with cone-like processes.

Ellisella (Fig. 21.23). Colonies are bushy, with whip-like branches, repeatedly branched in a dichotomous manner, generally <0.7 m in diameter. Polyps are not retractile, usually folded up against the branch surface. Sclerites are double heads and waisted spindles. Colour: red, yellow or white. Uncommon but widely distributed.

Ctenocella (Fig. 21.24). Members of this genus have comb- or lyre-shaped colonies that can grow to >1.5 m in size. Otherwise they are the much the same as *Ellisella*, although the waisted spindles are very short. Colour: red (usually). They generally occur in low light.

Junceella (Fig. 21.25). Unbranched whip-like colonies up to >2 m in length, occur in this genus. The polyps are not retractile, usually folded up against the branch surface. Surface sclerites are clubs, subsurface sclerites are double stars. Colour: white, cream, yellow or red. Members of this genus are occasionally found in dense patches, and are quite common and widely distributed. Similar genus: *Viminella*.

Family Isididae. Colonies in this family have a distinctly segmented axis free of sclerites. Although the calcareous internodes are usually solid, in some species

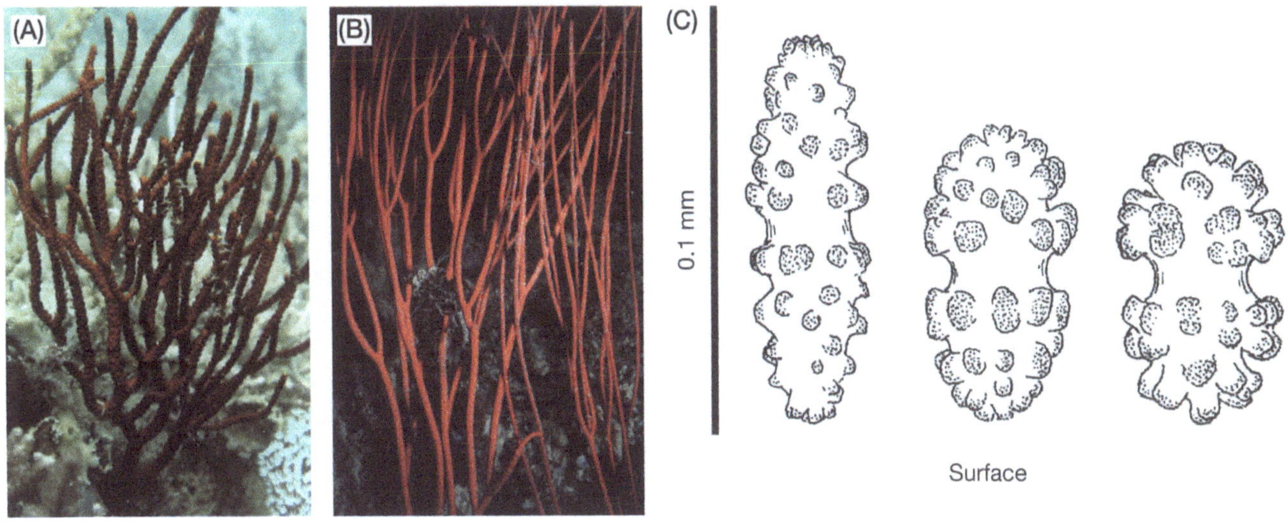

Figure 21.23 *Ellisella* sp. (family Ellisellidae). *A–B*, colonies; *C*, some of the sclerites representative of this genus. (Photos: K. Fabricius; illustration: P. Alderslade.)

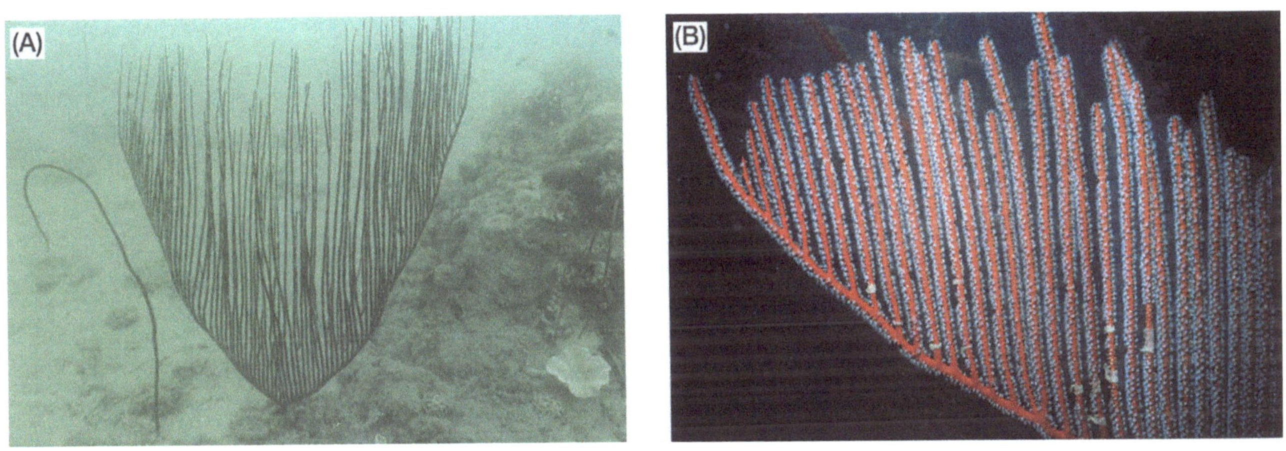

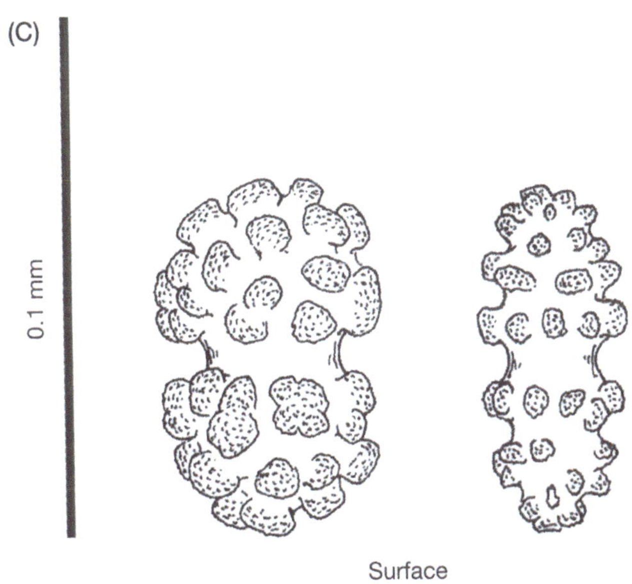

Figure 21.24 *Ctenocella* sp. (family Ellisellidae). *A–B*, colonies; *C*, some of the sclerites representative of this genus. (Photos: K. Fabricius; illustration: P. Alderslade.)

Figure 21.25 *Junceella* sp. (family Ellisellidae). *A–B*, colonies; *C*, some of the sclerites representative of this genus. (Photos: K. Fabricius; illustration: P. Alderslade.)

Figure 21.26 *Isis hippuris* (family Isididae). *A–B*, colonies; *C*, some of the sclerites representative of this genus. (Photos: K. Fabricius; illustration: P. Alderslade.)

they are hollow, but this tubular nature is not to be confused with the soft, cross-chambered central core of the holaxonians. Internodes can be coloured, and although they may be quite smooth, they are often ornamented with prickles and ridges. The nodes that alternate with the internodes consist of pure gorgonin. Colonies can be whip-like, but are usually profusely branched, bushy or fan-like, but rarely net-like. Polyps can be retractile or non-retractile. If non-retractile, they are commonly covered with broad scales, narrow rods or needles. If retractile, the polyps either have no sclerites or are armed with spindles in a collaret and points arrangement. Sclerites in the rest of the colony can be of many forms. The majority of species in this family are deep-water inhabitants; *Isis hippuris* and *Jasminisis cavatica* are the only species found in shallow waters of the GBR.

Isis (Fig. 21.26). Members of this genus are planar to bushy, generally 20–40 cm tall. The axial skeleton consists of alternating horny nodes and non-spicular, opaque, solid calcium carbonate internodes. Polyps are small and retractile into a very thick coenenchyme containing abundant sclerites in the form of capstans and clubs. Colour: bright yellow to green or brown. Common in clear waters. Similar genus: none.

ADDITIONAL READING

Alderslade, P. (2001). Six new genera and six new species of soft coral, and some proposed familial and subfamilial changes within the Alcyonacea (Coelenterata: Octocorallia). *Bulletin of the Biological Society of Washington* **10**, 15–65.

Bayer, F. M. (1981). Key to the genera of Octocorallia exclusive of Pennatulacea (Coelenterata: Anthozoa), with diagnoses of new taxa. *Proceedings of the Biological Society of Washington* **94**, 901–947.

Dinesen, Z. D. (1983). Patterns in the distribution of soft corals across the central Great Barrier Reef. *Coral Reefs* **1**, 229–236.

Fabricius, K., and Alderslade, P. (2001). 'Soft Corals and Sea Fans: A Comprehensive Guide to the Tropical Shallow Water Genera of the Central-West Pacific, the Indian Ocean and the Red Sea.' (Australian Institute of Marine Science: Townsville.)

Fabricius, K. E. and De'ath, G. (2008). Photosynthetic symbionts and energy supply determine octocoral biodiversity in coral reefs. *Ecology* (in press).

Grasshoff, M. (1999). The shallow water gorgonians of New Caledonia and adjacent islands (Coelenterata, Octocorallia). *Senckenbergiana Biologica* **78**, 1–121.

Hyman, L. H. (1940). 'The Invertebrates. Protozoa through Ctenophora.' (McGraw-Hill: New York.)

Paulay, G., Puglisi, M. P., and Starmer, J. A. (2003). The non-scleractinian Anthozoa (Cnidaria) of the Mariana Islands. *Micronesica* **35–36**, 138–155.

van Oppen, M. J. H., Mieog, J. C., Sánchez, C. A., and Fabricius, K. E. (2005). Diversity of algal endosymbionts (zooxanthellae) in octocorals: the roles of geography and host relationships. *Molecular Ecology* **14**, 2403–2417.

Verseveldt, J. (1977). Australian Octocorallia (Coelenterata). *Australian Journal of Marine and Freshwater Research* **28**, 171–240.

Williams, G. C. (1995). Living genera of sea pens (Coelenterata: Pennatulacea): illustrated key and synopses. *Zoological Journal of the Linnean Society* **113**, 93–140.

22. Worms

P. A. Hutchings

The common name 'worms' can refer to any of several groups of animals, including polychaetes (segmented worms), nemerteans (ribbon worms), sipunculans (peanut worms), echiuroids (spoon worms), myzostomes, nematodes (that may be free living or parasitic), and the parasitic worms, the platyhelminthes, which includes the cestodes (tapeworms), trematodes (liver flukes) and the free living turbellarians (flatworms). All these groups are abundant on the Great Barrier Reef (GBR), but the amount of information that is known about them varies according to the group. Each of these groups will now be discussed in terms of their diversity, where they occur on the reef and their feeding and reproductive ecologies. At the end of this chapter a series of references are given that will facilitate the identification of each group and provide additional information about them.

POLYCHAETES

Polychaetes are the most diverse and abundant of the worm groups and originally were referred to as a class within the Phylum Annelida, together with the earthworms (Class Oligochaeta) and the leeches (Class Hirudinea), and this classification will be found in many textbooks. Recent studies (both morphological and molecular) on the annelids have failed to show that they are a monophyletic group, and while the earthworms and leeches do form a single clade (the clitellates) their relationship to the polychaetes is still being debated. So, today the term 'polychaetes' is widely used rather than the term 'Annelida' and this group includes all the traditional polychaete families as well as the leeches, earthworms, and the siboglinids, sometimes known as 'beard worms'. This latter group until recently was considered either as one or two phyla and included the Vestiminifera, which are restricted to cold and hot vents, and the Pogonophora, or thin worms found in sediments. But recent studies have shown that these worms are closely related to each other and represent a family within the polychaetes. To date, no examples of siboglinids have been found in the GBR, but some may occur in deeper waters off the outer barrier reef. While it is widely accepted that the leeches, earthworms, siboglinids and probably the echiuroids and polychaetes are all closely related, their relationships are still being debated and numerous molecular and morphological studies have failed to completely resolve these issues.

Polychaetes consist of two presegmental regions, the prostomium and peristomium, a segmented trunk and postsegmental pygidium (Fig. 22.1*A*). The body wall consists of circular and longitudinal muscle layers enclosing a body coelom. Usually they have a well defined head with sensory and/or feeding appendages, followed by numerous body segments that may

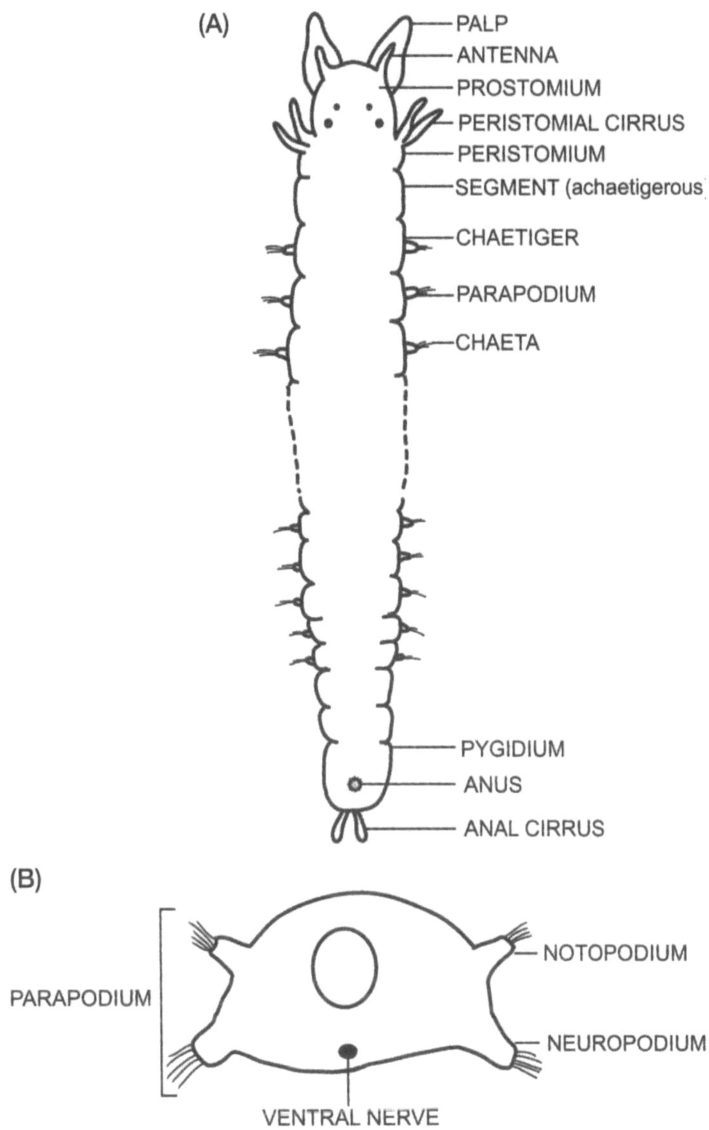

Figure 22.1 *A*, Stylised diagram showing major morphological characters of a generalised polychaete. *B*, Cross-section of the body of a stylised polychaete, showing structure of the parapodia and location of the gut and ventral nerve cord. (After Fauchald, 1977.)

be differentiated into thoracic and abdominal regions. Typically each segment has a pair of parapodia with chaetae (bristles) (Fig. 22.1*B*). The diversity of lifestyles exhibited by polychaetes is often reflected in their morphology. For example, species that burrow through sediments (like the capitellids) tend to have few, if any, sensory or feeding appendages and reduced parapodia; in contrast nereidids that crawl actively over the substratum have well developed sensory appendages and parapodia (Fig. 22.2*A*).

Traditionally, polychaetes consist of about 72 families and these are grouped into a series of clades; such a term is preferred to assigning them an actual taxonomic rank (such as order) as the relationships among clades, and in some cases even their exact composition, are still being debated. The problem arises because polychaetes lack a good fossil record as they are typically soft-bodied worms, but they are an old group with some fossils known from the Ediacaran period (around 580–545 Mya) being identified as polychaetes. So, since that time, presumably many polychaetes have evolved and became extinct, so today's polychaetes are descendants of a very ancient lineage. Many morphological characters, while closely resembling each other, may not necessarily indicate that species are closely related (i.e. are homologous structures), rather they may have evolved over long periods of time in response to the similar environments in which the species live. So, while similar terms have been used across the families to describe particular traits, they may not represent homologous structures, making coding of characters for any phylogenetic analysis difficult. In addition, many species have lost characters, which also makes coding difficult. In the 'Additional reading' at the end of the chapter references are given that will allow people interested in this area of research to find out the latest developments. Not only are the relationships within the polychaetes still being actively debated, but also the relationships of these segmented worms to other invertebrate groups. For many years these segmented worms were regarded as being closely related to the arthropods, based upon the presence of segmentation in both groups, but recently it has been shown that they are more closely related to molluscs. Polychaetes remain one of the last large groups of invertebrates for which a widely accepted classification is still not available.

Figure 22.2 Polychaetes: *A*, Family Nereididae, *Perinereis helleri* with its pharynx everted showing paragnaths, an important diagnostic character for this family. Size: up to 4–5 cm in length. (Photo: H. Nguyen.) *B*, Family Amphinomidae, *Chloeia flava* moving over the sediment. The chaetae easily break off and become embedded in fingers if the animal is picked up. Size: up to 100 mm in length. (Photo: R. Steene.) *C*, Family Amphinomidae, *Eurythoe complanata*, 'Fire worm'. These worms are commonly found under boulders intertidally and shallow subtidally. The dark red bushy branchiae are present adjacent to the parapodia all along body. Size: up to 60–80 mm in length. (Photo: K. Atkinson.) *D*, Family Amphinomidae, *Pherecardia* sp. crawling over the substrate in search of food. It uses its eversible muscular pharynx to feed on sponges, anemones, hydroids etc. (Photo: R. Steene.) *E*, Family Terebellidae, *Reteterebella queenslandia* deeply embedded within the coral substratum with highly extensile feeding tentacles spread out over the surface of the reef collecting food particles and moving them along the tentacles to the mouth. (Photo: P. Hutchings.) *F*, Family Serpulidae, *Spirobranchus corniculatus*-complex illustrating the diversity of colours on a single live colony of *Porites*. This species settles on a damaged polyp, secretes a calcareous tube and encourages the coral to grow around the worm. (Photo: R. Steene.) *(Continued)*

Figure 22.2 *(Continued)* *G*, Sabellidae, a colourful fan worm. Each radiole has numerous fine filaments along its length that are used to strain the water passing through the crown. Particles pass down the axis of each radiole to the mouth, where they are eaten, used for tube construction or rejected. The crown is 50–100 mm in diameter. (Photo: R. Steene.) *H*, Family Sabellidae, *Sabellastarte* sp. This worm is fully extended from its muddy tube. The tentacular crown 50–100 mm in diameter is used in filter feeding and for respiration. (Photo: R. Steene.) *I*, Family Eunicidae, *Marphysa* sp. These are commonly known as blood worms and are common in seagrass and muddy habitats adjacent to reefs. The animal is up to 200 mm in length. It has branched branchiae adjacent to the parapodia. (Photo: K. Atkinson.) *J*, Family Eunicidae, *Eunice aphroditois* emerges at night from its burrow to feed, keeping its jaws open for passing prey such as crustaceans and drift algae that are then grabbed and drawn down into the buccal cavity. They are sensitive to vibration and can rapidly withdraw into their burrows. They are found on shallow coral reefs and may reach 1–2 m in length. (Photo: R. Steene.) *K*, Family Phyllodocidae. Phyllodocid found in coral sediments, with expanded dorsal cirri all along its body. Size: 4–5 cm in length. (Photo: K. Atkinson.) *L*, Family Phyllodocidae, *Phyllodoce* sp. Mass spawning of these worms coincides with coral spawning. These worms normally live under rubble and swim up into the water column to spawn at around 2100 hrs. (Photo: P. Hutchings.)

Polychaetes occur throughout the world in all habitats from the supralittoral to the deepest parts of the ocean. They are predominantly marine or estuarine but a few species occur in moist terrestrial environments. Most are free living although some are commensal or parasitic. Currently over 13 000 species have been described worldwide and many more remain to be described. Many benthic marine communities are dominated by polychaetes not only in terms of the number of species but also by the number of individuals present. In estuarine environments with fluctuating salinity levels the diversity of polychaetes is typically low, although individuals may be abundant. They range from species a few millimetres or less in length with few segments to those many centimetres in length and hundreds of segments.

While polychaete families and many of the genera are known to occur worldwide or have very wide distributions, it is at the species level that one finds restricted distributions. While no comprehensive survey of the polychaetes of the GBR has been undertaken, they are diverse and abundant both in inter-reefal sediments and as borers or nestlers in dead coral substrate. Other families, such as the serpulids, always live in calcareous tubes that are firmly attached to hard substrates or to other organisms such as algae, seagrasses (Fig. 22.3*A*), molluscs or crab carapaces. Some adult pelagic polychaetes, as well as the larval stages of many species, are also found in the reefal plankton. In addition, species are attached to floating debris, as commensals on the undersurface of holothurians and as fouling organisms on buoys and hulls of ships (Fig. 22.3*B*, *C*). Few polychaetes have common names and scientific names must be used.

It has been suggested that there are over 500 species of polychaetes on the GBR. This is almost certainly a serious underestimate, although to date no comprehensive study has been undertaken. Reefal polychaetes are important at all levels of the ecosystem. Many of those living in the substratum are important in terms of bioturbation—breaking down the organic matter in the sediment as it passes through their bodies—and others shred plant material, making it more available to other organisms. They are abundant in seagrass beds and mangrove areas where large concentrations of organic matter accumulate from shed leaves. Such soft-bodied worms on intertidal reel flats are an important food source for wading birds at low tide and for fish and crustaceans as they move over such flats at high tide. Other polychaete species found on the reef are active carnivores, such as the large fire worms commonly found underneath boulders or rubble at the base of reefs. Fire worms belong to the Amphinomidae (Fig. 22.2*B–D*), and they should be handled with care as their long chaetae easily break off and lodge in your fingers. Apart from being an irritant there is a poison gland at the base of the chaetae and some people are quite allergic to this poison—hence their name. This family is well represented in tropical waters, and floating pieces of pumice from a distant volcanic eruption are often covered with them, together with goose barnacles. They range from a few millimetres to several centimetres in length, with robust, typically square-shaped bodies, and most are active carnivores although others are omnivores or opportunistic feeders.

Long white tentacles can often be seen spread out over the reef; these belong to a species commonly referred to as the 'spaghetti worm' (*Reteterebella queenslandia*), a member of the Terebellidae. When touched, the tentacles rapidly contract back into their tubes (Fig. 22.2*E*). These tubes are soft and flimsy, made of fine sediment cemented together with mucus, and are found either under boulders on the reef flat, or in deeper water, the tubes are deeply embedded in boulders well below the surface of the reef. It seems likely that these feeding tentacles contain noxious compounds as they are avoided by fish. This and other terebellid species are all surface deposit feeders. They spread their tentacles ('U' shaped in cross-section) out over the surface of the substratum. Longitudinal rows of cilia along the tentacles create currents that move fine sediments into the centre of the 'U', where it is trapped in mucus and formed into small bundles. These bundles are moved along the length of the tentacle to the mouth that is surrounded by a series of lips that sort these particles, the finest being swallowed, medium sized particles used for tube construction and maintenance, and the large particles are thrown out. The animal is not digesting the actual sediment but rather the algae and bacteria and organic matter that are attached to the surface of the particles. Another common species of terebellid is *Eupolymnia korangia* (Fig. 22.3*D*) that lives at the base of small bommies. The striped buccal tentacles are often seen spreading out over the sand and can rapidly retract back into the sandy tube in which this animal lives wedged between the coral.

Conspicuous 'Christmas tree' (*Spirobranchus corniculatus*-complex) worms occurring in a diverse range of colours (Fig. 22.2*F*) are common in live coral, especially the massive corals such as species of *Porites*. The tree-shaped branchial crown, when held upright in the water column, is able to effectively strain the water and small particles of food are trapped in mucus along the ciliated filaments; these particles are transported along the axis of the filaments down to the mouth at the base of the crown where they are eaten. The branchial crown has well developed eyes and they rapidly withdraw their crowns in response to shadows and vibrations, closing the calcareous tube with a modified branchial filament that forms a plug (or operculum). This reduces the opportunity for fish to eat the crown, but if it is eaten (and examination of fish guts suggests that this does occur) it can rapidly be regenerated. While these worms, which belong to the family Serpulidae, on first inspection appear to have bored into the live coral, they have not; rather they have stimulated the coral to grow around them. This happens when their pelagic larvae, which have spent several weeks in the plankton, settle on a damaged polyp on the living coral (to avoid being eaten), and rapidly secrete a thin calcareous tube. This stimulates the coral to grow around the tube and gradually the worm tube becomes firmly embedded in the coral. While commonly referred to as *S. giganteus* on the GBR this species is actually restricted to the Caribbean. The Australian species belong to the *Spirobranchus corniculatus*-complex and some undescribed species occur. The genus has been widely reported from all coral reef areas of the world. However, the colour of the branchial crown, which can vary from yellow, blue, red and purple, is not a useful character to distinguish species, and the structure of the opening tube as well as chaetal characteristics are important. Obviously, once the worm has become embedded in the coral it is entombed for life. This species-complex is restricted to living coral colonies, so presumably if the coral dies then the worms also die. Why this occurs is unknown, but a possible explanation is that the local water currents created by the coral when it feeds also are beneficial to the filter-feeding polychaete by continually renewing the water mass above them and its associated food particles. Many other species of serpulids are common on the reef and most have an operculum, the structure and ornamentation of which is critical for species and generic identification. Common on algal fronds and on blades of seagrasses are small calcareous tubes that are circular and comma-shaped; these all belong to the spirorbids (Fig. 22.3*A*), a group that is now regarded as belonging to the family Serpulidae. These small worms also have an operculum. In some species a brood chamber develops on the undersurface where the fertilised eggs develop, hatching as miniature adults. They may breed continuously over several months. Serpulids are commonly transported across oceans by drift algae and as fouling organisms on the hulls of ships (Fig. 22.3*B*, *C*). Some species are gregarious and larvae are attracted to settle close to adults of the same species.

Another group of filter-feeding polychaetes are the sabellids, or fan worms, which also have a branchial crown, but are always found in muddy/sandy tubes and lack an operculum (Fig. 22.2*G*, *H*). Again, a tremendous diversity is found in the sabellids, which range from fabriciniids a few millimetres in length living among algal turfs to large conspicuous fan worms with branchial crowns 2–3 cm in diameter. All these fan worms can retract rapidly back into their tubes in response to a passing shadow or vibration, but sometimes they are not quick enough and are eaten by predatory fish. Providing the worm can rapidly contract the body wall to prevent loss of coelomic body fluids, they can regenerate the crown. The larger sabellids, belonging to the sabellines, are found in sediments with the tubes anchored onto a small piece of hard substrate such as shell fragment; others actively bore into coral substrate and line their burrows with fine chitinous tubes.

Species of eunicids may be found underneath large boulders. They may reach several centimetres in length, have numerous segments, often an iridescent epithelium and, along the body on restricted segments, are bright red tufted branchiae (Fig. 22.2*I*). One species of eunciid, *Eunice aphroditois*, is found in the low intertidal zone deeply embedded in the reef crest, and at early morning low tides extends out of the reef crest and grazes on algae (Fig. 22.2*J*). All eunicids have a well developed set of jaws and are regarded as scavengers, feeding on both animal and plant matter. Some species bore into dead coral substrate using both mechanical and acid secretion to excavate their burrows deep within the coral, maintaining an opening to the exterior through which they obtain their food and oxygen from the water column, as well as discharging their gametes.

Another family, the nereidids or ragworms, are common on the reef, occurring in many habitats: in the

Figure 22.3 Polychaetes: *A*, Serpulidae, *Eulaeospira* sp., firmly attached to a blade of seagrass. These small animals filter feed and they are able to seal off their tubes as one of the branchial filaments is modified to form a plug or an operculum. In some species a brood chamber is developed underneath the operculum. (Photo: T. Macdonald.) *B*, Serpulidae, *Hydroides sanctaecrucis*. Close-up of calcareous tubes on a fouled yacht. (Photo: J. Lewis.) *C*, Family Serpulidae, *Hydroides sanctaecrucis*. This is an introduced species that has colonised a yacht in a marina in Cairns. (Photo: J. Lewis.) *D*, Family Terebellidae, *Eupolymnia koorangia*. This is a surface deposit feeder living under rocks or at the base of bommies. Size: 30–50 mm in length, excluding tentacles. (Photo: R. Steene.) *E*, Family Polynoidae. A scale worm, probably a species of *Iphione*, crawling over coarse sediment. Scale worms are active carnivores, catching prey using their eversible pharynx. Encrusting barnacles are present on some elytra in this specimen. (Photo: R. Steene.) *F*, Hesionidae, *Hesione* sp. or *Leocrates* sp., crawling over the substrate. Hesionids are active carnivores that feed by everting the pharynx and sucking up their prey. (Photo: R. Steene.) *(Continued)*

sediments, associated with filamentous and encrusting algae, as nestlers in dead coral substrate, and in the infralittoral zone underneath decaying vegetation. Members of this family have an eversible pharynx with well developed jaws that they use to collect prey and scavenge opportunistically (Fig. 22.2*A*). A closely related family, the Phyllodocidae, is also common and they can be abundant on reef flats at particular times of the year when the mature worms congregate on the surface to spawn, producing egg masses that are attached to pieces of algae or substrate (Fig. 22.2*K*). Other phyllodocids spawn in the water column (Fig. 22.2*L*).

Scaleworms, belonging to the polynoids, are common underneath coral substrate and are characterised by a series of overlapping scales (or elytra) that cover the dorsum and which may be brightly coloured, with distinct pigment patterns, and/or heavily ornamented (Fig. 22.3*E*). This group is highly speciose on the reef and one group lives commensally on the underside of some species of holothurians.

The hesionids (Hesionidae), like the scaleworms, are another group of active carnivores (Fig. 22.3*F*) found on the reef.

Common in the sediments both at the base of the reef and in lagoonal sediments are capitellids, which superficially resemble earthworms with poorly developed parapodia and a blunt head with no obvious sensory structures. They have an eversible pharynx that is used to swallow the sediment, obtaining nutrients from the algae and bacteria on the surface of the particles. These thread-like worms are typically bright red in colour due to the haemoglobin that is dissolved in their blood that allows them to absorb oxygen from the pore water in between the sediment particles.

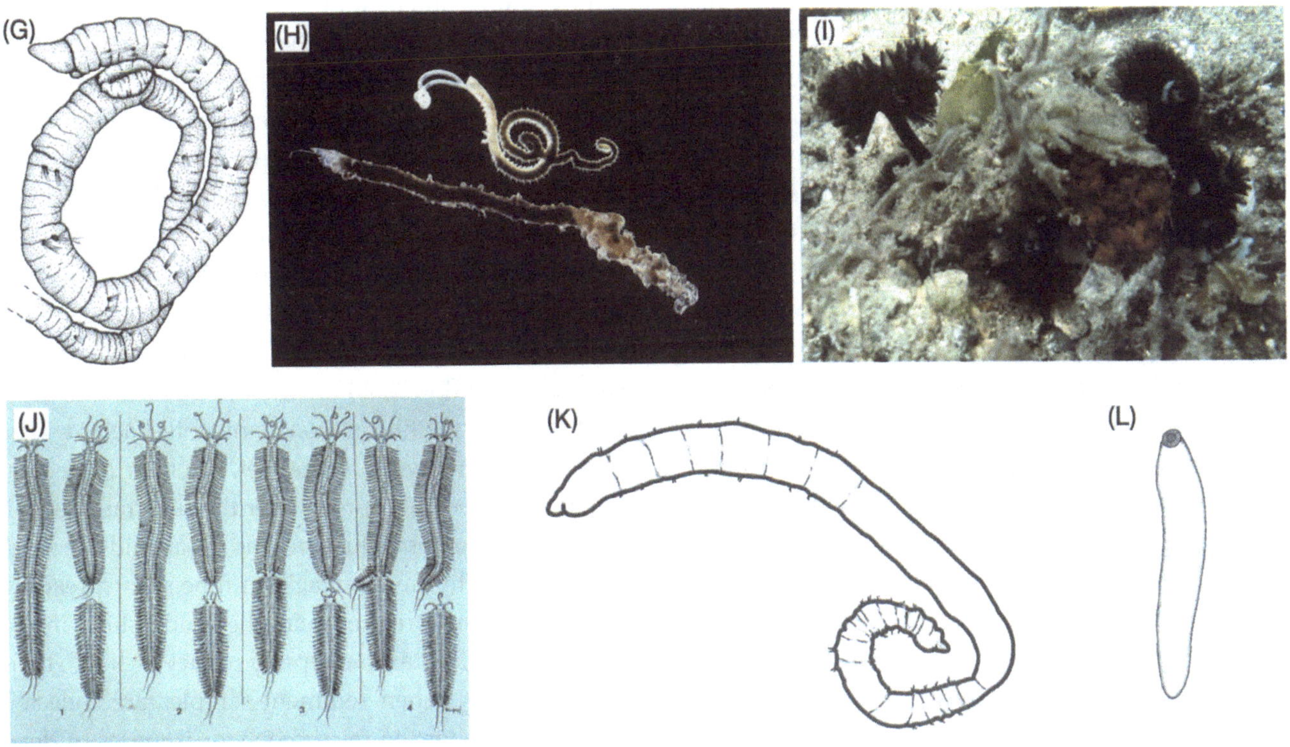

Figure 22.3 *(Continued)* *G*, Questidae, *Questa ersei*. Line drawing of anterior end of the animal, showing its simple body lacking parapodia, paired bundles of chaetae on all segments and the head reduced to a simple palpode with an eversible pharynx. Found living within sediment, it can reach lengths of up to 10 mm. (Figure reproduced with permission from Polychaetes and Allies (2000): A. Murray.) *H*, Family Chaetopteridae, *Mesochaetopterus* sp. One is removed from its tube and the other is still in its flimsy tube made of sand grains. These worms are common in reefal sediments. The anterior palps are used for feeding. The enlarged parapodial lobes in the mid body create water currents that pass through the tube and particles of food are collected. Size: up to 10–20 mm in length. (Photo: K. Atkinson.) *I*, Phoronid. These animals are always associated with burrowing sea anemones. (Photo: K. Attwood.) *J*, Line drawings of syllids undergoing asexual reproduction. (Source: after M. Durchon 1959). *K*, Line drawings of a marine oligochaete. (Artwork: K. Attwood.) *L*, Line drawing of digenetic trematode. (Artwork: K. Attwood.)

Sometimes occurring in the sediment, conical tubes made of cemented sand grains resembling 'ice cream cones' are found, inside which live pectinariids. Their golden spines, which are used to dig the sediment head down to collect food particles, cover the opening of the tube.

Representatives of one family, the Questidae, which are common in reefal sediments, have often been referred to as 'oligochaetoid polychaetes'. While they show some resemblance to marine oligochaetes, especially the Tubificidae, they share more characters with the polychaetes (Fig. 22.3G). One species, *Questa ersei*, has been recorded from Wistari Reef at Heron Island and also from the Houtman Abrohlos in Western Australia. These elongate and slender worms up to about 10 mm in length with up to 60 chaetigers are usually reddish-orange in colour. They have a blunt head without appendages and feed by swallowing sand grains using their ventral eversible pharynx and digesting the bacterial/diatom film on the surface of sand grains. This species is probably common throughout reefal sediments. Another family common in reefal sediments are members of the family Chaetopteridae, belonging to the genus *Mesochaetopterus* (Fig. 22.3*H*). These animals build sandy tubes that protrude from the surface of the sediment and water is pumped through these tubes, trapping particles that are used by the animal for either tube construction or are eaten.

The distribution of polychaetes is largely dependent on the type of substratum present; for example, the size and type of sediment, presence of suitable reefal substrate for the borers and nestlers, hard substrates for the encrusting species to settle on, and suitable algal substrates. Factors such as exposure and water currents are important for filter-feeding organisms. Species living in sediments need to have stable sediments so high energy beaches are typically low in the number of species and individuals. As all polychaetes are soft-bodied they need protection from predatory organisms, either by secreting a tube into which they can rapidly retract or habitats in which they can burrow and avoid predation. A few species have developed anti-predator strategies, for example, the numerous spiny chaetae of amphinomids (Fig. 22.2*B–D*) may make them unpalatable to fish and other predators. However, examination of fish and bird gut contents reveals that polychaetes are an important prey item for many species and obviously even when buried in sediments they can be preyed upon. Lacking an external skeleton or shell they provide an easy to digest source of food.

Polychaetes exhibit an amazing diversity of reproductive strategies, including both sexual and asexual reproduction. While most polychaetes are dioecious, that is, with separate males and females, some species are hermaphrodites, and others may be males at particular times of their life and then become females subsequently. Species may live for a few weeks or months to many years, some breed continuously over several months, whereas others are restricted to spawning on a single day. Some species actively mate, with the male fertilising the eggs as they are laid. In some the fertilised eggs are then placed under the scales. The eggs of some other species are brooded within their tubes, or placed in the chamber below the operculum, others lay their fertilised eggs in brood capsules that are then attached to the substrate where development occurs. In some cases, the developing embryos are cannibals, eating some of the other embryos, but in other cases the large yolky eggs have sufficient nutrients for them all to hatch as miniature adults. Other species are broadcast spawners with gametes being released into the water column where fertilisation occurs. In these cases, free swimming larvae, or trochophores, are produced that spend anything from a few hours to many weeks in the plankton before settling and metamorphosing into adult worms. Some larvae feed while in the plankton whereas other species, which produce large yolky eggs, do not. Most of the terebellids, for example, produce large yolky eggs and the pelagic larval stage is almost non-existent. Crawling benthonic larvae are produced that settle near where they hatch. In all cases it is important that males and females are ripe at the same time, and that spawning is synchronised—polychaetes have well developed endocrine systems to co-ordinate this. (Box 22.1 and Box 22.2.)

In summary, while not too many polychaetes will be observed when diving on the reef, on closer inspection tentacles can be seen spread over the substrate, as well as the expanded branchial crowns of sabellids and serpulids. At night more tentacles can be seen spreading out over the substrate or being held up fishing in the water column, but all can rapidly retract back into their burrows. Only when sediment samples are collected and sorted under the microscope, or if pieces of dead coral substrate are split open to reveal borers and nestlers, will the diversity of polychaetes be revealed. However, you must have a permit to collect such samples (see Chapter 12). Even though most of the diversity of this group is not readily visible, its members play an important part in the functioning of coral reef ecosystems, with roles including bioturbating sediments, breaking down organic matter, and settling and boring into coral substrate (see Chapter 8). They are also important prey items for a wide variety of organisms.

CLITELLATES

The clitellates consist of both the oligochaetes and the hirudineans. While oligochaetes are often thought of as being terrestrial, many species occur in both freshwater and marine sediments and while few studies have been carried out they are known to be common in marine sediments on the GBR. Species tend to be small, <1 cm in length, and thread-like (Fig. 22.3*K*), and often bright red in colour due to the blood pigment haemoglobin. They play an important role in meiofaunal communities in the breakdown of particulate matter and they themselves are predated upon by many animals.

BOX 22.1 MASS SPAWNING OF POLYCHAETE SPECIES

In the late 1980's when mass coral spawning was discovered, often polychaetes were also seen swimming in the water column and they too were spawning. Such mass spawning has been known and exploited by the Samoans for centuries. They could predict the night on which these sedentary eunicids, the 'Palolo' worms that live deep within the coral substrate, would swim up into the water column where their posterior ends would split open to release their gametes, triggered by the particular phase of the moon. The Samoans would collect these swimming sacs full of protein-rich eggs and sperm and dry them to produce a nutritious flour that they would make into biscuits or else trade with highlanders. On the GBR a suite of species spawn at particular phases of the moon during the summer months including species of nereidids, amphinomids, syllids and phyllodocids (Fig. 22.2*L*). In these species the entire worm develops swimming chaetae and large eyes, and leaves the substratrum and swarms in the water column where the males secrete a pherome that stimulates the females to spawn. By the end of the night the beaches are littered with the remains of their bodies. All the gametes are released into the water column, providing fish and other predators with an amazing feast. Almost certainly eunicids spawn on the GBR like in Samoa but to date this has not been documented. In all these cases, while water temperature and light levels are critical for co-ordinating spawning, other factors must be involved months before to ensure that the germinal epithelium of these worms proliferate the gametes and that there is sufficient time for the eggs and sperm to develop to maturity within the coelom. Again, laboratory studies have revealed that all these processes are controlled by an endocrine system that responds to external cues such as temperature and light.

BOX 22.2 ASEXUAL REPRODUCTION OF POLYCHAETES

Some species of polychaete can undertake asexual reproduction as well as sexual reproduction. *Dodecaceria* (family Cirratulidae) is common in dead coral substratum as it can bore into such habitats. An adult worm is capable of splitting into individual segments and each one develops a new head and tail to become a new individual. Another family that exhibits asexual reproduction is the syllids. In some cases the worm develops a series of stolons, resembling railway carriages, with the most posterior one the next one to be released as a new individual (Fig 22.3*J*); in other cases the stolons are proliferated off at angles to the adult worm.

Studies on the diversity of marine oligochaetes on the GBR are few and have tended to concentrate on soft sediment habitats close to research stations where facilities exist to carefully sort the sediment under a dissecting microscope, or elutriate the sediments under running seawater to allow the meiofauna to float off and be collected for study.

Oligochaetes are segmented worms without parapodia, but have segmental bundles of simple hook-like or capillary chaetae; often only a few are present, or they are sometimes absent. The prostomium is pointed, lacking any sensory appendages. They are hermaphrodites, and worms copulate to exchange gametes; fertilised gametes are laid in cocoons that are secreted by the clitellum. The cocoons are poorly developed, containing both the male and female gonopores and are deposited in the sediment. Miniature worms hatch out from the cocoons after variable periods of time. Some species undergo asexual reproduction by transverse division of the parent worm into two or more individuals. Marine oligochaetes feed on fine detritus, algae and other micro-organisms using their eversible pharynx. The diversity of the oligochaete fauna is a good indication of the 'health' of the sediment.

Marine leeches (Hirudinea) are present on the GBR. While some are free living (Fig. 22.4*A*, *B*) others rely upon an intermittent supply of blood from marine vertebrates, including turtles and fish. Some appear to be fairly host specific, whereas others are not. They use their anterior sucker to attach onto the host and inject an anticoagulant that allows the leech to suck out the blood without it clotting. A blood feeding leech will be two to three times as large after feeding than it was before. Leeches are typically dorsoventrally flattened and tapered at each end, with segments at both extremities modified to form suckers. Leeches are segmented, although external segmentation is obscured, and chaetae are absent. A clitellum is present and always formed by segments 9, 10 and 11, although these are only conspicuous during reproductive periods. Branchiae are present in the Pisciolidae as lateral leaf-like or branching outgrowths. In species without branchiae respiration occurs through the skin. Leeches are hermaphrodites and animals mate and exchange sperm that is then stored and used as required to fertilise the eggs, which are laid in cocoons from which miniature adults hatch out after some weeks. Relatively little is known about marine leeches but some act as vectors for protozoan parasites.

MYZOSTOMES

Myzostomes are a bizarre group of worm-like animals that are all obligate symbionts, mainly with crinoids and other echinoderms. They range from species that are mobile and move freely over the host stealing food, to others that are endoparasitic and live inside the host. Some of these stimulate the host to form cysts or galls around them. While myzostomes occur worldwide they are commonest in reefal areas where crinoids are abundant. The body forms of the mobile species range from flattened ovals to disc-like forms and some are ridged or have elongate extensions that resemble the pinnules on the crinoid arms (Fig. 22.4*C*, *D*). Many are coloured to resemble their hosts and they may be quite difficult to spot on the crinoid. The mobile forms basically have a fused body and appear unsegmented, the head is not a distinct structure, there are no eyes and the mouth is a ventral or terminal structure. When dissected, it can be seen that the body is clearly segmented with the five segments separated by septa and bearing five pairs of appendages. Each appendage has a stout emerging hook-shaped chaeta, supported by a single aciculum. These structures are far less obvious in the endoparasitic species. They are usually protandric hermaphrodites and are initially functional males that then become simultaneous hermaphrodites at maturity. Fertilisation is internal and fertilised eggs are spawned into the water column giving rise to planktonic, non feeding trochophores and they settle on a host within 5–8 days. The larvae have two bundles of long chaetae that resemble those found in the larvae of some polychaete families. How the endoparasitic and the gall-producing species reproduce still has not been documented.

The exact relationship of Myzostomes to other worm taxa is still being debated and while some studies suggest that they are embedded within the polychaetes, others support the concept of a separate group—the jury is still out.

PLATYHELMINTHES

This group of unsegmented worms includes the free living turbellarians and three classes of parasites. The endoparasitic cestodes (tapeworms, which are secondarily segmented) are common in many of the marine vertebrates, especially elasmobranchs; the trematodes (flukes) (Fig. 22.3*L*) occur in all vertebrate classes, and monogeneans are ectoparasites found on the gills of fish. Recent studies have revealed that all the para-

Figures 22.4 *A*, Hirudinae, *Trachelobdella* sp. (photo: R. Steene); *B*, Hirudinae, *Stibarobdella macrothela* that feeds exclusively on elasmobranchs (photo: R. Steene); *C*, Myzostome, *Hypomyzostoma dodecephalcis* on its host *Zygometra elegans* (a crinoid) from Lizard I. (photo: G. Rouse); *D*, *Myzostoma* sp. (may be new) on its host *Amphimetra tessellata* (a crinoid) from Lizard I. (photo: G. Rouse); *E*, marine flatworm gliding over the coral (photo: R. Steene); *F*, flatworm *Pseudobiceros hancocki* appears to be almost swimming over the reef (photo: R. Steene); *G*, Sipunculida, an unidentified species of *Aspidosiphon* (family Aspidosiphonidae), removed from its burrow deep within the coral substrate (photo: K. Atkinson); *H*, Sipunculida entombed in the coral (photo: P. Hutchings); *I*, Echuira extracted from its burrow, probably a species of *Bonellia* (photo: Australian Museum collection); *J*, Echuira, bifid tentacle of *Bonellia* spread out over the sediment at the base of a coral bommie at night (photo: R. Steene); *K*, line drawings of nematode (artwork: K. Attwood); *L*, nemertean, *Baseodiscus hemprichii* (photo: R. Steene).

sitic groups (cestodes and the trematodes) are really a subgroup, the Neodermata, of one of the subgroups, Rhabdocoela, of the old Class Turbellaria. So 'Turbellaria' as a classificatory group is no longer recognised, but for convenience, and in many publications the term is still widely used to distinguish between free living flatworms and their parasitic relatives.

Cestodes are at their most abundant in sharks and rays; on the GBR only rare individuals of these groups are without tapeworms. All tapeworms have complex multihost life cycles. All commence with an egg in the faeces of the final host. First intermediates are usually small crustaceans (although this has never been demonstrated on the GBR). Second intermediate hosts are a range of invertebrates, especially teleost fishes, that are often heavily infected with tapeworm metacestodes.

It has been estimated that there may be 20000 to 25000 species of parasitic trematodes worldwide and it has been suggested that the fish of the GBR may sustain 2270 digenetic trematodes, with about 10% of these known to science. Digenetic trematodes parasitise the gut or its outgrowth and they are common primarily in teleost fishes, as well as large marine reptiles and mammals. Green turtles can host 12 digenetic species and 16 species have been recorded from the dugong. The life cycle of digenetic trematodes is complex, involving two or three, and sometimes more, hosts and the first intermediate host is almost always a mollusc. To date only one complete life cycle of a trematode on the GBR is known: that of *Paucivitellosus fragilis*; the final hosts are blennies and mullet occurring on the Heron Island reef.

Many years of fish sampling around Heron Island suggests that the 1000 fish species may be infected with as many as 2000 monogenean species that are usually external parasites in the gill chambers or on the skin. Unlike digeneans their life cycle only involves a single host (Fig. 22.3*L*). It has been suggested that under normal circumstances the host lives in some sort of harmony with its parasites, however, under conditions of stress (as occurs in aquaculture), declining water quality with increased levels of fertiliser or pesticides can disrupt this balance with the health of the host species declining.

Free living flatworms are divided up into a number of groups, but they are all bilaterally symmetrical and unsegmented, with a body in which the organs are embedded in a solid cellular matrix of parenchymatous tissue rather than lying in a body cavity, the coelom, as occurs in the annelids. The gut is sac-like, unless it has been lost, and their nervous system has an anterior 'brain' and lateral nerve cords. The body is made up of three layers: the ectoderm, from which the epidermis and nervous tissue develops; endoderm, from which the gut arises, and mesoderm, from which the muscles and other organs arise. These soft-bodied worms regulate their body fluids by a complex series of channels in which specialised cells, the protonephridia, beat and propel fluids to the exterior. Flatworms, especially those living on coral reefs, are often brightly coloured (Fig. 22.4*E*, *F*) and are sometimes confused with nudibranch molluscs. They can be easily separated, however, as nudibranchs have a muscular foot, anterior rhinophores and posterior frilly gills, all of which are absent in flatworms.

Flatworms have anterior pseudotentacles and anterior pigment spots that are light sensitive, and a mouth that is not terminal and which may be quite small, leading to a muscular pharynx that enables these worms to suck up their prey as they glide over the substrate. Some animals glide on a sheet of mucus by the beating of the ciliary epidermis. Other species swim using well developed dorso-ventral muscles that allow the worm to send waves along the body to swim up into the water column. All flatworms can rapidly regenerate lost body tissues.

Flatworms occur on a variety of reefal habitats, and some live in close association with other invertebrates, especially species of echinoderms and soft corals. They are carnivores and may either injest their prey whole or just remove bits at a time. They evert their pharynx and secrete enzymes that begin to digest the prey tissue and the partially digested prey is sucked up into the gut. Target organisms include individual zooids of corals, bivalve molluscs, and colonial ascidians. As they lack an anus, any undigested particles must be ejected via the mouth. Nutrients diffuse into the body from the gut as there is no circulatory system.

Many of the species of flatworms on the reef are highly conspicuous and, as already mentioned, some can swim up into the water column by undulating their body margins, and yet they are almost universally ignored by predatory fish as their body tissue often contain toxins. The bright colours may also warn potential predators that they are poisonous to eat. Others mimic other poisonous animals so that while they are not themselves poisonous the animal that they are mimicking is.

While flatworms are hermaphrodites they appear not to self fertilise. When two mature individuals meet, one slowly glides over the other and raises the front half of the body. A penis armed with a hypodermic-like stylet is pushed into the body wall of the other worm and the sperm are injected into the animal, they then move through the body to find an egg in the oviduct where fertilisation occurs. In some other species the worms copulate tail to tail and the sperm are delivered into the female part of the reproductive system, being released into the oviduct as required. Fertilised eggs are placed in a capsule that is attached to the substratum and after a while miniature adults are produced, which just creep away from the capsule that then disintegrates.

Flatworms are abundant and diverse on the GBR, and range in size from very small (less than 1 mm) to several centimetres in length. Colour patterns are useful in species identification, although these are often lost when the material is preserved. Identification from good quality coloured photos is often possible, although many species still await formal description.

SIPUNCULANS

Sipunculans are soft-bodied, unsegmented coelomate worm-like animals, commonly called peanut worms (Fig. 22.4G). They consist of a muscular trunk or body and an anteriorly placed, more slender introvert that bears the mouth at the anterior extremity. The introvert is highly elastic, capable of considerable extension; at other times it is partially or completely retracted within the body cavity. The mouth leads to a long, recurved, spirally wound alimentary canal that lies within the coelom. Tentacles either surround the mouth or are associated with it. Chaetae are absent. One or two pairs of nephridia are present. The body wall is usually thick and often rubbery and has well developed circular and longitudinal muscles and occasionally a thin layer of oblique or diagonal muscle is also present. The skin often contains pigmented papillae and the surface of the introvert may be covered in hooks. Around the anus a caudal shield or calcareous knob or cone is often present. Sexes are separate, fertilisation is external, and a free swimming trochophore larva is produced.

While sipunculans are typically regarded as sedentary, some species, especially those that burrow in sand or silt, show considerable activity when they dig. When threatened, most species are able to rapidly retract their sensitive and probing introvert. Species respire through their body wall and tentacles, and cells in both the vascular and coelomic systems contain a red respiratory pigment, haemerythrin, which only gives up its oxygen at a very low partial pressure—much lower than that commonly found in sea water. It has been suggested that it functions only in adverse situations, such as may occur on intertidal reef flats on the reef especially during low tides in the middle of the day during summer.

Sipunculans are abundant on the reef, although not often seen unless sediment samples are collected or reef rock broken off (Fig. 22.4*H*). They live in characteristic shaped burrows in reefal substrate or in tubes in the sediment; others bore into solitary corals or inhabit empty gastropod shells. The solitary coral, *Heteropsammia michelini*, commonly contains the sipunculan *Aspidosiphon muelleri*. Rock boring species secrete an acid-like substance and after settling as larvae metamorphose and bore into the substrate, forming a flask-shaped burrow where they are effectively entombed whilst retaining a narrow passage to the exterior through which they obtain their oxygen and release their gametes. They are detritus feeders, using their tentacles and extensible introverts to collect algae, sediment and detritus. Sipunculans themselves are selectively fed upon by some molluscs and humans eat them in selected localities in the Indo-Pacific (although whether they were eaten by Australian Aborigines is not known).

Six families of sipunculans are known and numerous genera and species have been recorded from the GBR but many remain to be described. They appear to have phylogeographic affinities with other Indo-Pacific species. Juveniles are difficult to identify as they lack many of the adult characteristics.

ECHIURA

Echiurans are non-segmented, coelomate, bilaterally symmetrical worm-like marine animals (Fig. 22.4*I*). They have an elongate muscular trunk and an anterior extensible proboscis. They are commonly referred to as 'spoon worms', as their proboscis when extended is used to collect sediment from around its burrow.

The muscular trunk is usually light to dark green in colour, but sometimes reddish brown, and typically bears numerous flat or swollen glandular and sensory papillae. A pair of golden brown chaetae is usually present on the ventral surface of the trunk, just behind the mouth. The proboscis is usually flattened and ribbon-like but may be fleshy and spatulate. It is highly extensible and contractile but cannot be withdrawn completely into the body cavity like that of sipunculans. The distal end of the proboscis is usually truncate or bifid. It is this proboscis that is typically seen on the reef, especially on night dives, and when touched it retracts and is smooth and slippery (Fig. 22.4*J*). The mouth of an echuiran is present antero-ventrally at the base of the proboscis and the anus is at the posterior extremity of the trunk.

Echuirans are detritus feeders, except for species of *Urechis* that trap very fine particles by secreting a mucous net.

The sexes are separate, mature eggs and sperm pass out through the paired nephridia that range from 1–10 in number. Fertilisation is external, and a pelagic trochophore is produced. The only exceptions are species of *Bonellia* (Fig. 22.4*I*) that are common on the reef, where the male is minute and lives permanently within the female body.

Echiurans make burrows in sand and mud, live in the shells of molluscs and tests of sand dollars and in galleries made by other animals in the coral. There is no evidence that they themselves are capable of burrowing into the coral. They also occur on tropical mud flats intertidally.

To date only 13 species or subspecies of echiurans have been described from Australia and seven of these are endemic. While they are present on the GBR, it is not known how many species are present and undescribed species may occur.

NEMATODES

Nematodes include both parasitic and free living species that burrow in the sediment. They are characterised by a shiny iridescent cuticle and an unsegmented body (Fig. 22.4*K*). In many species, males and females exhibit no sexual dimorphism with streamlined bodies pointed at both ends, in other cases males have complex copulatory organs. Free living species are abundant in reefal sediments as a constituent of the meiofauna, and probably are extremely speciose but they have been poorly studied. Nematodes are important recyclers of nutrients in sediments and are an important food source for many other animals. Our knowledge of parasitic nematodes on the GBR is extremely limited, but they can be found in the guts of many marine vertebrate species.

NEMERTEANS

Many nematode species have long, thin, unsegmented bodies, sometimes brightly coloured, and covered in sticky mucous. The head end is typically pointed without any sensory appendages, although pigment spots may be visible in the surface epidermis. Their mouth is subterminal and they catch their prey by rapidly everting their proboscis. Only some of the large conspicuous species have been described from the GBR, including *Baseodiscus hemprichii* (Fig. 22.4*L*), and many more remain to be described. They can be seen crawling over the substrate, but many of the smaller species live in the sediments. They appear to be unpalatable to many species and it has been suggested that their mucus sheet on which they glide contains chemical compounds that renders them unpalatable.

PHORONIDS

Phoronids represent a small group of unsegmented, sedentary worm-like animals that live in tubes (Fig. 22.3*I*) that are often attached to the tubes of boring sea anemones, like species of *Cerianthus*. Superficially they resemble sabellid polychaetes but on closer examination the retractile tentacles (lophophores) are seen to be ciliated and arranged in a horseshoe shape around the mouth. They are hermaphrodites and gametes are released through their excretory ducts; the eggs drift up the tube and are trapped in the tentacles. After they have been fertilised, pelagic larvae are released into the water column and they spend several weeks in the plankton before settling and metamorphosing into adults.

ADDITIONAL READING

Identification of polychaetes

Wilson, R., Hutchings, P. A., and Glasby, C. J. (2003). 'Polychaetes: An Interactive Identification Guide'. (CD-ROM) (CSIRO Publishing: Melbourne.)

General information on polychaete families, myzostomes, echuira and sipuncula

Beesley, P. L., Ross, G. J. B., and Glasby, C. J. (2000) (Eds). 'Polychaetes and Allies. The Southern Synthesis. Fauna of Australia. Vol. 4A: Polychaeta, Myzostomida, Pogonophora, Echiura, Sipuncula.' (CSIRO Publishing: Melbourne.)

Rouse, G. W., and Pleijel, F. (2001). 'Polychaetes.' (Oxford University Press: Oxford.)

Flatworms

Newman L., and Cannon, L. (2003). 'Marine Flatworms: The World of Polyclads.' (CSIRO Publishing: Melbourne).

Newman, L., and Cannon, L. (2005). 'Fabulous Flatworms: A Guide to Marine Polyclads'. (CD-ROM) (CSIRO Publishing: Melbourne).

Parasitic worms

Rohde, K. (Ed). (2005). 'Marine Parasitology.' (CABI Publishing and CSIRO Publishing: Melbourne.)

Nemerteans

Gibson, R. (1997). Nemerteans (Phylum Nemertea). In 'Marine Invertebrates of Southern Australia. Part III'. (Eds S. A. Shepherd and M. Davies.) pp. 905–974. (South Australian Research and Development Institute: Adelaide.)

Gibson, R. (1983). Nemerteans of the Great Barrier Reef. 6. Enopla Hoplonemertea (Polystilifera: Reptantia). *Zoological Journal of the Linnean Society* **78**, 73–104.

23. Arthropoda: Crustaceans and Pycnogonids

S. T. Ahyong

CRUSTACEA

The Crustacea is the largest group of marine arthropods, earning it the title 'insects of the sea'. Being arthropods, crustaceans have a tough, chitinous, segmented exoskeleton, but are distinguished from other arthropods by having two pairs of antennae at some stage in their life cycle (antennae are lost in adult barnacles). Crustaceans are primarily marine species, but also occur in many terrestrial and freshwater habitats. They range in size from microscopic copepods to the Japanese giant spider crab (*Macrocheira*) with a 4 m leg span. The best-known crustaceans are the crabs, shrimps and lobsters for their large size and culinary qualities.

The crustacean body generally comprises three regions: the cephalon, thorax and abdomen (or pleon). Each region usually bears paired biramous limbs. The cephalon bears the compound eyes and two pairs of antennae. The limbs of the thorax form the mouthparts, grasping and locomotory legs; and those of the abdomen typically form pleopods (swimmerets) and uropods. Many crustaceans breathe through gills located either at the base of the thoracic limbs (e.g. decapods) or abdominal limbs (e.g. stomatopods).

Crustaceans continue to grow throughout their lives, unlike insects, which stop growing at maturity. The hard exoskeleton, however, cannot stretch or expand to accommodate growth. To increase in size crustaceans must periodically shed their old exoskeleton in a process known as moulting or ecdysis. In the lead-up to a moult, calcium and various other minerals are resorbed into the body from the old exoskeleton, beneath which a new, soft skin develops. The old exoskeleton 'loosens' at key locations and fractures along various defined points (called moult sutures). The animal extracts itself from the old exoskeleton, often within only a few minutes. Recently moulted animals, being soft and extremely vulnerable to predation, hide and remain inactive. At this time, calcium and other minerals from the old exoskeleton and surrounding seawater are slowly redeposited in the new soft exoskeleton. Within 24 to 48 hours, the new exoskeleton hardens and the animal can resume normal activity.

Sexes are usually separate in crustaceans and most have planktonic larvae. Planktonic larvae pass through many stages and can spend months in the plankton before settling on the bottom. Crustacean larvae are a significant component of the zooplankton.

On the GBR, crustaceans are abundant in all habitats, with around 1300 species recorded. In general, back-reef sites on mid-shelf reefs have the highest overall crustacean diversity, but most habitats await detailed faunal study. The most conspicuous crustaceans are the comparatively large and usually colourful decapods (crabs, shrimps, lobsters) and stomatopods (mantis shrimps), but small-bodied, speciose groups such as peracarids, ostracods and copepods, are also common. The major crustacean groups to be encountered are discussed below.

Ostracoda

Ostracods (Fig. 23.1*A*), commonly called seed shrimps, have a bivalved carapace that fully encloses the body. They range in length from less than 1 mm to about 20 mm, but most species seen on the GBR are 3–5 mm long. Ostracods use their two pairs of antennae for locomotion, either swimming or 'creeping', and their thoracic limbs are reduced to two pairs. Ostracods are divided into two major groups: the Podocopa and Myodocopa. Podocopans are typically elongate and peanut shaped whereas myodocopans are generally rounded or ovate with a notch in the margin. Almost any sediment or algal sample will contain ostracods; species may be detritivores, carnivores or scavengers. More than 40 species are known from Lizard Island Elsewhere on the GBR, ostracods have been scarcely studied but many more species almost certainly await discovery.

Copepoda

Copepods are abundant on the GBR and form a major component of zooplankton. Most are no longer than a few millimetres, though some parasitic species reach 300 millimetres. Copepods lack a carapace, and most free living species have a single median eye, a much-elongated pair of antennae that is held outwards, a broadened trunk area and slender abdomen. The most common copepods on the GBR are free living forms—the compact harpacticoids and barrel-shaped calanoids and cyclopoids. Harpacticoids are generally benthic, living under rocks, in algae and sediment. Calanoids and cyclopoids are common in the plankton where they may form dense swarms. Free living copepods feed variously on other zooplankton, diatoms, and algae, and are an extremely important food source for larval fish and decapods. Parasitic copepods feed on the blood or tissue of their hosts. Examples include the siphonostomatoids and poecilostomatoids that attach to the skin of sharks and fish, and certain cyclopoids, which have become worm-like as internal invertebrate parasites.

Cirripedia

The sessile habit and interlocking shell plates of barnacles led early investigators to believe they were molluscs. Not until the 1830s, when barnacle larvae were studied, were they correctly recognised as crustaceans. Barnacles are more abundant in temperate waters than in the tropics, but nevertheless range throughout the GBR on all hard substrata, including the carapace of sea turtles and the shells of other crustaceans. More than 40 species and 20 genera of barnacles are presently known from GBR waters. The body of most barnacles is enclosed by a somewhat tubular or conical opened topped chamber formed by a series of interlocking calcareous plates—usually known as the shell. The animal is positioned on its back, and thrusts its modified, feathery thoracic limbs, known as cirri, out of the shell to strain food particles from the water.

Unlike most other crustaceans, barnacles are hermaphrodites, meaning that each animal possesses male and female reproductive organs. The majority of barnacles in the reproductive phase develop a greatly elongated penis (up to 10 times their body length) that is used to reach out of the shell to fertilise a receptive neighbour. The fertilised eggs develop into free-swimming larvae that eventually settle on the substratum close to others of their species, identified by their unique chemical signature.

Three major types of barnacles will be encountered on the reef: balanomorphs (Fig. 23.1*B*), lepadomorphs (Fig. 23.1C) and rhizocephalans. In balanomorph barnacles, the so-called acorn barnacles, the shell is firmly fixed to the substratum, be it rock, coral, mangroves or wharf piles. Common balanomorphs on the reef flat rocks and boulders include species of *Balanus*, *Chthamalus*, *Megabalanus*, *Tesseropora*, *Tetraclita* and *Tetraclitella*. Not all barnacles are fixed to the substratum by the shell, however. The lepadamorphs, the goose-necked barnacles, are attached to the substratum by a flexible stalk, with common genera including *Lepas* and *Ibla*. *Conchoderma virgatum* is unusual in using sea snakes as a substratum. In contrast to the balanomorphs

Figure 23.1 *A*, ostracod (*Cypridinodes* sp.); *B*, balanomorph barnacle (*Chthamalus* sp.); *C*, lepadomorph barnacle (*Lepas anserifera*); *D*, opossum shrimp (Mysida); *E*, comma shrimp (Cumacea); *F*, apseudomorph tanaid (Tanaidacea); *G*, cirolanid isopod (*Cirolana curtensis*); *H*, sphaeromatid isopod (*Cerceis pravipalma*); *I*, parasitic isopod on fish (*Anilocra apogonae*); *J*, gammaridean amphipod (*Leucothoe* sp.); *K*, gammaridean amphipod (*Grandidierella* sp.); *L*, caprellidean amphipod (*Metaprotella* sp.). (Photos: *A, D–G, I–L,* R. Springthorpe, © Australian Museum; *B, C,* © GBRMPA; *H,* © N. D. Pentcheff.)

and lepadomorphs that simply use the substratum as an anchor point, the rhizocephalan barnacles have an entirely different lifestyle as partially internal and partially external parasites of decapod crustaceans. Rhizocephalans have lost any resemblance to typical adult barnacles, including loss of the calcareous plates. A common rhizocephalan, *Sacculina*, infects crabs and is visible externally only as a fleshy mass beneath the abdomen of the crab. Rhizocephalans interfere with the reproductive cycle of crabs and through the effect of hormones, cause male crabs to slowly develop female characteristics, an effect known as 'parasitic castration'.

Peracarida

The peracarids are a large group of malacostracan crustaceans recognisable by the possession of the *lacinia mobilis* (a blade like process on the mandible), and direct development of young, which are brooded in a pouch formed by lamellar outgrowths of the bases of the thoracic limbs (with the exception of the thermosbaenaceans, which brood their eggs under the carapace). In most cases, peracarids are minute, not exceeding 10 mm in length (many individual exceptions exist of course, such as the giant deep-sea isopod, *Bathynomus*). Despite their small size, however, peracarids are highly speciose and abundant in most habitats. Of the nine generally recognised peracarid orders, the most important on the GBR are the Mysidacea, Cumacea, Tanaidacea, Isopoda, and Amphipoda.

Mysidacea: opposum shrimps

The mysidaceans (Fig. 23.1*D*) resemble juvenile decapod shrimps, and usually do not exceed 20 mm in length. Though mysids superficially resemble decapod shrimps, an important distinction is the position of the statocyst (balance organ resembling a circular pore). The statocyst of mysids is located at the base of each branch of the tailfan, instead of the base of the first pair of antennae in decapods. Unlike other peracarids, mysidaceans have a well-developed carapace and stalked, instead of fixed, eyes. Most mysids are omnivorous suspension feeders, and form dense schools near the bottom or in the water column. Some species, however, have more particular associations, such as the red and white *Heteromysis harpaxoides*, which lives with hermit crabs inside the gastropod shell. More than 800 species in more than 120 genera occur worldwide, with around 50 species on the GBR.

Cumacea: comma shrimps

The cumaceans (Fig. 23.1*E*) are distinctive because of their bulbous carapace and slender abdomen that terminates in a pair of stick-like uropods. They burrow in reef sediments or seagrass beds, and feed on detritus, algae and microbes. At night, cumaceans ascend into the water column to moult and mate. Of the approximately 1300 species of cumacean worldwide, around 20 species are known from Queensland of which several are known from the GBR.

Tanaidacea

The tanaids (Fig. 23.1*F*), usually less than 10 mm length, resemble a slender isopod (below) with a pair of chelipeds. A unique feature of the tanaids is the fusion of the head with the first and second thoracic segments. Most tanaids fall into one of two major groups: the apseudomorphs, which have a somewhat flattened body, and tanaidomorphs, with a cylindrical body. Though seldom seen in the open, tanaids are abundant in reef sediments, seagrass, algae and hydrozoans where they burrow or build tubes from sand and fine particles. Despite their numerical abundance, however, the GBR tanaids are poorly known with only five of more than 800 known species formally recorded from the region.

Isopoda: slaters, pill bugs, fish lice

The isopods are a remarkably diverse group of peracarids that have colonised almost all habitats, from terrestrial leaf litter to abyssal depths. Isopods are characterised by fixed, immovable eyes, no carapace, seven pairs of thoracic legs, five pairs of pleopods, and a pleotelson (formed by fusion of the last two body segments). Most isopods have the familiar, flattened, oval body shape. The largest isopod, the deep-sea *Bathynomus giganteus*, reaches 35 cm in length, but on the GBR, most species are less than 10 mm long. Some

200 species of isopod are known from the GBR, and most obvious are the mostly free living flabelliferans, particularly species of the families Cirolanidae (Fig. 23.1*G*) and Sphaeromatidae (Fig. 23.1*H*). Cirolanids are efficient scavengers, emerging from the substratum to swim rapidly towards a food source such as a dead or maimed fish. They can also be voracious predators, sometimes biting humans. Sphaeromatids are perhaps less conspicuous than the cirolanids, but often have a highly sculptured pleotelson and trailing uropods. Some isopods vary markedly from the 'typical' form. These include the slender valviferans and anthurideans, and most notably the endoparasitic bopyroids that infect decapods. Bopyroids have lost many of the features of free living isopods, and being internal parasites, are seldom seen. A rounded bulge on the surface of the carapace of the decapod host is the only external clue that a bopryroid lies beneath. Other parasitic isopods such as the cymothoids (Fig. 23.1*I*) infect the skin, mouth, and gills, of fish, to which they cling with sharp, hooked 'feet'.

Amphipoda: sandhoppers, scuds and skeleton shrimps

The amphipods are the most speciose of the peracarids with more than 6000 species worldwide. They can be recognised by their (usually) laterally compressed body, biramous antennules and three pairs of uropods. On the GBR, about 200 species are known from depths shallower than 50 m, of which 80% are yet to be formally named, and this all from near Lizard Island in the far northern section of the Reef. Thus, many more species can be expected from the GBR overall. The three major Suborders of Amphipoda (Gammaridea, Hyperiidea and Caprellidea) each have a very distinctive appearance. The majority of amphipods are gammarideans (Fig. 23.1*J*, *K*), commonly known as scuds and sandhoppers (for terrestrial species), and these are the most familiar form. Gammarideans have a relatively deep, curved body that is obviously laterally compressed; they are the most accomplished swimmers of the Amphipoda, and can 'scuttle', 'skip' and 'flick' their bodies to escape capture. Gammarideans live in almost every habitat on the GBR—among rubble, boulders, coral and algae, and sometimes in association with hydroids, ascidians and sponges. Some are predatory, some scavenge, and others graze on algae and detritus. The hyperiideans resemble gammarideans with a swollen head and long slender legs. Unlike gammarideans, however, hyperiideans are seldom seen on the reef because they are oceanic creatures that live inside salps or on jellyfish. The caprellideans (Fig. 23.1*L*), or skeleton shrimps, are so named because of the very slender, stick-like body. They can be seen standing upright with forelimbs spread out whilst clinging by their posterior legs to algae or hydroids. They crawl with a somewhat looping motion by alternately using their anterior and posterior limbs. Though appearing very different from gammarideans, caprellideans probably represent a highly specialised lineage that arose from within the Gammaridea.

Stomatopoda

The mantis shrimps, Order Stomatopoda, are the most flamboyant crustaceans, and only distant relatives of the decapods. They are primarily shallow water tropical animals and are abundant throughout the GBR. They are active, aggressive predators and often vividly coloured. Adults range in size from less than 10 mm to almost 400 mm. The body is elongate, with a short, shield-like carapace that does not cover the last three thoracic somites. Unlike decapods, which either broadcast spawn or incubate the eggs on the pleopods, stomatopod females carry the egg mass with their mouthparts until hatching. The most characteristic feature of stomatopods, however, is the greatly enlarged raptorial forelimbs used to catch prey. The raptorial claw of stomatopods resembles that of the insect, the praying mantis, in which the last segment (dactylus) folds like a jackknife against the preceding segment (propodus). However, instead of grasping prey like the praying mantis, stomatopods strike violently with considerable speed and force. The raptorial strike of the stomatopod is one of the fastest known animal movements, being completed within five milliseconds. Obviously, large stomatopods must be handled with great care.

Stomatopods can be divided into two functional groups depending on how they use their raptorial claws during the strike. 'Spearers' (Fig. 23.2*A*, *B*, *C*) strike with an open dactylus, impaling prey—soft-bodied animals like fish, cephalopods and shrimps—on a row of sharp spines. 'Smashers' (Fig. 23.2*D*, *E*, *F*) strike with the dactylus closed against the propodus. The outer edge of the dactylus is a calcified heel, used like a hammer to smash open hard-bodied prey such as crabs and snails. The power of the raptorial strike can be lethal—the force of the impact may approach that of a small calibre bullet. So, in 'smashers', the raptorial claws are used not only for hunting, but also to avoid unnecessary confrontations. Many 'smashers' have evolved elaborate threat displays that involve lunging and spreading the limbs to display a coloured 'meral spot' on the inner surface of the raptorial claw (Fig. 23.2*F*). The intensity of the display and brightness of the 'meral spot' varies between species, and gives a potential opponent some indication of the ferocity that it might encounter.

The powerful raptorial claws of stomatopods are complemented by finely tuned vision. Each eye can be independently rotated and is divided in half by a central band of ommatidia. This provides the stomatopod with binocular vision from each eye, and the central band of ommatidia provides many of the reef species with colour vision and polarising filters. The combination of excellent vision and powerful raptorial claws make the stomatopod an efficient predator of both slow and fast moving prey.

Almost 500 species of stomatopod are known worldwide and more than 90 from GBR waters. Many new species await scientific documentation, especially small species that live deep inside rock and coral crevices. Around Lizard Island, almost 30 species are known. The most common stomatopods among rubble and corals are 'smashers' of the superfamily Gonodactyloidea. Common gonodactyloid genera on the reef flat are: *Gonodactylellus*, *Gonodactylus*, *Gonodactylaceus*, *Chorisquilla* and *Haptosquilla*. *Haptosquilla glyptocercus* and *H. trispinosa*, both with striped antennae and not exceeding about 40 mm length can be seen on the reef flat peering out of circular holes in reef rock. 'Spearers', primarily members of the Squilloidea and Lysiosquilloidea, burrow in the sand and sediment of the reef flat and on the seabed of the lagoon and between reefs. Several species of squilloid are commercially harvested, taken as trawl bycatch (e.g. *Harpiosquilla harpax* and *Oratosquillina* spp.). On the reef flat, the very large piscivore, *Lysiosquillina maculata* (up to 385 mm long), lives in mated pairs in deep, vertical burrows, with burrow entrances almost 100 mm across (Fig. 23.2*B*). *Lysiosquillina maculata* is the largest known stomatopod, with a body marked by bold, light and dark transverse bands. *Pseudosquilla ciliata*, another common spearer, burrows in seagrass beds or under rocks on the reef flat, and ranges in colour from bright yellow or dark green to mottled green and grey (Fig. 23.2*A*, *C*). On the reef slope, the number of species increases, and some of the most spectacularly coloured stomatopods, those of the 'smashing' genus *Odontodactylus*, can be seen walking out in the open (Fig. 23.2*E*).

Decapoda

The best known of all crustaceans are the decapods. They include the largest arthropods and many commercial species such as the mud crab (*Scylla serrata*), blue swimming crab (*Portunus armatus* (previously known as *P. pelagicus*)), red-spot king prawn (*Melicertus longistylus*), Moreton Bay bug (*Thenus* spp.) and spiny lobsters (*Panulirus* spp.). Decapods have a well-developed carapace covering the thorax, and five (rarely four) pairs of legs of which one or more pairs usually forms a cheliped (pincer). At least nine decapod infraorders are recognised: Dendrobranchiata (peneidean prawns), Caridea (shrimps), Stenopodidea (coral shrimps), Polychelida (deep-sea blind lobsters), Achelata (spiny-crayfish), Astacidea (clawed lobsters and freshwater crayfish), Thalassinidea (mud-shrimps, marine yabbies), Anomura (hermit crabs, squat lobsters, mole crabs), and Brachyura (true crabs). With the exception of the deep-sea polychelids, members of the other infraorders occur on the GBR.

Dendrobranchiata: prawns

The dendrobranchiates, collectively known as prawns in Australia, are possibly the oldest lineage of the decapods. They have the familiar elongate body with a slender, muscular abdomen and three pairs of small chelipeds. Unlike other decapods, female prawns do not carry eggs until hatching, but spawn their fertilised

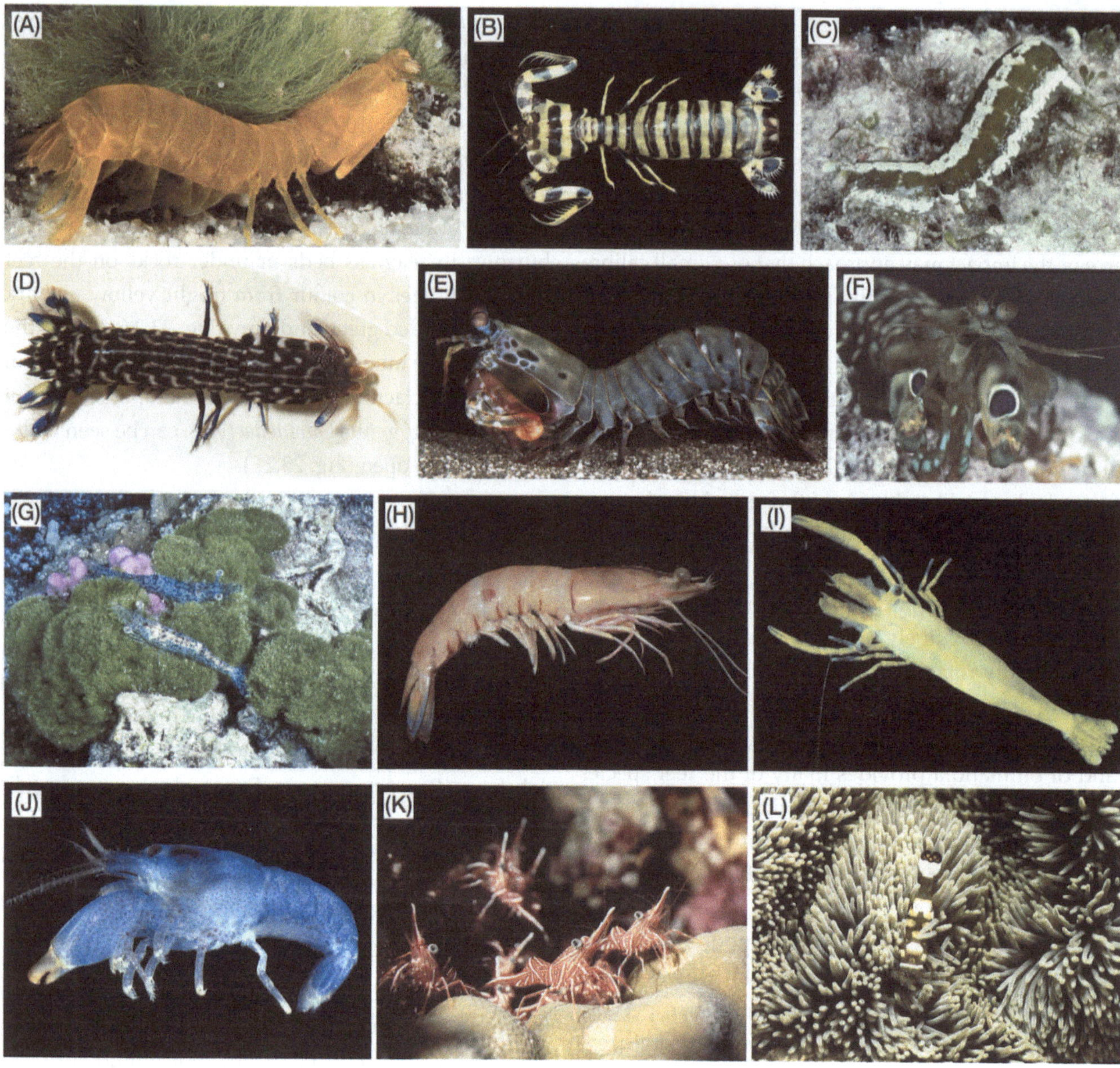

Figure 23.2 *A*, 'spearing' mantis shrimp (*Pseudosquilla ciliata*); *B*, 'spearing' mantis shrimp (*Lysiosquillina maculata*); *C*, 'spearing' mantis shrimp (*Pseudosquilla ciliata*); *D*, 'smashing' mantis shrimp (*Gonodactylaceus graphurus*); *E*, 'smashing' mantis shrimp (*Odontodactylus scyllarus*) carrying egg mass; *F*, 'smashing' mantis shrimp (*Gonodactylus smithii*) showing threat display; *G*, penaeid prawn (*Heteropenaeus longimanus*); *H*, red spot king prawn (*Melicertus longistylus*); *I*, caridean shrimp (*Periclimenes* sp.); *J*, snapping shrimp (*Synalpheus demani*); *K*, hingebeak shrimp (*Rhynchocinetes durbanensis*); *L*, caridean shrimp (*Periclimenes brevicarpalis*) on sea anemone. (Photos: *A, C–F*, © R. Caldwell; *B*, © S. Ahyong; *H*, © G. Ahyong; *G, K, L*, © GBRMPA; *I, J*, R. Springthorpe, © Australian Museum.)

eggs directly into the water column. Prawns are not generally common on corals or hard reef structures, though *Heteropenaeus longimanus* (Fig. 23.2*G*) and several species of *Metapenaeopsis*, *Sicyonia*, and *Acetes* have been recorded. Prawns are more abundant on the sediments of the lagoon and between reefs, where they support an important trawl fishery. One of the most important commercial species is the red-spot king prawn (Fig. 23.2*H*; *Melicertus longistylus*, previously known as *Penaeus longistylus*). Many commercial penaeids 'grow-out' in estuaries or nearshore habitats, and migrate offshore as adults, but red-spot king prawns spend their entire lives in the vicinity of the reef. Juveniles use shallow reef flats and lagoon areas as a nursery, and adults move to deeper parts of the lagoon to spawn.

Caridea: shrimps

The shrimps of the Infraorder Caridea, like the prawns, are accomplished swimmers. Unlike prawns, however, carideans incubate their eggs beneath the abdomen, have only two pairs of chelipeds and have the pleuron (lateral plate) of the second abdominal segment overlapping both those of the first and third segments. More than 200 species of caridean shrimp are presently known from the GBR, the most common of which are members of the Palaemonidae (subfamily Pontoninae) and Alpheidae. Most GBR pontonines are associates of other reef invertebrates: corals, anemones, nudibranchs, sponges, echinoderms and molluscs. Pontonines (Fig. 23.2*I–L*) tend to be delicate and retiring, and though common, must be searched for. One of the more easily found pontonines is *Periclimenes brevicarpalis* (Fig. 23.2*L*), which lives under the protection of the tentacles of sea anemones. Here, *P. brevicarpalis* rolls and sways its body, which is transparent except for its several large white patches, and orange and purple highlights. Interestingly, few pontonines have drab colouration; most are either brightly coloured or almost completely transparent. Alpheids, the snapping shrimps, are abundant throughout the GBR, but are more often heard than seen. Snapping shrimps live in burrows in sand, sponge, mud and coralline rock. The characteristic feature of the snapping shrimps is the greatly enlarged snapping claw, which, when snapped shut produces a loud, cracking sound with a localised shock wave sufficient to injure or kill small invertebrates and fish. Snapping shrimps usually live in pairs, but some form large social networks (e.g. *Synalpheus*; Fig. 23.2*J*), and many species of *Alpheus* cooperatively share a burrow with gobiid fish. Other conspicuous carideans on the reef include the cleaner shrimp, *Lysmata amboinensis*, the hinged-beak shrimps (*Rhynchocinetes durbanensis*; Fig. 23.2*K*) and the marbled shrimp (*Saron marmoratus*). One of the most unusual shrimps on the GBR is the harlequin shrimp (*Hymenocera picta*), which feeds exclusively on seastars. As a rule, caridean shrimps hide during the day and are best observed at night, when they emerge to feed.

Stenopodidea: coral shrimps

The stenopodidean shrimps are generally small, spiny shrimps characterised by having three pairs of chelipeds of which the third pair is the largest. Four species of stenopodidean are known from the GBR of which the red and white banded coral shrimp, *Stenopus hispidus* (Fig. 23.3*A*), is the largest and most familiar, reaching about 75 mm in length. Banded coral shrimps live in mated pairs in deep, rock or coral crevices and on the underside of overhangs. They are also known as 'fish cleaners'. Pairs of *S. hispidus* attract fish to their 'cleaning station' by waving their long, white antennae. Large fish approach the cleaning station, where the shrimps 'pick-off' dead skin and parasites. Other stenopodideans on the GBR include other species of *Stenopus* and the 15 mm *Microprosthema validum*.

Achelata: spiny and slipper lobsters

The spiny lobsters (Palinuridae) and slipper lobsters (Scyllaridae) are unusual decapods in lacking chelipeds. They are common on the reef flat and slope where they hide in crevices by day and emerge at night to feed. Palinurids include the large, edible species like the painted cray (*Panulirus versicolor*), ornate cray (*P. ornatus*) and spotted cray (*P. bispinosus* (Fig. 23.3*B*), often misidentified as *P. femoristraga*), for which a commercial fishery operates in northern reef waters between Cape York and Princess Charlotte Bay. Unlike their temperate water counterparts, species of *Panulirus* seldom enter traps, and are instead harvested by hand. Scyllarids are smaller than palinurids, and the antennae, rather than being long and whip-like, are formed into short flat plates. Whereas tropical palinurids are usually brightly coloured, scyllarids are often camouflaged to match their coralline surroundings. Scyllarids, such as *Scyllarides* and *Parribacus*, are common on the reef slope, among coral and rubble, whereas *Thenus* (Fig. 23.3*C*) and *Ibacus*, often called 'bugs', live on the muddy inter-reef sediments where they are commercially trawled. An unusual feature of *Thenus* and *Ibacus*,

Figure 23.3 *A*, banded coral shrimp (*Stenopus hispidus*); *B*, spotted spiny crayfish (*Panulirus bispinosus*); *C*, Moreton Bay bug (*Thenus australiensis*); *D*, dwarf reef lobster (*Enoplometopus occidentalis*); *E*, mud shrimp (*Strahlaxius glytocercus*); *F*, ghost shrimp (*Trypaea australiensis*); *G*, hermit crab (*Dardanus megistos*); *H*, squat lobster (*Allogalathea elegans*); *I*, porcelain crab (*Polyonyx* sp.); *J*, ghost crab (*Ocypode cordimana*); *K*, poisonous xanthid crab (*Lophozozymus erinnyes*); *L*, rubble crab (*Actaeomorpha scruposa*). (Photos: *A, D,* © S. Ahyong; *B, G, J, K,* © GBRMPA; *C,* © G. Ahyong; *E, F,* © C. Tudge; *H, I, L,* R. Springthorpe, © Australian Museum.)

and probably and adaptation to living on level soft sediments, is their very flat carapace that extends sideways, covering their legs.

Astacidea: clawed lobsters and freshwater crayfish

The Astacidea is the large group containing the marine clawed lobsters and freshwater crayfish. The most obvious feature of the astacideans, differentiating them from the palinurid lobsters, is the possession of at least one pair of enlarged, powerful chelipeds. In almost all cases, the large, first pair of chelipeds is followed by two pairs of smaller chelipeds. Freshwater crayfish, such as the yabby and marron (*Cherax*) are common on the mainland, and marine counterparts, such as scampi (*Metanephrops*, *Nephropsis* and *Thaumastochelopsis*) occur in deep, outer shelf waters. However, only a few species of astacideans live on the reef itself: the dwarf reef lobsters of the genus *Enoplometopus* (Fig. 23.3*D*) and family Enoplometopidae. Enoplometopidae, containing about 20 species worldwide, possibly represents the last remnant of a lineage of lobsters that otherwise went extinct in the Lower Jurassic Period. Species of *Enoplometopus* do not exceed 20 cm in length (more often less than 10 cm) but have vivid orange, red and purple colouration. Enoplometopids are very distinctive, but they are seldom seen because of their shy, nocturnal habit and lower reef slope habitat.

Thalassinidea: mud shrimps, ghost shrimps, marine yabbies

Thalassinideans are shrimp-like decapods closely related to the Anomura and Brachyura. They typically have at least one pair of enlarged chelipeds, small eyes, a much-reduced first abdominal segment, but long abdomen, and tailfan. The carapace and abdominal exoskeleton of thalassinideans is usually thin, translucent and somewhat soft, so not surprisingly, they rarely leave their burrows. Most thalassinideans burrow in sand and mud, but some burrow into sponges, live in holes or under coral rocks. Those burrowing in soft sediments, especially of species of the Callianassidae and Upogebiidae, are important bioturbators, often creating tall mounds of excavated sand and mud around the burrow entrances. The species most likely to be seen on the reef flat is the orange *Strahlaxius plectorhynchus*, living in vertical, rubble lined burrows on the inner reef flat. *Strahlaxius glyptocercus* (Fig. 23.3*E*) and *Trypaea australiensis* (Fig. 23.3*F*) occur in muddier or more fine-grained habitats, especially closer to the mainland coast. At least 38 species in nine families are known from the GBR. Thalassinids are usually difficult to capture because of their deep burrows, and no doubt many more species than presently recorded occur in the region.

Anomura: hermit crabs, squat lobsters, porcelain crabs and allies

The anomurans are the nearest relatives of the Brachyura, the true crabs, and include the hermit crabs (Paguroidea), squat lobsters and porcelain crabs (Galatheoidea) and mole crabs (Hippoidea). Unifying features of the Anomura include the very small, reduced last walking legs, and presence of an uncalcified groove along the sides of the carapace (the *linea anomurica*). The last pair of walking legs is so reduced, that at first sight, most anomurans appear to have only six instead of eight walking legs. The last pair of walking legs are usually folded under the abdomen or edges of the carapace, and used for grooming. Anomurans are abundant on the GBR, with the most obvious being the hermit crabs. Most hermit crabs have a soft, vulnerable abdomen, and usually use a gastropod shell for protection. Hermit crabs (Fig. 23.3*G*) are especially well adapted to occupying gastropod shells; the asymmetrical, coiled abdomen and asymmetry of walking legs and claws enable the animal to fit snugly inside. Species range from less than 10 mm to almost 300 mm in length. Common hermit crabs on the GBR fall into one of three major groups: the coenobitids, or terrestrial hermit crabs; the diogenids (the left handed hermit crabs), and pagurids (the right handed hermit crabs). Coenobitids, represented by *Coenobita* on the GBR, live on beaches above the tide line where they forage among flotsam and jetsam. To avoid desiccation, *Coenobita* carries water in its gastropod home and shelters in the shade or under driftwood during the hottest part of the day. Though they spend their entire adult life out of water, coenobitids must return to the sea to breed. Diogenids, with the left cheliped generally larger than the right, and the pagurids, with the opposite pattern, are common on all parts of the reef, from the intertidal beach rock to the reef flat, slope, and inter-reef areas. Hermit crabs are generally scavengers or herbivores, and large numbers can often be seen at night foraging on the reef flat or on the reef slope.

Other anomurans, such as the colourful squat lobsters and porcelain crabs, are common in most parts of the reef. Squat lobsters (Galatheidae) resemble a small, flattened lobster with numerous transverse groves on the carapace, the tail tucked under the body and chelipeds pointing forwards. Common galatheids include species of *Munida* that live in deeper, inter-reefal waters, species of *Galathea*, that live among corals, on bryozoans and sponges, and the conspicuously striped *Allogalathea elegans* (Fig. 23.3*H*) that lives in pairs on crinoids. Porcelain crabs (Porcellanidae) have a short, smooth or spiny carapace and chelipeds protruding sideways. Common porcellanid genera include the free living *Petrolisthes* and *Polyonyx* (Fig. 23.3*I*), and anemone associated *Neopetrolisthes*. Not more than 30 species each of Porcellanidae and Galatheidae are presently known from GBR waters.

The mole crabs, Hippoidea, are highly specialised for burrowing in sand. Their elongate bodies are flattened or oval in shape, and the legs are flattened into digging spades. They can burrow extremely rapidly and are difficult to detect, let alone capture. Most hippoids live in shallow water, but some live at depths beyond 200 m.

Brachyura: crabs

The Brachyura, the true crabs, are possibly the most successful of all decapods, with more the 6500 described species in more than 50 families worldwide. Almost 500 species in about 40 families occur on the GBR. The shallow water crabs of the region are relatively well documented, though many new species continue to be discovered. The name Brachyura, means 'short-tail' and refers to the characteristic feature of crabs—a short abdomen that is tucked beneath the body. Crabs occur in every reef habitat, and some species occasionally enter the water column as hitchhikers on jellyfish or among floating algae or debris.

On beaches above the tideline, the fast running stalk-eyed ghost crab (*Ocypode*; Fig. 23.3*J*) is conspicuous, particularly early and late in the day. Ghost crabs scavenge and opportunistically prey on whatever they can find on the beach, including the occasional turtle hatchling. Though ghost crabs live high up on the beach, their burrows usually reach down to the water table. On the reef flat, the slow moving xanthoid crabs, such as *Actaea* (Fig. 23.4*A*), *Atergatis*, *Carpilius*, *Chlorodiella*, *Eriphia*, *Etisus*, and the poisonous *Lophozozymus* (Fig. 23.3*K*) are common in and under crevices of coral and rocks. Xanthoids are well adapted to living in and among corals and rock, and are the most abundant and diverse of coral reef crabs. The xanthoid body is typically compact, often covered in small spines, bumps, or bristles, and the legs are relatively short and stout. The claws of many xanthoids have dark fingertips, hence the common name 'dark fingered crabs'. The parthenopoids also occur among the rubble of the reef flat, and common genera include *Actaeomorpha* (Fig. 23.3*L*) and *Daldorfia*. Parthenopoids usually have a somewhat triangular or rectangular carapace, with the whole surface of the body and claws covered in tubercles knobs and pits, effectively simulating their rubbly habitat. Unlike the xanthoids and parthenopoids, the swimming crabs (family Portunidae) are fast moving and aggressive. The hind legs of portunids form swimming paddles and the other legs are flattened to assist swimming and digging. The most common swimming crabs on the reef flat are of the genus *Thalamita* (Fig. 23.4*B*, *C*), recognisable by their widely spaced eyes. Species of *Thalamita* are powerful and aggressive, and can be seen foraging in the open over sand, rubble and seagrass.

The mud crab, *Scylla serrata* (Fig. 23.4*D*), is the largest known swimming crab, and is recreationally and commercially fished throughout tropical Australia. It usually lives along the coast in mangrove swamps, but can also be found on the soft offshore sediments between reefs, and amongst mangroves that sometimes occur on or near reefs flats. The claws are particularly powerful and can cause serious injury if mishandled. A much smaller, more 'gentle' crab of the mangroves is

Figure 23.4 *A*, xanthid crab (*Actaea peronii*); *B, C*, swimming crab (*Thalamita crenata*); *D*, pea crab (*Durckheimia lochi*); *E*, soldier crabs (*Mictyris longicarpus*); *F*, box crab (*Calappa lophos*); *G*, mud crab (*Scylla serrata*); *H*, coral crab (*Trapezia cymodoce*); *I*, a well camouflaged spider crab (Majidae); *J*, blue swimming crab (*Portunus armatus*); *K*, male sea spider carrying egg masses (*Anoplodactylus perissoporus*); *L*, sea spider (*Endeis mollis*) on *Millepora* coral. (Photos: *A, H*, R. Springthorpe, © Australian Museum; *B, E, I*, © GBRMPA; *C, F, G, J*, © S. Ahyong; *D*, I. Loch, © Australian Museum; *K, L*, © C. Arango.)

the soldier crab (*Mictyris*; Fig. 23.4*E*), which forages in large numbers on receding tides. Soldier crabs are easily recognised by their somewhat spherical, bluish body, and forward instead of sideways walk. They may march in the thousands, but at the first sign of danger, bury themselves in the sand and mud in a quick 'corkscrew' motion leaving only a small pock mark on the surface where they previously stood. Other conspicuous crabs on the open reef flat include the box crabs (*Calappa*; Fig. 23.4*F*). Box crabs hide, buried in the sand, and emerge at low tide to feed on molluscs, especially gastropods. Box crabs are often called 'shame-faced crabs' because the chelipeds are broad and high, almost entirely hiding the mouth and eyes. A peculiar feature of box crabs is a special 'peg-like' tooth on the side of the moveable finger of the right cheliped. The box crab uses this special tooth on the right cheliped like a can opener to literally 'peel' open gastropod shells.

Some crabs are very specific about their habitat. These include the turtle-grass crab (*Caphyra rotundifrons*), which lives only in the green alga *Chlorodesmis*; the swimming crab (*Lissocarcinus orbicularis*), which lives on and inside holothurians; and coral crabs (*Trapezia* (Fig. 23.4*H*) and *Tetralia* of the families Trapeziidae and Tetraliidae, respectively), which live in the branches of scleractinian corals such as *Acropora* and *Pocillopora*. Trapeziids and tetraliids feed mostly on coral mucous, and though not more than 20 mm wide, will attempt to defend their host against intruders including crown-of-thorns starfish (*Acanthaster planci*). The pinnotherid crabs, commonly called pea crabs, usually make their homes in another type of host, in most cases, bivalve molluscs. Species of *Nepinnotheres* and *Arcotheres* are common in pearl oysters, the rare *Durckheimia lochi* (Fig. 23.4*D*) lives in file shells (Limidae), and *Xanthasia murigera* lives in tridacnid clams. Spider crabs (Majidae; Fig. 23.4*I*) and sponge crabs (Dromiidae) are common on the reef, but they are seldom obvious because they move slowly and are well camouflaged. As their name implies, sponge crabs carry a sponge, held against the carapace by the last two pairs of legs. Spider crabs usually adorn themselves with algae, coral and sponge that are attached to Velcro-like hooked setae; they are often known as 'decorator crabs'. Over time, the sponge, algae or coral grows over the surface of the crab to the extent that the crab can hardly be detected unless it moves. As a small, mobile piece of 'reef', other hitchhikers including worms, amphipods and even other small crabs can be found on large majids. Numerous species of crabs occur on the soft sediments between reefs, particularly swimming crabs of the genera *Charybdis* and *Portunus*. Among these is the edible blue swimming crab, *Portunus armatus* (previously known as *P. pelagicus*) (Fig. 23.4*J*) and mud crab, *Scylla serrata* (Fig. 23.4*G*).

PYCNOGONIDA

Sea spider is an apt name for the pycnogonids (Fig. 23.4*K*, *L*), for they superficially resemble true spiders (Araneae), albeit with a more slender body and sometimes more than four pairs of legs (up to six pairs in some Antarctic species). The body of pycnogonids is small and slender. The head and thorax is fused to form the prosoma, and the abdomen (or opisthosoma) is small and unsegmented. Pycnogonids are also unusual in having the mouth placed at the end of a proboscis that may be as long as half the body. The pycnogonids are an ancient group dating back to the Devonian, probably representing the nearest relatives of the Chelicerata (spiders, mites, scorpions), or possibly the nearest relatives of the remaining Arthropoda. Around 1500 species of pycnogonid are known worldwide with about 30 species in 17 genera known from the GBR and adjacent Queensland coast. Some deep-sea pycnogonids have a leg span of 750 mm, but species on the GBR are much smaller, at less than 20 mm across. All are cryptic and slow moving, so detection can be difficult. Pycnogonids prey on bryozoans, polychaetes, hydroids, xoanthids and even scleractinian corals.

ADDITIONAL READING

Ahyong, S. T., and O'Meally, D. (2004). Phylogeny of the Decapoda Reptantia: resolution using three molecular loci and morphology. *Raffles Bulletin of Zoology* **52**(2), 673–693.

Arango, C. (2003). Sea spiders (Pycnogonida, Arthropoda) from the Great Barrier Reef, Australia: new species, new records and ecological annotations. *Journal of Natural History* **37**, 2723–2772.

Davie, P. J. F. (2002). 'Crustacea: Malacostraca: Eucarida (Part 1): Phyllocarida, Hoplocarida, Eucarida. Zoological Catalogue of Australia 19.3A.' (CSIRO Publishing: Melbourne.)

Davie, P. J. F. (2002). 'Crustacea: Malacostraca: Eucarida (Part2): Decapoda—Anomura,Brachyura.Zoological Catalogue of Australia 19.3B.' (CSIRO Publishing: Melbourne.)

Jones, D., and Morgan, G. (2002). 'A Field Guide to Crustaceans of Australian Waters.' 2nd edn. (Reed Holland: Sydney.)

Poore, G. C. B. (2004). 'Marine Decapod Crustacea of Southern Australia: A Guide to Their Identification.' (CSIRO Publishing: Melbourne.)

24. Mollusca

R. C. Willan

INTRODUCTION

Molluscs belong to a group of highly sophisticated invertebrates in terms of their morphology, having the greatest biodiversity of any phylum in the marine environment. This diversity is exceedingly high in the tropical waters of the Indo-Pacific Ocean, particularly in coral environments like those of the Great Barrier Reef (GBR), where it has been estimated that approximately 60% of all the marine invertebrate species are molluscs. There are as many as 3000 species of molluscs on the GBR.

On the GBR, molluscs vary in size from the impressive metre-long giant clams (Tridacnidae) down to micromolluscs (Rissoidae, Barleeidae, Stenothyridae) and nudibranchs (Hedylopsidae, Microhedylidae) less than 1 mm long. These tiny nudibranchs spend their lives between grains of coral sand. Molluscs exploit every possible kind of plant or animal as food. Herbivores and carnivores among them use a specialised feeding organ (radula) to scrape food off the substratum into their mouths. Some species graze algae or sponges from rocks, while others hunt down their quarry actively, detecting it either by smell (gastropods) or by acute eyesight (cephalopods). Many molluscs, particularly bivalves, make a living feeding off the soup of plankton suspended in the water by filtering out edible particles using mucous nets or gill sieves. There is also an ecologically important group of molluscs known as detrital feeders that separate as food the minute organic particles from the sediments. In doing so, these detrital feeders recycle nutrients that would otherwise be locked up in the sediments.

Not only are molluscs linked into every food chain on the GBR, but they are also integral for the very formation and destruction of the reef itself. Their shells serve as settlement sites for crust-forming red algae (Corallinaceae) that grow outward to weld the hard corals together, thus building up the reef. Other molluscs (especially Mytilidae, Pholadidae and Gastrochaenidae) actively bore into live (and dead) coral, weakening it, and eventually causing its collapse. Molluscs are fundamental to the economy of the GBR in terms of its productivity, for their eggs are laid in tens of millions and the floating larval stages form a very important component of the zooplankton in the waters over the reef.

During the life of a mollusc, its shell serves as a living space for other invertebrates that also take advantage of the protection afforded by the rightful inhabitant. In doing so, they reward the mollusc by helping to defend the shell from attackers like fishes and crabs. Examples of such commensal invertebrates are polychaetes that live inside the shell and hydroids that live on the outside. After the mollusc that manufactured the shell has died, the shell continues to be recycled by other animals like hermit crabs, tanaids and peanutworms as their 'home' until it finally breaks down completely. One species of hermit crab, *Trizopagurus strigatus*, is specially adapted to live only in empty cone shells; its body, claws and legs are all flattened to fit inside the narrow-mouthed shell. During its decomposition the shell also serves as a site of attachment for a multitude of other invertebrates, general and specific, like boring sponges, sea anemones, barnacles and foraminiferans (single celled animals enclosed within a tiny calcareous shell belonging to the Phylum Protista; indicated by an arrow in Fig. 24.1).

Molluscs are present in all habitats on the GBR, from coral outcrops and rocky shores, to sandy lagoons, and in the plankton that washes over the reef. Not only are molluscs present in all habitats, but the sheer abundance of some species like those of the Littorinidae can be staggeringly high. Some families (e.g. Pyramidellidae and Eulimidae) are almost all parasitic, on echinoderms in particular, and members of several other families (Cancellariidae, Galeommatidae, Tergipedidae) live commensally with a range of animals from other phyla.

One of the most significant points about the molluscs of the GBR is that there are definitely no non-native (i.e. introduced or exotic) species among them, not a single one. This situation is unique in the world today and it is one reason authorities need to be so wary when dealing with the shipping traffic moving up and down and through the GBR every day, because ships can easily transport non-native 'pests' on their hulls, in ballast water and in internal seawater compartments called 'sea chests'.

The larger gastropods and bivalves with external shells are the best known groups taxonomically on the GBR, so the level of accuracy of identification for them is relatively high compared to the smaller molluscs, particularly the micromolluscs, and those without shells. The basis for the classification of molluscs has shifted over the last century from the shell to the animal itself.

BODY PLAN

There is no standard molluscan shape and, indeed, there is no word in the English language that takes in the whole of the Phylum Mollusca. The term 'mollusc' means soft-bodied and all molluscs lack an internal supporting skeleton, therefore, in its absence, there is no 'typical' shape that is characteristic for the phylum and the shapes of molluscan animals are very variable. In fact, the shape of the molluscan animal is largely determined by the shape of the external shell it manufactures for protection. These shells are hard and three-layered, consisting of crystals of calcium carbonate, either calcite or aragonite, set in a matrix of protein material called conchiolin. An additional layer of a horny substance (periostracum) is laid over the exterior of the shell and it is also used as a plug to seal the aperture in gastropods (operculum).

The body of all molluscs consists of two main parts, the head/foot and the visceral mass. The former consists of a muscular foot in combination with the head with its array of sensory appendages. The visceral mass is essentially a sac containing the organs responsible for digestion, respiration, circulation and reproduction all encased in a thin membrane (the mantle). The mantle produces the shell from its outer edge.

But not all molluscs have external shells. Each of the classes of molluscs has representatives with simplified and reduced shells. It seems that evolution of modern-day molluscs has proceeded since the Palaeocene era (65 million years ago) from forms with external shells to forms with reduced shells or no shell at all. This process of shell reduction and loss has produced examples of parallel evolution many times in different groups (e.g. the families of 'limpets', most of which are unrelated to each other: Lottiidae, Patellidae, Nacellidae, Hipponicidae, Phenacolepadidae, Capulidae, Calyptraeidae and Siphonariidae).

The Mollusca is a very ancient phylum and its members separated early in the Palaeozoic era (540 million years ago) into several branches that are accorded the rank of classes in the taxonomic hierarchy. Seven such classes exist today. The univalves (Gastropoda, Monoplacophora and Scaphopoda) have a single shell encasing the animal, or none at all. Although squid, cuttlefish and octopus (Cephalopoda) have no external shell, it is clear from their long-distant relative, the pearly nautilus, which is still surviving as a 'living fossil' today that they evolved from ancestors with a single coiled shell. The bivalves (Bivalvia) have two shells, one on either side of the body, with an elastic ligament as a hinge at the top. The chitons (Polyplacophora) have eight shell plates roofing the body, all bound together by a leathery girdle. The gastroverms (Aplacophora) are simple and worm-like with their skin impregnated with tiny rods of calcium (spicules). All these classes, except the most primitive Monoplacophora and Aplacophora are represented on the GBR. In fact, aplacophans probably do live on the GBR too, but nobody has searched for them because they are inconspicuous and very difficult to locate.

In terms of their body plan, the chitons (Fig. 24.5*A*) are the most primitive and simplest molluscs one can encounter on the GBR (given that no aplacophorans have been found there yet, see above). Chitons are dorso-ventrally flattened, bilaterally symmetrical molluscs. There is a large flat foot ventrally and the body is covered dorsally with eight separate, usually articulating, shell plates bearing sensory aesthetes (simple light-detecting organs). The shell plates are surrounded by a muscular girdle that is either naked or covered with calcareous plates like a suit of armour. The head lacks eyes and tentacles. There are about 30 species of chitons on the GBR and they all live on hard substrata.

In terms of numbers of species, the most diverse molluscs on the GBR are the gastropods (Gastropoda) (Figs 24.2, 24.3, 24.4, 24.5*A–J*, 24.6*A*). There are about 2500 species of gastropods on the GBR. Gastropods have a multitude of different body forms, encompassing sea snails and slugs, limpets, sea hares and nudibranchs. Gastropods are united by the dramatic way in which the veliger larva metamorphoses. In the veliger, the cavity containing the gills faces backwards and ventrally, and the viscera are massed above the head/foot. At metamorphosis, an asymmetrical retractor muscle pulls the visceral mass through 180°. As a result, the mantle cavity becomes dorsal behind the head and faces forwards. The digestive system and its associated nerve fibres also become twisted during this process called torsion. In all gastropods but the true limpets, this process of torsion is accompanied by spiral coiling of the shell as it grows. Gastropod shells are asymmetric because growth takes place in a clockwise direction resulting in a shell with the aperture situated to the right of the longitudinal axis. Only a few gastropods (like those of the family Triphoridae, one of which is arrowed in Fig. 24.1), consistently coil in an anticlockwise direction.

The foot of gastropods has a flat sole for creeping, but it may be modified for swimming, digging or blocking the aperture of the shell. The foot is last to be drawn into the spacious final coil of the shell (the body whorl) and the aperture is finally plugged with the operculum, a horny or calcareous plate, borne on top of the foot.

The head/foot region, as the complex above the foot is known, extends from beneath the shell when the gastropod is active. By contrast, the visceral mass is permanently contained within the upper coils of the shell (the spire) and is covered by the mantle. A large part of the visceral mass is occupied by the digestive gland that extends nearly to its apex and is part of the digestive system. The rest of this system consists of a mouth at the end of a snout or retractable proboscis, a feeding organ (radula) and jaws within the pharynx, salivary glands, oesophagus, stomach (that opens into the digestive gland), intestine, rectum and anus. Because of torsion, faecal pellets are discharged over the head. The structure of the radula is a fundamental diagnostic feature used in the classification of the Gastropoda.

Bivalves (Bivalvia) (Figs 24.6*B–E*) are laterally compressed, bilaterally symmetrical molluscs. There are about 500 species of bivalves on the GBR. The shell consists of two valves joined dorsally by an elastic ligament and held together by two large muscles attached to both shell valves (adductor muscles). The mantle is often fused around the edges and extended posteriorly into retractable siphons. The head is simple, lacking eyes and tentacles, but there is a flap of tissue (labial palp) on either side of the mouth. The foot is axe-shaped and often bears a byssal gland that secretes a bundle of threads (byssus) for attachment to the substratum. The foot is retracted into the shell by a special pedal retractor muscle. The gills are much enlarged and serve more for feeding than they do for respiration. The gut includes a complex stomach.

The watering pots (family Penicillidae) are the most bizarre and aberrant of all bivalves on the GBR, albeit with only a few species. Juvenile watering pots, which are fully shelled, either burrow into the sediment or

Figure 24.1 Example of micromolluscs (including triphorid gastropods, one indicated by the black arrow) sorted from a single bag of coral sand from Arlington Reef. Note also the foraminiferan indicated by the white arrow. Scale: 10 mm. (Photo: U. Weinreich.)

Figure 24.2 Shells of representative taxa from the *Cymbiola pulchra* species-group (Volutidae) (from left to right): *C. pulchra woolacottae* from Heron I., 66.1 mm shell length; *C. pulchra houarti* from the Swain Reefs, 78.0 mm; *C. pulchra craecenta* from John Brewer Reef, 82.7 mm; *C. intruderi* from Halfmoon Reef, 71.5 mm; *C. excelsior* from Elusive Reef, 62.5 mm. (Photos: *C. excelsior*, A. Limpus; the rest, MAGNT.)

Figure 24.3 Triton's trumpet, *Charonia tritonis* (Ranellidae), *in situ*, about to devour a crown-of-thorns starfish, *Acanthaster planci*. (Photo: GBRMPA.)

Figure 24.4 This geography cone snail, *Conus geographus* (Conidae), has stung a small demoiselle fish and is expanding the anterior end of its digestive system in preparation for ingesting the fish. (Photo: U. Weinreich.)

Figure 24.5 *A*, The chiton, *Cryptoplax larvaeformis* (Cryptoplacidae), *in situ*, emerging at night to graze on algae. (Photo: GBRMPA.) *B*, Asses ear abalone, *Haliotis asinina* (Haliotidae), *in situ* showing the animal extended from its shell. (Photo: GBRMPA.) *C*, Gilbert's top snail, *Jujubinus gilberti* (Trochidae), showing the animal extended from its shell. (Photo: U. Weinreich.) *D*, Strawberry top snail, *Clanculus margaritarius margaritarius* (Trochidae), showing the animal extended from its shell. Note the remarkable similarity of the pattern of the animals' foot to its shell. (Photo: U. Weinreich.) *E*, Sea hare, *Aplysia dactylomela* (Aplysiidae), with its parapodia (upward-directed extensions from the foot) opened to show the mantle protecting the internal shell. (Photo: G. Cobb.) *F*, Tiger cowrie, *Cypraea tigris* (Cypraeidae), *in situ*, showing the animal extended from its shell. (Photo: GBRMPA.) *G*, Magnificent dorid nudibranch, *Chromodoris magnifica* (Chromodorididae), *in situ*. (Photo: R. C. Willan.) *H*, Much-desired aeolid nudibranch, *Flabellina expotata* (Flabellinidae), *in situ*. Note the specialised defensive sacs (cnidosacs; one indicated by an arrow) at the ends of outgrowths (cerata) on the dorsal surface. (Photo: R. C. Willan.) *I*, Parasitic snail, *Balcis* sp. (Eulimidae), *in situ* on host crinoid. (Photo: U. Weinreich.) *J*, This striated cone snail, *Conus striatus* (Conidae), has extended its proboscis (the narrower tube (arrowed) above the siphon) from its foregut in preparation for firing a toxin-loaded radular tooth into its prey. (Photo: U. Weinreich.)

Figure 24.6 *A*, Acorn dog whelk, *Nassarius glans* (Nassariidae), *in situ* showing the animal extended from its shell. (Photo: GBRMPA.) *B*, Flashing file clam, *Ctenoides ales* (Limidae), showing the animal extended from its shell. (Photo: U. Weinreich.) *C*, The pedum oyster, *Pedum spondyloideum* (Pectinidae), here photographed *in situ*, is an unusual scallop that lives permanently buried in brain corals. (Photo: J. G. Marshall.) *D*, Lilac venus clam, *Callista lilacina* (Veneridae), showing the animal extended from its shell. (Photo: U. Weinreich.) *E*, Giant clam, *Tridacna derasa* (Tridacnidae), *in situ* showing the brightly coloured mantle. (Photo: GBRMPA.) *F*, Cuttlefish, *Sepia* sp. (Sepiidae), *in situ* showing the head, eyes, arms and funnel. (Photo: GBRMPA.)

bore into soft calcareous rock. As they grow, the original shell valves of the juvenile fuse to the anterior end of a sealed, calcareous tube that is permanently buried in the sand. The anterior end of this tube is swollen like a balloon and perforated by numerous tiny tubes, like the holes of a watering can. Watering pots have lost the posterior adductor and pedal retractor muscles and their anterior equivalents are vestigial, so the animal is only attached to its tube by retractor muscles arising from the pallial line and by an array of muscular papillae. A foot is present, but it is only small. In its place, an enlarged pedal disc acts as an hydraulic pump to bring water into the mantle cavity from the interstitial water surrounding the anterior tube through the perforations.

Tusk snails (Scaphopoda) are very elongate, cylindrical, bilaterally symmetrical molluscs. There are about 25 species of scaphopods on the GBR. The mantle is fused mid-ventrally and the long tubular shell is open at both ends. The head bears a long snout and two groups of slender tentacles (captacula). The foot is cylindrical and pointed. Scaphopods have no gills, distinct blood vessels or auricles. They burrow in sediments and use the captacula to haul foraminiferans to the mouth.

Cephalopods (squid, cuttlefish, octopus) (Fig. 24.6*F*) are bilaterally symmetrical molluscs with a dorso-ventrally elongated body. There are about 35 described species of cephalopods on the GBR. In modern cephalopods the shell is internal and the visceral hump is covered by a muscular mantle, giving the body a rounded or streamlined shape. Speed, alertness and large body size are the keynotes of the cephalopods and they rival fishes in their locomotory prowess. They swim freely in the sea like fish, or move nimbly over the bottom. The head bears a pair of large, morphologically complex eyes. Cephalopods have a series of eight prehensile arms around the head so the mouth lies at their centre. These muscular arms bear rows of suckers and have been described as 'super lips'. Squid and cuttlefish additionally have two retractile tentacles bearing suckers at their distal tips. Cephalopods also have a specialised muscular organ, called a funnel, formed from part of the foot. Water from the mantle cavity is ejected via the funnel and is used as a means of jet propulsion in swimming.

BIOGEOGRAPHY

The northern GBR is the region of the Australian continent where the greatest number of molluscan species exist today (about 3000 species). Interestingly however, it is also the region with the lowest endemicity of molluscs, probably reflecting the youthfulness of its geological formation and fact that most species have planktonic larvae and so can mix with populations of the same species elsewhere in the tropical Indo-Pacific Ocean. No molluscan species is endemic to the GBR as such, though two gastropods only occur in the northern section: *Peristernia australiensis* (Fasciolariidae) and *Thecacera boyla* (Polyceridae), and *Corallastele allenae* (Trochidae) seems to be endemic to the Capricorn Group of islands. The diversity of molluscs does attenuate slightly toward the southern extremity of the GBR, but total biodiversity is little changed because of the presence of endemic temperate Australian taxa at the northern limits of their range. Here, the deeper water endemic taxa (Turritellidae, Cypraeidae and Volutidae) have short-lived or direct-developing larvae and there are permanent populations in the cooler inter-reefal areas and channels deeper than 100 metres. By contrast, the eastern Australian shallow-water endemic taxa that do occur on the GBR, like the nudibranch *Chromodoris splendida* (Chromodorididae), have planktonic larvae and it is not known if their sporadic occurrences on the southern GBR represent breeding populations or adults resulting from chance northward-flowing larvae that have successfully grown to adulthood.

One cannot cover the biogeography of molluscs of the GBR without mentioning the remarkable group of endemic Australian balers popularly known as Heron Island volutes. The genus *Cymbiola* (Volutidae) contains a complex of about 20 allopatric species, subspecies and forms with very different shaped shells and colour patterns (Fig. 24.2) that live along the eastern Australian coast between 16°S and 32°S and includes by far the majority on the GBR itself (12 taxa alone occur in the section between the Swain Reefs and Fraser Island). Taxa of this assemblage occur from the Ribbon Reefs east of Cape Flattery in the north and extend continuously, in both shallow water on the coral reefs themselves and the deep water channels throughout the GBR, to Lady Elliot Island. South of Lady Elliott Island they are only found in deeper water (i.e. greater than 100 m). Interestingly, the deeper water taxa are ancestral and they have diverged less than the shallow water forms. Like all balers, these taxa of *Cymbiola* hatch from their egg capsules as crawl-away juveniles, so there is no genetic mixing between them that reinforces their distinctiveness. Although the evolution of these balers must have taken place since the formation of the GBR in the Pleistocene era (2 million years ago) we are a long way from understanding their evolutionary mosaic.

FEEDING

Were it not for the combined efforts of all the gastropods and chitons grazing algae and algal sporelings off hard substrata, the whole GBR would be green with seaweed. Such herbivores have the foregut elaborated into a cuticularised rasping structure (radula) and specialised ciliary fields in the stomach. Examples of algal-grazing molluscs common on the GBR are snake skin chitons (Chitonidae), narrow-plated chitons (Cryptoplacidae) (Fig. 24.5*A*), limpets (Lottiidae, Patellidae, Siphonariidae), abalones (Haliotidae) (Fig. 24.5*B*), top and turban snails (Trochidae, Turbinidae, Phasianellidae) (Fig. 24.5*C*, *D*), nerites (Neritidae), clusterwinkles (Planaxidae), periwinkles (Littorinidae), a few cowries (Cypraeidae), bubble snails (Bullidae, Haminoeidae), sap-suckers (Juliidae, Plakobranchidae, Limapontiidae, Polybranchiidae) and sea hares (Aplysiidae) (Fig. 24.5*E*).

Even more gastropods graze on encrusting animals than they do on plants. Animal-grazing is accomplished by the radula as in the algal grazers. Probably sponges serve as the main food group within this category of 'meats' (if that term can be applied to such animal tissues). Examples of sponge-grazing molluscs common on the GBR are spikey chitons (Acanthochitonidae), keyhole- and slit-limpets (Fissurellidae), top snails (Trochidae), cerithiopsids (Cerithiopsidae), triphoras (Triphoridae) (Fig. 24.1), the majority of cowries (Cypraeidae) (Fig. 24.5*F*), side-gilled sea slugs (Pleurobranchidae), and many nudibranchs (Hexabranchidae, Aegiridae, Dorididae, Discodorididae, Chromodorididae, Actinocyclidae, etc.) (Fig. 24.5*G*). But there are lots of other groups of invertebrates that constitute the food for grazing gastropods apart from sponges. Examples of hydroid-grazers are top snails (Trochidae) and aeolid nudibranchs (Flabellinidae, Aeolidiidae, Tergipedidae, Facelinidae). Examples of hard coral-grazers are coral snails (Coralliophilidae), wentletraps (Epitoniidae) and nudibranchs (Tergipedidae). Examples of soft coral-grazers are egg cowries (Ovulidae) and nudibranchs (Tritoniidae). Examples of bryozoan-grazers are nudibranchs (Polyceridae, Goniodorididae). Examples of kamptozoan-grazers are nudibranchs (Goniodorididae). Examples of ascidian-grazers are lamellarias (Velutinidae), bean cowries (Triviidae) and nudibranchs (Polyceridae, Goniodorididae).

The most remarkable of all these 'meat'-grazing gastropods are the aeolid nudibranchs (Fig. 24.5*H*), all of which feed on cnidarians. After ingestion, the tissue of their cnidarian prey is separated mechanically by cilia into digestible material and indigestible stinging cells (nematocysts), the latter being prevented from firing through copious volumes of mucus. These stinging cells are moved (carefully!) along fine branches of the digestive gland by other types of cilia to specialised sacs (cnidosacs; indicated by an arrow in Fig. 24.5*H*) way out at the ends of finger-like outgrowths from the dorsal surface of the body (cerata) that otherwise serve for respiration. When a predatory fish takes a bite out of an aeolid, the nematocyst-loaded cerata are quickly pointed in the direction of the attacker and the nematocysts are fired off all at once. So the predator gets a face full of stinging cells instead of a meal! Aeolids provide the best example of animals using the defensive structures made by other organisms for their own defense (kleptoplasty).

The natural progression from grazing other animals is to feeding upon one's own kind, and indeed there are two families of nudibranchs (Polyceridae and Gymnodorididae) and one family of bubble snails (Agjajidae) that have members specialising for eating other sea slugs. This usually involves ingesting the slug's body whole.

Instead of grazing 'meat' unselectively, some gastropods have become specialised as parasites on specific tissues (like the blood) of their hosts. The best known 'blood-suckers' are pyramidellids (Pyramidellidae) that feed on a range of invertebrate hosts, eulimids (Eulimidae) (Fig. 24.5*I*) that parasitise echinoderms (either as ectoparasites or as endoparasites living entirely in 'galls' inside the host) and margin snails (Marginellidae) that suck the blood of sleeping fishes.

On the GBR, an enormous number of gastropods actively hunt down mobile prey. Those that prey specifically on polychaetes are cone snails (Conidae, Terebridae), drupes (Muricidae), vase snails (Turbinellidae), bubble snails (Acteonidae, Aplustridae) and nudibranchs (Vayssiereidae). Mitre snails (Mitridae, Costellariidae) prey only on peanut worms (sipunculans). Those that prey only on crustaceans are harp snails (Harpidae), rock snails (Muricidae) and nudibranchs (Tethydidae). Those that prey only on echinoderms are helmet snails (Cassidae, Tonnidae) and trumpet snails (Ranellidae) (Fig. 24.3). Those that only prey on other gastropods are frog snails (Bursidae), balers (Volutidae) and cone snails (Conidae). Those that only prey on bivalves are moon snails (Naticidae) and rock snails (Muricidae). Those that only prey on fishes are basket snails (Cancellariidae) and cone snails (Conidae) (Figs 24.4, 24.5*J*).

Most species of dove snails (Columbellidae) are carnivorous, eating crustaceans and polychaetes. However, a few species are scavengers, and, remarkably, two species (one of which, *Euplica scripta*, occurs on the GBR) are facultative herbivores, that is, they occasionally eat algae in addition to their normal diet of crustaceans.

Balers (Volutidae) live on soft substrata, occasionally foraging on hard substrata. They hunt down other gastropods (like top snails and turban snails, and even poisonous cone snails) and smother them with their large foot. Having entrapped the prey, balers often then 'larder' them in a special pouch under the foot. This pouch may contain several gastropods, still alive, that have been hunted down during a night's foraging.

For an essentially bottom-dwelling group, the gastropods have a large number of species that, although attached to reefal substrata, exploit the plankton in the water column above the reef for food. Examples of plankton-feeding gastropods are cap snails (Capulidae), slipper limpets (Calyptraeidae) and worm snails (Vermetidae, Siliquariidae). The worm snails, the most interesting group of these plankton-feeding gastropods and the commonest encountered on the GBR, use a mucus trap to feed; a gland on their foot secretes copious quantities of sticky mucous that form thin threads streamed into the water as a feeding web. The web is hauled back into the mouth with the aid of the radula and the plankton stuck onto it is ingested. To maximise the time spent feeding, at least some species of worm snails are able to produce a second web concurrently with the ingestion of the first one.

The most agile and beautiful planktonic-feeding gastropods are those that live permanently in the water column forming part of the macroplankton. The sea butterflies (Cavoliniidae and Limacinidae) have their foot modified into two enlargements anteriorly (the so-called 'wings') for swimming. Many species of sea butterflies use a transparent and delicate mucous 'fishing' web to strain other microscopic algal cells (i.e. diatoms and dinoflagellates) out of the plankton. Sea goddesses (Clionidae, Hydromylidae and Pneumodermatidae), another wholly planktonic group of gastropods, are active predators upon these sea butterflies, spotting them with their well-developed eyes. Sea goddesses capture sea butterflies with the suckered tentacles on their heads, these being attached to the inside of the aperture of the sea butterfly shell. The proboscis is then thrust into the shell of the sea butterfly and the body is ripped out whole.

When plankton falls down onto the sea floor, it joins other organic and inorganic material deposited there. A number of gastropods 'vacuum' up these nutritionally rich deposits from the surface of both hard and soft substrata. Examples of deposit-feeding gastropods are sand creepers (Cerithiidae), longbums (Potamididae), conchs and spider snails (Strombidae, Seraphsidae), carrier shells (Xenophoridae), onchidiids (Onchidiidae) and ear snails (Ellobiidae).

There is an array of gastropods dedicated to scavenging dead animals. Indeed, one family of scavengers, the dog whelks (Nassariidae) (Fig. 24.6*A*), has about 30 species on the GBR. These different species have partitioned the habitats neatly between themselves; some only live intertidally, others in shallow lagoons, and yet others in fine sediments on outer slopes. Other scavenging gastropods are a few species of cowrie (Cypraeidae) and whelks (Buccinidae).

Though sedentary, bivalves use their extremely large and complex gills as filters to strain the plankton. Long cilia on the gills draw a powerful current of sea water into the mantle cavity (i.e. the space between the shell valves) where it is passed through the gill sieve formed by other types of cilia. A typical bivalve filters 30 to 60 times its own volume of water every hour. Common examples of filter-feeding bivalves on the GBR are mussels (Mytilidae), ark clams (Arcidae, Cucullaeidae, Noetidae), bittersweet clams (Glycymerididae), pearl oysters (Pteriidae), hammer oysters (Malleidae), mangrove oysters (Isognomonidae), pen shells (Pinnidae), file clams (Limidae) (Fig. 24.6*B*), oysters (Gryphaeidae, Ostreidae), kittens paws (Plicatulidae), saucer scallops (Propeamusiidae), scallops and fan shells (Pectinidae) (Fig. 24.6*C*), thorny oysters (Spondylidae), jingle shells (Anomiidae), jewel boxes (Chamidae), yoyo clams (Galeommatidae), cardita clams (Carditidae), cockles (Cardiidae), giant clams (Tridacnidae), trough clams (Mactridae), razor clams (Solenidae, Pharidae), wedge clams (Donacidae), venus clams (Veneridae) (Fig. 24.6*D*), basket clams (Corbulidae), piddocks (Pholadidae) and watering pots (Penicillidae).

Giant clams (Fig. 24.6*E*) lie upside down in depressions on the coral. Their mantle lobes are extensively developed and 'farm' dense colonies of microscopic dinoflagellates (*Symbiodinium* sp., popularly called zooxanthellae). These zooxanthellae appear to supply the majority of older clams' carbon requirements, most importantly glucose, through their own photosynthesis.

The largest family of bivalves numerically, the wafer clams (Tellinidae), all feed on organic particles deposited on the sea floor, thus recycling nutrients in the way of the deposit-feeding gastropods mentioned above. They have a very long and flexible inhalant siphon that 'vacuums' these deposits off the top layer of sediment. The ingested sediment is sucked into the mantle cavity where it is sorted by the enormous labial palps into edible particles and waste matter. Two other families of bivalves closely related to the wafer clams are also deposit feeders, the sunset clams (Psammobiidae) and semele clams (Semelidae).

The lucine clams (Lucinidae) are a particularly specialised family of bivalves whose shells are quite commonly encountered on the GBR. Lucines lack an inhalant siphon and have an inhalant opening instead for feeding and respiration, through which they communicate with the water column by means of an inhalant tube constructed up through the sediment by the foot. The foot is highly extensible, with the tip in the form of a pointed bulb capable of secreting mucus that lines the inhalant tube. The gills have a rich, resident bacterial flora contained in large vacuoles; these bacteria undertake sulphide-oxidising reactions.

The cephalopods are all carnivorous, rivalling the fishes in their ability to see and hunt down active prey like fishes and swimming crustaceans. Common examples of free-swimming cephalopods on the GBR are cuttlefish (Sepiidae) (Fig. 24.6*F*) and calamari squid (Loliginidae). These catch their prey by rapidly shooting out their pair of feeding tentacles. These tentacles move so fast that they are difficult to see.

Dumpling squid (Sepiolidae) and octopuses (Octopodidae) spend most of their time on the sea floor, feeding on less active prey like benthic crustaceans, bivalves and sleeping fishes. However, their ability to see and hunt down their prey is every bit as good as their free-swimming relatives. All octopuses have strong toxins that quickly immobilise their prey.

REPRODUCTION

Most molluscan species have separate sexes and there are generally no external differences between males and females of the same species. However, the spider snail *Lambis lambis* (Strombidae), shows some dimorphism when the shells are fully grown, with females having larger shells with long upward-curved spines and males having smaller shells with short horizontally-pointing spines. Other notable examples are the very small yoyo clams (Galeommatidae), where dwarf males live inside the mantle cavity of the female.

Many bivalves (most notably Ostreidae, Tridacnidae and Galeommatidae) change sex from male to female as they grow; that is, they are protandric hermaphrodites. All the sea slugs are functionally hermaphrodites, producing both male and female gametes and having both male and female reproductive organs in the same body. In other words, they are simultaneous hermaphrodites. However, they never fertilise themselves (although this is physically possible), and instead are able to mate with any other mature animal of the same species they encounter, both individuals acting as sperm donors and egg recipients at the time of copulation. The sea hares (Aplysiidae) (Fig. 24.5*E*) have an even more extreme kind of partnering where they form long mating chains; the individual at the rear of such a chain acts as a male to the one in front of it by delivering sperm. This second-to-last individual acts as a female by receiving the sperm and, simultaneously, as a male by delivering sperm to the partner in front of it, and so on through the chain. Indeed, there is one report in the literature of the animals at the front and rear of such a chain coming together to form a complete mating ring!

Some sea butterflies (Cavoliniidae) are believed to possess remarkable asexual reproduction. Under laboratory conditions, an unusual-looking skinny individual is formed by transverse fission of the body of a normal individual. The new individual then detaches itself from its shell as a naked animal, transporting only gonads, and grows independently into a fully-shelled separate individual.

Most molluscs simply shed their gametes (eggs and sperm) directly into the seawater where fertilisation occurs, but fertilisation is internal in some gastropods and all cephalopods. In those taxa with an intermediate stage, the resulting larva has a shell like a tiny cap, and lobes on its foot for swimming and capturing plankton for feeding. A larva of this type, termed a veliger, is typical of all molluscs. Veligers drift with the currents until they come across the adult food source at which time they break off their swimming lobes and become crawling juveniles. Those molluscs that have internal fertilisation lay their eggs in tough cases. Either, veligers hatch from these cases and join the plankton, or a miniature version of the adult crawls out. In the first case, the female produces a relatively large number of small eggs, each with a small amount of yolk (lecithotrophic development), and this is the most common pattern found in warmer waters like those of the GBR. In the second case, which is termed direct development, the female produces only a relatively few large eggs, each with a rich supply of yolk, and this occurs mainly in molluscs living in cold waters, although some examples are found on the GBR.

Sex reaches new heights in cephalopods, in terms of both mating and brood protection. The sexes are always separate. Many shallow water cephalopods come together in large numbers to spawn. They have elaborate courtship displays with the males 'dancing' and rapidly changing their colour and pattern at the same time to impress the females. The males of some octopuses display specially enlarged suckers to females as a sign of sexual maturity, with males of one species on the GBR, *Octopus cyanea*, waving a raised and coiled modified arm tip in the direction of the female during its courtship display. The sperm duct of cephalopods has become exceedingly specialised for the manufacture of sperm bundles (spermatophores). Each spermatophore is a narrow, hard, torpedo-shaped tube containing a dense mass of sperm. While a penis is used in some species to transfer spermatophores directly to the female, in most cephalopods, the male uses a specially modified arm (the hectocotylus) to transfer spermatophores from the terminal opening of the genital duct to the female (the cephalopod equivalent of copulation), sometimes even leaving his arm holding the bunch of spermatophores inside the female's mantle cavity. Males of other cephalopods bite small holes in the female's skin into which they insert the spermatophores. These implanted spermatophores resemble small, white, parasitic worms under the skin. Cephalopod embryos are special among the molluscs from the moment after fertilisation. The eggs are comparatively large and yolky, and do not completely cleave so that the embryo is built up from a smaller disc of cells on the upper pole of the egg, and the larger part of the egg goes to form a yolk sac from which the young animal is nourished. Female octopuses brood their eggs until the embryos hatch, fiercely defending them from predators.

Although brood protection is best developed in octopuses, it is by no means unique to octopuses or just cephalopods. Females of several gastropods (Cypraeidae, Ranellidae and Facelinidae) remain with their spawn masses until the young hatch out and it is a common sight to see a female cowrie 'sitting on her eggs' when one turns over a dead coral slab on the GBR. It is one of the reasons people should always replace rocks and coral slabs if they turn them over; if not, the female and all the embryos she is guarding will die from desiccation.

Females of a few bivalves (most notably Galeommatidae) take brood protection to the extreme by incubating their eggs, and even the developing embryos, inside their mantle cavities. These little incubatory yoyo clams only produce a small number of embryos (up to a dozen) because of the considerable energy that needs to be invested in such maternal care.

DANGEROUS MOLLUSCS FOUND ON THE GREAT BARRIER REEF

Only a very few molluscs living on the GBR are deadly to humans, and they cause harm accidentally by defending themselves.

The blue-ringed octopus, *Hapalochlaena lunulata*, has particularly potent saliva that quickly paralyses its prey (normally crustaceans). One component of the saliva is tetrodotoxin, which is produced by bacteria in the salivary glands, and to which the human nervous system is particularly susceptible. This chemical prevents messages coming from the brain from reaching the muscles, so that the human victim is paralysed. Interestingly, tetrodotoxin only paralyses voluntary muscles and involuntary muscles, such as the heart, iris of the eye and the gut, continue to function normally. The victim scarcely feels the bite of a blue-ringed octopus, but after a short time he has difficulty in breathing. This is followed by nausea and vomiting, complete cessation of breathing and collapse. The victim remains fully conscious and may die from lack of oxygen. There have been no deaths from blue-ringed octopuses on the GBR, but in 1954 a man in Darwin died within two hours of being bitten by one.

A victim definitely feels a sharp pain if he has been stung by a cone snail. There are actually more than 100 species of cone snails (*Conus*) in all habitats on the GBR, but only three of them living among live coral and coral rubble (*Conus geographus*, *C. tulipa* and *C. obscura*) are capable of killing humans. These cone snails are all fish-eating (piscivorous) species and have an enlarged aperture to the shell into which the dead fish can be hauled for digestion (Fig. 24.4). The sting is caused by the piercing of the skin by a single radular tooth resembling a tiny harpoon. The tooth, which is shot out forcefully from the tip of a long flexible proboscis, injects a powerful neurotoxin (contoxin) as it is driven into the flesh of the prey. Cone snails have an elaborate venom apparatus consisting of a muscular bulb and a tubular secretory duct opening into the mouth cavity. A modified proboscis that is tubular and muscular like an elephant's trunk (Fig. 24.4) actually fires out this radular tooth. A human victim of a sting from a cone snail can lose vision, or have hearing or speech affected and he may become partially or completely paralysed within half an hour if he has been unlucky enough to have been stung by one of the deadly species. Indeed, there is one recorded death in 1935 of a man on Hayman Island from the sting of a *Conus geographus*. A sting by a species of cone snail other than these three may cause pain, swelling and discolouration of the area near the puncture, but not death. So one should walk carefully over the reef and not pick up a cone shell, even if it appears to be empty, because its animal may be fully retracted within the shell.

FUTURE RESEARCH

At present the most urgent need is for taxonomic research to establish what species of molluscs live on the GBR and where they live. At present the figure of 3000 species is a best guess and the number of adequately identified species is rapidly increasing as more and more groups of micromolluscs like those shown in Fig. 24.1 are studied. An indication of the richness of these micromolluscs on the GBR can be obtained from just one family, the curious left-twisted triphoras (Triphoridae; a triphora shell is arrowed in Fig. 24.1), where a single sample of coral sand is known to have contained 80 species of triphoras, many of them undescribed scientifically.

Genetic technology is becoming not only an important tool for taxonomic research, but it is also highly significant for assessing the relationships between the different taxa. Genetics has great power to separate taxa that do not offer sufficient morphological characters to distinguish them, or present a confusing array of morphological characters. The prime candidate for this novel genetic technology would be the beautiful balers of the genus *Cymbiola* (Volutidae) mentioned above and shown in Fig. 24.2.

Despite the very large number of species of molluscs on the GBR, research on their behaviour and physiology has barely begun. As shown above, we have general knowledge about the feeding types of molluscs, but there has hardly been any research on individual species. For example, although we know some chitons have 'home' sites that they return to after feeding excursions, we do not know how these chitons find their way 'home' or what makes a particularly good 'home' site. Similarly, we know that some members of the mussel genus *Lithophaga* (Mytilidae) burrow into coral using a chelating agent secreted by pallial glands, but we do not know what turns this agent off when individual mussels attain maturity and stop burrowing, or how they avoid being overgrown by living corals.

As the advent of scuba diving on the reefs themselves has opened our eyes to the nudibranch fauna of the GBR, drift diving in mid-water has just started to reveal some astonishing information about the behaviour of the holoplanktonic molluscs. For example, the very peculiar sea goddess *Hydromyles globulosa*, which is moderately common on the GBR, has its entire animal encased in a transparent and flexible cuticle. When threatened, an animal can retract completely into this cuticle and seal the slit-like opening with a fold in the cuticle, thus turning itself into a completely impervious sphere.

If one investigates the body of any mollusc closely, one is likely to find it harbours numerous parasites (e.g. ciliated protists, nematodes, trematodes, isopod crustaceans), both externally and internally. For example, on Heron Island, the common clusterwinkle *Planaxis sulcatus* (Planaxidae) acts as an intermediate host for a number of trematodes, including *Austrobilharzia terrigalensis*, which is responsible for 'swimmer's itch'.

Nudibranchs have come to symbolise all the molluscs on the GBR because of their fragile bodies, bizarre shapes, bright colours and remarkable behaviours. Nowadays no diver can have a holiday on the GBR without taking at least some digital images of nudibranchs. Photography is certainly a form of 'collecting', albeit one that does not involve the removal of any live animals themselves, though damage is inadvertently caused to the reef by trampling, boat's anchors and incorrect weighting. In the past, shell collectors received much criticism by outsiders for their activities. One hears anecdotes of the 'devastation' caused by shell collectors and one reads emotive passages like: 'By their thoughtless depredation the whole productivity of that reef flat and the natural habitat of thousands upon thousands of its inhabitants can literally be ruined in a matter of hours' Bennett (1986: 126). But such hysterical statements have never been tested on the GBR. Humans have collected molluscs for ages, for food, trade and decoration, and detailed studies on population ecology have shown many species are highly resilient to collecting. Responsible shell collectors only take a few individuals and never remove immature specimens or females guarding eggs. Furthermore, they are aware of the ecological consequences of not turning over dead coral slabs. It is the shell collectors who have provided us scientists with the specimens for our comparative studies and museum reference collections. In recognising shell collectors as the least threatening of a host of (natural and unnatural) processes that could affect the GBR—human trampling, bottom trawling, commercial fishing, collection for consumption by Aboriginal and non-indigenous people, eutrophication from land-derived nutrients, cyclones and rising sea levels—the Great Barrier Reef Marine Park Authority currently allows recreational collection of molluscs on certain reefs on the GBR and along the adjacent coast without a permit. This collecting is limited to five individuals of a species (i.e. a total that comprises living animals and/or their shells, but not more than five of any species) to be taken in one day. For all other purposes, a permit is required from GBRMPA (see Chapter 12).

REFERENCES

Taxonomy of balers of the *Cymbiola pulchra* species-group with hypotheses on their evolution

Bail, P., and Limpus, A. (1998). 'Revision of *Cymbiola* (Cymbiolacca) from the East Australian coast the '*pulchra* complex'.' (Evolver Publications: Rome.)

Detailed treatment of all mollusc families occurring in Australia

Beesley, P. L., Ross, G. J. B., and Wells, A. E. (Eds) (1998). 'Mollusca: The Southern Synthesis. Fauna of Australia. Vol. 5.' (CSIRO Publishing: Melbourne.)

Species present in Australian waters with geographical distributions

Gowlett-Holmes, K. (2001). Polyplacophora. In 'Zoological Catalogue of Australia 17.2 Mollusca; Aplacophora, Polyplacophora, Scaphopoda, Cephalopoda'. (Eds A. Wells and W. W. K. Houston.) pp. 10–84. (CSIRO Publishing: Melbourne.)

Lamprell, K. L., and Healy, J. M. (1998). A revision of the Scaphopoda from Australian waters. *Records of the Australian Museum Supplement* **24**, 1–189.

Lu, C. C. (2001). Cephalopoda. In 'Zoological catalogue of Australia 17.2 Mollusca; Aplacophora, Polyplacophora, Scaphopoda, Cephalopoda'. (Eds A. Wells and W. W. K. Houston.) pp. 129–308. (CSIRO Publishing: Melbourne.)

Detailed descriptions and illustrations of all the species from Heron Island plus a listing of species present on the GBR

Marshall, J. G., and Willan, R. C. (1999). 'Nudibranchs of Heron Island, Great Barrier Reef: A Survey of the Opisthobranchia (Sea Slugs) of Heron and Wistari Reefs.' (Backhuys Publishers: Leiden.)

■ WHAT ARE BRYOZOANS?

Bennett, I. (1986). The Great Barrier Reef. (Lansdown Press: Dee Why West, NSW.)

25. Bryozoa

D. P. Gordon & P. E. Bock

Anyone seeing bryozoans for the first time might mistake them for turfing seaweeds, hydroids, or tiny corals. Bryozoans are zoologically unrelated to reef corals, of course, but their hard, calcareous crustose, mounded, and branching colonies superficially resemble those of cnidarians. Whereas in the tropics, bryozoans are mostly dwarfed by stony corals, in cooler temperate waters they come into their own, and can form bioherms and mini-reefs.

Bushy bryozoans used to be called moss animals (the literal meaning of Bryozoa) and the flat encrusting ones sea mats. Lace corals, or lace bryozoans, are the descriptor for some lacy fenestrate (mesh-like) forms and even the fragile encrusters can be said to create lacy sheets over seaweeds, stones and shell, and under coral rubble. No one common name is adequate for all, so 'bryozoan' is probably the best all-round term. (Box. 25.1.)

In the 1800s, the Bryozoa were held to comprise two very distinct groups, respectively named Ectoprocta (bryozoans in the strict sense) and Entoprocta (also called Kamptozoa), each of which was raised to phylum rank. Ectoprocta as a phylum name may be found in some American textbooks, but the International Bryozoology Association has formally adopted only the name Bryozoa. Entoprocts/kamptozoans (variously known as goblet worms or nodding animals) appear

BOX 25.1 HOW ARE THEY CONSTRUCTED?

Colonies are made up mostly of feeding zooids (autozooids), but bryozoans are justifiably famous for exhibiting a higher degree of polymorphism ('many forms') than almost any other invertebrate group, having different kinds of non-feeding individuals. The most famous are avicularia, named by Charles Darwin. They have various roles (cleaning, defence, locomotion). Then there are kenozooids, simple chambers modified to serve as attachment rootlets (rhizoids, stalks), spines, stolons, space-fillers, or part of the support structure for large erect colonies. To feed, bryozoans evert a funnel-shaped plume of tentacles (called a lophophore) through an opening (orifice) in the body wall that merely puckers inwards upon closure or is variously modified as a pleated collar, as pair of stiff lip-like folds, or as a lid-like operculum.

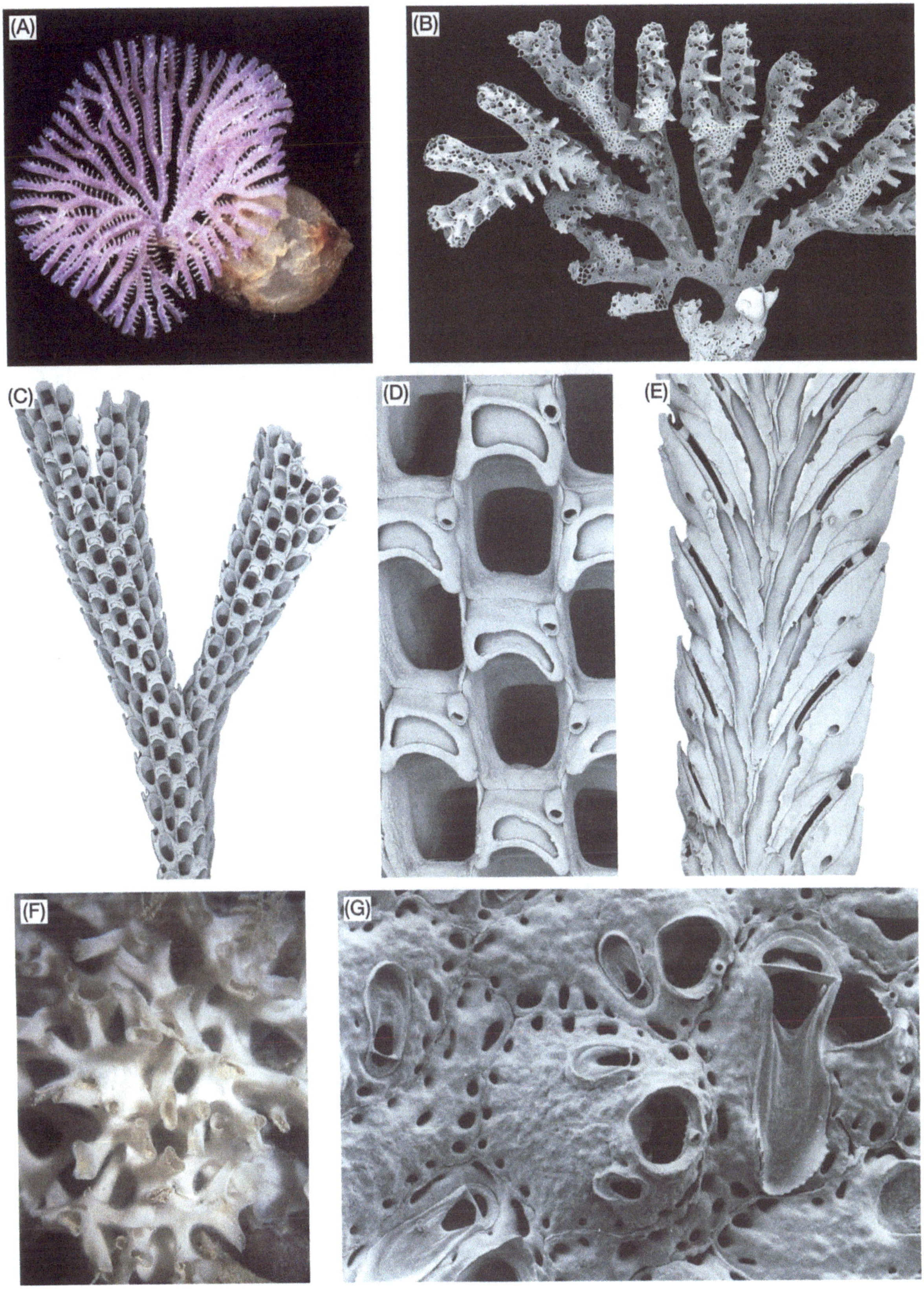

Figure 25.1 The cyclostome bryozoan *Mesonea radians*, showing a dried colony (×3) in *A*, and a scanning electron micrograph of a colony *B*, with several brood chambers (the porous structures below the branch bifurcations). Scanning electron micrographs of the cheilostome bryozoan *Caberea dichotoma*, showing a branch at *C*, (×15), with close-ups of zooids with ovicells and spine bases *D*, and the dorsal side of a branch with vibracular chambers *E*. The cheilostome bryozoan *Parasmittina vacuramosa*, showing a colony of tubular branches at *F*, (×1.5) and a scanning electron micrograph of zooids with orifices and avicularia of different sizes *G*. (Photos: *A, B*, P. Bock; *C, D, E, G*, D. Gordon; *F*, K. Gowlett-Holmes).

not be closely related to genuine bryozoans but the evolutionary origins of both groups is very obscure. Only six species of Entoprocta have been reported from the Great Barrier Reef (GBR); there are many more to discover. They are not further discussed in this account.

With the exception of a few species that consist of a single individual for part or most of their lives, bryozoans are colonies of a few to millions of individuals (zooids), each of which is produced by a form of asexual budding (Box 25.2). Hence, bryozoans can be thought of as modular or clonal organisms. Adult colonies range in size from less than a millimetre to about a metre in diameter. Most are one to a few centimetres in width (if encrusting) or height (if erect).

Bryozoans have an excellent fossil record but, curiously, Bryozoa is the only major phylum not found fossilised from the Cambrian. There are about 15 000 fossil species and nearly 6000 living species.

■ HISTORY OF COLLECTING AND TAXONOMIC DIVERSITY OF GREAT BARRIER REEF BRYOZOA

As the British biologist John Ryland noted, the most outstanding thing that can be said about research on GBR Bryozoa is its paucity, given the undoubted richness of the fauna, although the number of studies on particular taxa have been increasing in recent years. Ryland summarised the history of collecting from the 1840s to the 1970s and wrote a short account of the Bryozoa of the GBR for the 1984 *Coral Reef Handbook* based on his personal collecting. This collection, mostly from Heron Island, formed the basis of two important papers published by him and another British colleague in the 1990s. (For information on collecting bryozoans, see Box 25.3.)

Current known bryozoan diversity in the Great Barrier Reef Province from Torres Strait to the southernmost section is based on distributional information, species lists, and descriptions provided by 20 bryozoologists and two ecologists from 1852 to 2006. The tally from all sources, after accounting for synonyms, is 319 species for the entire Great Barrier Reef Province (see Table 25.1).

This figure for the number of species is probably extremely conservative, especially given the geographic area covered by the GBR, as well as the wide variety of habitats and niches that can be occupied by the range of colony form that bryozoans can exhibit (see below). The neighbouring New Caledonian bryofauna comprises 407 species, of which 178 species are found in the first 100 metres and 232 species range into or are found only at depths greater than 100 m. Some 178 species were recently reported in the Solomon Islands fauna, of

BOX 25.2 REPRODUCTION

Usually zooids are hermaphroditic, their sperm maturing before their ova. Some bryozoan species have separate female and male zooids that differ in shape, size, and anatomy. Purely reproductive zooids may lack a gut and have a few tiny non-ciliated tentacles whose sole function is to release sperm. This surprising use of tentacles as vasa deferentia (sperm ducts) appears to be true for all bryozoans. Since the lophophore also serves as a gill for respiration, it is a truly a multipurpose structure.

Fertilisation in bryozoans is internal. Sperm from one colony are captured by a recipient colony's tentacles, to which they first adhere then move downwards towards a duct (intertentacular organ) or, more commonly, a pore that allows entry into the body cavity. Fertilised eggs may be incubated in a special sac within the zooid, within a modified zooid (gonozooid or brood chamber), or, as in the majority of species, in a hood-like ovicell distal to the maternal orifice.

BOX 25.3 COLLECTING

Erect bushy, stick-like, and lace-like bryozoans can easily be detached from their substratum. Encrusting bryozoans are harder to collect unless they are on pieces of shell, seaweed, or coral. A thin blade can be used to lift or scrape some off the substratum. Details regarding permits required for collecting in the various zones of the GBR are provided in Chapter 12.

Preservation

Uncalcified and delicate bryozoans are best preserved in alcohol, usually 70% ethanol, but 90–100% ethanol is necessary for DNA studies. The easiest way to treat most encrusting and robust erect species is to let them dry out. There is very little smell.

Preparation for study

Identification is based mostly on details of the zooid, especially the skeleton in those forms (the majority) that have one. For this, a microscope is necessary. (Most bryozoologists tend to use a scanning electron microscope as it reveals the greatest details and quickly yields good digital photos.) To see the skeleton clearly, bryozoologists soak small pieces of colony in liquid domestic bleach (hypochlorite solution) to remove soft tissues and cuticular structures.

Identification

Information on bryozoans is scattered in monographs and scientific papers and there are very few popular guides (none for the GBR). However, the internet is becoming a good source of information. The best entry point, especially for photographs of Australian bryozoans, is http://bryozoa.net [Verified 21 March 2008]. The official website of the International Bryozoology Association is www.nhm.ac.uk/hosted_sites/iba/ [Verified 21 March 2008].

which 72 (40%) were new, giving an indication of the level of taxonomic novelty that tends to occur whenever a tropical island group is explored bryozoologically intensively and for the first time. Some 725 species are known for the Philippine-Indonesian region (i.e. the 'Coral Triangle') which is the most biodiverse marine area globally. Even this region, however, has been explored relatively superficially, both in shallow water and at depth, and one can expect this tally to be doubled. Equally, one should expect more than 1000 bryozoan species within the boundaries of the entire Great Barrier Reef Province. While there is considerable overlap of species in the Coral Triangle, a number of species appear, at this juncture, to have limited distributions.

THE IMPORTANCE OF BRYOZOANS ON CORAL REEFS

Bryozoan roles

Bryozoan diversity can be very high in tropical reef environments. Apart from some turfing, robustly encrusting, or prominent erect species, however, most bryozoans are small and cryptic, living on the undersides of coral heads or eroded coral boulders, often in the spaces between branches or in excavations provided by other organisms. On inner reef flats, conditions are unsuitable for bryozoans but on the outer flat where boulders and rubble are common, bryozoans can be found in abundance, especially at the reef edge

Table 25.1 Bryozoan taxonomy diversity in the Great Barrier Reef province

Taxon	No. of families	No. of genera	No. of species
Class Stenolaemata	10	13	16
Order Cyclostomata	10	13	16
Class Gymolaemata	68	127	301
Order Ctenostomata	6	7	10
Order Cheilostomata	62	120	291
Totals	78	140	317

at low water and on the seaward slope. Where space allows (as on the underside of concave surfaces) relatively large, erect colonies can grow. Bryozoans may be prolific on windward reef slopes below the main level of wave action.

Although bryozoans in tropical reefs may not generally achieve the structural dominance that occurs in some temperate settings, they can nevertheless play several important roles in reef environments, as hidden encrusters, cavity dwellers, cavity fillers, and dead-coral veneerers; between reefs they can act as sediment binders, stabilisers, and trappers. Hence, bryozoans can contribute to reef build-up, especially those species that engage in self-overgrowth or which grow over other organisms. Such colonies tend to become a permanent part of the calcareous mass that comprises a reef, acting as accessory frame-builders. Alternatively, those bryozoans that break up upon the death of the colony can contribute to the sediment that forms between clumps and reefs, just as they do in more temperate settings.

Bryozoan growth forms

Bryozoans exhibit an impressive range of colony form, a fact noted and much used by non-bryozoologists (as well as bryozoologists), particularly paleontologists who have sought to correlate colony form with environment in fossil assemblages in which bryozoans have been found in rock-forming abundance. Colonies can be two-dimensional encrusters (the majority), or, through frontal budding and/or self-overgrowth, these can become mounded (well exemplified by species of *Celleporaria*) or erect. Hence, erect bryozoans may be firmly fixed to the substratum (like lace corals of the family Phidoloporidae), but many are not, having root-like rhizoids that attach them to the substratum. Owing to the fact that rhizoids can spread out, and in a few cases even become finely divided and attach to sand grains, some species are adapted to live on soft sediments. Some soft-sediment dwellers can be tiny conical forms, scarcely more than a millimetre in size and may have only a single anchoring rootlet. In some others, the rootlet is a robust tube that anchors a large 'platey' colony above the sediment. Bryozoan colonies may be thickly calcified, lightly calcified, or (in the case of ctenostomes) uncalcified. If erect colonies are lightly calcified they can bend in a current and are said to be flexibly erect (members of the families Bugulidae and Candidae). But well-calcified colonies can also be flexible if they are articulated by uncalcified joints (Cellariidae, Catenicellidae, Poricellariidae, Quadricellariidae, Savignyellidae, Tetraplariidae) or/and are basally rooted. (Examples of all these forms can be seen on the website in Box 25.3.)

Apart from encrusting, fixed-erect, and flexible-erect colonies, there are two other main categories. One is free living (vagrant), that is, the colony is unattached to the substratum and, owing to bristle-like avicularian mandibles around the periphery of the colony (cap- or discus-shaped), has the power of mobility (lunulariids, otionellids, and selenariids). One other distinctive category is that of shell-borer/shell-eroder (found among several families of Ctenostomata). None has yet been reported from the GBR but they must exist. Their small size, transparency, and cryptic habit render them easily overlooked, however.

For the entire GBR Province, 55% of the bryozoan species are two-dimensional encrusters, 20% are flexibly erect, and 15% are firm three-dimensional structures that are rigidly attached to the substratum. These categories of colonial morphology are very broad. They can be subdivided, and more than 20 such categories have been recognised and named (for example, encrusting colonies may be runners, spots, or sheets; erect colonies may be sticks, trees, or fronds). Such categories based on colony form potentially allow for the possibility of paleoenvironmental analysis, but it must be said that correlations between morphology and environmental conditions in the literature have typically been inferential and not backed up by experimental or ecological observations based on living colonies. For this reason, a new, integrated classification scheme based on growth-habit characteristics and the disposition of modules (zooids) has been developed to allow a testable common ground for systematic comparison of character states among varied bryozoan growth habits. It would be fruitful applying this system to the varied range of colony forms represented in the GBR Province.

ADDITIONAL READING

Cuffey, R. J. (1977). Bryozoan contributions to reefs and bioherms through geologic time. In 'Reefs and Related Carbonates—Ecology and Sedimentology'. (Eds S. H. Frost, M. P. Weiss and J. B. Saunders.) pp. 181–194. [AAPG Studies in Geology No. 4.] (The American Association of Petroleum Geologists: Tulsa.)

Hageman, S. J.; Bock, P. E.; Bone, Y., and McGowran, B. (1998). Bryozoan growth habits: classification and analysis. *Journal of Paleontology* **72**, 418–436.

Hayward, P. J. (2004). Taxonomic studies on some Indo-West Pacific Phidoloporidae (Bryozoa: Cheilostomata). *Systematics and Biodiversity* **1**, 305–326.

Hayward, P. J., and Ryland, J. S. (1995). Bryozoa from Heron Island, Great Barrier Reef, 2. *Memoirs of the Queensland Museum* **38**, 533–573.

Mather, P., and Bennett, I. (Eds) (1984). 'A Coral Reef Handbook.' 2nd edn. (The Australian Coral Reef Society: Brisbane.)

Ryland, J. S. (1974). Bryozoa in the Great Barrier Reef Province. In 'Proceedings of the Second International Reef Symposium, Vol. 2'. (Eds A. M. Cameron, B. M. Campbell, A. B. Cribb, R. Endean, J. S. Jell, O. A. Jones, P. Mather and F. H. Talbot.) pp. 341–348. (The Great Barrier Reef Committee: Brisbane.)

Ryland, J. S. (1984). Phylum Bryozoa—lace corals and their relatives. In 'A Coral Reef Handbook' 2nd edn. (Eds P. Mather and I. Bennett.) pp. 68–75. (The Australian Coral Reef Society: Brisbane.)

Ryland, J. S., and Hayward, P. J. (1992). Bryozoa from Heron Island, Great Barrier Reef. *Memoirs of the Queensland Museum* **32**, 223–301.

Tilbrook, K. J. (2006). Cheilostomatous Bryozoa from the Solomon Islands. In 'Santa Barbara Museum of Natural History Monographs 4: Studies in Biodiversity 3'. (Ed. H. W. Chaney.) pp. 1–385. (Santa Barbara Museum of Natural History: California.)

26. Echinodermata

M. Byrne

■ INTRODUCTION

Echinoderms are a conspicuous and diverse component of the invertebrate fauna of the Great Barrier Reef (GBR) and have been reviewed in a number of taxonomic and biogeographic studies (see Additional reading). Most echinoderms from tropical Australia have a broad distribution in the Indo-Pacific Ocean. The 630 species of echinoderms recorded from the GBR are from the five classes as follows: sea stars (Asteroidea) 137 species; brittle stars (Ophiuroidea) 166 species; sea urchins (Echinoidea) 110 species; sea cucumbers (Holothuroidea) 127 species; and feather stars (Crinoidea) 90 species. Echinoderms are a conspicuous and ecologically important component of reef communities. They are often the dominant organisms on the sea floor and this is particularly true of tropical Holothuroidea, the elongate sausage-shaped animals (e.g. *Holothuria*) seen on reef flats and sandy areas. Some of the more common echinoderm species on the GBR are illustrated in Figs 26.1–26.8.

The body symmetry of adult echinoderms is secondarily radial and based on pentamery. There are usually five radii (e.g. arms of sea stars), although multiarmed (6+ arms) asteroids and brittle stars are common and some burrowing sea urchins have four radii. Echinoderm classes are easily identified due to their distinctive body profiles. Asteroids are star-shaped with the arms (five or more) tapering from the disc (e.g. *Linckia, Nardoa*) or a cushion-like pentagon shape lacking arms (e.g. *Culcita*) (Figs 26.1; 26.2*A–F*). Holothuroids by contrast are elongate (Figs 26.2*F–I*; 26.3–26.5). Ophiuroids have a round central disc and slender flexible arms that are sharply set-off from the body (Figs 26.6, 26.7). Their body can have a simple (brittle stars or serpent stars) or branched (basket stars) profile. Echinoids have a rigid, globose body (= test) covered by spines of varying length (Fig. 26.8*A–F*). Crinoids have an array of feather-like arms that range in number from five to several hundred (Fig. 26.8*G–I*).

Some aspects of the biology of echinoderms, for instance reproduction, are general to the phylum, while other features such as feeding are class specific. Class specific features are dealt with below with a focus on species that occur on the GBR.

■ REPRODUCTION AND LIFE HISTORY

For the most part, there are no external morphological differences between males and females. Most echinoderms are sexual reproducers, spawning copious numbers of eggs. The crown-of-thorns starfish, *Acanthaster planci*, is estimated to release up to 60 million eggs per year (Box 26.1). Many echinoderms from the GBR spawn in the spring and summer and can have a very long spawning period. Mass spawning of echinoderms is common and many species spawn at the same time as the corals on the GBR. Release of gametes

Figure 26.1 Asteroidea. *A, Acanthaster planci* (× 0.07) (photo: J. Keesing); *B, Linckia laevigata* (× 0.25); *C, Linckia guildingii* (× 0.30); *D, Linckia multifora* (× 0.25); *E, Nardoa novaecaledoniae* (× 0.33); *F, Culcita novaeguineae* adult (× 0.18); *G, Culcita novaeguineae* juvenile (× 1.00) (photo: S. Walker); *H, Echinaster luzonicus* (× 0.40); *I, Fromia milleporella*. (× 0.80) (Photos: M. Byrne, unless noted.)

by erect sea cucumbers during summer evenings is a sight often seen by divers. Reproductive periodicity of some echinoderms along the GBR changes with latitude. *Archaster typicus*, a widespread and abundant sea star on muddy sand flats throughout the Indo-Pacific, is unusual in showing pairing of males and females during spawning. As the breeding season approaches the male climbs on the female's aboral surface. This pairing behaviour undoubtedly enhances fertilisation success.

Most echinoderms are free spawners and have dispersive larvae. These larvae are beautiful and distinct for each class. In contrast to adults, the larvae have bilateral symmetry. Some need to feed in the plankton (planktotrophic larvae) and others are sustained by egg nutrients (lecithotrophic larvae). Some echinoderms like the asteroid *Aquilonastra byrneae* (Figs 26.2C) have benthic development and consequently lack a dispersive stage. Brooding echinoderms such as the asteroid *Cryptasterina hystera* (Fig. 26.2*A*) and the ophiuroid *Ophiopeza spinosa* (Fig. 26.7C) care for their young and give rise to mobile crawling juveniles. These two species are unusual in having a pelagic-type larva in the brood chamber and so have potential to brood and broadcast their young, but it is not known if they do so.

Asexual reproduction is also common in the echinoderms of the GBR and occurs in asteroids, ophiuroids and holothuroids. This involves the animals breaking in half or fragmenting a part of their body. As a result these echinoderms often have an unusual body profile (Figs 26.1*D*, 26.2*D*). This process is followed by regeneration of each portion to form a complete individual. On the GBR fission (splitting in half) is particularly common in brittle stars and sea cucumbers. Some small asterinid sea stars are fissiparous (Fig. 26.2*D*). *Holothuria atra* and *Stichopus chloronotus* are well known for asexual reproduction by fission. The asteroid *Linckia multifora* has an impressive capacity to propagate asexually and is often seen with arms at different stages of regeneration (Fig. 26.1*D*). Whole stars can regenerate from a single autotomised arm.

ECHINODERM DIVERSITY

Asteroidea

Asteroids (Figs 26.1, 26.2*A–F*), although diverse, are generally not abundant, with the exception of the spectacular outbreaks of *Acanthaster planci* (Fig. 26.1*A*) (Box 26.1). They occur in reef, soft sediment and rubble habitats. *Luidia* and *Astropecten* species are found in lagoon and inter-reefal sediment habitats. The Ophidiasteridae is a large family of tropical sea stars including the distinct blue sea star *Linckia laevigata* (Fig. 26.1*B*) and other common species *Fromia milleporella* (26.1*I*), *Nardoa novaecaledoniae* (26.1*E*) and *Ophidiaster granifer*. The Asterinidae (Fig. 26.2*A–E*) is a species rich family that is particularly challenging with respect to species identification due to the presence of morphospecies

BOX 26.1 FAMILY ACANTHASTERIDAE

Acanthaster planci

The crown-of-thorns (COTS) starfish, *Acanthaster planci* (Fig. 26.1*A*) is probably the best-studied asteroid in the world because of the effect that periodic outbreaks of this species have on coral reefs (see Additional reading). We have comprehensive knowledge of the ecology, reproduction, and feeding biology of *A. planci*. This species is an unusual asteroid in being a specialist corallivore and outbreaks result in marked decrease in live coral cover to below 1–5% in some reports. *Acanthaster planci* generally prefers acroporid and pocilloporid corals, but other corals are also consumed. The outbreak ends when coral prey is exhausted. Reef recovery following intense predation by COTS is variable with some reefs not recovering for 10–15 years. The eggs of *A. planci* contain saponins, a chemical toxic to fish and presumed to protect the eggs and embryos from predation.

There are three hypotheses on the causes of *A. planci* outbreaks: (1) the predator removal hypothesis proposes that overfishing of COTS predators (triton shells, fishes) influences the increase in numbers; (2) the natural phenomenon hypothesis, and (3) the terrestrial runoff hypothesis that proposes that nutrient runoff from the land leads to an increase in the phytoplankton food of *A. planci* larvae, leading to enhanced recruitment. Recent data analyses provide strong support for the nutrient hypotheses, emphasising the need to control nutrient runoff from anthropogenic sources into the GBR lagoon (see Chapter 11). While it was thought that COTS outbreaks might be due to a major pulse of recruitment, it now appears that they are comprised of individuals from multiple recruitment years.

Figure 26.2 Asteroidea. *A*, *Cryptasterina hystera* (× 1.67) (photo: from Byrne and Walker 2007); *B*, *Cryptasterina pentagona* (× 1.67) (photo: from Dartnall *et al.* 2003); *C*, *Aquilonastra byrneae* (× 1.67) (photo: from Byrne and Walker 2007); *D*, *Ailsastra* sp (× 0.30) ; *E*, *Disasterina* sp. Holothuroidea (× 1.25). *F*, *Afrocucumis africana* (× 0.92); *G*, *Polyplectana kefersteinii* (× 1.10); *H*, *Euapta godeffroyi* (× 0.10); *I*, *Synapta maculata* (× 0.23). (Photos: M. Byrne, unless noted.)

complexes comprising a number of species. A number of cryptic species have been recently identified in tropical Queensland based on differences in reproduction and development.

Echinaster luzonicus (Fig. 26.1*H*) is a common species in the GBR. This multiarmed (6+ arms) species often has an irregular profile with unequal portions due to its propensity to autotomise distal portions of its arms followed by regeneration. The pincushion star *Culcita novaeguineae* is also common (Fig. 26.1*F*, *G*). Juvenile *Culcita* are flatter and have more distinct rays than the adult (Figs 26.1*G*).

Figure 26.3 Holothuroidea. *A, Holothuria atra* (× 0.27); *B, Holothuria atra* with commensal crabs (× 0.50); *C, Holothuria atra* with commensal worm (× 0.67); *D, Holothuria edulis* (× 0.38); *E, Holothuria leucospilota* (× 0.17); *F, Holothuria dofleinii* (× 0.21); *G, Holothuria difficilis* (× 0.63); *H, Holothuria fuscogilva* (× 0.19); *I. Holothuria whitmaei* (× 0.15) (photo: H. Eriksson). (Photos: M. Byrne, unless noted.)

The feeding biology of sea stars varies from specific to general diets. Many species are predators on molluscs, sponges, corals and other echinoderms. Other species are surface grazers and scavengers. These asteroids predominantly feed off hard substrata such as coral rubble, eating sponges, microscopic organisms, or coral mucus. *Culcita novaeguineae* feeds on corals, other invertebrates and benthic films. Coral mortality caused by individual *Culcita* is less than that caused by individual *Acanthaster*, but its selective feeding habits may influence the relative abundance of some coral species.

The Holothuroidea (Figs 26.2*F–I*, 26.3–26.5) are diverse and common all along the GBR and there are three major orders: (1) the Aspidochirotida, the common deposit feeders that graze on the surface of the sediment; (2) the Dendrochirotida, suspension feeders with an array of branching tentacles and; (3) the Apodida, which are also deposit feeders. Several surveys of holothuroid diversity on the GBR have been undertaken (see Additional reading).

Aspidochirotids are by far the most abundant and

BOX 26.2 BÊCHE-DE-MER SPECIES

Approximately 15 species comprise the bêche-de-mer fisheries across northern Australia and elsewhere in the Indo-Pacific. Several Australian fisheries for the high value species collapsed in short order, repeating the pattern seen elsewhere in the Indo-Pacific. As illustrated by the teatfish (*Holothuria* [*Microthele*] *whitmaei* and *H.* [*M.*] *fuscogilva*) complex, there are serious conservation concerns for populations of commercial species that have been on the decline for some time or have completely disappeared from some localities (see Additional Reading). Holothuroids are particularly susceptible to overfishing because of their limited mobility, slow growth, late maturity, poor recruitment and density-dependent propagation. Local areas are quickly stripped of valuable species. Currently, fishers in Australia are moving on to less valuable or less accessible (remote areas, deeper waters) stock, repeating the 'fishing-down' pattern seen in other jurisdictions.

One of the first species targeted by fishers on the GBR was the high value species, the black teatfish (BTF) *Holothuria whitmaei*. This species is still listed as *H. nobilis* in some guides, but this is an Indian Ocean species. The GBR fishery for the BTF was short-lived as the stock of this shallow water species was quickly depleted. Unfished reefs and green zones support high densities of bêche-de-mer species. The large populations of *H. whitmaei* at Raine Reef and other reserves are crucial for conservation of the species.

The general lack of juveniles in sea cucumber populations indicates that recruitment will be slow. The first principle of reproduction, gamete contact, is likely to be compromised by the current low densities of spawning individuals on some reefs.

conspicuous holothuroids on the GBR (Figs 26.3–26.5). They occur in a variety of habitats from the reef dwelling *Actinopyga* (surf red fish, Fig. 26.5*F*), to the intertidal and deep water aspidochirotids *Actinopyga*, *Bohadschia*, *Holothuria*, and *Stichopus* species. The most abundant species on the GBR are capable of asexual and sexual reproduction. These include *H. atra* (Fig. 26.3*A*), *H. difficilis* (Fig. 26.3*G*), *H. edulis* (Fig. 26.3*D*), *H. hilla* (Fig. 26.4*C*) and *S. chloronotus* (Fig. 26.5*A*). They are fissiparous, splitting in half and subsequently regenerating the anterior and posterior halves to make two complete individuals. Although these species also reproduce sexually, spawning gametes and have dispersive larvae, it appears that fission is important in maintaining the populations. Many species provide habitat to commensal crabs and scale worms. These are commonly seen on the body wall of *H. atra* (Fig. 26.3*B*, *C*). Parasitic eulimid snails are also commonly seen on the body surface of some holothuroids (Fig. 26.4*F*). These snails extend their proboscis through the body wall and into the coelom where they utilise nutrients in their host body fluids.

Aspidochirotids are deposit feeders and are prominent members of the biota of soft sediment environments. They feed with their spatula-like tentacles (Fig. 26.5*G*) scooping up sand, algal films and detritus. Aspidochirotids provide important ecosystem services by enhancing local productivity through their bioturbation and feeding/digestive activity. Burrowing species are particularly important in bioturbation of the nutrient poor carbonate sediments that dominate much of the GBR. Taxonomically, these sea cucumbers have long been a challenge, with some aspidochirotids well characterised, but others being intractable to traditional taxonomy. Several commercial species of *Stichopus* and *Actinopyga* appear to be morphospecies complexes and the identity of species currently being fished needs to be assessed (Box 26.2). Synaptids (Fig. 26.2*G–I*) are less familiar because many are small and some are noctur-

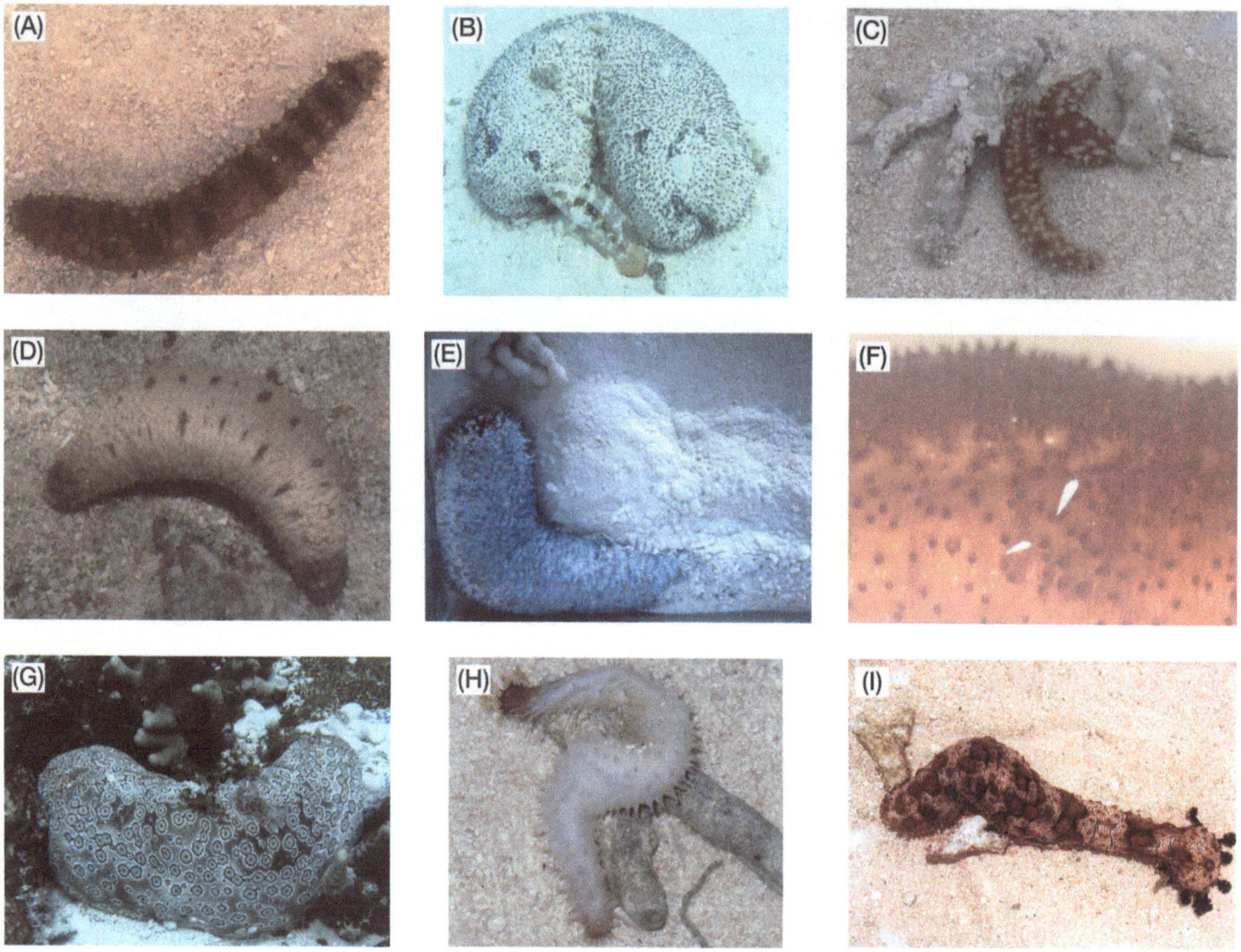

Figure 26.4 Holothuroidea. *A, Holothuria impatiens* (× 0.23); *B, Holothuria lessoni* (× 0.20) (photo: H Eriksson); *C, Holothuria hilla* (× 0.14); *D, Holothuria arenicola* (× 0.25); *E, Holothuria isuga* (× 0.12); *F, Holothuria isuga* with parasitic snails (× 1.00); *G, Bohadschia argus* (× 0.17) (photo: H. Eriksson); *H, Labidodemas semperianum* (× 0.40); *I, Personothuria graeffei* (× 0.16). (Photos: M. Byrne, unless noted.)

nal. The suspension feeding dendrochirotids are less common. *Afrocucumis africana* (Fig. 26.2*F*), a small, dark purple dendrochirotid is often seen under boulders and rubble in the intertidal zone. Other dendrochirotids burrow in sediment or live in the reef infrastructure or algal turfs and so are rarely seen. Their tree-like dendritic tentacles that they use for filter feeding can be seen emerging from the substrate.

Ophiuroidea

The species diversity of ophiuroids on the GBR is impressive, with species in the families Ophiocomidae, Ophiotrichidae and Ophiodermatidae being well represented (Figs 26.6, 26.7). *Ophiocoma* and *Macrophiothrix* species are a diverse assemblage of large species. Several genera are in need of revision, with cryptic species evident. Ophiuroids are often common shoreward of coral habitats where many species are sympatric, aggregating under slabs of coral rubble and in crevices. Along the GBR these include *Ophiarachnella gorgonia* (Fig. 26.7*B*), *Ophiolepis elegans* (Fig. 26.7*D*), *Ophiocoma dentata* (Fig. 26.6*A*), *O. erinaceus* (Fig. 26.6*C*), *O. scolopendrina* (Fig. 26.6*B*), *Ophionereis porrecta* (Fig. 26.7*F*) and *Macrophiothrix* species (Fig. 26.7*G, H*). Ophiuroids also inhabit the reef infrastructure and soft sediments. *Ophiocoma scolopendrina* (Fig. 26.6*B*) is the most con-

Figure 26.5 Holothuroidea. *A*, *Stichopus chloronotus* (× 0.15); *B*, *Stichopus herrmanni* (× 0.17); *C*, *Stichopus monotuberculatus* (× 0.20); *D*, *Stichopus vastus* (× 0.18); *E*, *Thelenota ananas* (× 0.14); *F*, *Actinopyga* sp. (surf red fish) (× 0.17) (photo: S. Walker); *G*, *Actinopyga* tentacles (× 0.50); *H*, *Actinopyga echinites* (× 0.13); *I*, *Actinopyga miliaris* (× 0.11). (Photos: M. Byrne, unless noted.)

spicuous ophiuroid in the Indo-Pacific. This species specialises in shallow habitats that are emersed at low tide where it reaches densities up to 320 m^{-2} and is often seen feeding on surface scum at low tide.

Ophiuroids have varied diets with many being filter feeders, extending their arms from their hiding places in the reef infrastructure to feed. This is often at night. Burrowing amphiurid ophiuroids keep their disc below the surface of the sediment and extend their arms into the water column to filter feed. Ophiodermatids are predators and scavengers. *Ophiarachna incrassata* (Fig. 26.7*A*), a spectacular ophiodermatid, traps fishes under its arms and also feeds on carrion left behind by major predators. It has been seen scavenging turtle flesh left behind by tiger sharks.

Echinoidea

The Echinoidea includes the familiar sea urchins and irregular urchins (sand dollars and spatangoids). On the GBR most sea urchins are cryptic during the day, with species such as *Diadema savignyi* (Fig. 26.8*A*) and *Echinometra mathei* (Fig. 26.8C) emerging at night. *Echinostrephus* species excavate holes in coral to form a permanent residence from which its spines emerge. The

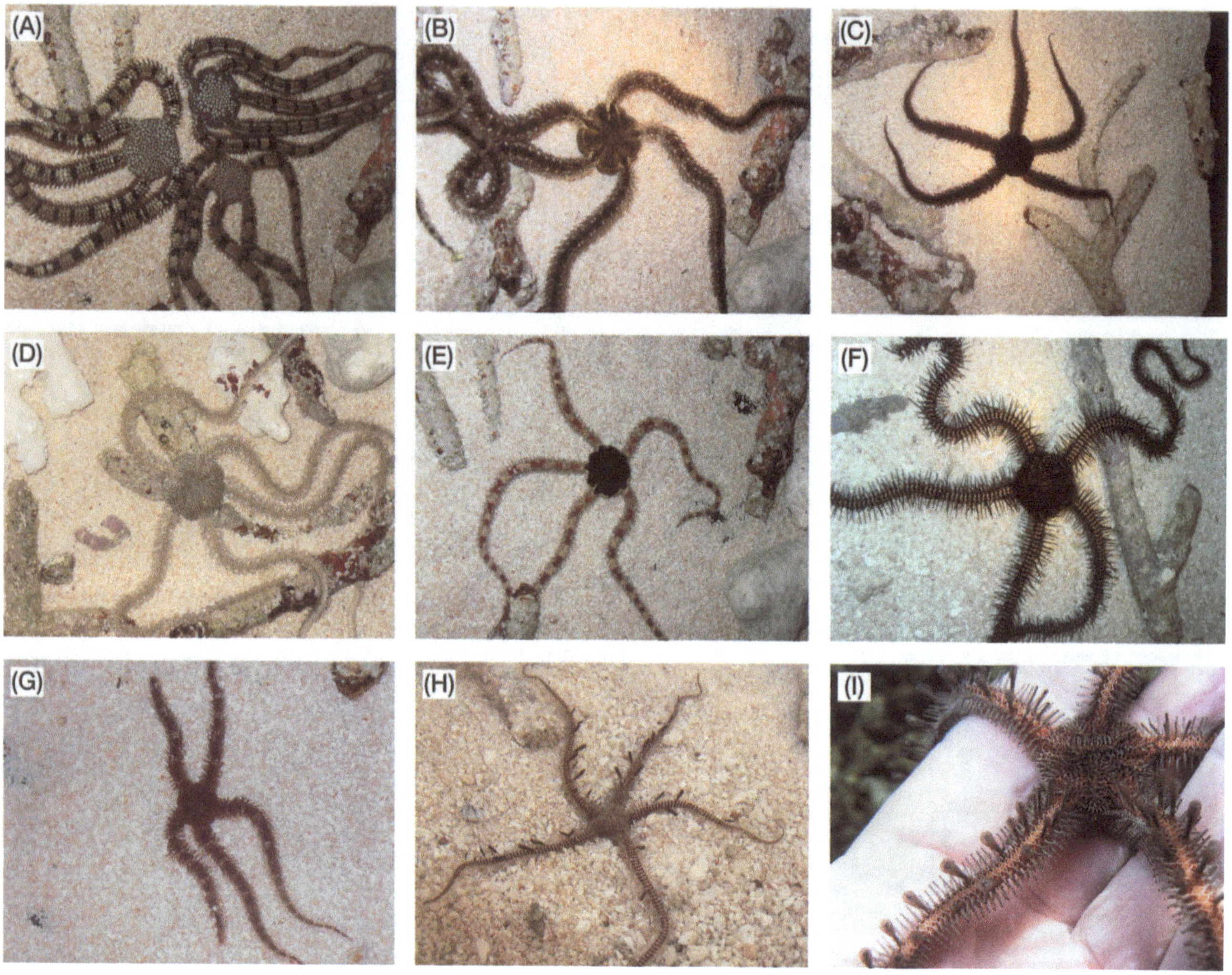

Figure 26.6 Ophiuroidea. *A, Ophiocoma dentata* (× 0.50); *B, Ophiocoma scolopendrina* (× 0.50); *C, Ophiocoma erinaceus* (× 0.33); *D, Ophiarthrum pictum* (× 0.38); *E, Ophiarthrum elegans* (× 0.50); *F, Ophiomastix janualis* (× 0.38); *G, Ophiomastix mixta* (× 0.50); *H, Ophiomastix caryophyllata* (× 0.50); *I, Ophiomastix annulosa* (× 1.00) (photo: S. Uthicke). (Photos: M. Byrne, unless noted.)

commercial urchin *Tripneustes gratilla* (Fig. 26.8*D*) is common in sea grass areas. Common irregular echinoids include the sand dollar *Peronella lesueuri* that occur in abundance along the Queensland coast and the spatangoid urchin *Breynia australasiae* that burrows in sandy inter-reefal areas and can be locally abundant. The burrowing activity of echinoids, particularly by *Echinometra* species, is important in bioerosion of coral reef and intertidal habitats (see Chapter 8).

Most sea urchins are grazers, using their hard calcareous teeth to remove algae and encrusting organisms from the surface. As a member of the grazing guild on coral reefs, sea urchins contribute to keeping the biomass of algae low, thereby preventing overgrowth of algae over coral substrate. Sand dollars are suspension feeders, collecting particles from flow and transferring these by specialised tube feet down the food groves on the surface of the test to the mouth. Spatangoids have a secondary bilateral profile associated with their burrowing life style. They are deposit feeders removing organic matter from ingested sand as they propel themselves through sediments using their specialised 'rowing' spines.

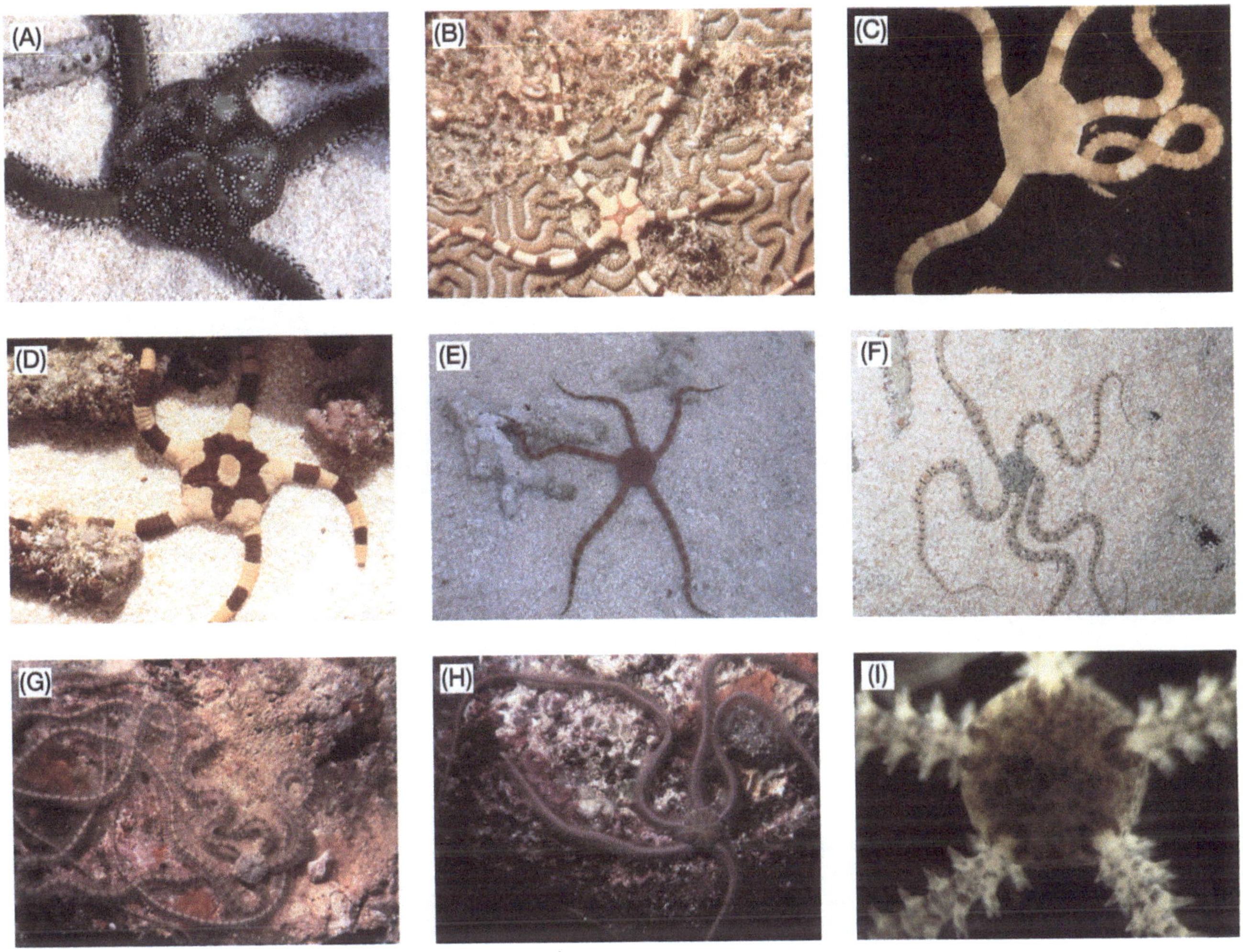

Figure 26.7 Ophiuroidea. *A, Ophiarachna incrassata* (× 0.71); *B, Ophiarachnella gorgonia* (× 0.50) (photo: J. Keesing); *C, Ophiopeza spinosa* (× 2.50); *D, Ophiolepis elegans* (× 0.63); *E, Ophiomyxa australis* (× 0.25); *F, Ophionereis porrecta* (× 0.50); *G, Macrophiothrix lorioli* (× 0.33) (photo: J. Keesing); *H, Macrophiothrix nereidina* (× 0.33) (photo: J. Keesing); *I, Amphipholis squamata* (× 8.33). (Photos: M. Byrne, unless noted.)

Crinoidea

Most crinoids on the GBR are subtidal and a few are abundant in lower intertidal areas. Reefs along the GBR have a diverse assemblage of crinoid species. Some species occupy the surface of soft sediments. There are several taxonomic reviews and surveys of the crinoid fauna of the GBR (see Additional reading). The Comasteridae is a major family of tropical crinoids including *Comanthus* (Fig. 26.8G), *Comaster* (Fig. 26.8*H*), *Comatella*, *Comatula*, *Oxycomanthus* and *Phanogenia* species. The Mariametridae also has a number of genera on the GBR including *Dichrometra*, *Lamprometra*, *Mariametra*, *Himermetra* (Fig. 26.8*I*), and *Stephanometra*.

Crinoids are locally abundant on the GBR, as shown in a number of investigations of their population ecology. They extend their feather-like arms into the water column and may or may not have cirri, basal claw-like appendages, to cling to the substrate. Those that lack cirri use their arms for attachment. Many crinoids are fully or partially concealed within the reef infrastructure during the day and emerge at night to feed. Other crinoids are fully exposed at all times. The cryptic behaviour of crinoids may be to avoid predation by fishes, and several species have defensive chemicals that may deter predators. When disturbed from their perch, crinoids exhibit a striking

Figure 26.8 Echinoidea. *A, Diadema savignyi* (× 0.14) (photo: A. Hoggett); *B, Echinothrix diadema* (× 0.23); *C, Echinometra mathaei* (× 0.54); *D, Tripneustes gratilla* (× 0.50) (photo: U. Bové); *E, Mespilia gobulus* (× 1.00); *F, Eucidaris metularia* (× 1.00) (photo: A. Hoggett). Crinoidea. *G, Comanthus alternans* (× 0.16) (photo: A. Hoggett); *H, Comaster schlegelii* (× 0.20) (photo: A. Hoggett); *I, Himerometra robustipinna* (× 0.10) (photo: A. Hoggett). (Photos: M. Byrne, unless noted.)

swimming behaviour with co-ordinated movement of the arms.

Crinoids are filter feeders, extending their feather-like arms above the substrate to capture food particles on their small tube feet that line the arms. The food is captured from flow and so crinoids take up positions on the reef and adjust the arrangement of the arms to take advantage of the ambient water movement. Food particles are conveyed to the mouth in the centre of the disc on food tracts.

Any collecting of echinoderms requires a permit (see Chapter 12).

ADDITIONAL READING

General references

Byrne, M. (2001). Echinodermata. In 'Invertebrate Zoology'. (Ed. D. T. Anderson.) pp. 366–395. (Oxford University Press: Sydney.)

Byrne, M., Cisternas, P., and O'Hara, T. (2008). Brooding of a pelagic-type larva in *Ophiopeza spinosa:* reproduction and development in a tropical ophiodermatid brittlestar. *Invertebrate Biology* **127**, 98–107.

Byrne, M., Cisternas, P., Hoggett, A., O'Hara, T., and Uthicke, S. (2004). Diversity of echinoderms at Raine Island, Great Barrier Reef. In 'Echinoderms: München'. (Eds T. Heinzeller and J. H. Nebelsick.) pp. 159–164. (Taylor and Francis Group: London.)

Clark, A. M., and Rowe, F. W. E. (1971). 'Monograph of Shallow-Water Indo-West Pacific Echinoderms.' (Trustees of the British Museum (Natural History): London.)

Messing, C. G., Meyer, D. L., Siebeck, U. E., Jermin, L. S., Vaney, D. I., and Rouse, G. W. (2006). A modern soft-bottom, shallow-water crinoid fauna (Echinodermata) from the Great Barrier Reef, Australia. *Coral Reefs* **25**, 164–168.

Rowe, F. W. E., and Gates, J. (1995). 'Echinodermata. Zoological Catalogue of Australia. Vol. 33.' (CSIRO Publishing: Melbourne.)

Crown-of-thorns sea star

Brodie, J., Fabricius, K., De'ath, G., and Okaji, K. (2005). Are increased nutrient inputs responsible for more outbreaks of crown-of-thorns starfish? An appraisal of the evidence. *Marine Pollution Bulletin* **51**, 266–278.

DeVantier, L. M., and Done, T. J. (2007). Inferring past outbreaks of crown-of-thorns seastar from scar patterns on coral heads. In 'Geological Applications to Coral Reef Ecology'. (Eds. R. Aronson and R. Beer.) pp. 85–125. (Springer: New York.)

Uthicke, S., Schaffelke, B., and Byrne, M. (In Press). A boom or bust phylum? Ecological and evolutionary consequences of large population density. *Ecological Monographs* (in press).

Bêche-de-mer holothuroids

Conand, C., and Byrne, M. (1993). Recent developments in the bêche-de-mer fishery in the Indo-Pacific. *Marine Fisheries Review* **55**, 1–13.

Uthicke, S., O'Hara, T. D., and Byrne, M. (2004). Species composition and molecular phylogeny of the Indo-Pacific teatfish (Echinodermata: Holothuroidea) bêche-de-mer fishery. *Marine and Freshwater Research* **55**, 1–12.

Uthicke, S., Welch, D., and Benzie, J. A. H. (2004). Slow growth and lack of recovery in overfished holothurians on the Great Barrier Reef: Evidence from DNA fingerprints and repeated large-scale surveys. *Conservation Biology,* **18**, 1395–1404.

Photo credits

Byrne, M., and Walker, S. J. (2007). Distribution and reproduction of intertidal species of *Aquilonastra* and *Cryptasterina* (Asterinidae) from One Tree Reef, Southern Great Barrier Reef. *Bulletin of Marine Science,* **81**: 209–218.

Dartnall, A. J., Byrne, M., Collins J., and Hart, M. W. (2003). A new viviparous species of asterinid (Echinodermata, Asteroidea, Asterinidae) and a new genus to accommodate the species of pan-tropical exiguoid sea stars. *Zootaxa* **359**, 1–14.

27. Tunicata

P. Kott

Tunicata once were thought to be a subphylum of the Chordata (containing all vertebrates: fish, amphibians, reptiles, birds and mammals). Now they are recognised as a separate phylum, sharing a common ancestor with chordates. At some time in their lives, tunicates (like chordates) have a dorsal nerve cord, a notochord (a rod of turgid cells that, in vertebrates, becomes the backbone), perforations through the pharyngeal wall to the exterior (like gills in fish), and a ventral thoracic gland (the endostyle) that secretes thyroxin (like the vertebrate thyroid gland). The notochord and muscles associated with it are in a larval tail that, in the majority of tunicates, is lost during metamorphosis (as a tadpole larva loses its tail and becomes a frog).

Table 27.1 A possible chordate phylogeny (adapted from Schaeffer 1987).

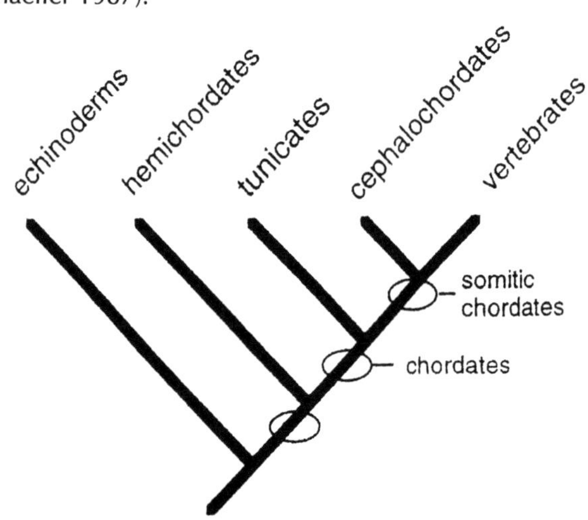

In addition to their chordate characters, a special feature of tunicates (from which they derive their name) is the acellular coat (the test) of tunicin. Tunicin is a material similar to plant cellulose and the only example of such a substance outside the plant kingdom. Also, intracellular reducing systems in Tunicata (generating acid on cell lysis) are unique in living organisms, as is the presence of large concentrations of vanadium in a significant number of species. This unique chemistry, together with molecular data, support the hypothesis of a long and independent evolution following an origin from an evolving chordate stem that successively gave rise to echinoderms, hemichordates, tunicates and cephalochordates.

Tunicates, found at intertidal to hadal depths in all the world's oceans, are significant in marine ecosystems as powerful filter feeders, straining minute organic particles and bacteria from water they drive through filtering membranes. Their relationship to chordates confers on them a value as experimental organisms for investigations on chordate processes (such as immune reactions). Their unique chemistry has attracted the interest of the pharmacological industry as a possible source of novel biologically active molecules with useful clinical applications. This makes them the target of well funded bioexploration.

The three very distinct classes of tunicates (see Table 27.2) are the sessile Ascidiacea and the planktonic Thaliacea and Appendicularia. Ascidiacea and

Table 27.2 Characters of the Protochordata phyla, Tunicata classes and higher taxon Ascidiacea.

PROTOCHORDATA
An informal group of bilaterally symmetrical coelomate animals thought to be related to chordate ancestors, having a perforated pharynx, elements of a dorsal nerve cord, but lacking an internal cartilaginous skeleton

PHYLUM HEMICHORDATA	**PHYLUM TUNICATA**	**PHYLUM CHORDATA SUBPHYLUM CEPHALOCHORDATA**
Sessile, solitary or colonial, collagen branchial skeleton, ciliated tornaria larva sub-dermal nerve net.	Sessile or planktonic, ectoderm secretes cellulose-related tunicin, no branchial skeleton, notochord-like cells in larval tail, groove in ventral mid-line, metabolising iodine, dorsal ganglion	Active swimmers, myomeres, collagen branchial skeleton, dorsal notochord, no larva, groove in ventral mid-line, metabolising iodine, dorsal nerve chord.
acorn worms, pterobranchs	*salps, doliolids, appendicularians, sea squirts*	*lancelots*

THALIACEA	**ASCIDIACEA**	**APPENDICULARIA**
Planktonic, sexual and vegetative generations alternate, tailed larvae, filter-feed through pharyngeal perforations, atrium behind pharynx	Adults fixed, solitary or colonial, free-swimming tailed larvae, filter-feed through pharyngeal perforations, atrium around pharynx	Planktonic, solitary, development direct, tail present through life, filter-feed through mucous house, no atrium, pharynx reduced
salps and doliolids	*sea squirts*	*appendicularians or larvaceans*

ENTEROGONA	**PLEUROGONA**
Gut behind or alongside pharynx, one gonad of each sex, paired atrial invaginations	Gut and gonads alongside pharynx, median atrial invagination

APLOUSOBRANCHIA	**PHLEBOBRANCHIA**	**STOLIDOBRANCHIA**
Epicardium regenerative, gut and gonads behind pharynx	Epicardium excretory, gut alongside pharynx, gonads in gut loop, branchial sac flat	Epicardium excretory, gonads on both sides of body, branchial sac flat

Thaliacea have an outer coat of the characteristic tunicin, a larval tail lost on metamorphosis and feeding currents driven by pharyngeal cilia through the filtering membrane (a mucous sheet covering the inner wall of the pharynx). In the minute Appendicularia, tunicin occurs only in small quantities in some embryos, the tail persists through life and its beating drives the feeding current through a filtering membrane in a mucous house that encloses the animal's trunk.

■ CLASS THALIACEA (SALPS AND DOLIOLIDS) (FIG. 27.1A–C)

These transparent, jelly-like planktonic tunicates swarm at certain times, becoming so crowded that they exclude most other zooplankton from coastal waters. They swim by a sort of jet propulsion and have the mouth at one end of a barrel-shaped body and the atrial cavity and its excurrent opening at the other. Rings of muscle, their number

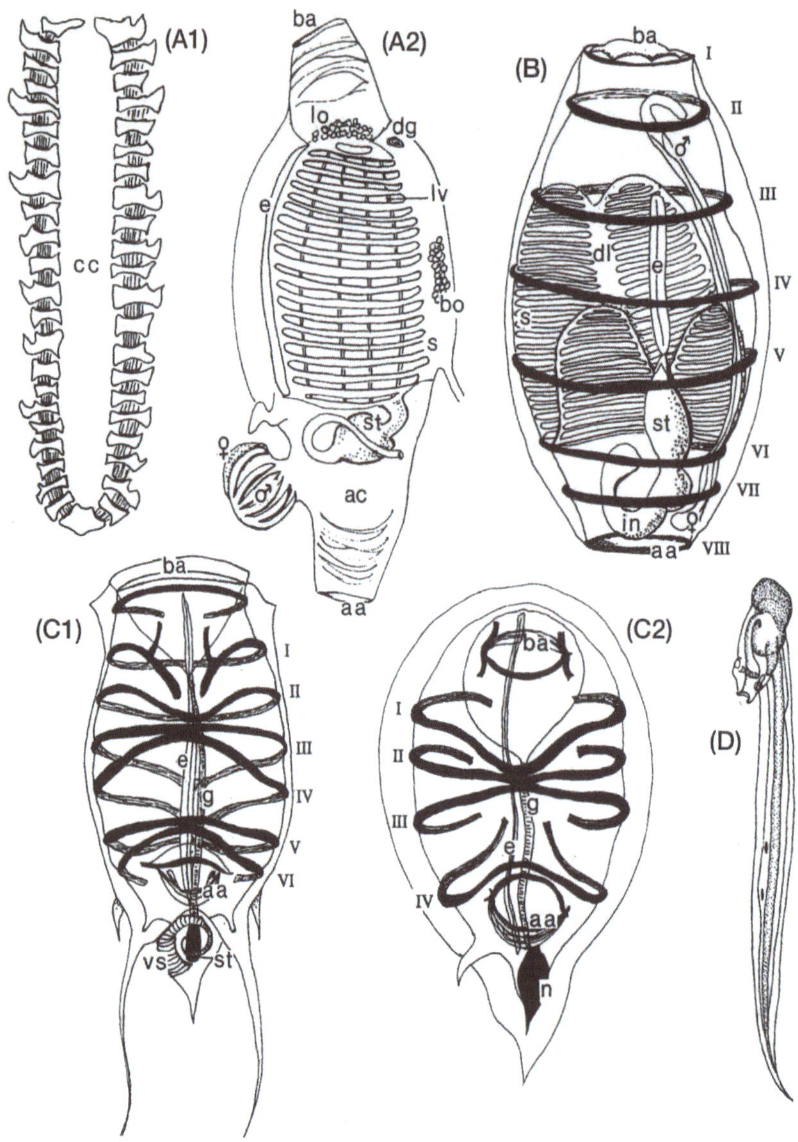

Figure 27.1 Thaliacea (*A–C*) and Appendicularia (*D*). *A*1, *Pyrosoma*, section through colony; *A*2, zooid; *B*, *Doliolum* gegenbauri; *C*, *Thalia democratic*; *C*1, solitary vegetative zooid; *C*2, colonial sexual zooid; *D*, *Oikopleura dioica*. aa, atrial opening; ac, atrial cavity; ba, mouth; bo, light organ; cc, common cloacal cavity; dg, dorsal ganglion; dl, dorsal lamina; g, gill; lo, blood forming organ; lv, longitudinal vessel; in, intestine; s, stigmata; st, stomach; vg, vegetative stolon. (Figure: after Thompson 1948.)

and arrangement unique to each species, encircle the body. The largest known tunicate (colonies up to 20 m) is a *Pyrosoma* species, a genus also famous for bioluminescent organs that produce brilliant displays visible from space.

All Thaliacea have complex life histories of alternating sexual and asexual generations. In salps, colonies consisting of chains of connected sexual zooids are produced by cloning from a stolon in the solitary individuals of the alternating asexual generation. Fertilisation is internal and embryos are incubated in the sexual colonial zooids. Only purse salps (genus *Pyrosoma*) are exclusively colonial. In doliolids the life histories are even more complex than in salps, involving sexually produced larvae that metamorphose into asexual individuals. The latter multiply to form polymorphic complexes of gastrozooids and phorozooids that are eventually freed and from which the sexual gonozooids eventually detach.

■ CLASS APPENDICULARIA (FIGS 27.1*D*)

This class consists of planktonic free living tunicates that have a trunk seldom more than 0.5 cm long and a permanent tail four or five times that length. Development is direct, without the intervention of a larva. Generally, individuals are protandrous, with the testis (with a duct to the exterior) maturing first. Eggs, externally fertilised, are released later by rupture of the adult trunk before the animal degenerates and dies. The mucous house, which encloses the trunk, is discarded as the animal grows and generates a new house. The discarded house, with its load of entangled organic material rejected from the feeding stream, is a source of food for other organisms. The pharyngeal perforations are reduced to two tubes (spiracles) from the floor of the pharynx to the ventral surface of the trunk. The endostyle is relatively short.

Only about 60 species are known worldwide and, being planktonic, they are spread around the world in ocean currents and are carried into coastal waters with other oceanic species usually in spring and summer.

■ CLASS ASCIDIACEA (SEA SQUIRTS) (FIGS 27.2–27.9)

The Ascidiacea, commonly known as sea squirts or ascidians, are the largest and most diverse class of the Tunicata, containing at least 3000 species worldwide. In size they range from colonial zooids about 1 mm long to large solitary individuals of 20 cm or more. A large perforated pharynx supports the filtering mucous sheet secreted by the mid-ventral thoracic groove called the endostyle. The atrial (peribranchial) cavity almost completely surrounds the pharynx and the gut loop is embedded in the body wall, either behind or folded up in the pallial body wall beside the pharynx (outside the atrial cavity). Ascidans are always hermaphrodite, although either androgynous or protogynous, thus avoiding self-fertilisation.

Ascidian reproduction and larvae

Not surprisingly, these fixed organisms almost always have free-swimming larvae that find places to settle and ensure recruitment and gene flow between populations. The larval tail, with a complex array of muscles and nerves, stiffened by the rod of notochord cells and guided by pigmented sense organs, propels the larva in specified directions (rather than passively drifting around as do the ciliated larvae of most other invertebrates). Responding to gravity and light, the larvae first move up to the light and away from the sea floor for dispersal, and then down toward the sea floor and into shaded places for settlement and metamorphosis (when the tail is withdrawn into the larval haemocoele and is resorbed).

Fertilisation (either internal or external) is probably synchronised by pheromones. Life histories patterns contribute to the success of sexual reproduction, gene flow and population maintenance. In some respects, these patterns are different in solitary and colonial species. Solitary forms (with very few exceptions) have large gonads, produce large numbers of gametes and are fertilised externally. Their larvae are free-swimming for relatively long periods and are subject to dispersal away from the parents. In colonies, large numbers of replicated zooids, each with gonads, compensates for their small size (compared to the gonads in solitary species) and ensures adequate concentrations of gametes and opportunities for fertilisation. In an adaptation possibly associated directly with internal fertilisation and incubation of embryos, colonial zooids have relatively large testes and small ovaries. Retention of tailed larvae within the parent so their free swimming life is short (often less than 10 minutes) ensures settlement close to

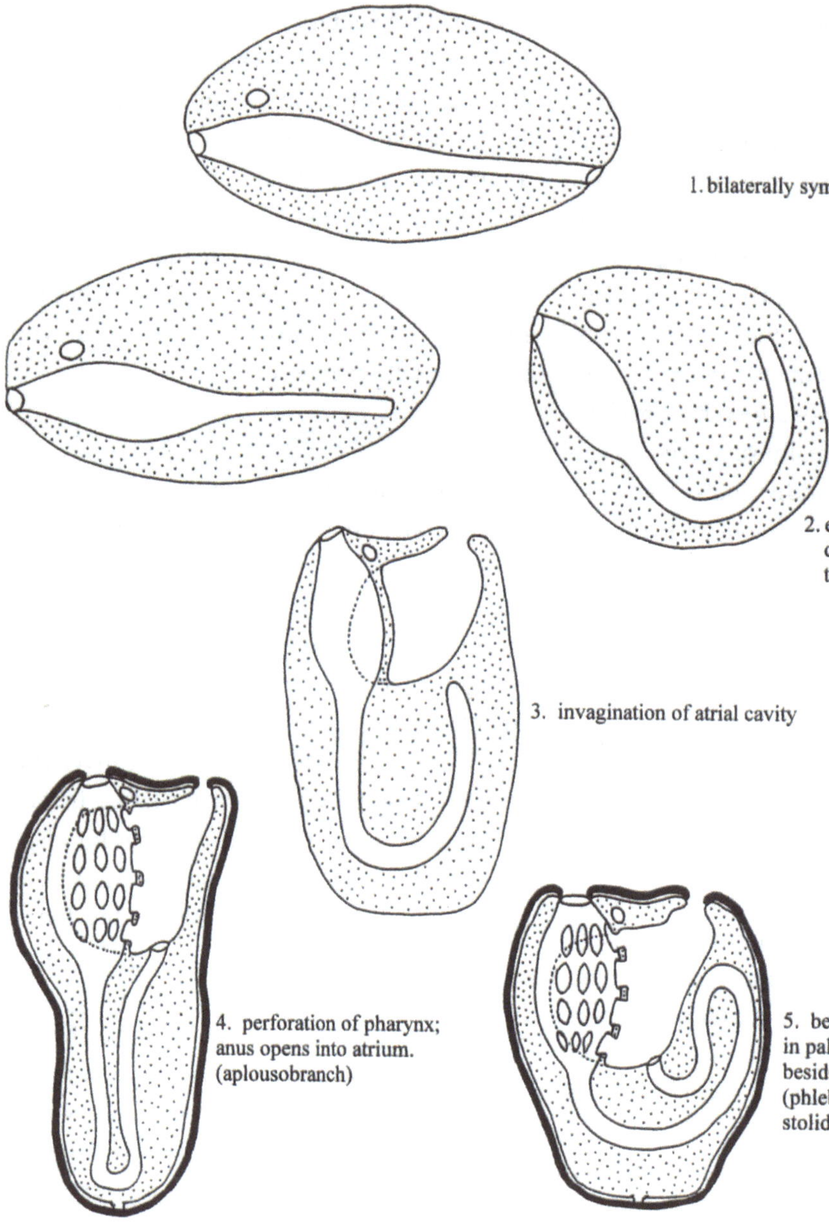

Figure 27.2 Diagrammatic representation of the development of the ascidian body from an imaginary bilaterally symmetrical ancestor. (Figure: P. Kott.)

parent colonies. This strategy maintains relatively large populations and hence opportunities for fertilisation (as does the profusion of suitable habitats in tropical locations).

Despite pressures to maintain populations by reducing the length of larval life, tailed larvae persist in most ascidian species and maintain gene flow by chains of recruitment through the vast geographic ranges that appear to characterise the tropical ascidian fauna. Direct development, without the intervention of tailed larvae occurs only in some species in open sea floor habitats.

Ascidian habits and interactions with the environment

Both colonial and solitary ascidians can be compared to the occupants of a house without doors that do all

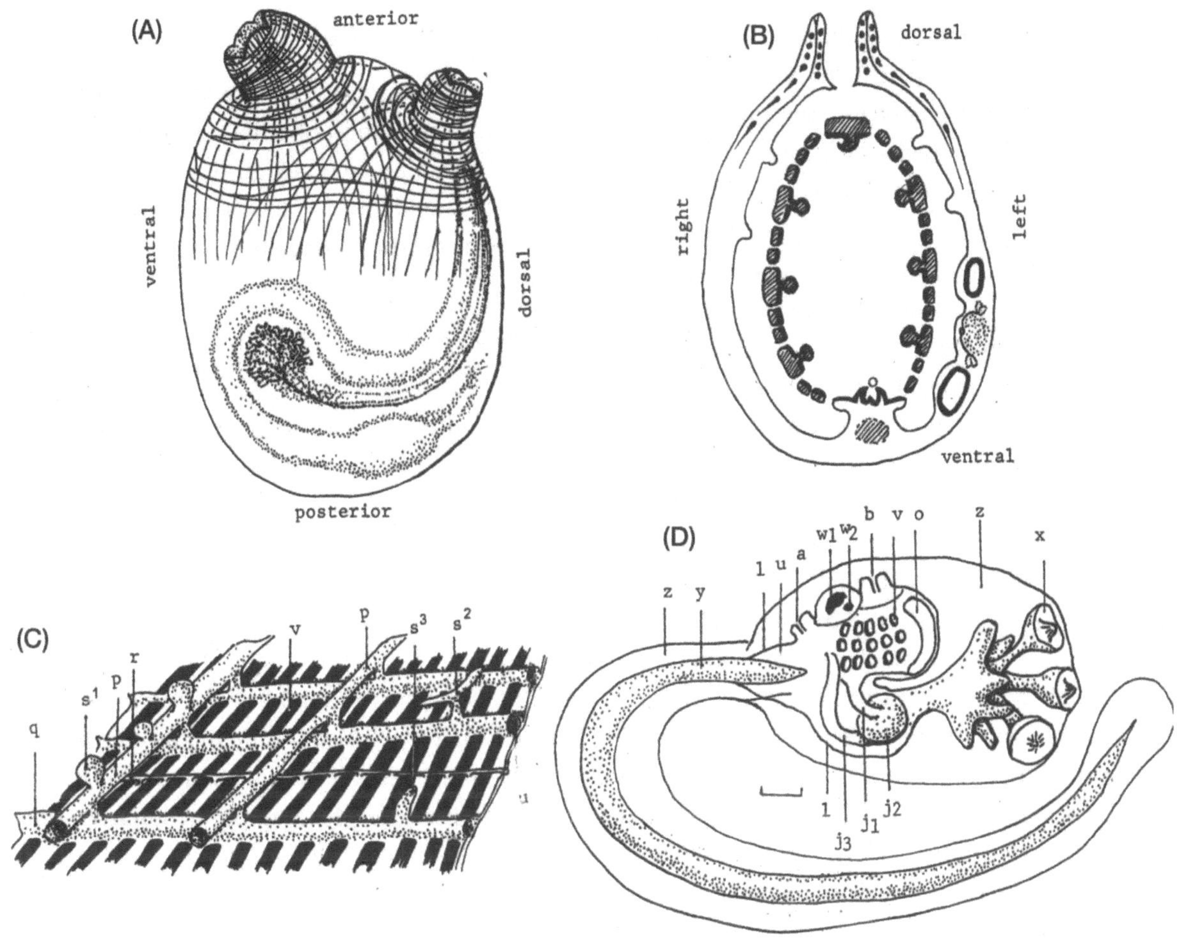

Figure 27.3 Aspects of ascidian morphology (semidiagrammatic). *A–B*, phlebobranch ascidian: *A*, from left side; *B*, oblique section through atrial aperture and gonads. *C*, inner wall of branchial sac showing: q, transverse vessel; r, parastigmatic vessel; s1–3 (secondary, bifid and simple, respectively); v, stigmata. *D*, ascidian larva showing: a, excurrent aperture; b, incurrent aperture; j 1–3 oesophagus, stomach and rectum, respectively; l, trunk epithelium; o, endostyle; u, larval haemocoele; v, stigmata; w 1–2, ocellus and otolith, respectively; x, adhesive organ; y, larval tail; z, larval test. (Figure: Kott 1993.)

their business with the outside world that flows past their house through a front and a back window. Adult ascidians are fixed to a substrate, or are rooted in, or lie immobile on, sea floor sediments. Each individual has an incurrent and excurrent opening. Colonial zooids sometimes (like solitary species) have both apertures opening separately to the exterior and are either joined to one another by strands of test or are partially or completely embedded in common test. More integrated colonies with embedded zooids have excurrent openings into common cloacal cavities or canals inside the colonial test. In these integrated colonies, individuals are most isolated from the exterior, their incurrent (branchial) apertures being the only zooid openings with direct access to the external environment.

Ascidians squirt water out of the pharynx as the muscular body contracts, hence their common name 'sea squirt'. This activity has been found to be rhythmic in some species, probably ensuring that the branchial cavity is cleaned and irrigated. Also, in a response known as the 'cross reflex', stimulation of the inside of either aperture (to emulate an intrusive organism or irritant) will cause the other to close while water is ejected to wash away the intruder. However, in life, both solitary and colonial ascidians are almost continuously engaged in filtering water. Although cryptic and often obsured by epibionts, they can be detected by their open apertures. Pharyngeal perforations, the anus, and gonoducts all open into the peribranchial cavity and faeces and either gametes or larvae are released in the excurrent stream of filtered water propelled away by the positive pressure built up inside the organism. Thus, the excurent water is separated from, and does not pollute, the incurrent water. Pressure gradients in individuals with both openings direct to the exterior are maintained by the smaller diameter of the separate excurrent openings relative to the incurrent one.

In species with zooids arranged around common cloacal cavities, some of the internal pressure is transferred from the zooid into the common cloacal spaces in the colony. This will be affected by the number and size of the common cloacal openings in relation to the number of zooids and the sum of their individual excurrent flows into the common cloacal cavity. Sometimes separately opening zooids are arranged in circular rudimentary systems with excurrent openings crowded in the centre of the circle of zooids. Excurrent flows from each zooid combine into a strong current ejected away from the incurrent apertures spread around the periphery of the circle. Some of the advantages of true cloacal systems in reinforcing the separation of incurrent and excurrent water thus are available to less integrated colonies. This tends to support the view that, at the depths so far explored, advantages of a colonial habit include the isolation of incurrent and spent water. This is additional to the usual advantages associated with coloniality, *viz.* insulation of zooids from the environment, internal fertilisation, maintenance of populations by incubation of embryos inside the colony and flexibility in the growth form of colonies compared with solitary organisms. Internal pressure gradients in living solitary individuals as well as in colonies are the principle means by which they maintain their shape.

Colonial ascidian species are about five times as numerous as solitary ones in both temperate and tropical waters, and colonial species with well integrated cloacal systems are more numerous in the tropics than in temperate waters. It is possible that the best integrated colonies have some particular advantage in tropical habitats. These include:

- rapid two-dimensional growth to occupy space and internal incubation of embryos to maintain populations (ascidian colonies grow as a result of replication of zooids and colonies with the most prolific replication tend to have the greatest rate of growth); and
- a decrease in zooid size as a result of the interruption to growth of individuals during replication.

These effects are conspicuous in the family Didemnidae, where prolific replication and zooid size reduction are at their greatest and colony growth is rapid and two dimensional, covering expanses of substrate and excluding other sedentary organisms.

The ascidian test

Ascidians are surrounded by an inert acellular coat of tunicin (the test). It endows them with the flexibility to exploit different habits and occupy a diversity of habitats. In addition to affording protection from predators and physical forces, it has adhesive properties that attach the organism to the substrate. Its intrinsic strength and protective qualities are often enhanced by the development of hairs and roots that allow attachment to, or incorporation in, a wide range of sediments. It can also develop stalks to raise individuals or colonies above the sediments and allow them to take advantage of currents that flow by. Tests form a framework for colonies, connect zooids to each other, or form a matrix in which they are embedded and in which embryos are incubated. It can be soft and gelatinous, flexible, firm and rigid, delicate, brittle, or tough and leathery, thick or thin. In several taxa calcareous spicules produced by the ectoderm are embedded in it. These are sometimes so crowded that they form an internal skeletal support for large branching colonies. In most solitary stolidobranch species, the test lining the body openings is armed with microscopic overlapping spines that point outwards and appear to prevent other organisms from intruding into the open apertures. Ectodermal vessels projecting from the body wall synthesise and nourish the test and help to anchor the body to it. Otherwise, the usually muscular ascidian body is attached only by

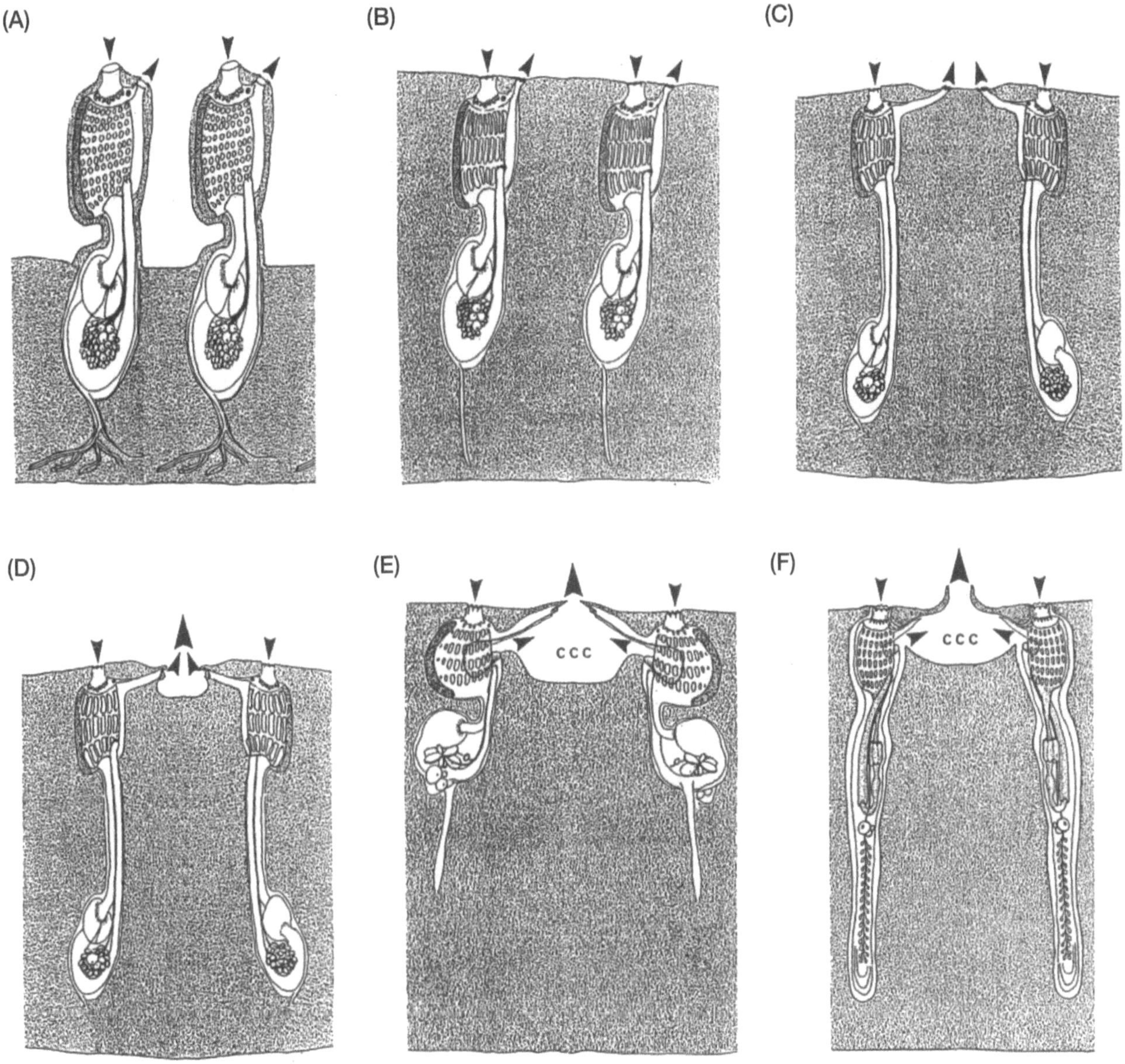

Figure 27.4 Evolution of aplousobranch ascidian colonial systems. *A–C*, zooids independently opening. *A*, partially and *B*, completely embedded; *C*, rudimentary systems; *D–F*, zooids arranged around common cloacal cavities; *D*, simple circular systems (e.g. Polycitoridae); *E–F*, extensive common cloacal canals and cavities (gonads in gut loop, e.g. Didemnidae; or in a posterior abdomen, e.g. Polyclinidae). (Figure: Kott 1990.)

invaginations of the test that line its siphons. The test often shelters other organisms such as crustaceans and molluscs that bore into it. In some didemnid species, symbionts embedded in the test or attached to the lining of common cloacal cavities contribute to the nutrition of their hosts. The test also contributes to ascidian interactions with the environment by differential growth that affects the orientation of apertures, offering incurrent openings to oncoming (food laden) currents and away from falling sediments and from the excurrent (spent) water from the same or adjacent individuals or colonies.

Biogeography

Seven hundred and twenty ascidian species are now known from Australian waters, of which only 150 are

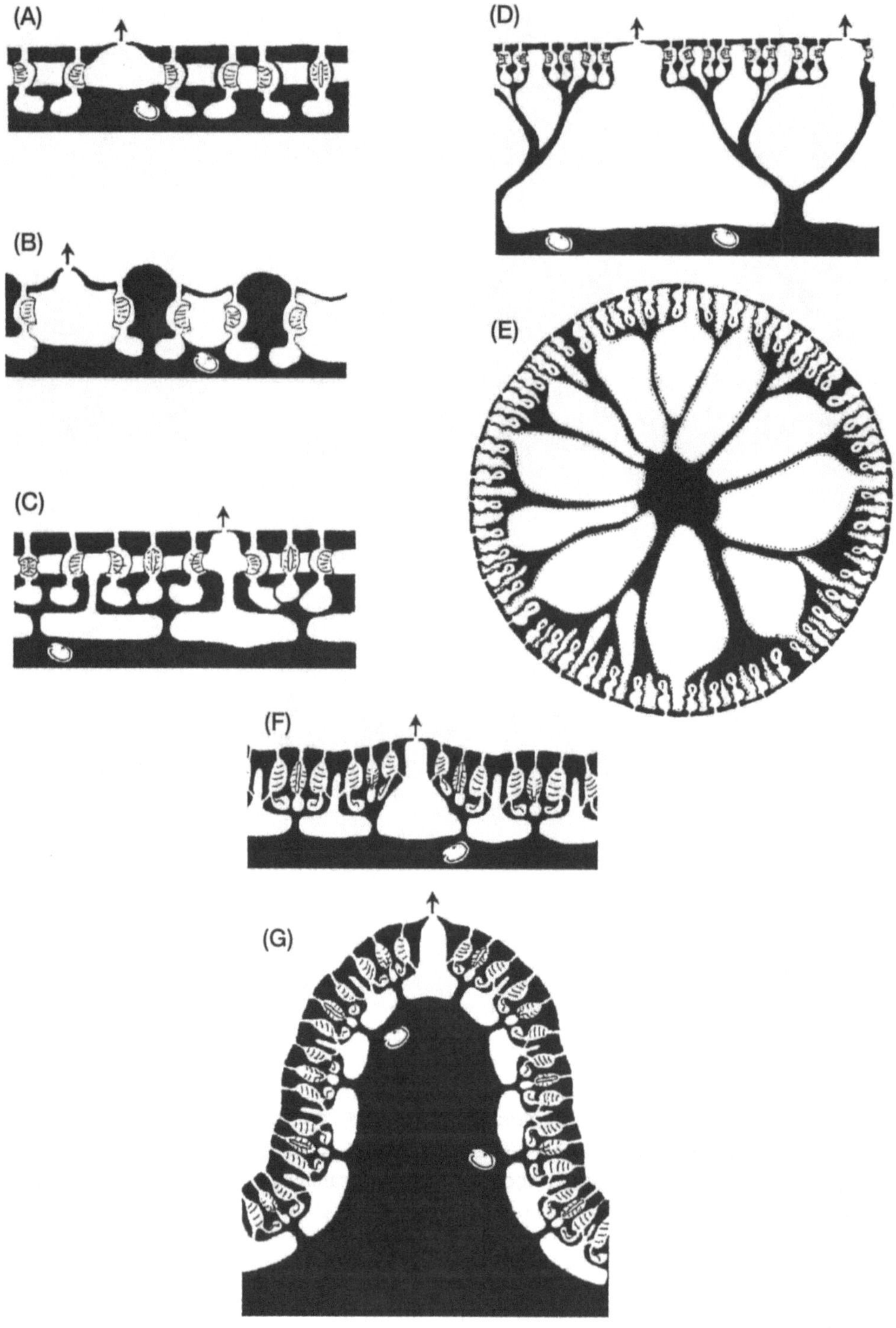

Figure 27.5 Common cloacal systems of Didemnidae: *A*, horizontal thoracic cavities (*Didemnum, Polysyncraton*); *B*, deep horizontal cavities (*Didemnum, Polysyncraton, Lissoclinum*); *C*, three dimensional thoracic and posterior abdominal canals; *D*, zooids cross extensive cavities between surface and basal layers of test in test connectives (*Lissoclinum, Diplosoma*); *E*, flask-shaped colony with central test core surrounded by cavity (*Didemnum, Polysyncraton, Lissoclinum*); *F–G*, zooids with atrial siphon and three-dimensional oesophageal and posterior abdominal canals, *F*, encrusting; *G*, upright colony lobe (*Trididemnum, Leptoclinides*). *Note:* The type of system is often found in, but not exclusive to, the genus indicated. (Figure: Kott 2001.)

solitary. These are both indigenous species not known beyond Australia and non-indigenous species that almost invariably are tropical, especially recorded from the Great Barrier Reef (GBR), and with a range extending into some part of the wider Indo-West Pacific coralline region. Tropical and temperate locations around the Australian continent contain almost equal numbers of ascidian species. However, they have significantly different biogeographical affinities. Although indigenous and non-indigenous species are found in more or less equal numbers in the tropics, the number of both solitary and colonial indigenous species is more than twice the non-indigenous component of temperate waters. In tropical waters ascidian species are more often non-indigenous. The great species diversity seen in the GBR can be partly attributed to its location in the centre of the vast Indo-West Pacific coralline region. Speciation producing a unique fauna is more apparent in waters across the southern part of the Australian continent. Nevertheless, it is probable that the GBR, with its great latitudinal extent, has acted as a bridge between the tropics and the temperate waters of the continent, conveying genetic material from the tropics into the southern waters where it is now incorporated into the indigenous fauna. So far, a reduction in species diversity from north to south in the GBR has not been detected for ascidians. The most conspicuous difference between the ascidian populations of northern and southern reefs is the absence of the great mats of didemnid/algal symbioses from southern reef flats where diurnal temperature changes in winter may inhibit these species.

Symbioses

It is not unusual for tropical organisms to take advantage of sunlight, one of the most celebrated examples being the intra-cellular zooxanthellae providing photosynthates that nourish corals and allow many species to be largely independent of ingested food (autotrophic, see Chapter 7). Certain species of the colonial ascidian family Didemnidae have also taken advantage of abundant tropical sunlight developing symbioses with chlorophyll containing symbionts and have become autotrophic. Development of these symbioses is enhanced by a colony form that characteristically has habitats for symbionts in large spaces inside the colony as well as in the test itself. Further, species of Didemnidae contain calcareous spicules in the test that, arranged (by the ascidian host), shield from or expose the contained symbionts to sunlight to regulate the rate that photosynthates are delivered to the host and to reduce photoinhibition (see Chapter 7). This regulation can be differentially imposed so that part of a colony can respond independently to the light falling on it. A colony growing around the side of a rock is green (without spicules) on the shaded underside of the rock; white (with spicules increasingly crowded in the surface) where it grows around the side, and pink (with carotenoid pigment added) on the part of the colony in full sunlight on top of the rock.

Some (*Lissoclinum timorense* and *L. bistratum*) didemnid/*Prochloron* symbioses form vast mosaics covering reef flats, especially in the northern part of the GBR and high tropics of the Indo-West Pacific (where diurnal temperature ranges are not as great as they are in the south). The open reef flat is an unusual place for ascidians, which usually are more cryptic, however, the habitat has obvious advantages for autotrophic species able to regulate the light their symbionts are exposed to. Further, occupation of this habitat is enhanced by the capacity of colonies to divide and move to space themselves evenly, maintaining the mosaic-like pattern on the reef flat.

The vase shaped *Didemnum molle* is another common species with a remarkable geographic range from the west Indian Ocean to Fiji. It also subdivides and moves, climbing up its substratum (staghorn coral skeletons, aquarium walls and the sides of reefs) toward the light by drawing in tendrils of test from the base of the colony. *Didemnum simile*, a conspicuous species with a similar geographic range, neither subdivides nor moves. However, it forms extensive sheets and encrusts vast areas of undercut, growing through gaps and holes in the reef flat and binding rubble together. Its colour, affected by its contained *Prochloron*, varies from bright sapphire blue to emerald green.

The existence of green plant cells in certain tropical species of the Didemnidae was well known by the 1950s although their identity had not been determined, being referred variously to zoochlorellae or zooxanthellae or Chlorophyta. In 1973, aboard the American research vessel *Alpha Helix* working in waters off Lizard Island, Eldon Newcomb of the University of Wisconsin recognised the green prokaryotic cells in many ascidian colonies as Cyanophyta (blue-green algae). Subsequently a new division of the algae, Prochlorophyta, containing a single genus *Prochloron* was erected to accommodate these organisms. There was some support for an hypothesis that they were ancestors of green plant chloroplasts, having chlorophyll *b* in addition to chlorophyll *a* and lacking phycobiliproteins. However, depending on the ascidian species, the symbionts are either in the common cloacal cavity or embedded in the colony test and are not endosymbionts (like coral zooxanthellae) as they were claimed to be. In fact they are never even in the ascidian zooid. Later *Prochloron* was confirmed as a genus of the Cyanophyta and now is grouped (as a 'prochlorophyte') with two other genera containing chlorophyll *b*.

The number of species in *Prochloron* is not known and it has not been cultured outside the ascidian host. Obligate symbioses are known in four of the eight known genera of the Didemnidae. Except for symbioses in closely related species of *Diplosoma*, adaptations of the larval ascidian test to carry the symbionts to the next host generation are diverse. It is most likely that the symbiosis arose at least once and possibly more than once in each genus. About 25 obligate symbioses are now known. Sometimes non-obligate symbionts are on the surface of the colony and are readily brushed off. Also there are chlorophyte symbionts in some didemnid species both on their own and with *Prochloron*.

Evolution of ascidians

Evolution of extant ascidians appears to have proceeded in two streams from some large solitary ancestor like *Ciona intestinalis*, a solitary aplousobranch ascidian. Endodermal tissue of paired epicardial sacs, outgrowths of the gut at the posterior end of the pharynx, is the regenerative tissue in *Ciona*. This appears to be homologous with its role in most related aplousobranch families where replicates are formed by horizontal division of the abdomen or posterior abdomen involving epicardial tissue. The other main evolutionary pathway appears to be associated with the use of pericardial sacs for an excretory purpose in the largely solitary phlebobranch and stolidobranch ascidians.

Apart from *Ciona*, Aplousobranchia are almost exclusively colonial. Colonial taxa, however, have evolved in otherwise primarily solitary Suborders Phlebobranchia and Stolidobranchia by stolonial or pallial budding, involving only ectodermal tissue. These colonies are convergent with those of Aplousobranchia, suggesting that similar selective pressures are operative. Colonial zooids of all suborders are small and simplified, with small branchial sacs and without digestive diverticulae of the gut. Ovaries are small and testes are large, a condition possibly associated with internal fertilisation and intracolonial incubation of embryos. In colonial species special organs for excretion have been detected only in colonial Styelidae where they persist as relicts of their stolidobranch ancestors.

Solitary phlebobranch and stolidobranch species display similar evidence of convergent evolution in general adaptations of organs to serve larger individuals. For instance, conspicuous in solitary taxa are large branchial sacs and increases in filtering area (e.g. the formation of folds in Stolidobranchia and stigmata spiralling around conical prominences in the pharyngeal wall in both Phlebobranchia and Stolidobranchia). The gut is large with complex digestive organs branching off it. Most species have special excretory cells derived from the epicardium, and in the stolidobranch family Molgulidae a large kidney is formed from epicardial tissue. Also, with few exceptions, male and female gonads are large, species are oviparous, fertilised externally and seldom adaptated to retain developing embryos in the large pharynx and surrounding peribranchial sac.

Collection, narcotisation, fixation, preservation

Plastic bags and buckets, always with enough seawater to cover the specimens, are convenient ways to collect. Change the water from time to time to avoid temperature increases. If possible, photograph the specimen before collecting it and ensure that you have a reliable system to match the specimen to the photograph (because after you have collected it that ascidian will never look the same again). The colours of living specimens

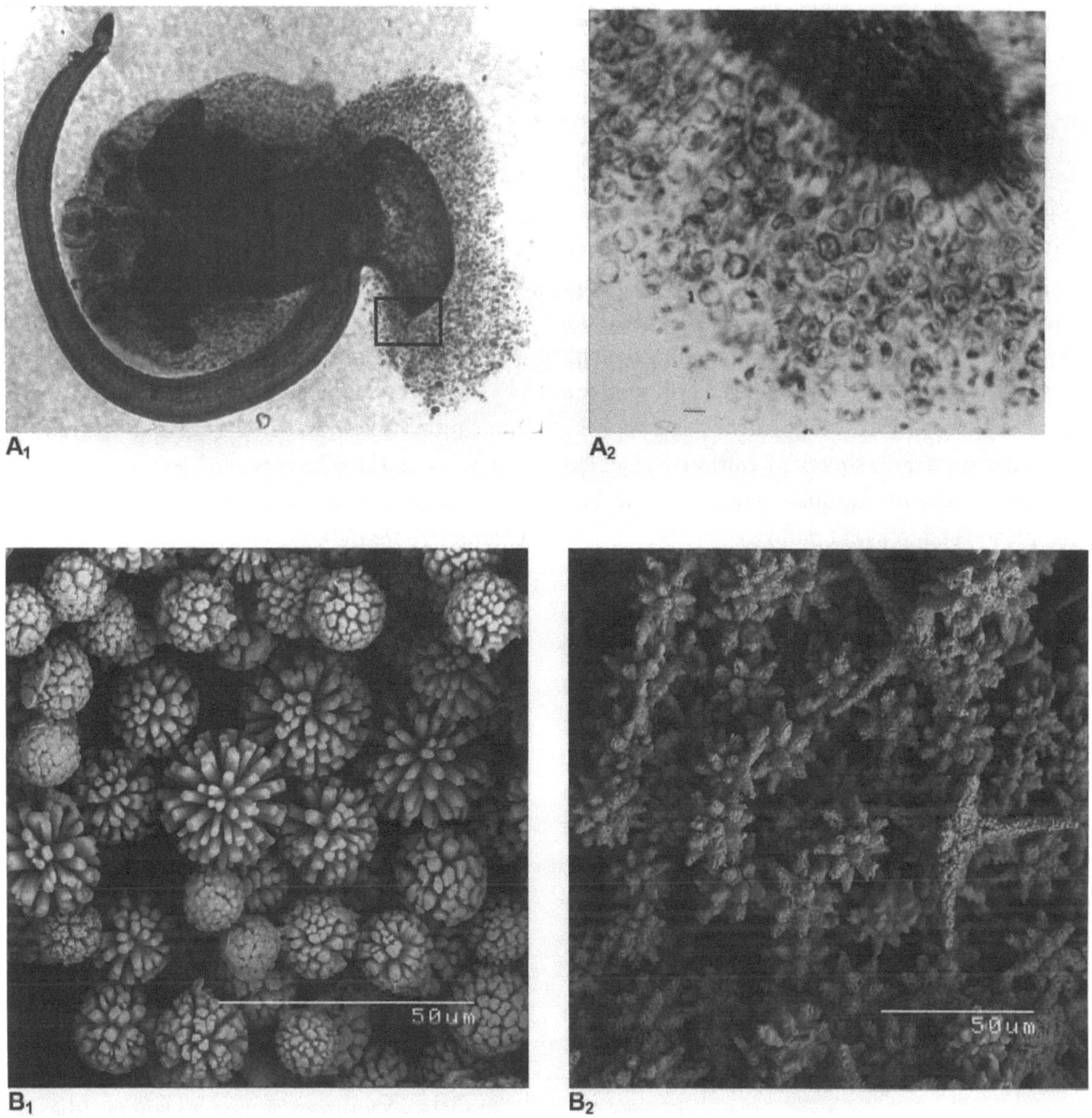

Figure 27.6 A_{1-2}, *Diplosoma simile* larva. A_1, immature larva with the tail curved around the ventrum of trunk, antero-median adhesive organs, the rastrum (a T-shaped outgrowth from the larval haemocoel dorsal to the base of the tail at the posterior end of the trunk) differentiated into hairs; A_2, tip of the right horn of a mature larval rastrum (indicated in A_1) showing *Prochloron* cells entangled in hairs differentiated from the thick larval test covering the rastrum. B_{1-2}, Didemnid $CaCO_3$ (aragonite) spicules: *Didemnum albopunctatum* (globular spicules with rod-shaped rays), *Didemnum membranaceum* (stellate spicules with conical rays and some giant spicules). (Figure: light micrographs, Kott 1982; SE micrographs, Kott 2001.)

are conspicuous, especially in colonies where the surface is not obscured by epibionts and sand. However, there is a great deal of intraspecific variation in colour and usually it changes when specimens are removed from their substrate. In the field, ascidians can often be confused with equally brightly coloured sponges although, unlike sponges, the apertures of the ascidian close tightly when it is disturbed.

Field notes should be made *in situ* before the animal is disturbed and any changes resulting from disturbance, including its removal from the substrate, should be noted together with colour, size, shape and details of the zooid arrangement in the colonies and notes on the habitat, location, collecting method and depth. Make the notes in pencil on good quality waterproof paper and put them into a plastic bag with the specimen; or in a field notebook, carefully cross-referencing each entry to the number that you keep with the specimen. It is a good idea to take a colour chart with you when you collect ascidians. (Make a replicate of any reliable standard by matching its colours with colour chips from the paint charts you can pick up at any hardware store; then paste up the matched chips on a few sheets of cardboard, waterproof or seal them in plastic and carry them with you in a black plastic bag to avoid fading.)

Dislodge specimens carefully from their substrate using a sharp fishing knife. Sample the colonies too large to collect (it is not necessary to take the whole specimen). However, draw the whole specimen in a field notebook or on the back of the label, note its general dimensions and shade or indicate in some way the portion of it that you have sampled. Take the sample so that it is representative of the colony, that is, cut a wedge from the centre of the colony to its outer margin, or from a large common cloacal aperture to the outer margin, or take a whole lobe or branch. Always cut vertically, from surface to the base, parallel to the zooids. Never cut an ascidian colony horizontally or you are likely to cut zooids in half. If you cut into a didemnid colony, do it underwater as damage causes cells to generate acid that can dissolve calcareous spicules. The sea water will help to neutralise the acid quickly.

Narcotise specimens with menthol crystals for up to three hours (colonies) to five or more hours (large solitary specimens) by using just enough water to cover the specimen and closing the bag or vessel as closely as possible. The specimen is narcotised if the siphons do not close when stimulated. As menthol is a bacteriocide, the specimen remains in good condition despite the long time taken for narcotisation. Also, menthol is expensive so save the menthol crystals when narcotisation is complete. Do not on any account use an alcoholic solution of menthol or you will merely kill the specimen before it is narcotised.

Fix the specimen by quickly adding to it one part of 40% formaldehyde to the nine parts represented by the specimen, plus enough seawater to cover it. Do not add formalin already made up to 4%; if the specimen is the same volume as the preservative you will end up with only 2% formalin. In other words think of the specimen as water (which it mostly is). Keep the preservative neutral with a little calcium carbonate in the water. (*Note*. Formalin is generally used to describe a dilution of 40% formaldehyde, so that 10% formalin is actually 4% formaldehyde.)

Preserve the formalin-fixed specimens after at least three months by transferring them to 70% ethanol with a dash of glycerol.

Permits are required for collecting ascidians, see Chapter 12 for details.

Examination of specimens

- Observe the entire individual or colony, noting its size, shape, colour, position, orientation and form of the openings, nature of the test and its inclusions and the arrangement of colonial zooids.
- If the test is transparent it may be possible to observe the body through it. However, usually it is necessary to remove the individual or zooid from the test. Solitary specimens should be opened around the ventral midline (along the endostyle), either by opening the whole body including the test, or by removing the body from the test before making the incision. The cut is made (with sharp scissors, one blade inside the branchial sac) from siphon to siphon (the long way around). Specimens can be pinned out, and the branchial sac examined before gently removing it from the body wall, by severing the connectives, so that gut and gonads and other structures embedded in or on the pallial body wall can be seen (see Fig. 27.7).
- Colonial zooids can be examined in water or in a drop of glycerol on a slide after their removal from the test. The pharynx can often be opened along the endostyle by inserting the very sharp tip of fine scissors or watchmaker's forceps into the branchial sac and cutting it open against a hard surface. A drop of stain in the water or glycerol often makes structures less obscure. Smaller aplousobranch zooids may

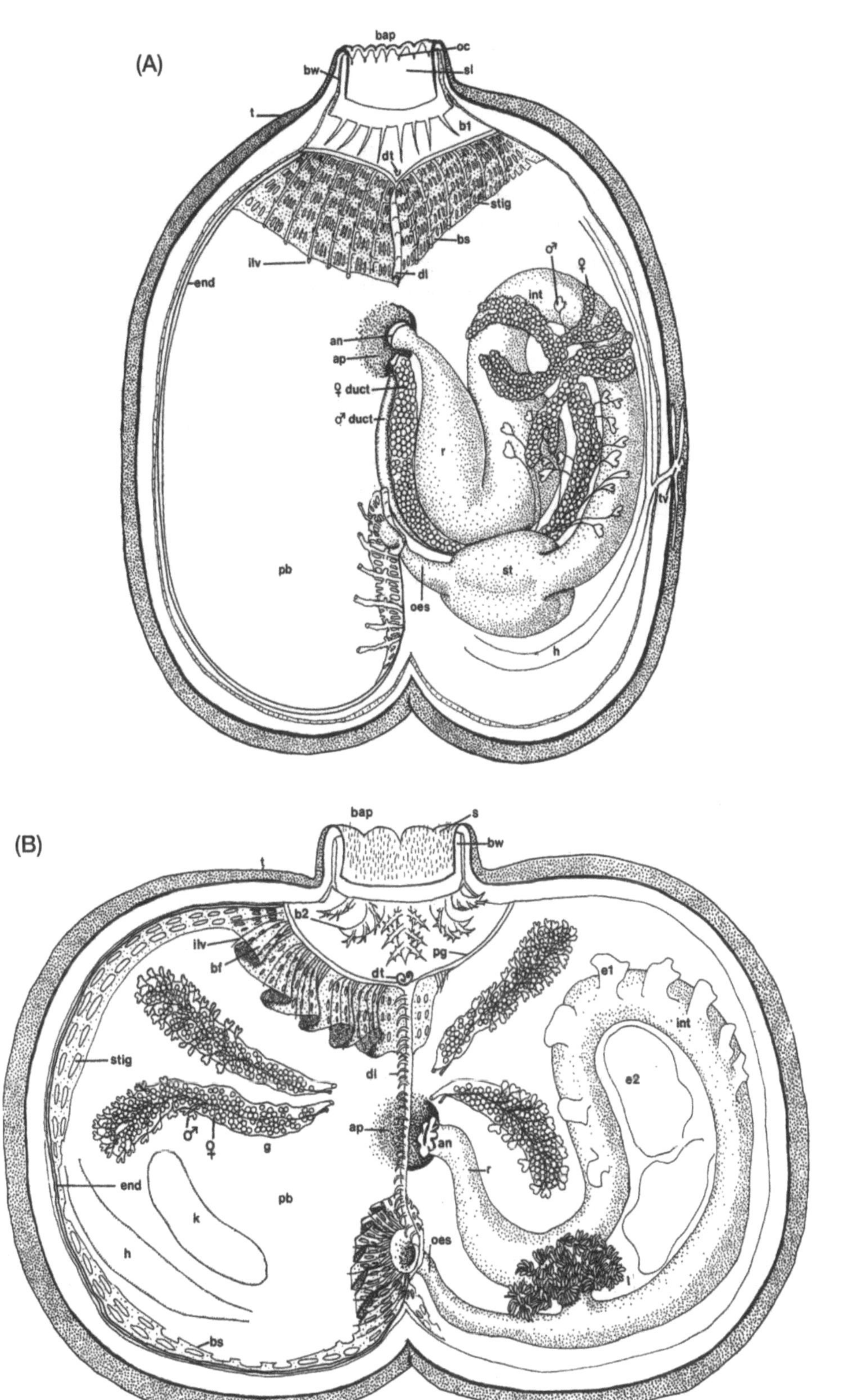

Figure 27.7 Morphology of a phlebobranch and stolidobranch ascidian, *A–B*, respectively. Each opened along the mid ventral line and with the branchial sac largely removed to show structures embedded in the pallial body wall (semidiagrammatic): an, anus; ap, atrial aperture; b1–2, branchial tentacles; bap, branchial aperture; bf, branchial fold; bs, branchial sac; bw, body wall; dt, dorsal tubercle with opening of neural gland; dl, dorsal lamina; e1–2, endocarp; end, endostyle; h, heart; ilv, internal longitudinal vessel; int, intestine; k, kidney; l, liver lobules; oes, oesophagus; pb, perobranchial cavity; r, rectum; st, stomach; t, test; tv, test vessel. (Figure: Kott 1985.)

need to be stained, cleared and mounted for light microscopy.

- Often it is easier to remove zooids from a section cut vertically from the colony and laid on its side so that the arrangement of the zooids and the common cloacal cavities can be observed. Vertical handcut sections of didemnid species will need to be decalcified (in 3% HCl). When all the spicules are dissolved, the whole section, with the zooids in place, can be washed (to neutralise it), stained, cleared and mounted in the usual way.
- Colonial species of all suborders are viviparous and the larvae being brooded in the colonies often are found in the atrial cavities of the zooids, or in brood pouches attached to the thorax or abdomen of zooids or lying free in the test. Very few solitary (phlebobranch or stolidobranch) species are viviparous. The exceptions are some species of *Agnezia*, *Molgula* and *Polycarpa*. Larvae need to be stained, cleared and mounted for their structures to be seen.

Calcareous spicules of Didemnidae and some Polycitoridae are best examined by scanning electron microscopy. Prepare the stub by wiping a thin smear of glue over it with a finger tip and letting it dry thoroughly. Then dehydrate a small strip of test containing spicules (remove any zooids or other foreign matter) by holding it in the tip of stainless steel forceps and incinerating it in a hot flame (e.g. over a bunsen burner). When it is completely white, dip the completely dehydrated strip into a drop of absolute alcohol. The spicules will be spread by surface tension in the drop of alcohol, which evaporates immediately leaving the spicules spread over the surface of the stub. This stage is very rapid, because the alcohol must not dissolve the glue before it evaporates.

Identification

The characters useful for identifying ascidians and understanding them and their phylogenetic relationships include the consistency of the test and surface markings, position and orientation of the apertures, habit of the organism, organisation of zooids in the colony, relative sizes of the thorax, abdomen, and posterior abdomen (if present), branchial tentacles, opening of the neural gland on the dorsal tubercle, dorsal lamina (plain membrane or languets), siphonal spines, muscles (position, number, inner and outer layers), branchial sac and internal longitudinal vessels, branchial (pharyngeal) folds, the shape of the perforations (stigmata), the position and course of the gut and its sub-divisions, the nature of any gut diverticulae (liver), anal border, condition of the stomach wall, presence of a kidney, the position and type of gonads, and aspects of the larvae including adhesive organs, ectodermal ampullae, trunk size, and the presence of blastozooids.

Ascidians are contractile and contraction directly affects the appearance and consistency of the test and shape of both solitary individuals and colonies. Conformation of cloacal cavities and zooid systems (in life maintained by positive internal pressure of incoming ciliary currents) are obscured as colonies collapse on fixation. Colour also is lost following removal from the substrate and as fixative and preservative are added. The morphology of many internal organs, especially in small colonial zooids are obscured, particularly the number and appearance of muscle bands, branchial vessels and stigmata. Gonads are sometimes ephemeral and age and maturity can dramatically affect their appearance. Also, in these organisms there is a wide range of intraspecific variation and both individuals and colonies and are affected by environmental factors. For these reasons ascidians are often difficult to identify.

Equipment

- Dissecting instruments: a pair of strong scissors, scalpels (with removable blades), two pairs of watchmaker's forceps, one round-tipped forceps, one fine (retina) scissors.
- Light microscopes: stereo (dissecting) microscope with direct and substage light and an accessory objective lens to give a magnification of about ×80; compound microscope to magnification of ×400 with eyepiece and stage micrometer. Drawing tubes are desirable on both microscopes.
- Access to a scanning electron microscope is desirable.

KEYS TO FAMILIES OF THE ASCIDIACEA

For keys to genera and species see Additional reading.

KEY TO THE APLOUSOBRANCHIA

Usually colonial (exceptions *Ciona*, and some species of Diazonidae, Clavelinidae and Euherdmaniidae); gut loop posterior to pharynx; gonads in, or posterior to, gut loop.

1. Gut loop horizontal (*Ciona*) .... Cionidae
   – Gut loop vertical 2
2. Apertures smooth-rimmed 3
   – Apertures not smooth-rimmed 4
3. Abdomen about twice the length of the large thorax (*Clavelina, Nephtheis*) .... Clavelinidae
   – Abdomen many times the length of the thorax (*Pycnoclavella*) .... Pycnoclavellidae
4. Internal longitudinal branchial vessels or vestiges of them present; gonads in the abdomen (*Diazona, Rhopalaea*) .... Diazonidae
   – Lacking internal longitudinal branchial vessels or vestiges of them except when gonads in a posterior abdomen 5
5. Transverse thoracic muscles forming a continuous coat (*Polycitor, Eudistoma*) .... Polycitoridae
   – Transverse thoracic muscles not forming a continuous coat 6
6. Stigmata in only three rows (*Pseudodistoma, Sigillina*) .... Pseudodistomidae
   – Stigmata in more than three rows 7
7. Gonads in the abdomen 8
   – Gonads in the posterior abdomen 10
8. Excurrent apertures of zooids opening directly to the exterior; apertures with fringed lobes (*Stomozoa*) .... Stomozoidae
   – Excurrent apertures not opening directly to the exterior; lobes of the apertures not fringed 9
9. Replicates by horizontal division of a posterior abdominal stolon (*Distaplia, Sycozoa, Hypsistozoa*) .... Holozoidae
   – Replicates by oesophageal budding (*Atriolum, Leptoclinides, Polysyncraton, Didemnum, Trididemnum, Lissoclinum, Didemnidae*) .... Didemnidae
10. Excurrent apertures open directly to the exterior 11
   – Excurrent apertures do not open directly to exterior (*Polyclinum, Synoicum, Aplidium, Aplidiopsis* .... Polyclinidae
11. Stomach at the end of a relatively long abdomen (*Euherdmania*) .... Euherdmaniidae
   – Stomach in the middle of a relatively short abdomen 12

12. Posterior abdomen short tapering, with bunched testis follicles) (*Pseudodiazona, Monniotus, Protopolyclinum, Condominium*)..Protopolyclinidae
   – Posterior abdomen long, testis follicles not bunched (*Ritterella*) ....Ritterellidae

KEY TO THE PHLEBOBRANCHIA

Branchial sac without folds; gut loop folded up to right or left of pharynx; gonads relatively large, one of each sex only, present on one side only (in gut loop).

1. Gonads in everted pouch of body wall embedded in the test (*Microgastra, Plurella*) .... Plurellidae
   – Gonads not in everted pouch of body wall embedded in the test............................2
2. Solitary individuals 3
   – Colonies (*Perophora, Ecteinascidia*) .... Perophoridae
3. Gut on right side of the branchial sac (*Rhodosoma, Corella*) .... Corellidae
   – Gut on left side of the branchial sac (*Phallusia, Ascidia*) ....Ascidiidae

KEY TO THE STOLIDOBRANCHIA

Branchial sac folded (exceptions only when colonial habit is associated with significant reductions in size and simplification of zooids); gut loop folded up at the left of the pharynx; gonads paired, present on each side of the body.

1. Branchial tentacles branched..2
   – Branchial tentacles not branched (Styelinae-solitary: *Styela, Cnemidocarpa, Polycarpa*; Polyzoinae-colonial: *Polyandrocarpa, Oculinaria, Stolonica, Symplegma*, Botryllinae-colonial systems: *Botrylloides, Botryllus*)................ Styelidae
2. Stigmata rectangular, no kidney vesicle on right (*Herdmania, Microcosmus, Pyura, Halocynthia, Hartmeyeria*)... Pyuridae
   – Stigmata coiled: kidney vesicle on right (*Molgula, Eugyra*) ....Molgulidae

ADDITIONAL READING

Christen, R., and Braconnet, J. C. (1998). Molecular phylogeny of tunicate, a preliminary study using 28S ribosomal RNA partial sequences: implications in terms of evolution and ecology. In 'The Biology of Pelagic Tunicates'. (Ed. Q. Bone.) pp. 265–271. (Oxford University Press: Oxford.)

Fenaux, R. (1998). Anatomy and functional morphology of the appendicularia. In 'The Biology of Pelagic Tunicates'. (Ed. Q. Bone.) pp. 25–34. (Oxford University Press: Oxford.)

Godeaux, J. E. A. (1989). Functions of the endostyle in the tunicates. *Bulletin of Marine Science* **45**(2), 228–242.

Godeaux, J. E. A. (1998). The relationships and systematics of the Thaliacea with keys for identification. In 'The Biology of Pelagic Tunicates'. (Ed. Q. Bone.) pp. 272–274. (Oxford University Press: Oxford.)

Godeaux, J., Bone, Q., and Braconnet, J. C. (1998). The anatomy of Thaliacea. In 'The Biology of Tunicates'. (Ed. Q. Bone.) pp. 1–23. (Oxford University Press: Oxford.)

Figure 27.8 *A*, *Rhopalaea crassa* (Diazonidae); *B*, *Clavelina moluccensis* (Clavelinidae); *C*, *Aplidium multiplicatum* (Polyclinidae); *D*, *Aplidium tabascum* (Polyclinidae); *E–H*, Didemnid/Prochloron symbioses: *E*, *Trididemnum cyclops*; *F*, *Lissoclinum bistratum* with carotenoids in surface test of exposed colonies; *G*, *Didemnum molle* colonies; *H*, *Diplosoma virens* colonies. (Photos: *A*, R. Steene; *B*, N. Holmes; *C*, K. Gowlett-Holmes; *D*, *H*, N. Coleman; *E*, D. Parry; *F*, E. Lovell; *G*, J. Hooper.)

Goodbody, I. (1974). The physiology of ascidians. *Advances in Marine Biology* **12**, 1–129.

Kott, P. (1982). Didemnid/algal symbiosis: algal transfer to a new host generation. In 'Proceedings Fourth International Coral Reef Symposium, Manila 1981'. Vol. 2, pp. 721–23. (University of the Philippines: Quezon City.)

Kott, P. (1985). The Australian Ascidiacea Pt 1, Phlebobranchia and Stolidobranchia. *Memoirs of the Queensland Museum* **23**, 1–440.

Kott, P. (1990). The Australian Ascidiacea Pt 2, Aplousobranchia (1). *Memoirs of the Queensland Museum* **29**(1), 1–266.

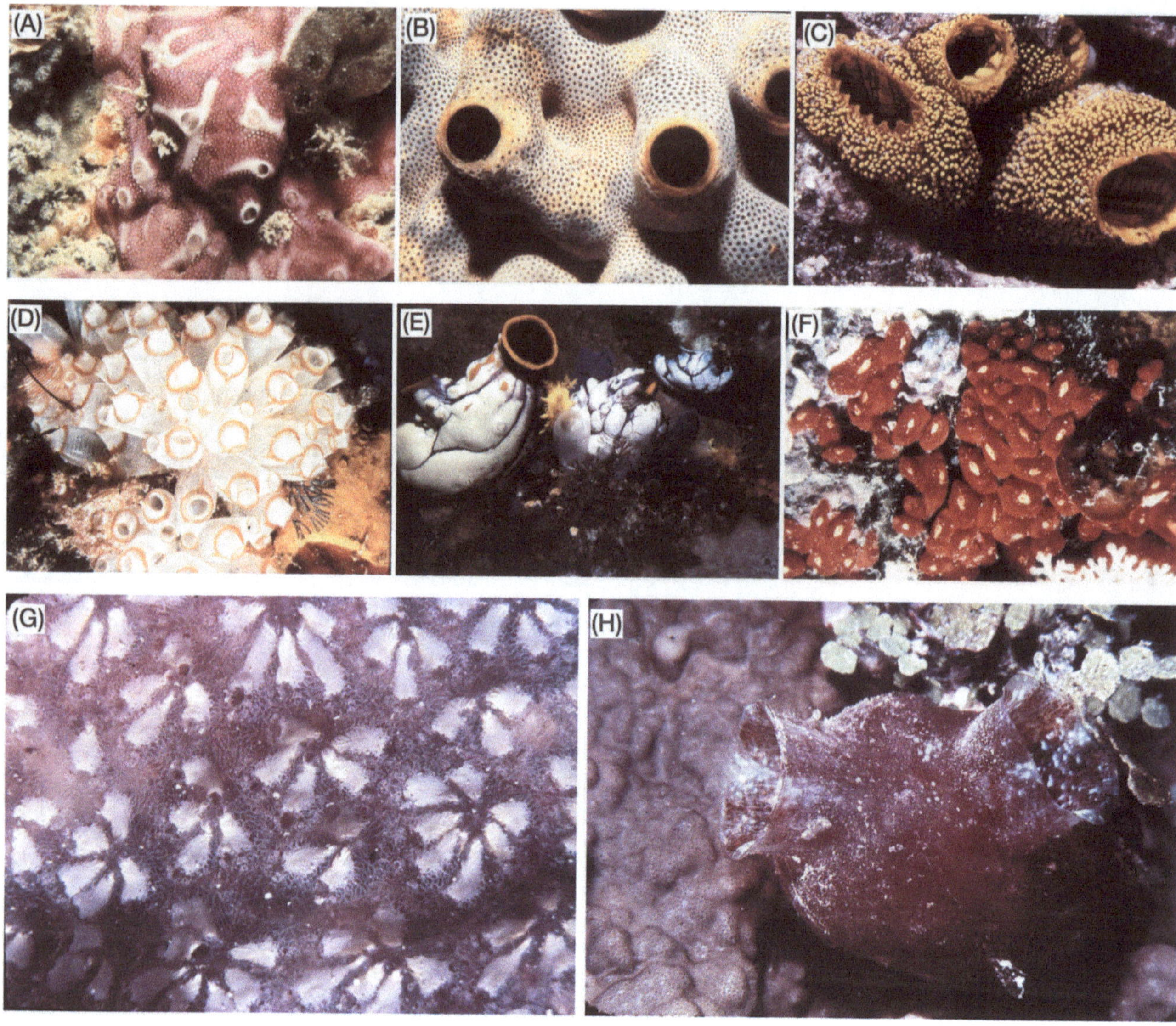

Figure 27.9 *A*, *Didemnum membranaceum* (Didemnidae); *B*, *Lissoclinum badium* (Didemnidae); *C*, *Phallusia julinea* (showing apertures of two individuals); *D*, *Ecteinascidia diaphanis* (Perophoridae); *E*, *Polycarpa aurata* (Styelidae, Styelinae); *F*, *Eusynstyela latericius* (Styelidae, Polyzoinae); *G*, *Symplegma teruaki* (Styelidae, Polyzoinae); *H*, *Herdmania momus* (Pyuridae). (Photos: *A*, *B*, *D*, *F*, N. Coleman; *C*, *E*, *G*, *H*, R. and V. Taylor.)

Kott, P. (1992). The Australian Ascidiacea, Pt 3, Aplousobranchia (2). *Memoirs of the Queensland Museum* **32**(2), 377–620.

Kott, P. (1993). Phylum Tunicata. In 'A Coral Reef Handbook'. (Eds P. Mather and I. Bennett.) (Surrey Beatty & Sons: Sydney.)

Kott, P. (2001). The Australian Ascidiacea Pt 4, Didemnidae. *Memoirs of the Queensland Museum* **47**(1), 1–410.

Schaeffer, R. (1987). Deuterosome monophyly and phylogeny. *Evolutionary Biology* **21**, 179–235.

Thompson, H. (1948). 'Pelagic Tunicates of Australia'. (Commonwealth Council for Scientific and Industrial Research Australia: Melbourne.)

28. The Fish Assemblages of the Great Barrier Reef: Their Diversity and Origin

J. H. Choat & B. C. Russell

INTRODUCTION

There are three critical features of coral reef fish faunas. First, they represent the most diverse assemblages of vertebrates on the planet. Second, this diversity may be seen at very local scales; hundreds of species co-occur within relatively small areas. Third, most species have broad geographical distributions. Thus, the observer moving across a small area of reef will encounter many different species of fish. Moving over relatively small areas (500 m^2) of reef habitat can reveal up to 100 species of reef fishes, far more vertebrates than would be encountered in any terrestrial habitat. If the scale of observation is increased to cover geographically distant reefs within the same ocean basin similar diversities may be encountered and in many instances the same species will be observed. However, species diversity and species identity do vary on geographic scales in ways that reflect the location and the evolutionary history of the reef habitat. In order to understand the processes that underlie this incredible diversity of fishes we must also appreciate the forces that have modified the reef habitat over time.

In this chapter we provide an introduction to some of the important species groups and to the functional diversity of fishes on the Great Barrier Reef (GBR). We do not try to provide a catalogue or identification guide to all the different groups. There are many studies that do this (Box 28.1). Instead we wish to pose some general questions about the GBR reef fish fauna. What is the relationship of the GBR fishes to other reef fish faunas? Do they have any unusual or unique features? Are there distinctive features of the GBR fauna that reflect the location, structure and history of the GBR itself? For this we must describe not only the fishes but also the geological history and oceanographic processes that have contributed to the formation of their habitat, the largest and most complex reef structure in the world.

We also provide examples of different groups of fishes that illustrate some of the important ecological features that help define the GBR and reflect its history. These examples provide an insight into some of the latest research initiatives on GBR fishes and identify future directions this may take.

FISHES – THE MOST DIVERSE VERTEBRATES

Fish are the largest and most diverse group of vertebrates on the planet. With an estimated 32 500 species they constitute approximately half of all the known

BOX 28.1 CATALOGUES AND KEY REFERENCES FOR IDENTIFICATION OF GREAT BARRIER REEF FISHES

Carpenter, K. E., and Niem. V. H. (Eds) (1998). 'FAO Species Identification Guide for Fishery Purposes. The living marine resources of the Western Central Pacific. Vols 1–6.' (FAO: Rome.)

Eschmeyer, W. N. (Ed.) (1998). Catalog of Fishes. Center for Biodiversity Research and Information, Special Publication 1. California Academy of Sciences. Available at www.calacademy.org/research/ichthyology/catalog/fishcatsearch.html.

Froese, R., and Pauly, D. (Eds) (2007). FishBase. World Wide Web electronic publication. Avaible at www.fishbase.org.

Hoese, D. F, Bray, D. J., Allen, G. R., and Paxton, J. (2006). Fishes. In 'Zoological Catalogue of Australia. Vol. 35.' (Eds. Beesley, P.L. and A. Wells.) 2248 pp. (ABRS and CSIRO Publishing: Australia.)

Lowe, G. R., and Russell, B. C. (1994). Additions and revisions to the checklist of fishes of the Capricorn-Bunker Group Great Barrier Reef Australia. Technical Memoir GBRMPA-TM-19, Great Barrier Reef Marine Park Authority, Townsville.

Nelson, J. S. (2006). 'Fishes of the World.' 4th edn. (John Wiley and Sons, Inc.: New York.)

Randall, J. E., Allen, G. R., and Steene, R. C. (1990). 'Fishes of the Great Barrier Reef and Coral Sea.' (Crawford House Press: Bathurst.)

Russell, B. C. (1983). 'Annotated checklist of the coral reef fishes in the Capricorn–Bunker Group Great Barrier Reef Australia.' Great Barrier Reef Marine Park Authority. Special Publication No. 1.

species of vertebrates. Moreover, the potential for discovering new species is far greater in fishes than any of the other groups of vertebrates. Even after three centuries the rate of discovery of new species remains undiminished and fish experts estimate about 5000 marine species remain to be discovered and identified.

Although the diversity of fishes is relatively high, their distribution through the biosphere is taxonomically biased. For example the 11 952 fishes recorded from fresh water environments are strongly represented by just three groups, the carps, characins and catfishes that constitute 65% of the diverse fauna. Similar biases occur in the marine environment. Coral reefs are dominated by a single but complex group of bony fishes, the perciformes. These are perch-like fishes with spiny fin rays characterised by modifications to the feeding and locomotory apparatus that allow them to exploit a variety of food items ranging from microscopic sessile organisms to highly motile invertebrates and fishes.

The distribution of fish through the aquatic biosphere is also heavily biased. The majority of fish species occur in warm water with high local productivity, on shallow tropical marine reefs and in tropical lakes and streams. Tropical reefs make up less than 1% of the total marine habitat and freshwater species occupy less than 1% of the world's aquatic habitat. Thus, the greatest diversity of fishes occurs within a tiny fraction of the Earth's aquatic habitats, in shallow productive waters at low latitudes.

The diversity of marine reef fishes is reflected in the following figures. The total number of species is approximately 16 000 with 10 000 occurring in shallow tropical waters. The diversity of coral reef fishes has a strong geographical focus, with about 2400 species occurring in the Indonesian and Philippine archipelagos. The majority of these are perciformes that have an extended evolutionary history to the start of the Cenozoic. Although some lineages were present in the Cretaceous period at 80–100 million years ago (My) the great morphological variety that we see in the present day was not established until the Eocene period, approximately 50 My.

The most speciose groups of present day coral reef fishes are gobies (family Gobiidae), wrasses (Labridae), groupers (Serranidae), damselfishes (Pomacentridae), cardinalfishes (Apogonidae) and blennies (Blenniidae). In addition, many groups such as butterflyfishes (Chaetodontidae), surgeonfishes (Acanthuridae), parrotfishes (Scaridae) and snappers (Lutjanidae) are also conspicuous and diverse in terms of species numbers, and with wrasses and damselfishes make up a majority of the individual fishes observed on coral reefs. In terms of size, coral reef fishes are highly skewed towards the lower end of the size range, with the majority of reef fishes being less than 20 cm in length. In addition, many reef fish species are locally rare, and a characteristic of fish communities on coral reefs is that they are dominated by a few very abundant species and numerous small rare species.

CORAL REEFS AS FISH HABITATS – AN ECOLOGICAL AND HISTORICAL PERSPECTIVE

Coral reefs are biological formations resulting from constructive processes that produce calcium carbonate and erosive processes that reduce this to sediment. Because important metabolic processes of corals are dependent on light, coral reefs are shallow water phenomena. Coral reefs are important fish habitats as they provide shelter, especially for the juveniles of many species. In shallow, clear water, coral reefs support complexes of small turfing algae capable of rapid growth and turnover. These algae trap organic detritus and provide sites for bacterial growth. These highly productive algal complexes and the associated detritus and microbes are a major source of readily accessible primary and secondary productivity for grazing animals. The calcium carbonate substrate is relatively soft and porous and subject to colonisation by boring organisms. These in turn provide habitats for microorganisms that represent an important protein source for those reef fishes that can excavate the substratum and process the material to extract invertebrates and living plants. The feeding activities of grazing and excavating fishes and the subsequent passage of ingested material through the alimentary tract produce a rain of fine sediment and enriched detrital material. The detritus serves as the primary food source for some of the most abundant groups of grazing fishes. Lastly, the currents in the vicinity of coral reefs form complex eddies and accumulate large volumes of planktonic and small nektonic organisms that serve as a food source for numerous species of plankton feeding reef fishes that are preyed upon by larger pelagic predators.

Understanding reef fish faunas for the purposes of management and conservation requires an appreciation of how these faunas differ between reefs and regions. Reef history, location and environmental influences may all have profound effects on fish assemblages. The GBR has had a distinctive history that in association with its regional location has left its imprint on the fish fauna. The following sections will deal with the location of the GBR with respect to other large reef systems, its history and the geological processes as they relate to the present day fish assemblages and their habitats.

GBR REEF FISHES

Patterns in space

Some 1625 species, including trawl fishes, are recorded from the GBR, of which 1468 are coral reef species. This is a relatively high value and may be explained by two critical influences.

First, there is the geographical location of the reef and its setting in terms of the hydrodynamic environment. The position of the GBR relative to the Indonesian and Philippine archipelagoes to the north and the close proximity to the western Pacific reefs has resulted in a diverse fauna of reef fishes. Colonisation by larval reef fish from sources to the north and west are reflected in the strong biogeographical affinities of the GBR fish fauna with widespread tropical Indo-West Pacific elements.

Second, there is the size, configuration and habitat structure of the reef. The GBR is exceptional in terms of

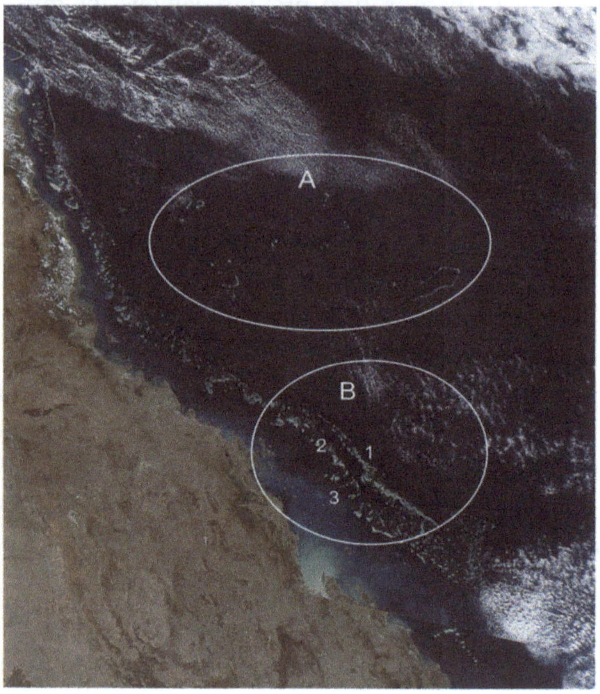

Figure 28.1 A satellite image of the north-east Australian coast showing the GBR and the characteristic structure from the coast to the outer barrier with inner, mid and outer shelf reefs across a longitudinal gradient. Reefs of the Queensland Plateau represent habitats from which reef fishes could recolonise the GBR during periods of rising sea level. *A*, Coral Sea reefs of the Queensland Plateau; *B*, Structure of the GBR: 1, Outer reefs exposed to oceanic influences; 2, Mid shelf reefs; 3, Inner reefs and islands merging with mid shelf reefs. (Image by NASA.)

its size (350 000 km^2), extending along the east Australian coastline from 10°S to 24°S, a distance of 1200 km (Fig. 28.1). The main structure of the reef terminates just south of the tropic of Capricorn. Beyond this there are a number of isolated coral reefs (Elizabeth and Middleton Reefs and Lord Howe Island) extending as far south as 31°30′S. These support faunas of coral reef fishes and also representatives of subtropical and temperate fish groups. A recent survey of Elizabeth and Middleton identified 322 species of reef associated fishes including 26 groupers, 25 butterfly fishes, 21 benthic feeding damselfishes, 22 parrotfishes and 21 surgeonfishes. Some species are endemic to these southern regions. Similar data from the northern end of the reef (Lizard Island) identifies a major shift in diversity, with approximately 900 species of reef associated fishes recorded. For the same groups the diversities were: groupers 41, butterfly fishes 38, benthic feeding damselfishes 48, parrotfishes 25 and surgeonfishes 31. However, the southern reefs also harbour species with warm temperate water affinities including morwongs (Cheilodactylidae), ludericks or sea chubs (*Kyphosus, Girella*), large pomacentrids (*Parma*), wrasses (*Pseudolabrus*) and surgeonfishes (*Prionurus*). These species contribute to the overall diversity of the reef, providing a small but distinctive southern element to the reef fauna (Fig. 28.2).

There are also strong longitudinal trends in reef fish diversity. Comparison of equivalent habitats from inner coastal reefs and mid-shelf and outer barrier reefs at the central region of the GBR revealed the following

Figure 28.2 *A*, The large surgeon fish *Prionurus maculatus* schooling on the Middleton reef crest. *B*, The kyphosid *Kyphosus pacificus* that frequently schools with *P. maculatus* on the Middleton reef crest. Both species use fermentation to digest their primary food source: macroscopic algae. These species extend into the southern GBR. *C*, The girellid fish *Girella cyanea* schooling with *K. pacificus*. *Girella cyanea* feeds on a higher proportion of animal material than the kyphosid. (Photos: A. M. Ayling, Sea Research.)

Figure 28.3 *A*, Inner reef feeding group of grazing fishes. The group consists of two species of parrotfishes *Scarus ghobban*, often found in non-reef environments and a predominantly inshore species *S. rivulatus*. Macroscopic algae is usually present at such sites. *B*, A mixed school of grazing fishes on a mid-shelf reef. These diverse groups of grazing fishes are characteristic of mid-shelf reefs. The example consists of eight species of parrotfish and four species of surgeonfish. *C*, The exposed outer barrier reef crest. This group is dominated by the large browsing surgeonfish *Naso tonganus* and plankton feeding *Acanthurus mata*. (Photos: *A*, *B*, A. Lewis, Tevene'I Marine; *C*, JCU.)

differences in species diversities: butterfly fishes: inner 7, outer 15; damselfishes: inner 10, outer 26; parrotfishes: inner 8, outer 20; surgeonfishes: inner 4, outer 15. An important aspect of this distribution is that some inshore species have restricted distributions that reflect some of the unusual habitat features of inshore reefs (Fig. 28.3). An important summary point is that the GBR maintains a very high diversity of fishes for two reasons. The first is its sheer size, comprising 2000 reefs spread over 350 000 km^2. This provides a massive target with a high probability of suitable habitats for larval fishes that may disperse from other regions to the north and east. The second is the variety of habitats, including inshore coastal reefs extending out to reefs exposed to fully oceanic conditions.

The continuous nature of the reef, its latitudinal distribution from tropical to subtropical environments and the strong longitudinal gradient of habitat structure from the coasts to the Coral Sea are factors that underlie the notable diversity of reef fish species. In addition, the unique evolutionary history of the Australian coastal fauna has made an important contribution to the diversity of the present day GBR fish assemblage.

Patterns in time

Although reef fish faunas vary according to their location, the most profound and interesting changes are those that occur over time. Reef fish diversities vary in response to processes that occur at different timescales.

- Intergeneration changes driven by the variation in the recruitment of juvenile fishes. Reef fish populations are subject to an open water dispersive stage before commencing life on the reef. This influences the number and type of fish that constitute the next generation of reef life.
- Decadal-scale changes associated with climatic variation such as that encountered in El Niño years, cyclonic activity and in some cases biological agents such as the crown-of-thorns starfish, *Acanthaster* (see Chapter 26). Pulses of temperature increase *Acanthaster* feeding, and destructive cyclones result in declines of living coral with concomitant changes in the numbers of many small reef fishes.
- Century-scale changes associated with longer term shifts in ocean temperature or current systems as exemplified by the little ice age (1200–1800 CE). Such climatic shifts will modify distributional patterns of fish, including migration into warmer sections of their range.
- Geological and oceanographic processes and long term climatic trends that usually operate over thousands or millions of years. These include changes to the geographical location of reefs (plate tectonics) and drastic modification of sea level (glaciation cycles).

The most informative approach to understanding the present day GBR fish fauna is to summarise the history of the reef through geological time. A comprehensive description of the geological history of the GBR is provided in Chapters 2 and 3.

At the start of reef fish history, 55 My, much of eastern Australia lay well south of the tropics. Over the next 50 million years tectonic processes moved the Australian continent northwards with the northern boundary of the reef reaching present tropical latitudes about 25 My. However, the entire extent of the reef was not wholly tropical until 3 My. Three things are important about this historical pattern. First, unlike the continents of the northern hemisphere, the Australian coast was not subject to episodes of extensive glaciation during the mid to late Cenozoic. In contrast to the temperate coasts of north America and Europe, the southern Australian fish fauna underwent periods of extensive diversification resulting in lineages of reef fish especially within the wrasses, morwongs and leather jackets (Monacanthidae) that are unique to the southern hemisphere. Some of these temperate-water groups have been able to penetrate tropical environments.

Second, over this period temperatures of the surrounding oceans have varied substantially. From 60 to 45 My the GBR was subject to water temperatures ranging from 9°C to 19°C, thus inhibiting coral reef formation. The combination of northwards continental movement and increasing oceanic temperatures provided a period of 17 My to the present day when an increasing proportion of the reef enjoyed temperatures that permitted coral reef growth.

A third factor, however, further inhibited tropical reef formation. Glaciation cycles drive major fluctuations in sea level and over the last 1.8 My 32 glaciations have been recorded. Over the last 430 thousand years, there has been an increase in the magnitude of cycles, resulting in four episodes of rapid sea level variation with maximum amplitudes of 120–140 m. Cycles of sea level fluctuation reduce and alter habitats that in turn modify reef fish populations. For the GBR, the most dramatic changes have occurred over the last 130 ka. At that time sea levels were equivalent to those of the present day. Sea levels then declined in a series of steps to 125 m below present day levels 20 ka ago. There followed an abrupt rise to present levels commencing 16–18 ka ago.

Given the configuration of the GBR and association with the continental shelf over much of the previous 130 ka there would have been no reef formation on the north-east Australian coast other than a fringing reef at the continental margin. The characteristic mid- and outer-shelf reefs that harbour most of the fish species would have been nonexistent. The best estimates of the rate of sea level rise place the age of the GBR in its present configuration as less than 7 ka old.

In summary, the present GBR is surprisingly young and has been subject to enormous changes over the last 200 ka. Over this period, characterised by cyclic changes in sea level, the reef has ceased to exist during low stands and then been reconstituted through the rising seas. In the periods of rising seas the reef must have been recolonised by reef fishes from the reef systems to the east and north. The present-day configuration with the system of midshelf platform reefs is only 6–8 ka old, although the fish species themselves are far more ancient.

FUNCTIONALLY IMPORTANT GROUPS OF FISHES ON THE GBR

Grazing fishes–parrotfishes and surgeon fishes

What defines a reef fish fauna? This debate has tended to focus on the taxonomic structure of reef fish faunas and their history. One emerging conclusion is that present day reef fish did not arise on coral reefs. The taxonomic debate has been influenced by the observation that the most abundant groups of coral reef fishes also have representatives in temperate waters. It may be easier to define reef fishes in terms of their functional attributes and ecological features. The strongest associations with coral reefs are seen in groups such as parrotfishes, surgeonfishes, butterflyfishes, blennies and many benthic-feeding damselfishes. This association has strong links with their nutritional ecology.

A study of these groups reveals many similar species that co-occur within small areas. This is especially true of the grazing fishes that characterise reef crests and flats. This begs the question as to how they share resources. To understand this it is necessary to examine how space is used in foraging and feeding activities on

Figure 28.4 *A*, An example of a scraping parrotfish, *Scarus oviceps*. Feeding is predominantly by removal of the top 2–3 mm of the calcareous substratum. *B*, An example of an excavating parrotfish, *Chlorurus microrhinos*. Note the deep profile of the head that incorporates the massive oral jaws and associated musculature. Feeding is predominantly by excavating the calcareous substratum. *C*, *Acanthurus olivaceus*, an abundant detrital feeding surgeonfish. Detritus and sediment are triturated in a muscular gizzard before digestion. (Photos: A. Lewis, Tevene'l Marine.)

reefs. In terms of fish real estate, reefs consist of contrasting areas of complex structure provided by living corals and extensive areas of carbonate rock and associated coral debris. Living corals are used extensively by smaller reef fishes and recruits of larger species as shelter. Extensive carbonate flats and fields of coral rubble support complexes of turfing and encrusting algae, fine sediment and detritus. It is the extensive areas of apparently bare calcareous substratum that provide a key to the diversity and dynamics of many types of reef fishes. If this substratum is examined microscopically then a complex 'tangled bank' harbouring a great variety of small plants and micro-organisms, which (to quote Darwin) are 'different from each other, and dependent upon each other in so complex a manner' is revealed. Very large numbers of fishes graze on this 'bank' with the primary groups being parrotfishes, surgeonfishes and rabbitfishes (Siganidae).

Although these groups are usually classed as 'herbivores' their trophic biology is more complex than just eating living plants (Fig. 28.4). Many, including the most abundant species, feed on detrital material, bacterial aggregates and small invertebrates. Others, and especially some species of surgeonfishes, target turfing and filamentous algae, but most of the species that constitute the mixed foraging schools of grazing perciformes that are so characteristic of reefs feed on detrital and bacterial aggregates and cannot be defined as herbivores. This is especially true of parrotfishes that have the capacity to remove calcareous material with scraping (Fig. 28.4*A*) or excavating (Fig. 28.4*B*) oral jaws to extract the microbial assemblages that occur within the porous upper substratum. Many species of surgeonfishes graze mainly on detritus and sediment (Fig. 28.4*C*). Moreover, analysis of feeding and foraging of smaller fish, including many damselfishes and most blennies, shows that they are also targeting detrital resources.

The foraging behaviour of abundant grazing fishes provides a consistent ecological signature for reef fish assemblages. The initial impression is one of uniform feeding by large multispecific schools of grazing fishes moving over tracts of reef. However, both feeding behaviour and the resources targeted are more complex than an initial impression suggests. Some species (exemplified by the aggressive surgeonfish *Acanthurus lineatus*) actively defend territories that support dense growths of turfing algae, their primary food source (Fig. 28.5*A*). Similar behaviour is exhibited by territorial damselfishes of the genus *Stegastes* (Fig. 28.5*B*) where defence of feeding territories against much larger grazers is common. In some reef habitats, such as reef crests, territories may cover up to 70% of the available reef substratum. Other groups of grazing fishes, especially most parrotfishes and many surgeonfishes, form mobile schools that graze over extensive areas, with feeding episodes continuously disrupted by territorial species.

The complexity of grazing behaviour is illustrated by the variety of diets, or the 'nutritional ecology' of the different species. The range includes (1) acanthurids

Figure 28.5 *A*, The highly aggressive surgeon fish *Acanthurus lineatus*. This species defends territories from grazing fishes and enhances the growth of red algae, the primary food source. *B*, A territorial damselfish of the genus *Stegastes*, which employs active defence of an algal garden against grazing fishes. (Photos: A. Lewis, Tevene'l Marine.)

that feed on filamentous algae, processing food via acid digestion (*Acanthurus lineatus*); (2) species that concentrate on large brown algae, through bacterial fermentation (*Naso lituratus*); (3) species that selectively harvest smaller turfing algae and also use bacterial fermentation (*Zebrasoma scopas*); (4) species that feed on both algae and plankton (*Naso brevirostris, N. vlamingii*); (5) species that feed exclusively on detritus (*Ctenochaetus striatus*); (6) species that feed on mixtures of detritus, animal matter and algae (*Acanthurus blochii*); and (7) species that scrape or excavate calcareous surfaces to extract detritus, animal material and turfing algae (*Scarus frenatus, Chlorurus microrhinos*). All of these groups may combine to form multi-species foraging schools that are frequently joined by rabbitfishes and species of goatfish (Mullidae) and wrasses that forage on small invertebrates disturbed by group feeding. At the upper end of the size spectrum of grazing fishes are the very large excavating parrotfishes (*Bolbometopon, Chlorurus*) that are capable of removing hundreds of tonnes of solid calcareous material per year and redistributing it as sediment on reef faces.

These groups have traditionally been classified as herbivores, placed at the bottom of the food chain and attributed the dual roles of enhancing coral growth through removal of algae and acting as a major conduit of carbon through reef systems by linking plant production to carnivores that consume the abundant herbivorous fish. The removal of algae by grazing fish has been demonstrated experimentally. However, the widespread consumption of bacteria, small invertebrates and a cosmopolitan mix of detrital material demonstrates that food chains and the flow of carbon through coral reef communities are likely to be far more complex than usually considered in trophic schemes. The apparent uniformity of grazing activity in mixed schools masks a much greater diversity of feeding behaviour, dietary targets and food processing than is currently recognised.

One of the best examples of the diversity of feeding behaviour in 'herbivorous' fishes is provided by an analysis of the evolutionary relationship in the surgeonfish genus *Naso* (unicorn fishes) carried out using gene sequences extracted from mitochondrial DNA. This provides a means of constructing a robust picture of the evolutionary relations among the 19 currently recognised species. This group is of interest on account of their morphological variation, including slow moving reef grazers and plankton feeders closely associated with shallow reef habitats, often with large frontal horns or head-bumps, to rapidly swimming pelagic tuna-like species. These species have a wide variety of diets including benthic algae, benthic invertebrates, macroplankton and small rapidly swimming nekton such as small fishes (Fig. 28.6). The unicorn fishes have a long evolutionary history, with closely related genera occurring as well preserved fossils in Eocene (50 My) shallow water marine deposits. The purpose of the evolutionary analysis was to determine the evolutionary pathways of the different feeding modes. Were the herbivorous species basal to the other groups or did they emerge much later? Is plant feeding restricted to certain evolutionary groups? What is the basis of the extraordinary morphological diversity of the unicorn fishes?

The answers (Box 28.2) were surprising. A number of well supported monophyletic evolutionary groupings (clades) were identified. The ancestral group was represented by the small, pelagic, fast swimming species of the *Naso minor* complex, not the more abundant reef associated species. Particular types of feeding behaviour and nutritional ecology (algal grazing, plankton feeding, pelagic foraging on nekton, fermentative digestion) did not occur in cohesive evolutionary groupings but were scattered through the different clades. The trademark morphology of the group, the extended frontal horn, occurred in unrelated species in different clades (e.g. *Naso annulatus* and *Naso brevirostris*). This was surprising, as such distinctive morphologies had strongly suggested natural groupings in past studies. The basic message is one of greater evolutionary flexibility in terms of feeding behaviour and morphology than anticipated from ecological studies. In summary, the large groups of grazing fishes that characterise the GBR are deserving of more detailed study.

Predatory guilds of reef fishes

It is a natural progression from grazing fishes to the predatory fishes that inhabit the reef. Many species are piscivores and tend to concentrate on parrotfishes, damselfishes and near-reef pelagic groups such as fusiliers (Caesionidae). Not surprisingly, species with well developed dorsal and anal fin spines (butterflyfishes)

BOX 28.2 EVOLUTIONARY RELATIONSHIPS IN FORAGING AND FEEDING MODES IN THE GENUS *NASO* (UNICORN FISHES)

Figure 28.6 illustrates the evolutionary relationships among the different species of the genus *Naso* in the form of a phylogenetic tree based on sequences from mitochondrial and nuclear genes. Members of the genus display a variety of behaviours including pelagic foraging in the water column, benthic foraging over the reef surface and a mixture of pelagic and benthic foraging modes. Species that swim slowly in open water develop extended frontal horns, fast swimming pelagic foragers have a tuna-like morphology. Different foraging modes are associated with different feeding patterns: macroplankton, benthic animal material, brown algae or mixtures of green and red algae. Some species feed on both planktonic and benthic animals; others feed on algae and plankton. The phylogenetic reconstruction shows that the two most speciose evolutionary groupings or clades (1 and 2) each show examples of the different foraging modes. The distinctive morphologies such as frontal horns and tuna-like body shapes have developed independently in the different clades. The ancestral groups are represented by small pelagic species in the subgenus *Axinurus*. The genus is a relatively ancient widely distributed group of reef fishes with the distribution ranges: IPO, Indo-Pacific; PO, Pacific; IO, Indian Ocean, WIO, West Indian Ocean indicated. The numbers represent the robustness of the tree structure using Bayesian, MP and ML analyses.

Further details are available in Klanten *et al.* (2004). *Molecular Phylogenetics and Evolution* **32**, 221–235.

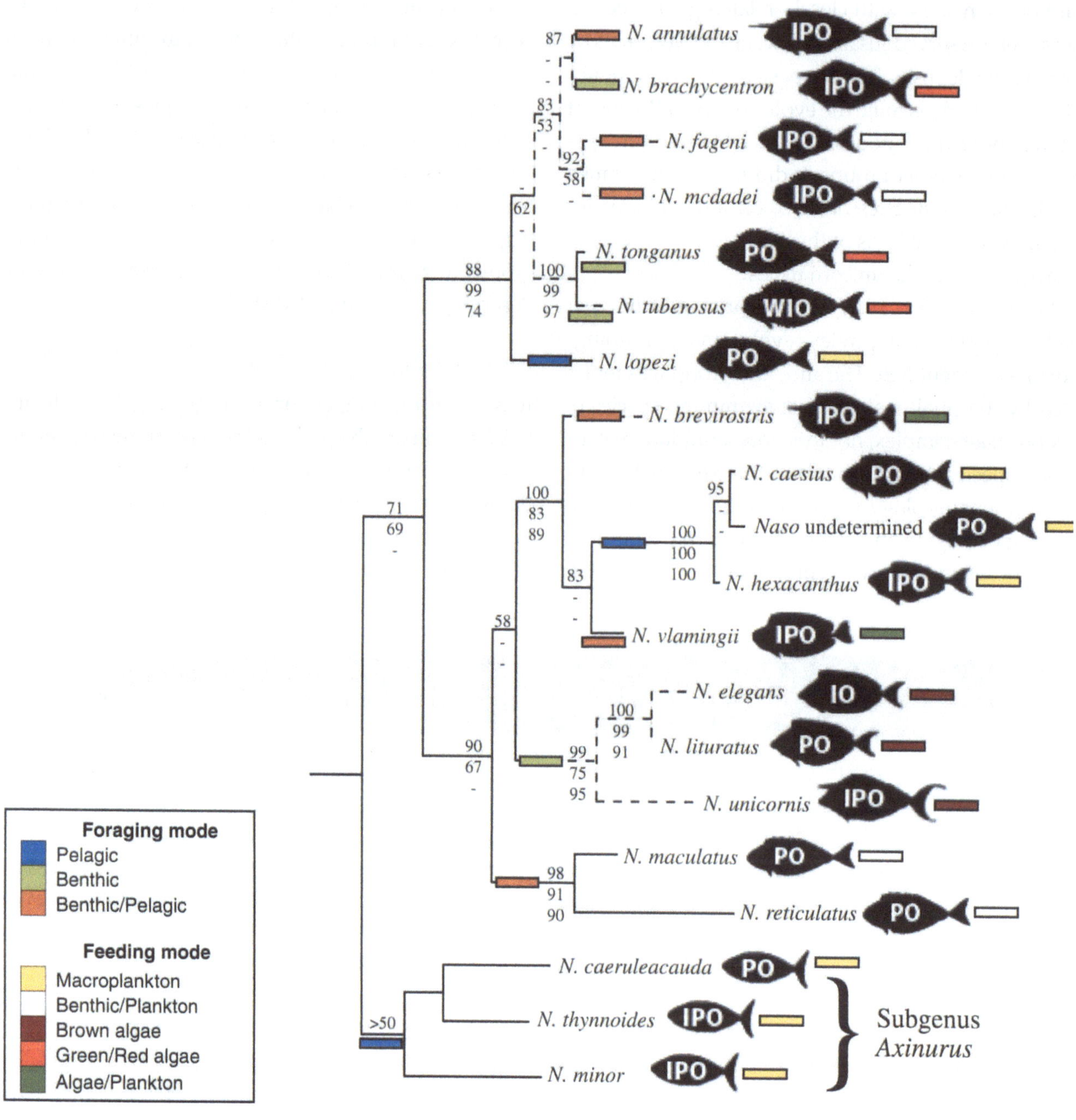

Figure 28.6 Evolutionary relationships in foraging and feeding patterns in *Naso*. (Image: S. Klanten JCU.)

and with caudal knives (surgeonfishes) occur very infrequently in the stomach contents. There are numerous different types of predatory fishes including pelagic species such as mackerels (Scombridae) and large trevallies (Carangidae) but the dominant predators on coral reefs are usually groupers. These are dominated by four genera, *Epinephelus*, *Cephalopholis*, *Plectropomus* and *Variola* and may be partitioned into ecological and foraging groupings. The largest genus is *Epinephelus* with 27 species recorded from the GBR followed by *Cephalopholis*, 10; *Plectropomus*, five and *Variola*, two.

Epinephelus is notable in that the species cover a very wide size range from *E. lanceolatus*, which reaches 2.5 m total length (TL), to small cryptic species such as *E. merra* at 0.25 m TL. The genus may be ecologically subdivided by size, with large species between 0.6–2.0 m TL often being mobile species but usually associated with bottom cover. Smaller species are more numerous than the larger species, with GBR grouper assemblages being dominated by fish 15–25 cm TL. Size is related to foraging activities. Small groupers, including members of the genus *Cephalopholis*, small species of *Epinephelus* and juveniles of the larger species are invariably cryptic, associated with the shelter of coral growths and debris. The larger species of *Epinephelus* are more mobile and may move considerable distances when foraging, but are strongly associated with complex reef structures and usually seek shelter when disturbed. Members of the genus are usually ambush predators preying on other reef fish. In contrast to the genera *Cephalopholis* and *Epinephelus*, members of the other two genera, *Plectropomus* and *Variola*, have a roving habit and usually forage above the substratum, especially on reef fronts, slopes and deeper reef bases. For this reason, they are more visible to the observer than other groupers. There are two unifying features of groupers. First, juveniles are invariably secretive and associated with areas of high cover. For this reason it is difficult to estimate recruitment patterns in reef groupers and to obtain information on the early growth stages. Second, members of this family frequently aggregate to spawn, with some species of *Epinephelus* reaching very high local densities at spawning sites. Other genera such as *Plectropomus* also form spawning aggregations, but with numerous local groups with relatively low numbers of fish.

The GBR grouper fauna has some unusual features. Most reef areas are dominated by smaller cryptic species with the larger mobile species being comparatively rare, even in areas protected from fishing. Consequently, observers see few groupers during normal activities. For example, the dominant groupers on relatively undisturbed reefs in the southern Seychelles were *Cephalopholis urodeta* and *Epinephelus fasciatus*, two small species that made up 70% of the grouper fauna. Larger species, such as *E. polyphekadion* and *E. fuscoguttatus*, occurred at less than one individual per 1000 m^2. Although four small species of *Cephalopholis* made up 47% of the GBR grouper fauna, two species of *Plectropomus*, *P. leopardus* and *P. laevis*, made up 33% of the fauna. The presence of these large, actively foraging predators on reef crests and reef fronts are a unique feature of the GBR predator fauna. By comparison, this genus was very rare in the southern Seychelles. This is not a reflection of fishing as other groupers of great commercial importance were recorded in high numbers from the monitoring sites. This simply reflects the natural distribution of *Plectropomus*, a genus confined mainly to the south-west Pacific and the Indo-Australian region. The ecological equivalent was *Variola louti*, a species that shows a similar above-bottom foraging habit to *Plectropomus*, although it made up only 6% of the serranid fauna. The most abundant populations of *Plectropomus* occur on the Australian plate and especially on the north-eastern Australian coasts and reefs where they also support extensive fisheries. This high concentration of roving and highly visible predators is very much an Australian phenomenon.

Predation is intense on coral reefs and the influence of predators such as groupers on reef fishes may be manifested in behavioural and morphological responses in the prey species. Many reef fishes have evolved colour patterns and behaviours as a means of avoiding predation, or in some cases to enhance their ability as predators. The most common way of avoiding predation is through protective resemblance or camouflage, where the fish closely resembles a part of a substratum, a plant, or a sedentary animal such as a sponge or soft coral to avoid predation (Fig. 28.7; Box 28.3).

■ THE RECENT HISTORY OF THE GBR – HOW HAS THIS INFLUENCED THE FISH FAUNA?

The complex structure of the GBR gives an impression of stability and permanence. The tracts of coastal fringing reefs, mid-shelf platform reefs and ribbon reefs of the outer barrier provide a parallel series of distinctive habitats extending along a north-south axis for 1000 km. The different habitats support distinctive fish communities. At a number of localities along the reef the habitats intergrade, especially in the vicinity of large coastal islands (Fig. 28.1). At these localities, elements of the inshore and mid-shelf reef assemblages may mix.

The geological history of the GBR shows that this structural and biological partitioning of the reef has

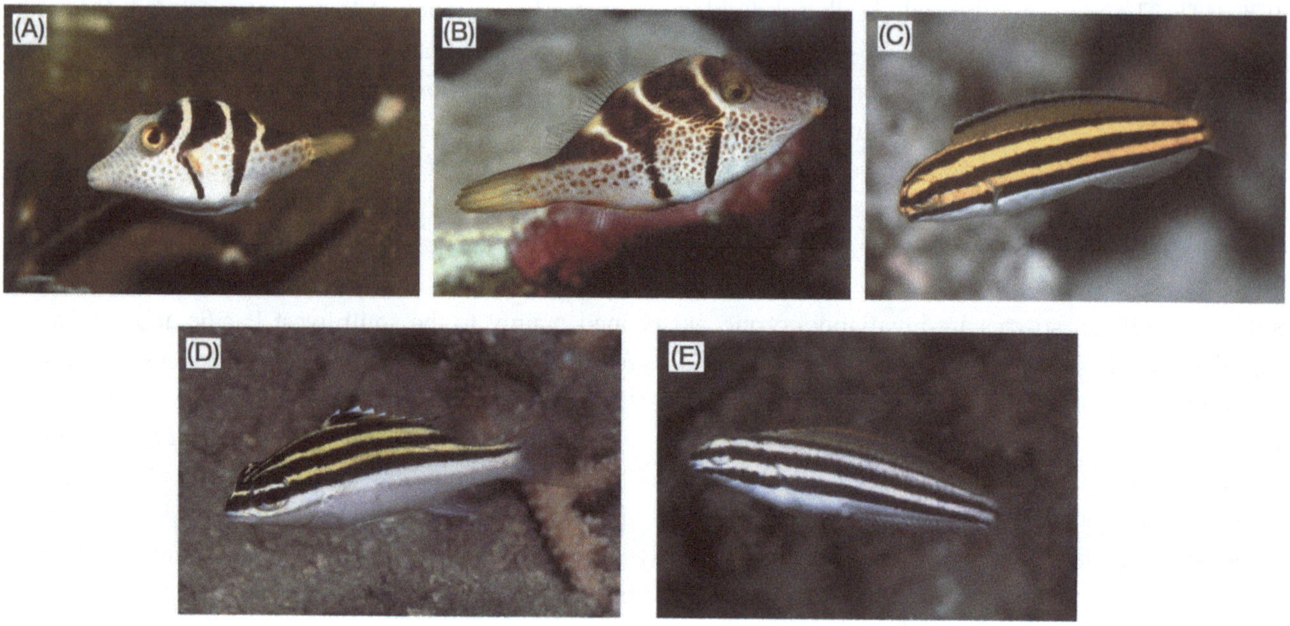

Figure 28.7 Mimicry in GBR Fishes. *A*, noxious tetradontid *Canthigaster valentini*; *B*, monocanthid *Paraleuteres prionurus* that mimics *C. valentini*; *C*, poison fang blenny *Meiacanthus lineatus*; *D*, *E*, mimics of *M. lineatus*, *Scolopsis bilineatus* (*D*) and *Petroscirtes fallax* (*E*). (Photos: *A*, *B*, *D*, Gerry Allen; *C*, *E*, Roger Steene.)

occurred very recently, and that sea level changes associated with glacial cycles have profoundly modified Australian shallow water reef systems. For the last 130 ka sea levels have fallen on a global basis from levels that were similar to those of the present day to a low of 140 m below present levels only 20 ka. The biogeographical consequences of this were profound, with isolation of the north-east Australian coast from the west due to the closure of the Torres Strait from 116 ka to 30 ka, and then by the emergence of the exposed coasts of Papua New Guinea to the north and east. Rapid sea level rises linked to the termination of the last glacial period 19 ka ago flooded continental shelves so that reef habitats increased from 50 000 to 225 000 km^2 at present. Thus, over the last 6000 years the north-eastern Australian coast has been transformed from a shelf habitat dominated by fringing reefs to one supporting the extensive and ecologically diverse system of the present day GBR. One consequence of the recent emergence of a new reef system is that the majority of the fish fauna must be the product of a rapid colonisation process from external sources. Where did the present day fish fauna come from and are there any evolutionary and ecological signatures of this colonisation process?

To explore this question we will consider two of the most prominent members of the reef fauna, the bar cheeked coral trout (*Plectropomus maculatus*) characteristic of inshore reefs and the leopard coral trout (*Plectropomus leopardus*) common on mid-shelf reefs. The geographical source of these populations and their recent evolutionary history was investigated by examining and comparing sequences of molecules amplified from mitochondrial genes. Such sequences are passed on through female parents and as they are inherited they provide a method for disentangling patterns of evolutionary descent in the sampled population. They also provide a species identification code and allow assessment of the degree of relatedness among groups of species.

An analysis of mitochondrial sequences of GBR *Plectropomus* species was carried out to establish the pattern and degree of relatedness of three species. These were *P. leopardus*, *P. maculatus* and *P. laevis* that are very similar in appearance and colour pattern (Fig. 28.8). In addition, the historical biogeography of the most abundant species, *P. leopardus*, which occurs on both the eastern and western tropical coasts of Australia, was carried out. The distribution of this species is of considerable

BOX 28.3 DISGUISE, DEFENCE AND AGGRESSION

Mimicry is a special kind of resemblance that involves co-evolution of colour and morphology, and even behaviour to enhance the deception. Mimicry among coral reef fishes, once thought to be rare, appears to be a general and widespread phenomenon, with about 100 cases now reported. Many of the known cases of interspecific mimicry in fishes involve one or more species of the family Blenniidae. Mimicry also appears to be particularly important during juvenile stages, with more than 25% of mimic species losing their mimic colouration when they outgrow their models and become less vulnerable to predation.

Most of the cases of interspecific mimicry reported so far can be classified as Batesian, Müllerian or Aggressive mimicry. Batesian mimicry is the resemblance of a harmless or palatable species to a harmful or unpalatable one. An example of Batesian mimicry among GBR fishes is that between the noxious *Cathigaster valentini* (Fig. 28.7*A*) and the triggerfish *Paraleuteres prionurus* (Fig. 28.7*B*). In Müllerian mimicry both species possess some undesirable qualities. This type of mimicry appears to be rare among fishes although it may contribute to the mimetic complexes involving members of the blenniid tribe Nemophini. Aggressive mimicry is the resemblance of a predatory species to a harmless or non-predatory form. Aggressive mimicry is the most prevalent type of mimicry in coral reef fishes, constituting about half of all known cases. An example among GBR fishes is the aggressive mimic blenny *Aspidontus taeniatus*, which closely resembles the colour and behaviour of the cleaner wrasse *Labroides dimidiatus*, and uses this deceit to closely approach and bite pieces from unsuspecting prey fishes.

In some cases, where two or more species of fishes are involved in a mimetic complex, elements of all three types of mimicry may be present. The spatial distribution of mimics also appears to be limited by that of their model species, although some mimic different models or different colour morphs of the same model in different habitats or in different parts of their range. For example, the juvenile coral bream *Scolopsis bilineatus* (Fig. 28.7*D*) and the blenny *Petroscirtes fallax* (Fig. 28.7*E*) both mimic the yellow and black-striped poison-fang blenny *Meiacanthus lineatus* (Fig. 28.7C) on the GBR, but in Fiji where an all yellow morph of *M. lineatus* occurs, it is mimicked by an unusual yellow colour form of the coral bream and also by a yellow form of the aggressive sabretooth blenny *Plagiotremus laudandus*, which preys on the soft tissue of other fishes. These examples suggest a high degree of phenotypic plasticity in mimetic colouration and little genetic differentiation among different mimics of the same species. For further reading see:

Randall, J. E. (2005). A review of mimicry in marine fishes. *Zoological Studies* **44**(3), 299–328.

Moland, E., Eagle, J. V., and Jones, G. P. (2005). Ecology and evolution of mimicry in coral reef fishes. *Oceanography and Marine Biology: An Annual Review* **43**, 455–482.

interest as it occurs on the Australian tectonic plate, the western and southern Pacific (Papua New Guinea, Solomons, New Caledonia) and the Indo-Philippine archipelagos, Taiwan and southern Japan. It achieves high abundances on the Australian plate, including New Caledonia, but abundances decline through the reef at lower latitudes, then increase in higher northern latitudes, which suggests an anti-tropical distribution.

Analysis of the geographical structure of *P. leopardus* populations includes some surprises. Eastern and western Australian populations are distinct, a reflection of the long period of closure of the Torres Strait and the sparse reef environment in the present day Arafura Sea. In fact, the closest relatives of the west Australian population appear to be from Taiwan. The GBR populations had their strongest affinities with New Caledonian fish, and the analysis of larval migration patterns strongly suggests an east to west gene flow. This provides a key to the question of the rapid colonisation of the GBR reef over the last 6–7 ka. During the period of low sea level stands the Coral Sea was characterised by large shallow areas of actively growing reefs—the Queensland and Marion plateaus. These are now inundated and well below the level of active reef growth and are now represented by only scattered groups of Coral Sea reefs and islands. However, during periods of low sea level, driven by glaciation cycles over the last 500 ka, these reefs and those of New Caledonia further to the east, would have served as recruitment sources when rising sea levels provided the opportunity for recolonisation of the GBR.

One of the most important messages is that the process of colonisation of the newly forming GBR and the partitioning of species into different habitats happened over a very short period. Are there any genetic signatures of these events? Genetic analysis of the relationships among species of *Plectropomus* shows that *P. leopardus, P. maculatus* and *P. laevis* are closely related, something that is reflected in their colour patterns. On the west coast of Australia *P. leopardus* and *P. maculatus* form distinct monophyletic groups or clades. However, on the east coast *P. maculatus* has the same mitochondrial genetic signature or haplotype as *P. leopardus*. Using mitochondrial genes, *Plectropomus maculatus* from the GBR cannot be distinguished from *P. leopardus,* which is in striking contrast to the pattern observed in Western Australia. The most parsimonious explanation for these distinctive coastal patterns is that on the east coast the species have a history of hybridisation, with male *P. maculatus* joining spawning groups of *P. leopardus* (Fig. 28.8; Box 28.4). The opportunity for the mixing and overlap of the species populations on the east coast was greatly enhanced by the turbulent history of the reef, with very rapid episodes of colonisation of new reef structures and sorting among the newly formed habitats. In contrast, the coast of Western Australia has had a far more stable history, with limited effects of sea level change and with populations of each species clearly partitioned between coastal and oceanic offshore reefs.

The distinctive genetic structures of *P. leopardus* and *P. maculatus* populations on the GBR, with evidence of past episodes of hybridisation, are in effect a signature of a distinctive geological and evolutionary history of this reef system. It is highly probable that other species will show similar evidence of a reef system subject to rapid structural and biological change.

THE FUTURE FOR REEF FISH RESEARCH ON THE GBR

Given the diversity of the fishes and the ecological complexity of the GBR, there are a multitude of novel research issues to follow up. Much of what we hear about the GBR concerns issues of habitat disturbance and loss and the consequences of overfishing. These are legitimate issues and the global picture of the status of coral reefs demonstrates that although the GBR is relatively healthy, the consequences of natural and anthropogenic disturbances can be manifested very quickly. The most effective way to evaluate the potential influences of disturbance and exploitation is through an understanding of the biology of the organisms themselves. Reef animals, and especially the fishes, vary in the way they respond to stress and environmental change. It is difficult to generalise from one group to another and the best approach is to understand the biological mechanisms, both evolutionary and ecological, that underlies the way different groups respond to change.

We identify three areas of future studies of GBR fishes that will yield the type of information discussed above. This list is not exhaustive or prioritised, but it captures some of the excitement of ongoing reef fish research.

Depth distributions of reef fishes. How deep does the GBR go? Using baited video cameras it is now possible to obtain visual samples of the diversity of reef fishes to a depth of 200 m, below the level of active coral growth. These surveys have revealed that a number of species

BOX 28.4 GENETIC STRUCTURE OF *PLECTROPOMUS* POPULATIONS

The genetic structure of *Plectropomus* on the east and west coasts of Australia. The diagram represents a phylogenetic tree based on molecular sequences from the D-loop region of the mitochondrial (mt) genome of *P. leopardus* and *P. maculatus* collected from the tropical coasts of eastern and western Australia. Sequences from *P. laevis* constitute an outgroup. The colours represent mt sequences characteristic of each species (blue, *P. leopardus*; green, *P. maculatus*).

Plectropomus leopardus sequences grouped as a single clade (A), with populations from the two coasts occurring as two genetically distinct sister clades. *Plectropomus maculatus* populations from the west coast form a distinct clade (B). However, on the east coast *P. leopardus* and *P. maculatus* do not separate into species-specific clades and are genetically indistinguishable with the mitochondrial marker used in this study. The failure of the mitochondrial marker to distinguish the two species is attributable to interspecific hybridisation between these species on the GBR. The structure of the phylogenetic tree is strongly supported in ML and Bayesian analyses. Further details of this study are provided in L. van Herwerden *et al.* (2006). *Molecular Phylogenetics and Evolution*, **41**, 420–435.

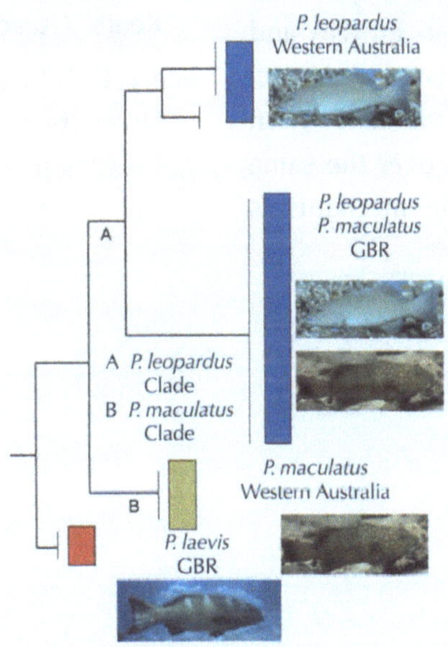

Figure 28.8 Genetic structure of *Plectropomus* populations. (Image: L. van Herwerden, JCU.)

that were considered to be shallow water fishes extend to 100 m depth. This work requires a reevaluation of the habitat association for some common reef species and also identifies species that are able to extend their depth range in response to shallow water disturbances, such as sudden increases in temperature or fishing effects.

Genetic analysis of population structures. Is there a consistent genetic signature associated with the very recent colonisation of the GBR by reef fishes including further examples of hybridisation between related species? Surveys of reef fish genetic structure are also important with respect to the detection of 'cryptic' species. However, this carries with it an important caveat;

it is unwise to use a genetic criterion such as differences in mitochondrial sequences to differentiate fish species, as the *Plectropomus* example shows. Two things are essential for the confirmation of cryptic species. At least two lines of evidence such as morphological and/or structural distinctions as well as genetic differences are required. More importantly, it is critical to have the capacity to correctly identify reef fishes in the field. This capacity is being eroded as taxonomic research in our museums winds down. Collecting tissues from reef fishes for such schemes as the Barcode of Life without proper taxonomic quality control is a recipe for disaster.

Multi-scale analysis of life history features. After a slow start, there are a number of study programs investigating the age-based life history features of coral reef fishes using age information extracted from reef fish otoliths. This is important for a number of reasons, including the analysis of stock structure. Many species of reef fish show significant differences in growth and mortality rates between localities. This is very conspicuous in GBR fishes. Such studies are valuable when linked to analyses of genetic structure over the same scale as it provides insights into the mechanisms underlying spatial differences in demography. Preliminary work suggests that reef fishes show great plasticity with respect to growth rates, reproductive outputs and age-structure, but it is unclear to what extent this has a genetic basis. When examined on a broader geographic scale, GBR reef fishes show some unique demographic features including far greater life spans than those in adjacent reef systems although the significance of this is not clear at present.

ADDITIONAL READING

Allen, G. R. (2007). Conservation hotspots of biodiversity and endemism for Indo-Pacific coral reef fishes. *Aquatic Conservation: Marine and Freshwater Ecosystems*. Available at http://dx.doi.org/10.1002/aqc.880 [Verified 21 March 2008].

Sale, P. F. (Ed.) (1991). 'The Ecology of Fishes on Coral Reefs.' (Academic Press: San Diego.)

Sale, P. F. (Ed.) (2006). 'Coral Reef Fishes. Dynamics and Diversity in a Coral Reef Ecosystem.' (Academic Press: San Diego.)

29. Reptiles

H. Heatwole & V. Lukoschek

SEA TURTLES

Six species of sea turtles occur in the waters of the Great Barrier Reef (GBR). Five of these belong to the family Cheloniidae: green turtle (*Chelonia mydas*) (Fig. 29.1*A*), flatback (*Natator depressus*), olive ridley (*Lepidochelys olivacea*), hawksbill (*Eretmochelys imbricata*) (Fig. 29.1*B*) and loggerhead (*Caretta caretta*). The remaining species, the leatherback or luth (*Dermochelys coriacea*), is a member of the family Dermochelidae.

Each of these species has its own feeding and nesting grounds but there is broad overlap among species. In the non-breeding season the turtles disperse widely to feeding grounds within and well beyond the GBR region. Those nesting on or near the GBR may go as far as Indonesia, New Guinea, the Solomon Is, Vanuatu, New Caledonia and Fiji. Only the flatback is endemic to Australia.

Green turtles nest on the Capricorn-Bunker islands and on the cays of the outer barrier from the Swain Reefs north to Princess Charlotte Bay. Loggerheads (Fig. 29.1C) nest on the cays of the Swain Reefs and Capricorn-Bunker islands as well as on beaches of the adjacent Queensland coast. Hawksbills lay their eggs mostly on the cays of the inner shelf north to Princess Charlotte Bay. The flatback has a restricted nesting area on the inshore continental islands between Gladstone and Mackay, and leatherbacks lay their eggs on the mainland coast just south of Bundaberg. The olive ridley does not appear to nest in the GBR region. Except for the flatback, all species also nest well beyond the GBR, and indeed are distributed worldwide in subtropical and tropical waters. The leatherback also ranges into cooler waters.

Female turtles return repeatedly to lay their eggs on beaches in the area from which they hatched. In the GBR region, sea turtles aggregate and begin to court near nesting beaches in spring (September to November) and females come ashore and dig pits in the sand in which they lay their eggs between October and March. Eggs incubate in the sand and the temperature of the nest determines the sex of the hatchlings: cooler nests produce males whereas warmer nests produce females. Hatchlings emerge in summer and autumn (January to May) and go to sea where they drift with currents and feed on macroscopic plankton. After 5–10 years they move into their traditional feeding grounds and adopt the adult diet.

Adult green turtles feed mainly on algae and seagrass on coral reefs and inshore seagrass flats. The loggerhead eats mainly molluscs and crabs in sandy lagoons on the reef or in inshore bays and estuaries. The olive ridley consumes small crabs in inter-reefal areas. Hawksbills eat algae, seagrasses, sponges, ascidians, bryozoans and molluscs, mainly on coral reefs. The flatback feeds on soft corals and other soft-bodied animals on soft bottoms in inter-reefal areas.

Figure 29.1 *A*, green turtle (*Chelonia mydas*) (photo: GBRMPA); *B*, hawksbill turtle (*Eretmochelys imbricata*) (photo: GBRMPA); *C*, loggerhead sea turtle (*Caretta caretta*) digging a nest pit on Heron I., Great Barrier Reef (photo: H. Heawole); *D*, head of olive sea snake (*Aipysurus laevis*) showing the two semicircular nostril valves (closed) (photo: V. Lukoschek); *E*, various sea snake species in an aquarium illustrating the differences in size, shape, and colour pattern within the group. On the left with saddle-blotches is *Astrotia stokesii* and the banded one is a species of *Hydrophis* (note that the head is small and protruding above the water and the paddle-shaped tail is large and submerged). The three on the right are *Aipysurus laevis* (note their marked differences in colour). Colour patterns vary greatly within many species, especially from one locality to another. (Photo: H. Heatwole.); *F*, Two turtleheaded sea snakes, *Emydocephalus annulatus*, courting. Male is the smaller (dark) snake pursing the larger female. There is enormous colour pattern variation in both sexes of this species ranging from entirely black (melanistic) to very pale salmon. (Photo: M. Berger.); *G*, Juvenile olive sea snake, *Aipysurus laevis*, showing strongly banded pattern of this species in its first year of life. (Photo: V. Lukoschek.); *H*, *Aipysurus duboisii* swimming over a sea grass bed in sandy habitat adjacent to a coral reef. (Photo: V. Lukoschek.); *I*, Two turtleheaded sea snakes, *Emydocephalus annulatus*, courting. The male is the smaller snake on top of the larger female. (Photo: M. Kospartov.); *J*, Adult olive sea snake, *Aipysurus laevis*, showing typical colour pattern (olive head and uniform grey to brown body) found on the Great Barrier Reef. (Photo: V. Lukoschek.); *K*, *Astrotia stokesii*, the bulkiest of all sea snake species, may reach 2 m in length and has a large head and an exceptionally wide gape. Its fangs are long and it is one of the few species capable of penetrating a wet suit. (Photo: AIMS.)

The leatherback eats jellyfish and salps, mainly in deeper water away from reefs.

A number of threatening processes affect marine turtles at all stages of their life cycles and in different parts of their geographic ranges. Because of this, marine turtle conservation requires a multifaceted approach and international co-operation. Throughout the world key threatening processes include the loss of nesting habitat on key nesting beaches, artificial lights at nesting beaches that attract hatchlings towards the light and away from the ocean, direct harvest of eggs for human consumption, increased predation on nests by feral animals (such as wild pigs), direct harvest of adults for consumption of meat and for 'tortoiseshell', tourism, boat strike, ingestion and/or entanglement in marine debris, and incidental capture in fisheries gear. In the GBR region (as with many other places), numerous conservation initiatives are in place, which aim to reduce or eliminate these impacts. These initiatives include the protection of key nesting beaches and important internesting and foraging habitats, assisting indigenous hunters (traditional owners) to manage their harvests within sustainable limits, increasing awareness of the need for boats to 'go slow' in key turtle habitats, increasing awareness of the impacts of pollution and litter in the marine environment, and a legal requirement that turtle exclusion devices (TEDs) be used by trawl fishers at all times. Ecotourism ventures in Queensland allow visitors the opportunity to observe females dig their nests and lay eggs (Fig. 29.1C) and to participate in the release of hatchlings; these activities aim to educate people about the fascinating lives of marine turtles.

SEA SNAKES

Sea snakes are true reptiles closely related to Australian venomous terrestrial snakes. Indeed, both groups are included in a single subfamily, Hydrophiinae, by most modern herpetologists. The nearest relatives of the hydrophiines are the Asian members of the subfamily, Elapinae. Hydrophiine and elapine snakes are united in the family Elapidae and defined by having fixed front fangs. Sea snakes differ from their terrestrial relatives by numerous marine adaptations, including smaller belly scales, a paddle-shaped tail used in swimming, nostril valves that close from the inside and keep out water while diving (Fig. 29.1*D*), and a salt gland that is located beneath the tongue sheath and excretes excess salt. Sea snakes are extraordinary divers, being able to descend to depths of 100 metres and remain submerged for two hours or more. They surface to breathe air but supplement their oxygen and carbon dioxide exchange by cutaneous respiration while submerged.

There are two kinds of sea snakes, sea kraits and true sea snakes (Fig. 29.1*E*). The sea kraits are egg-layers and come out on land to oviposit and to rest and mate, whereas all true sea snakes give birth to live young and never voluntarily leave the sea. Although sea kraits are abundant in New Guinea and the island chain of the Solomon Is, Vanuatu, New Caledonia, Tonga and Fiji, curiously they are unknown from Australia except as rare waifs. By contrast, Australia has a rich diversity of 32 species of true sea snakes, of which 14 species maintain permanent breeding populations in the GBR region. In general, courtship (Fig. 29.1*F*) and mating take place in winter and live young are born in summer. Clutch sizes vary considerably among true sea snake species, ranging from just two to four per clutch to around 20 per clutch. Juveniles generally have colour patterns that differ from those of adults of the same species (Fig. 29.1G).

Sea snake species occur in a variety of habitats: some are true reefal species, whereas others occur in deeper inter-reefal areas, or on rocky or muddy substrates between the GBR and the Queensland coast (Table 29.1). Of the fourteen species that occur in the GBR region, two are restricted to reefs (*Aipysurus duboisii* (Fig. 29.1*H*) and *Emydocephalus annulatus* (Fig. 29.1*I*)) and two are characteristic of reefs but also are found in other habitats (*Aipysurus laevis* (Fig. 29.1*J*) and *Astrotia stokesii* (Fig. 29.1*K*)); three are found in inter-reefal areas (*Acalyptophis peronii*, *Disteira kingii* and *Disteira major*); five are mainly from coastal habitats that are muddy or rocky; and one, *Pelamis platurus*, is a pelagic species that lives at the surface in slicks. *Pelamis platurus* is the only species that is not benthic. There is a progressive decrease in species richness along the GBR from north to south.

Table 29.1 Sea snake fauna of the Great Barrier Reef and Coral Sea and the habitats where they most commonly occur

Species	Habitat	Water depth	
Acalyptophis peronii	Inter-reefal	Deep	(30–64 m)
Aipysurus duboisii	Coral reefs, over sea grass or sandy habitats	Variable*	
Aipysurus eydouxii	Muddy bottoms, also in rivers and estuaries	Shallow	
Aipysurus laevis	Coral reefs, rocky coasts and soft sediments	Variable	(1–55 m)
Astrotia stokesii	Eurytopic	Variable	
Disteira kingii	Inter-reefal	Deep	
Disteira major	Inter-reefal over sandy and muddy habitats	Variable	(1–43 m)
Emydocephalus annulatus	Coral reefs	Shallow	
Enhydrina schistosa	Muddy & sandy habitats; estuaries and creeks	Shallow	
Hydrophis elegans	Sandy & muddy habitats; also in estuaries	Shallow	(1–18 m)
Hydrophis ornatus	–	Deep	(18–55 m)
Hydrophis sp.**	Sandy habitats	Deep	(15–40 m)
*Lapemis curtus****	Muddy and sandy habitats in turbid waters	Variable	(1–40 m)
Pelamis platurus	Pelagic	Water surface	

* This species variously reported (1) as a shallow-water form occurring in water 2 m to 10 m deep, based on direct observation of this species on coral reefs and (2) as occurring at depths of 10-30 fathoms (18-55 m) based on trawling catches.
** Sometimes erroneously called *Microcephalophis gracilis* in the literature.
*** Often listed as *Lapemis hardwickii* (a synonym).

The diets of sea snake species range from generalist to extremely specialised. *Aipysurus eydouxii* and *E. annulatus* have the most specialised diets and eat only fish eggs, while *A. peronii* feeds principally on goby fish, and a number of *Hydrophis* species are eel-specialists; *H. elegans* may be such a species. Other species, such as *Lapemis curtus, A. duboisii, A. laevis, D. major, P. platurus, E. schistosa* and probably *D. kingii*, are more general in their diet and eat fish from a variety of families (some even take an occasional invertebrate). The diets of the other sea snake species from the GBR are poorly known.

Most sea snake species feed on the bottom. They poke their heads into burrows and crevices and flick out the tongue to capture odours whereby they identify their prey. They do not recognise prey by sight and do not attempt to catch fish in open water. At least three species from the GBR are exceptions. *Enhydrina schistosa* lives in turbid water where visibility is low and it feeds by snapping at fish that bump into it in the muddy water. *Pelamis platurus* catches its prey while floating on the surface. It slowly bends its head toward vibrations of the water made by small fish and captures its prey using a fast lateral strike, sometimes accompanied by backward swimming. *Lapemis curtus* may feed in the water column as well as on the bottom. All species manipulate their prey until they reach the head and prey items are swallowed headfirst.

Sea snakes elicit fear in many people and indeed in Asia fishermen are bitten and die. Most bites are sustained, however, either by people using hand seines in muddy estuaries where they tread, bare-footed, on *E. schistosa* or when attempting to extract snakes by hand from their nets. About 90% of serious or fatal bites suffered by humans are attributable to *E. schistosa* and the rest are mostly inflicted by *Hydrophis cyanocinctus*, a species not present in Australia. Sea snakes do not constitute a significant risk on reefs where most sport divers concentrate their activities, as long as people exercise good sense and caution; that is, do not molest the snakes, and do not approach courting pairs. There are very few verified examples of unprovoked sea snake attacks. Although a wet suit is good protection against most species, *A. stokesii* (Fig. 29.1*K*) and larger female *A. laevis* (Fig. 29.1*J*) are exceptions; their fangs are long enough to penetrate the thickness of neoprene used in tropical wet suits. An anti-venom specific for the treatment of bites by *E. schistosa* is available and, in combination with anti-venom against tiger snakes, *Notechis scutatus*, is effective against other species of sea snakes.

Threatening processes affecting sea snakes are less well understood than those affecting sea turtles. In South-East Asia, many sea snake species are harvested for food and leather whereas, in Australia, all sea snake species are protected under the Environment Protection and Biodiversity Conservation Act, 1999 and direct harvesting is banned. There have been some reports of declining sea snake numbers on some reefs in Australian waters, possibly attributable to the degradation of tropical, shallow-water marine habitats due to overfishing, pollution, coral bleaching and disease. In addition, inter-reefal species comprise a significant component of commercial trawl bycatch. Considerable effort is being made to reduce this impact by developing and evaluating the effectiveness of bycatch reduction devices for sea snakes.

KEY TO THE SPECIES OF MARINE REPTILES OF THE GREAT BARRIER REEF

1. Animal with four flippers and encased in bony shell 2
   – Animal legless and not encased in bony shell 7
2. Carapace with distinct longitudinal ridges; no large flat plates .... *Dermochelys coriacea* (leatherback; luth)
   – Carapace lacking longitudinal ridges and with large flat plates 3
3. Four pairs of costal scales (row of plates flanking each side of central row of plates) 4
   – More than four pairs of costal plates 6
4. Plates on carapace overlapping .... *Eretomochelys imbricata* (hawksbill)
   – Plates on carapace juxtaposed 5
5. Carapace low-domed with edges upturned .... *Natator depressus* (flatback)
   – Carapace strongly arched with tapering edges .... *Chelonia mydas* (green turtle)
6. Carapace longer than wide and with five (rarely six) pairs of costal scales .... *Caretta caretta* (loggerhead)
   – Carapace oval, not much longer than wide and with six or more pairs of costal scales .... *Lepidochelys olivacea* (olive ridley)
7. Belly scales broad and conspicuous, at least three times as broad as adjacent body scales 8
   – Belly scales small and inconspicuous, much less than three times as broad as adjacent body scales 11

8. Three scales on upper lip of each side of mouth, the second one very long and clearly extended .... *Emydocephalus annulatus* (turtleheaded sea snake)
   – Six or more scales on upper lip of each side of mouth 9
9. Scales on parietal region of head irregular and small, about equal in size to scales on neck .... *Aipysurus duboisii* (Dubois' sea snake)
   – Scales on parietal region of head larger than those on neck 10
10. Parietal shields on head fragmented and asymmetrical on the two sides of the head .... *Aipysurus laevis* (olive sea snake)
   – Parietal shields on head not fragmented and are symmetrical on the two sides of the head .... *Aipysurus eydouxii* (Eydoux' sea snake)
11. Scales above eye projecting as spines .... *Acalyptophis peronii* (horned sea snake)
   – Scales above eyes not projecting or spine-like 12
12. Median longitudinal groove on underside of chin 13
   – No median longitudinal groove on underside of chin 14
13. 47 or more rows of scales around middle of body; black dorsal (upper body) colour sharply demarcated from lighter (yellow or brown) of the underside .... *Pelamis platursus* (yellow-bellied sea snake)
   – Fewer than 47 rows of scales around middle of body; hexagonal or squarish body scales with raised tubercles (spines) .... *Lapemis curtus* (spine-bellied sea snake)
14. Median anterior chin-scale triangular, broader than long 15
   – Median anterior chin-scale dagger-shaped and much longer than broad .... *Enhydrina schistosa* (beaked sea snake)
15. Ventral scales leaf-like in double row forming a keel.. .... *Astrotia stokesii* (Stoke's sea snake)
   – Ventral scales flat, not leaf-like 16
16. Anterior chin-shields longer than wide and separated by less than half their length from first pair of infralabials 17
   – Anterior chin-shields wider than long and separated by about half their length from first pair of infralabials 19
17. Head about as wide as rest of body; have ten or more maxillary teeth following fang .... *Hydrophis ornatus* (ornate sea snake)
   – Head narrower than rest of body; fewer than ten maxillary teeth following fang... 18
18. Ventral scales more than 300 .... *Hydrophis elegans* (elegant sea snake)
   – Ventral scales less than 300 ....*Hydrophis mcdowelli* (McDowell's sea snake)
19. Ventral scales more than 300 .... *Disteira kingii* (king's sea snake)
   – Ventral scales less than 300 .... *Disteira major* (greater sea snake)

TERRESTRIAL REPTILES

Land reptiles are known from at least 53 islands and cays of the GBR and include at least 31 species of lizards and nine species of terrestrial snakes. Detailed treatment of this fauna is beyond the scope of this chapter and only a summary is presented here.

The fauna of the continental islands is a subset of that of the Queensland coast. The geology, soils and vegetation of continental islands resemble conditions on the adjacent mainland and in many cases the islands and mainland were connected as recently as 10 000 years ago during the most recent Pleistocene glacial maximum, when sea levels were over 100 metres lower than present levels. Some reptiles are moderately adept at over-water dispersal and, although populations may have been separated since rising sea levels isolated continental islands, it is likely that at least some gene flow has persisted since then. Consequently, most of the populations of lizards and snakes on continental islands of the GBR region are not insular endemics, but are conspecific with mainland populations. The herpetofauna of continental islands encompasses a variety of lizards including mainly skinks (Scincidae) and geckos (Gekkonidae) as well as some flap-footed lizards (Pygopodidae) and goannas (Varanidae). Indeed, Lizard Island was named after its population of goannas. Dragons (Agamidae) are notably lacking. Small numbers of species of all the terrestrial families of Australian snakes are represented on continental islands in the GBR region.

The coral cays have a lower species richness of terrestrial reptiles than do the continental islands. Their fauna is composed of two elements, skinks and geckos that are readily dispersed over water naturally as waifs and which also are widely dispersed on islands well beyond the GBR region, and skinks and geckos that have been transported to the cays by human agency. In addition, there have been a few species of dragons inadvertently introduced by humans; these were ephemeral and did not establish permanent populations.

ADDITIONAL READING

Cogger, H. G. (2000). 'Reptiles and Amphibians of Australia.' 6th edn. (Reed New Holland: Sydney.) [Contains coloured photographs, diagrams showing names of scales and scale patterns used in key, and keys for all Australian species.]

Dunson, W. A. (Ed.) (1975). 'The Biology of Sea Snakes.' (University Park Press: Baltimore.) [This book contains chapters by Dunson, Heatwole, and Limpus dealing with the sea snake fauna of different geographic regions within Australia.]

Heatwole, H. (1997). Marine snakes: are they are a sustainable resource? *Wildlife Society Bulletin* **25**, 766–722.

Heatwole, H. (1999). 'Sea Snakes.' (University of New South Wales Press: Sydney.)

Heatwole, H., and Cogger, H. (1994). 'Sea snakes of Australia.' In 'Sea Snake Toxicology'. Chapter 5. (Ed. P. Gopalakrishnakone.) pp. 167–205. (Singapore University Press: Singapore.) [This chapter contains keys to all sea snakes that occur in Australia.]

Lukoschek, V., Heatwole, H., Grech, A., Burns, G., and Marsh, H. (2007). Distribution of two species of marine snakes, *Aipysurus laevis* and *Emydocephalus annulatus*, in the southern Great Barrier Reef: metapopulation dynamics, marine protected areas and conservation. *Coral Reefs* **26**, 291–307.

Lutz, P. L., and Musick, J. A. (Eds.) (1996). 'The Biology of Sea Turtles.' (CRC Press: Boca Raton.)

Lutz, P. L, Musick, J. A., and Wyneken, J. (Eds) (2003). 'The Biology of Sea Turtles. Vol. II.' (CRC Press: Boca Raton.)

30. Marine Mammals

H. Marsh

INTRODUCTION

The Great Barrier Reef World Heritage Area supports a diverse marine mammal fauna. As listed in Table 30.1, our knowledge of their distributions suggests that more than 30 species of marine mammals spend at least part of their lives in the region. Almost all of these animals are members of the Order Cetacea (whales and dolphins). The region also supports globally significant populations of one member of the Order Sirenia (sea cows), the dugong, *Dugong dugon.* Both cetaceans and sirenians spend their entire lives in the water.

There are two major groups (suborders) of cetaceans, the Mysticeti or baleen whales and the Odontoceti or toothed whales. Baleen whales lack teeth and have baleen plates hanging from roof of their mouths. Toothed whales generally capture their prey, typically fish or squid, one at a time. Their most usual dentition is a large number (up to 100) of peg-like teeth, all similar in form in each of the upper and lower jaws.

Each cetacean suborder contains several families, which in turn contain one or more genera. Two baleen whales, the humpback, *Megaptera novaengliae,* and the dwarf minke, *Balaenoptera acutorostrata,* are commonly seen in the GBR region during the winter and are important tourist attractions. The coastal odontocetes include the rare Indo-Pacific hump-backed dolphins, *Sousa chinensis* (Fig. 30.1), and Australian snubfin dolphins, *Orcaella heinsohni* (Fig. 30.2) (the only cetacean endemic to Australia and Papua New Guinea waters).

Some species such as killer whales and common dolphins are known to occur in the GBR region but are rarely reported. Other species have never been seen alive in the region but are known from strandings on the adjacent Queensland coast. One species, Longman's beaked whale, *Indopacetus pacificus,* was considered to be the world's rarest cetacean until recently. A skull found in Mackay, Queensland, was the basis for the initial description of this species.

The dugong is of great cultural and dietary value to indigenous Australians living adjacent to the GBR region. The importance of the populations of dugongs occurring in the GBR were part of the rationale for the region's World Heritage Listing.

IDENTIFYING MARINE MAMMALS

Identifying marine mammals at sea is extremely difficult, especially if only a small portion of an animal's body is seen fleetingly as it surfaces to breathe. High-quality photographs of an animal breaching (leaping from the water) or underwater with most or all of the body in the picture, or of stranded animals (alive or dead) are an important aid to identification and can be used by researchers to identify species and even individuals of some species. Sketches are also helpful.

Identifying stranded animals is also difficult, especially when they are decomposing and external fea-

Figure 30.1 Indo-Pacific humpback dolphin, *Sousa chinensis*. (Photo: G. Parra.)

Figure 30.2 Australian Snubfin dolphin, *Orcaella heinsohni*. (Photo: G. Parra.)

tures are used. However, if the skull is prepared, identification is much more certain. Stranded marine mammals in the GBR region should be reported through the stranding hotline at 1300 130 372.

An untrained observer is unlikely to be able to identify other than the most common and distinctive species of dolphins and whales. Beaked whales are particularly challenging because many species are rare and there are few reference specimens. Some species have been observed only as dead and rotting carcasses, so that the real appearance of the living animal is not known. Certain features, if observed and recorded carefully, are most helpful in species identification, when one has access to reference books or the opinion of experienced observers.

The observations that are most likely to be helpful in identifying a marine mammal at sea are:

(1) length of the animal;

(2) colour, including especially colour pattern (if any) and other markings or scars;

(3) presence or absence of a dorsal fin;

(4) if present, the size, shape and position relative to the distance between snout and tail flukes of the dorsal fin;

(5) the shape of the head (e.g. broad or narrow, square, round, bulbous or flat, beaked or snub-nosed);

(6) shape and height of the 'blow' (the cloud of vapour from the blowhole as the animal breathes out);

(7) observed behaviour (such as frequency of surfacing, how much of the back is seen when the animal surfaces, leaping, spinning in the air, slapping the water surface);

(8) the estimated number and composition of a group (e.g. are adults, calves or juveniles present, are all about the same size?)

Figure 30.3 is a sighting sheet that will help identify the species of marine mammals most likely to be seen in the GBR region; the species listed are not exhaustive. There are several good identification guides that provide more detail (see Additional reading).

There is still much to learn about cetacean distribution and the reported distribution maps for some species are probably inaccurate. A single report, or a small number of reports, of stranded individuals does not necessarily reflect the normal distribution and range of a species.

Some of the more well known marine mammals of the GBR region are introduced briefly below.

Humpback whale

The humpback whales that are born in, and migrate along, waters off the east coast of Australia form part of the Group E breeding stock. These whales generally feed in Area V in Antarctica and migrate along the eastern Australian coast to the GBR to mate and give birth. This population was severely depleted to a few hundred individuals or less by commercial whaling opera-

Table 30.1 List of marine mammals that are known or considered likely to occur in the Great Barrier Reef region from GBRMPA (2000). Generic distributional data from Bryden *et al.* (1998) and Steve Van Dyck, Queensland Museum.

Scientific name	Common name	Relevant generic habitat requirements	Known habitats in the Great Barrier Reef region
Order Cetacea, Suborder Mysticeti, Baleen whales			
Balaenoptera acutorostrata	Dwarf minke whale*	Probably throughout region in winter	Swain Reefs to Cape Grenville; especially between Lizard Island and Ribbon Reef No. 10, between March and October, particularly June–July. See Fig. 30.3*C*
Balaenoptera bonearensis	Antarctic minke whale*	Possibly throughout region in winter	More likely in south of region; most northerly sighting inside Ribbon Reef No. 5 but much less common in north than dwarf minke.
Balaenoptera edeni	Bryde's whale**	Throughout region all year; may be seen close to coast	
Balaenoptera musculus	Blue whale*	Oceanic possibly throughout region in winter	See Fig. 30.3*A*
Balaenoptera physalus	Fin whale	Oceanic; possibly in southern parts of region in winter	See Fig. 30.3*A*
Megaptera novaeangliae	Humpback whale*	Coastal and island waters in winter and spring; breeding grounds between about 15° and 20°S	Especially Whitsunday and Mackay regions in winter; seen at northern end of the Great Barrier Reef (10°30'S) between October and January, possibly all year. See Fig. 30.3*A*
Order Cetacea, Suborder Odontoceti Toothed Whales and Dolphins			
Delphinus delphis	Short-beaked common dolphin**	Pelagic and neritic waters throughout region	See Fig. 30.3*D*
Feresa attenuata	Pygmy killer whale**	Possibly throughout region	See Fig. 30.3*B*
Globicephala macrorhynchus	Short-finned pilot whale*	Possibly throughout region in open ocean and continental shelf waters	See Fig. 30.3*C*
Grampus griseus	Risso's dolphin*	Possibly throughout region both inshore and offshore	
Indopacetus pacificus	Longman's beaked whale*	Possibly beyond continental shelf throughout region	Confirmed standing in Mackay region

(continued)

Table 30.1 *(continued)*

Scientific name	Common name	Relevant generic habitat requirements	Known habitats in the Great Barrier Reef region
Order Cetacea, Suborder Mysticeti, Baleen whales			
Balaenoptera acutorostrata	Dwarf minke whale*	Probably throughout region in winter	Swain Reefs to Cape Grenville; especially between Lizard Island and Ribbon Reef No. 10, between March and October, particularly June–July. See Fig. 30.3*C*
Balaenoptera bonearensis	Antarctic minke whale*	Possibly throughout region in winter	More likely in south of region; most northerly sighting inside Ribbon Reef No. 5 but much less common in north than dwarf minke.
Balaenoptera edeni	Bryde's whale**	Throughout region all year; may be seen close to coast	
Balaenoptera musculus	Blue whale*	Oceanic possibly throughout region in winter	See Fig. 30.3*A*
Balaenoptera physalus	Fin whale	Oceanic; possibly in southern parts of region in winter	See Fig. 30.3*A*
Megaptera novaeangliae	Humpback whale*	Coastal and island waters in winter and spring; breeding grounds between about 15° and 20°S	Especially Whitsunday and Mackay regions in winter; seen at northern end of the Great Barrier Reef (10°30′S) between October and January, possibly all year. See Fig. 30.3*A*
Order Cetacea, Suborder Odontoceti Toothed Whales and Dolphins			
Delphinus delphis	Short-beaked common dolphin**	Pelagic and neritic waters throughout region	See Fig. 30.3*D*
Feresa attenuata	Pygmy killer whale**	Possibly throughout region	See Fig. 30.3*B*
Globicephala macrorhynchus	Short-finned pilot whale*	Possibly throughout region in open ocean and continental shelf waters	See Fig. 30.3*C*
Grampus griseus	Risso's dolphin*	Possibly throughout region both inshore and offshore	
Indopacetus pacificus	Longman's beaked whale*	Possibly beyond continental shelf throughout region	Confirmed standing in Mackay region
Order Cetacea, Suborder Odontoceti Toothed Whales and Dolphins			
Kogia breviceps	Pygmy sperm whale**	Possibly throughout region in oceanic waters	

(continued)

Table 30.1 *(continued)*

Scientific name	Common name	Relevant generic habitat requirements	Known habitats in the Great Barrier Reef region
Stenella longirostris	Spinner dolphin*	Throughout region, primarily pelagic but nearshore in some regions particularly around islands	See Fig. 30.3*D*
Steno bredanensis	Rough-toothed dolphin	Possibly throughout region, usually far offshore	See Fig. 30.3*D*
Tursiops spp.	Bottlenose dolphin*	Widely distributed in both coastal (*T. aduncus*) and pelagic waters (generally *T. truncatus*)	Seen in coastal waters particularly near rocky headlands and offshore waters. See Fig. 30.3*D*
Ziphius cavirostris	Cuvier's beaked whale*	Possibly throughout region	
Order Sirenia, Family Dugongidae, sea cows			
Dugong dugon	Dugong*	Coastal and island waters throughout region, especially seagrass meadows	Especially in protected bays, offshore in northern GBR region in summer. See Fig. 30.3*B*

*confirmed from GBR region from strandings;
+confirmed for GBR region from sightings only;
**confirmed from Queensland south of GBR from strandings; Long-fined pilot whale *Globicephala melas* and Sei whale *Balaenoptera borealis* also confirmed from South-East Queensland from strandings.

tions in the 20th century, but has increased rapidly since then. The most recent population estimate (for 2004) is around 7000 animals with a long term rate of increase of between 10–11% annually.

Information about humpback whales on their breeding grounds in the GBR is limited and mostly predates the recent increase in population size. The areas of highest concentrations of humpback whales, and in particular mothers with calves, seem to be around the Whitsunday Islands and the Mackay region but animals with calves are seen at least as far north as Cairns. Whaling records suggest that some humpbacks breed in the reefs to the east of the Coral Sea such as the Chesterfields. Humpbacks have also been sighted in the northern end of the GBR (10°31′S) between October and January, after the end of the main north-south migration. The significance of such sightings is unknown.

Humpback whales are distinguished by their very long flippers that may extend up to a third of their body length. The head, jaws and flippers bear a series of protuberances, the dermal tubercles, which give these parts of the animal a knobbly appearance. The small dorsal fin varies in shape from falcate to slightly rounded. The head is rounder than in the other baleen whales occurring in the GBR region and the general body somewhat stouter. The roundish blow rises to 2–3 m. Humpbacks are often active at the surface and spectacular behaviours such as flipper and tail slaps, spy-hopping and breaching are common.

Dwarf minke whale

Although the dwarf minke whale is known only from the southern hemisphere, it seems more closely related to the northern hemisphere minke whale, *Balaenoptera acutorostrata*, than to the Antarctic minke whale, *Balaenoptera bonaerensis*. It is currently regarded as an un-named subspecies of *B. acutorostrata*, but may be a distinct species. Both Antarctic and dwarf minke whales are found in GBR waters but only one Antarctic minke has been observed over the continental shelf in the northern region. The two species can be distinguished by their size and colouration: (1) female dwarf minkes have a maximum size of about 7.8 m, on average about 2 m shorter than Antarctic minkes, and (2) dwarf minkes have a white shoulder blaze and flipper base, with a dark grey tip on the flipper, in contrast to Antarctic minkes that have a light to dark grey shoulder and a uniformly paler grey flipper.

In the GBR region, dwarf minkes have been recorded from north of Cape Grenville Island to the Swain reefs. This distribution may reflect the pattern of human use of the region rather than the actual distribution of the whales. Dwarf minke whales are seen in the northern GBR between March and November, with over 90% of sightings in June and July. The outer shelf region from Ribbon Reef 10 (near Lizard Island) south along the Ribbon Reefs to Agincourt Reef is the main focus for minke whale tourism. People are allowed to swim with minke whales from permitted vessels in the Great Barrier Reef Marine Park, but only if the whales initiate the encounter. Regulations govern people's behaviour during such encounters. The colour patterns of dwarf minke whales are the most complex of any baleen whale and are used to identify individuals by researchers, with particular whales being sighted repeatedly over periods of up to eight years.

Coastal dolphins

Several species of coastal dolphins occur in the GBR region. Bottlenose dolphins *Tursiops* spp. occur in both coastal and pelagic waters in the GBR region. The Indo-Pacific humpback dolphin, *Sousa chinensis* (Fig. 30.1), and the Australian snubfin dolphin, *Orcaella heinsohni,* occur in small populations mainly close to the coast and estuaries.

Recent morphological and genetic studies of the genus *Orcaella* have revealed that Australian snubfin dolphins populations (Fig. 30.2) are a separate species from the Asian *O. brevirostris*. The species level taxonomy of humpback dolphins is unresolved and the humpback dolphins that occur in northern Australia are likely to join the Australian snubfin dolphin as the only cetaceans endemic to Australian waters. Thus, both species have extremely high biodiversity value at a national and international level. However, comprehensive research on these species in Australia has only been undertaken in Queensland, particularly in Cleveland Bay near Townsville.

The species are best distinguished by: (1) the shape of their head: the Australian snubfin dolphin lacks the beak characteristic of bottlenose and humpback dolphins, and the humpback dolphin has a longer and more defined beak than the bottlenose dolphin; (2) the shape of the dorsal fin: high and hooked in bottlenose dolphins, low and triangular in the humpback dolphin and small and triangular in the snubfin dolphin; (3) location: snubfin and humpback dolphins are likely to be in waters less than 10 m deep and up to 6 km offshore (sightings of humpback dolphins up to some 50+ km from the coast have been recorded in the northern GBR region, probably due to the physiography of the coastlines and continental shelves in this area), bottlenose dolphins occur in more open water or close to rocky headlands, (4) colouration: snubfin dolphins vary from different tones of pale grey to brownish grey; humpback dolphins are uniformly grey, with flanks shading to off-white and spotting towards the ventral surface; in some animals the dorsal fin, rostrum and melon whiten with age, while the rest of the dorsal surface remains pale grey; bottlenose dolphins are mainly dark grey (but mature individual of the *aduncus* form is spotted ventrally), and (5) school size: snubfin dolphins are mainly found in schools of 5–8 individuals (schools of one to 21 individuals have been observed in the wild), humpback dolphins form smaller schools of usually 2–3 individuals (school size ranges from one to 12 individuals); Bottlenose dolphins occur in schools of various sizes ranging from single animals to several individuals (>20).

Dugong

The dugong (*Dugong dugon*) looks rather like a cross between a rotund dolphin and a walrus. Its body, flippers and fluke resemble those of a dolphin without a dorsal fin. Its head looks somewhat like that of a wal-

A

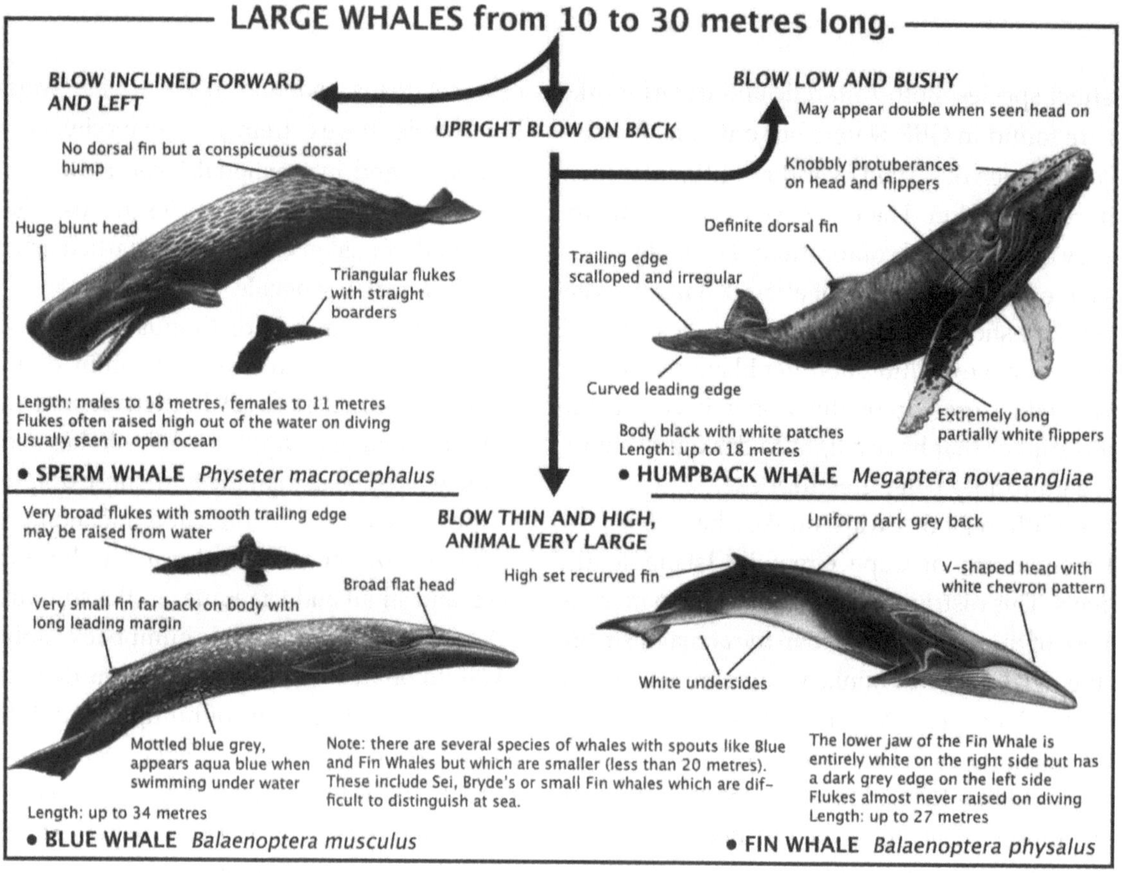
LARGE WHALES from 10 to 30 metres long.
BLOW INCLINED FORWARD AND LEFT
UPRIGHT BLOW ON BACK
BLOW LOW AND BUSHY
May appear double when seen head on
No dorsal fin but a conspicuous dorsal hump
Huge blunt head
Triangular flukes with straight boarders
Length: males to 18 metres, females to 11 metres
Flukes often raised high out of the water on diving
Usually seen in open ocean
• SPERM WHALE Physeter macrocephalus
Knobbly protuberances on head and flippers
Definite dorsal fin
Trailing edge scalloped and irregular
Curved leading edge
Extremely long partially white flippers
Body black with white patches
Length: up to 18 metres
• HUMPBACK WHALE Megaptera novaeangliae
Very broad flukes with smooth trailing edge may be raised from water
BLOW THIN AND HIGH, ANIMAL VERY LARGE
Broad flat head
Very small fin far back on body with long leading margin
Mottled blue grey, appears aqua blue when swimming under water
Length: up to 34 metres
• BLUE WHALE Balaenoptera musculus
Note: there are several species of whales with spouts like Blue and Fin Whales but which are smaller (less than 20 metres). These include Sei, Bryde's or small Fin whales which are difficult to distinguish at sea.
Uniform dark grey back
High set recurved fin
V-shaped head with white chevron pattern
White undersides
The lower jaw of the Fin Whale is entirely white on the right side but has a dark grey edge on the left side
Flukes almost never raised on diving
Length: up to 27 metres
• FIN WHALE Balaenoptera physalus

B

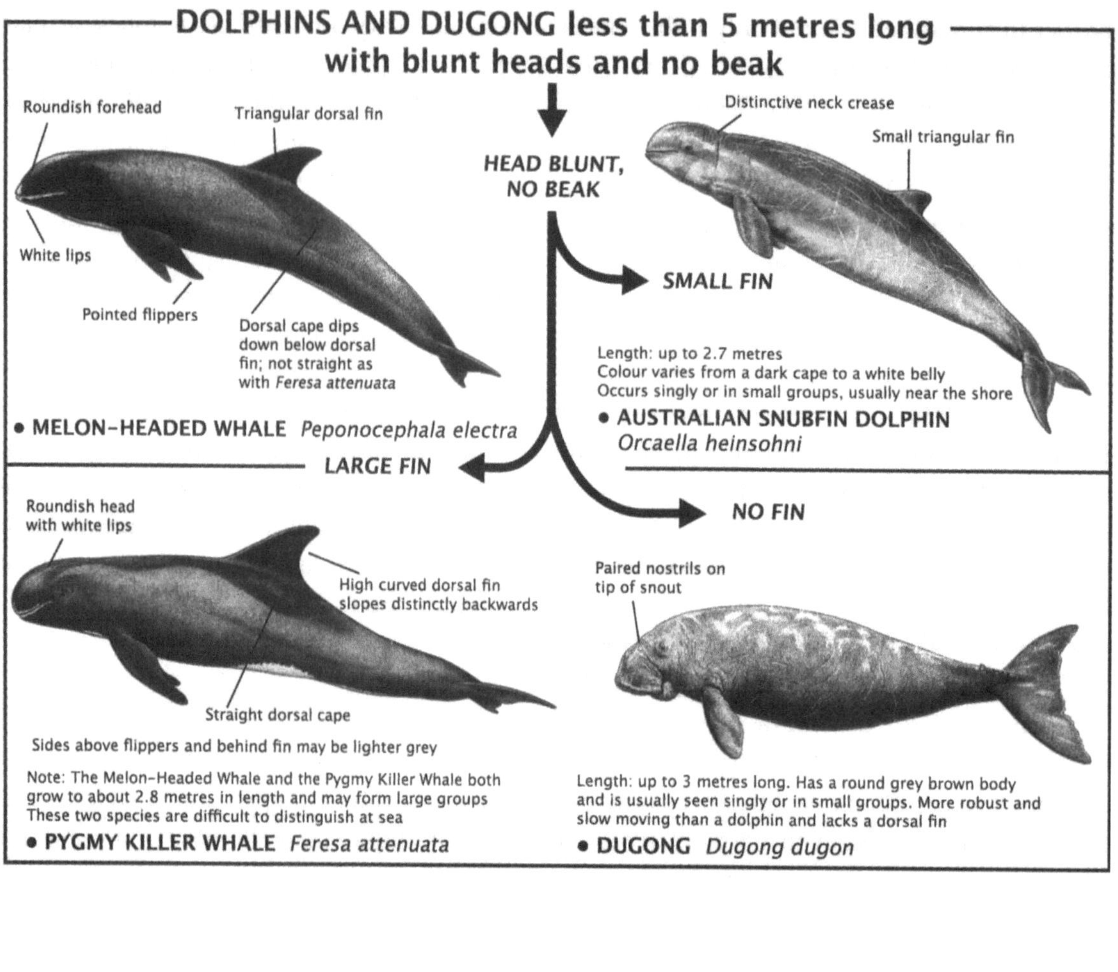
DOLPHINS AND DUGONG less than 5 metres long with blunt heads and no beak
HEAD BLUNT, NO BEAK
Roundish forehead
Triangular dorsal fin
White lips
Pointed flippers
Dorsal cape dips down below dorsal fin; not straight as with Feresa attenuata
• MELON-HEADED WHALE Peponocephala electra
Distinctive neck crease
Small triangular fin
SMALL FIN
Length: up to 2.7 metres
Colour varies from a dark cape to a white belly
Occurs singly or in small groups, usually near the shore
• AUSTRALIAN SNUBFIN DOLPHIN Orcaella heinsohni
LARGE FIN
NO FIN
Roundish head with white lips
High curved dorsal fin slopes distinctly backwards
Straight dorsal cape
Sides above flippers and behind fin may be lighter grey
Note: The Melon-Headed Whale and the Pygmy Killer Whale both grow to about 2.8 metres in length and may form large groups
These two species are difficult to distinguish at sea
• PYGMY KILLER WHALE Feresa attenuata
Paired nostrils on tip of snout
Length: up to 3 metres long. Has a round grey brown body and is usually seen singly or in small groups. More robust and slow moving than a dolphin and lacks a dorsal fin
• DUGONG Dugong dugon

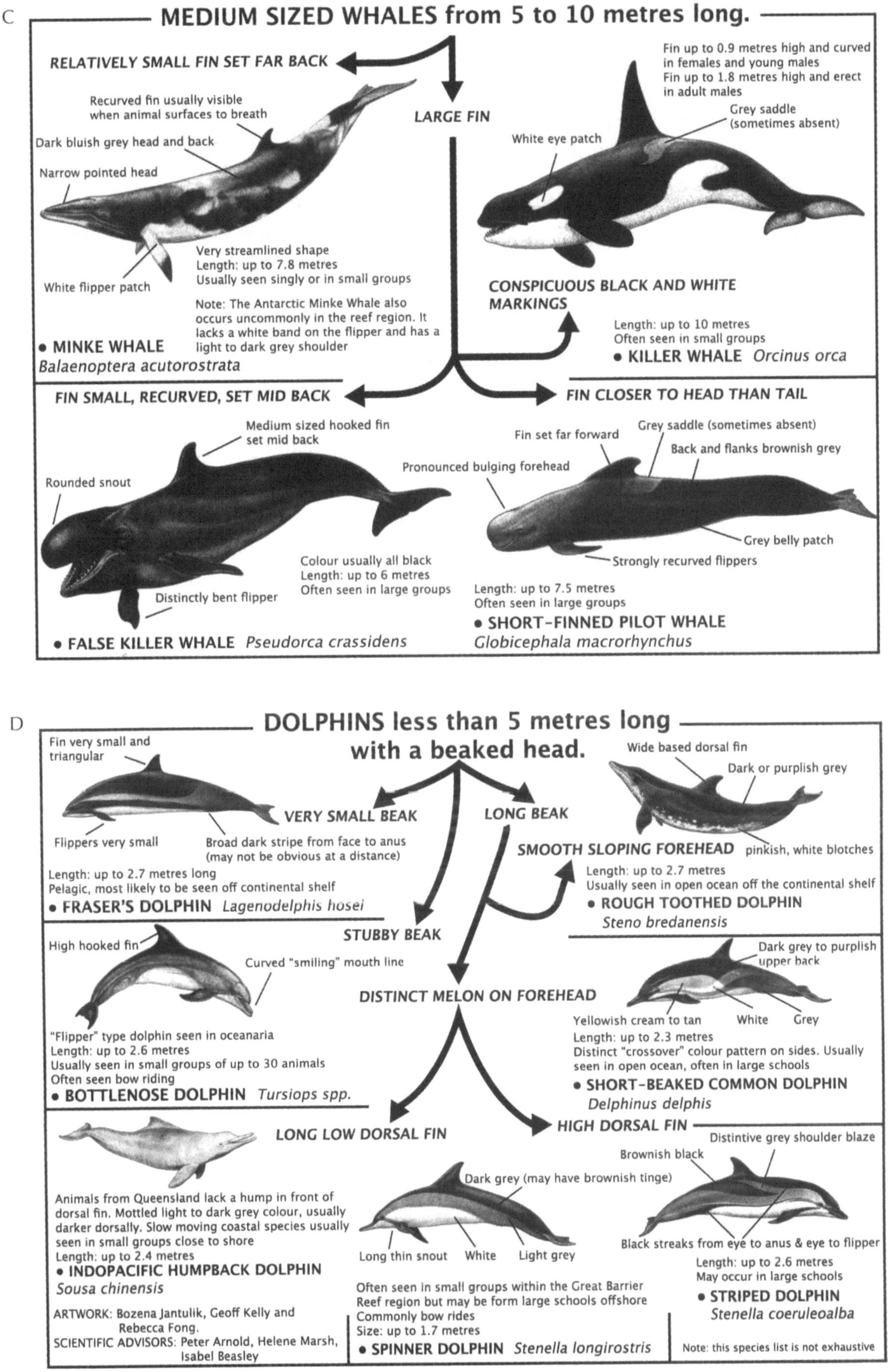

Figure 30.3
A–D, Identification guide to the marine mammals most likely to the seen in the Great Barrier Reef region. Not all animals from the region are illustrated. Table 30.1 provides a more complete list. (Figure: Bozena Jantulik, Geoff Kelly and Rebecca Fong.)

rus without the long tusks. Growing to a length of up to about 3 m, the dugong is the only extant plant-eating mammal that spends all its life in the sea. Dugongs can be difficult to distinguish from Australian snubfin dolphins in the wild, especially as both species often occur in inshore turbid waters. Dugongs surface very discreetly, often with only their nostrils showing above the water. Dugongs tend to move more slowly than dolphins and the lack of a dorsal fin is their most distinguishing characteristic for observers at sea.

Adults are grey in colour but may appear brown from the air or from a boat. Older 'scarback' individuals may have a large area of unpigmented skin on the back above the pectoral fins. The dugong's head is distinctive with the mouth opening ventrally beneath a broad, flat muzzle. The tusks of mature males and some old females erupt on either side of the head. There are two mammary glands, each opening via a single teat situated in the 'armpit' or axilla. The mammaries are somewhat reminiscent of the breasts of human females, which probably explains the legendary links between mermaids and sirenians. The tail of the dugong is triangular like that of a whale.

The dugong mainly occurs in the coastal waters of the GBR lagoon where its distribution is broadly coincident with that of its seagrass food. It is seen up to about 100 km offshore inside the reef in the northern GBR region in the summer.

ADDITIONAL READING

Baker, A. N. (1999). 'Whales and Dolphins of New Zealand and Australia: An Identification Guide. (Wellington: Victoria University Press.)

Bryden, M. M., Marsh, H., and Shaughnessy, P. (1998). 'Dugongs, Whales, Dolphins and Seals – A Guide to the Sea Mammals of Australia.' (Allen and Unwin: St Leonards, Australia.)

GBRMPA (2000). Whale and dolphin conservation in the Great Barrier Reef Marine Park: policy document. (GBRMPA: Townsville.) Available at http://www.gbrmpa.gov.au/corp_site/info_services/publications/misc_pub/whale_dolphin [Verified 17 March 2008].

Geraci, J. R., and Lounsbury, V. J. (2005). 'Marine Mammals Ashore: A Field Guide for Strandings.' 2nd edn. (National Aquarium in Baltimore: Baltimore, MD.)

Perrin, W, Wursig, B., and Thewissen, J. (Eds.) (2002). 'Encyclopedia of Marine Mammals.' (Academic Press: California.)

31. Seabirds

B. C. Congdon

GENERAL LIFE-HISTORY CHARACTERISTICS

Seabirds are highly visible, charismatic predators in marine ecosystems that feed primarily or exclusively at sea. They have a range of relatively unique life-history characteristics associated with this predominantly marine lifestyle. Many of these characteristics are directly linked with having to forage over large distances to obtain sufficient food to breed.

In general, seabirds are long-lived species that have deferred sexual maturity, small clutch sizes, slow chick growth rates and extended fledgling periods. For example, crested terns (*Sterna bergii*, Fig. 31.1*A*) may live 18–20 years, sooty terns (*S. fuscata*, Fig. 31.1*B*) up to 32 years and larger species such as boobies and frigatebirds (Fig. 31.1*C*) even longer. Most seabird species do not become sexually mature or return to breed for between 5 to 12 years after fledging and the nestling period for some species, such as frigatebirds, can be up to six months long.

When nesting, care by both parents is usually required to successfully rear chicks. Following hatching, adults take turns brooding chicks until they are able to regulate their own body temperature. After this, chicks are left alone at the colony while both parents forage for food. At this stage the chicks of some ground-nesting tern species are mobile and gather together in crèches for protection from predators (Fig. 31.1*D*). Chicks of other species usually remain at the nest site.

Chicks of many seabirds also accumulate extensive bodyfat reserves to buffer themselves against long periods between adult feeding visits. Greatest mortality occurs during early life-history stages, such as eggs, nestlings and in the postfledging and prereproductive periods. Predation levels in seabird colonies are often significant and any disturbance by humans usually leads to increased levels of predation on exposed eggs or chicks.

Most seabirds breed on relatively remote islands, different species tending to nest in specific habitat types. Seabirds of the Great Barrier Reef (GBR) prefer coral cays to other island types. Most nest in dense colonies on the ground using either bare scrape nests or nests built among low lying vegetation. Other species use rocky crevices or burrows dug among vegetation dense enough to stabilise the soil structure and still others nest in trees on platform nests built of leaves or sticks (Table 31.1).

DISTRIBUTION AND ABUNDANCE ON THE GBR

Australia's GBR supports a significant and diverse seabird community. Twenty different species (Table 31.1) comprising 1.7 millon individuals and more than 25% of Australia's tropical seabirds nest within the GBR region. This includes greater than 50% of Australia's roseate terns (*S. dougallii*), lesser-crested terns (*S. bengalensis*,

Figure 31.1 *A*, crested tern; *B*, sooty tern; *C*, least frigatebird; *D*, bridled tern; *E*, lesser crested tern; *F*, black-naped tern; *G*, black noddy; *H*, wedge-tailed shearwater; *I*, brown booby; *J*, masked booby; *K*, common noddy. (Photos: C. Erwin, B. Congdon, D. Peck.)

Table 31.1 Seabird species/approximate population estimates for the GBR only/nesting habitat/foraging guild

Species		GBR population	Nesting habitat	Foraging guild
Herald petrel	*Pterodroma arminjoniana*	3	Ground: scrape	Pelagic
Wedge-tailed shearwater	*Puffinus pacificus*	560 000	Burrow	Pelagic
Australian pelican	*Pelicanus conspicillatus*	1024	Ground: scrape	Coastal
Red-footed booby	*Sula sula*	172	Trees/shrubs/ground	Pelagic
Brown booby	*Sula leucogaster*	18 500	Ground: low grass/scrape	Offshore
Masked booby	*Sula dactylatra*	1100	Ground: scrape	Pelagic
Great frigatebird	*Fregata minor*	20	Trees/ground	Pelagic
Least frigatebird	*Fregata ariel*	2500	Shrubs/ground	Pelagic
Red-tailed tropicbird	*Paethon rubricauda*	100	Holes/crevices/tree roots	Pelagic
Caspian tern	*Sterna caspia*	70	Ground: low grass/scrape	Coastal
Roseate tern	*Sterna dougallii*	6000	Ground: scrape	Inshore
Black-naped tern	*Sterna sumatrana*	3900	Ground: scrape	Inshore
Little tern	*Sterna albifrons*	51 (possibly up to 1000)	Ground: scrape	Coastal
Sooty tern	*Sterna fuscata*	48 000	Ground: low grass/scrape	Pelagic
Bridled tern	*Sterna anaethetus*	13 900	Ground: low grass/scrape	Offshore
Crested tern	*Sterna bergii*	26 000	Ground: scrape	Inshore
Lesser-crested tern	*Sterna bengalensis*	6300	Ground: scrape	Inshore
Common noddy	*Anous stolidus*	46 000	Ground: low grass/scrape	Offshore
Black noddy	*Anous minutus*	300 000	Trees	Offshore
Silver Gull	*Larus novaehollandiae*	750	Ground: low grass/scrape	Coastal

Fig. 31.1*E*), black-naped terns (*S. sumatrana*, Fig. 31.1*F*), and black noddies (*Anous minutus*, Fig. 31.1*G*), and about 25% of the wedge-tailed shearwater (*Puffinus pacificus*, Fig. 31.1*H*), brown booby (*Sula leucogaster*, Fig. 31.1*I*), masked booby (*S. dactylatra*, Fig. 31.1*J*) and red-tailed tropicbirds (*Phaeton rubricauda*).

In general, seabirds are colonial nesters. Throughout the GBR there are 56 major and 20 minor seabird colonies (Fig. 31.2). Most major colonies are located in either the far northern, northern or far southern regions of the GBR. Raine Island in the far northern sector is one of the largest and most significant tropical seabird breeding sites in Australia. Of the 24 seabird species recorded as breeding in Queensland, 14 breed at Raine Island. A comprehensive review of the trends in seabird numbers at this site since the beginning of last century has recently been undertaken (see Additional reading).

Michaelmas Cay in the northern section of the GBR is a tropical seabird colony rated as the second most important bird nesting site in the GBR. The island constitutes a major nesting site for sooty terns and common noddies (*A. stolidus*, Fig. 31.1*K*), as well as for

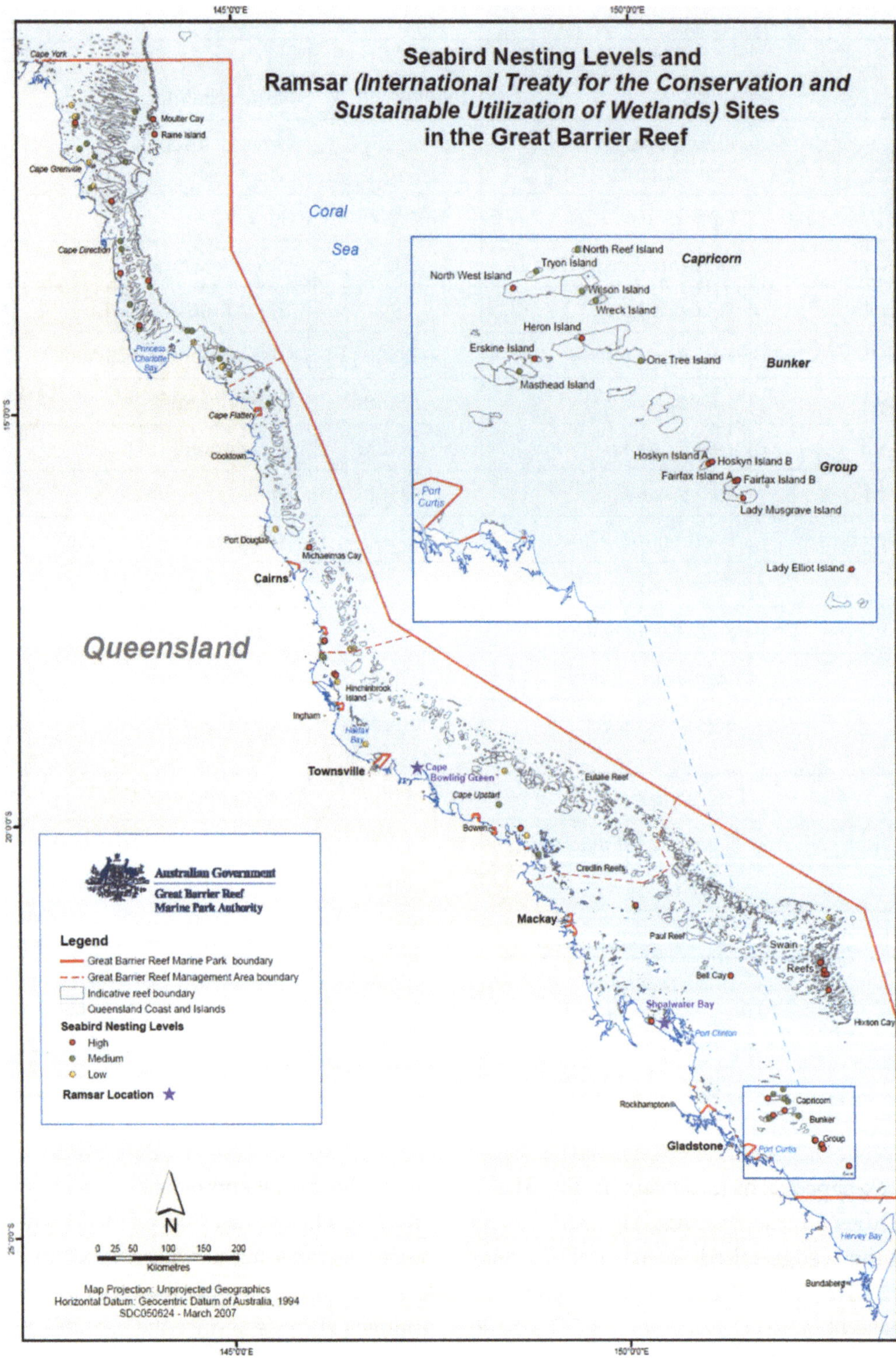

Figure 31.2 Seabird colony distribution on the GBR. (Map: Spatial Data Centre, GBRMPA.)

crested and lesser-crested terns. The islands of the Swain reefs in the far south-eastern region of the GBR also constitute one of six core seabird breeding areas.

Finally, the Capricorn-Bunker group of islands in the far southern GBR contains nationally and internationally significant seabird breeding populations. This island group supports the Pacific Ocean's largest breeding colony of wedge-tailed shearwaters and over 97% of the black noddy populations of the GBR. The Capricornia Cays also contain ~75% of the seabird biomass of the GBR World Heritage Area.

Many seabird species that breed within the GBR and in adjacent areas are considered migratory and/or threatened species and are listed under the Australian Environmental Protection and Biodiversity Conservation (EPBC) Act 1999 in a variety of categories. Many are also variously protected under international agreements such as: the China-Australia Migratory Bird Agreement (CAMBA), Japan-Australia Migratory Bird Agreement (JAMBA) and the Convention on the Conservation of Migratory Species of Wild Animals (Bonn Convention 1979). Additionally, the GBR region hosts migrating populations of some northern hemisphere breeding species such as common tern (*Sterna hirundo*) and much of the Asian population of roseate tern.

Foraging modes

Seabirds of the GBR can be classified into four major foraging guilds; coastal, inshore, offshore and open-ocean or pelagic. Species in each of these guilds travel different distances to access food and are away from young for different periods of time. Offshore and pelagic species may travel very long distances to obtain food (see Box 31.1) and because of this usually regurgitate food for young. Inshore foragers travel much shorter distances, forage in and around reefs adjacent to colonies and carry whole fish in their bill to feed young. Foraging mode also affects both the number of eggs laid and chick growth rates, with more pelagic foraging species having single egg clutches and much slower growing chicks.

Inshore, offshore and pelagic foragers show consistent differences in the size of breeding colonies. Inshore foragers usually have small, more densely packed colonies of up to a few thousand birds, while pelagic species can have enormous colonies spread over multiple closely spaced islands. For example, ~500 000 wedge-tailed shearwaters breed in the 13 islands of the Capricorn and Bunker groups.

Such high population abundance means that where large seabird colonies occur the quantity of marine resources consumed by seabirds is significant. Such high consumption rates also mean that seabirds play a number of important functional roles in marine ecosystems. Most seabird prey items are either pelagic fish or squid. Therefore, seabirds play a major role in transferring nutrients from offshore and pelagic areas to islands and reefs. For example, black noddies deposit an estimated 4 tonnes of guano on Heron Island per annum. Some species of seabirds also play an important role in seed dispersal and in distributing organic matter into lower parts of the developing soil profile (e.g. burrow-nesting species such as shearwaters).

Seabirds can forage alone, but most forage in large mixed species aggregations or flocks. The likely reason for this is that flocking offers advantages in both finding and accessing prey. Different species forage in different ways. Species that capture prey within a metre or so of the ocean surface usually just dip their beaks underwater either after landing, or while still in the air. Other species that forage to depths of up to 20 metres may plunge dive at speed using just the momentum of the dive to carry them to prey, or they may extend the pursuit of prey underwater using slight movements of their wings. A third group more actively pursues prey by swimming underwater using their wings or feet for propulsion. Regardless of foraging mode most seabirds are also heavily reliant on the activity of subsurface predators to drive prey within reach. For this reason the presence and activity of large predatory fishes such as tuna and mackerel, as well as other subsurface predators, are extremely important to seabirds. This is particularly so in the relatively resource poor foraging environment of the tropics.

Timing of breeding

The timing of seabird breeding activity on the GBR is complex and can vary with location, year and/or species. In general, both northern and southern colonies

BOX 31.1 CRITICAL FORAGING LOCATIONS FOR WEDGE-TAILED SHEARWATERS

Recently it has been shown that adult shearwaters in the southern GBR cannot simultaneously replenish their own body reserves and maintain chick development using only prey caught near breeding colonies. Because of this adults use a specialised dual-foraging strategy where they alternate multiple short foraging trips in resource-poor, near-colony waters with long trips to highly productive areas at large distances from colonies. During near-colony trips adults sacrifice their own body condition to satisfy chick energy requirements. Then at some critical body mass they perform an extended trip to regain body condition. This foraging strategy is well known in southern ocean tube-nosed seabirds but has not previously been seen in the tropics.

Using satellite transmitters, shearwaters from Heron Island have been observed travelling over 2000 km on a single 'at-distance' foraging trip to visit multiple seamounts in the Coral Sea. Still others have been observed foraging off the northern coast of New South Wales. At these locations extensive upwelling of nutrients supports very high levels of prey abundance. In both areas shearwaters were observed foraging along the edges of these upwelling zones. The conservation significance of such key foraging refuges cannot be overstated. The breeding success of shearwaters and probably other seabird species on a regional scale is likely to be totally dependent on the continued stability of upwelling at these sites. How climate change will affect productivity at these locations is currently unknown.

tend to have specific seasonal peaks in activity. Most breeding peaks occur over summer from October through to April each year, but for some species and especially in northern colonies breeding activity can occur at all times of the year. Different species may also use the same areas at different times of year. For example, masked booby breeding in the Swains Reefs peaks in winter, while brown booby breeding at the same locations peaks during the austral summer.

Threats to seabirds

There are a range of identified anthropogenic threats to the long term population viability of seabirds of the GBR. By far the biggest contemporary threat is climate change and its associated impacts on food supply, breeding habitat, and so on. Seabirds' total dependence on marine resources makes them key upper trophic level predators in marine systems. This means that seabird demographics and reproductive parameters are strongly impacted by, and closely reflect, changing food availability and oceanographic conditions. For this reason they are widely considered important indicator species in marine ecosystems. Therefore, understanding how seabird population dynamics and reproductive ecology are impacted by changing oceanography leads directly to important insights into the potential future impacts of climate change; not only on seabirds, but also on other functionally important components of reef ecosystems in general (see Box 31.2).

Other important identified threats to seabirds include, but are not limited to; commercial fishing, direct disturbance by visitors to islands, breeding habitat destruction, the introduction of exotic plants and animals, as well as pollution and water quality degradation along with their associated trophic disturbance.

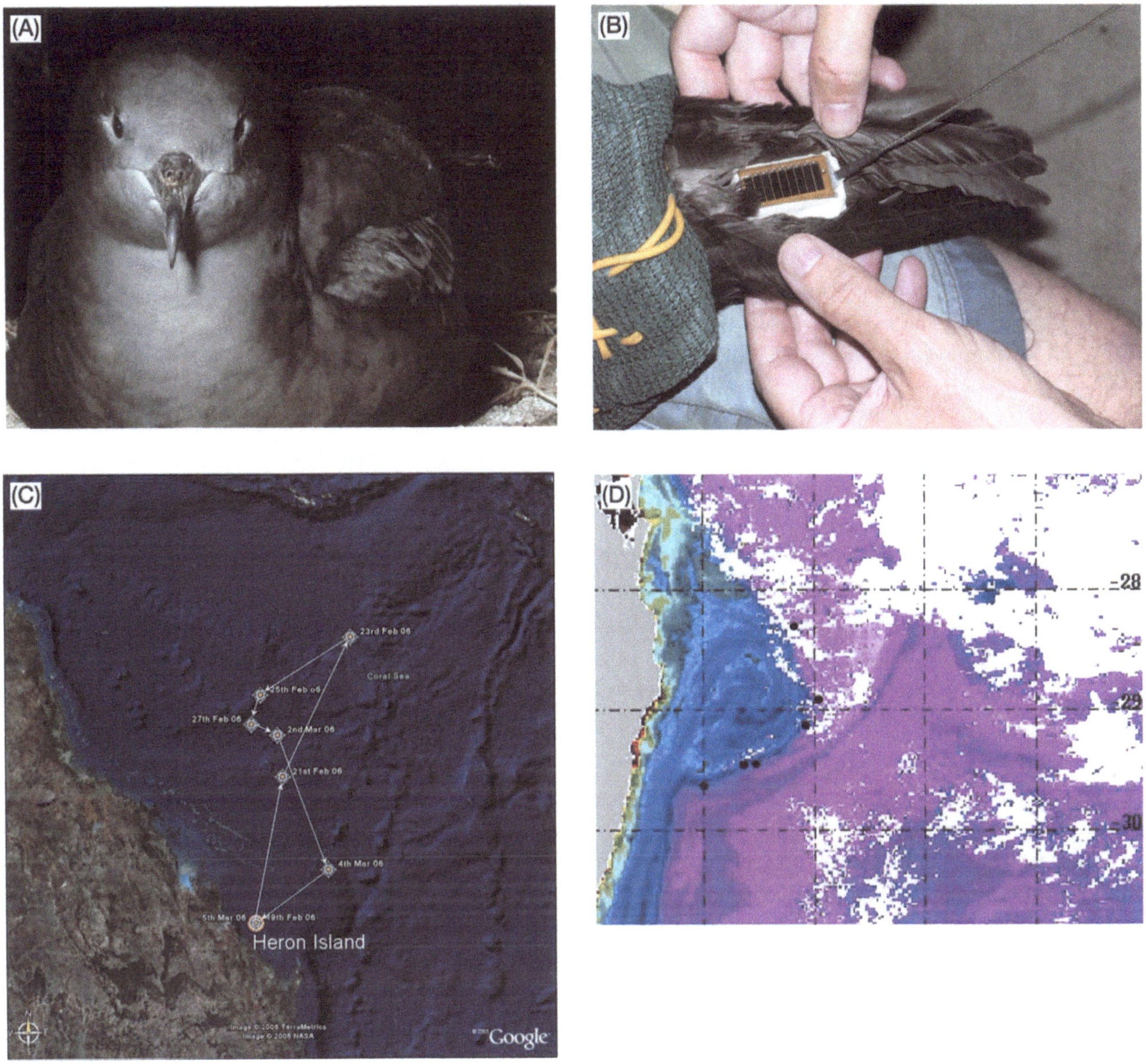

Figure 31.3 *A*, wedge-tailed shearwater; *B*, PTT-satellite transmitter; *C*, PTT foraging track for a single wedge-tailed shearwater adult during February–March 2006; *D*, wedge-tailed shearwater foraging positions (•) along the edge of an oceanographic frontal system of the northern NSW coast. (Photos: B. Congdon, S. Weeks.)

■ NON-MARINE BIRDS

Many other terrestrial, coastal and migratory bird species also use the continental islands and cays of the GBR. The numbers and types are too extensive to detail here, but these bird communities often resemble those occurring in similar habitats on the mainland. Some of the more common non-migratory species include the buff-banded rail (*Gallirallus philippensis*), Australian pelican (*Pelecanus conspicillatus*), eastern reef egret (*Egretta sacra*), pied (Torresian) imperial pigeon (*Ducula bicolor*), sooty oyster catcher (*Haematopus fuliginosus*) and beach stone-curlew (*Esacus neglectus*). Three species of fish-eating raptors also inhabit islands in the GBR, the osprey (*Pandion haliaetus*), brahminy kite (*Haliastur indus*) and white-bellied sea eagle (*Haliaeetus leucogaster*).

BOX 31.2 SEABIRDS AND CLIMATE CHANGE ON THE GBR

Climate change is predicted to raise sea-surface temperatures (SST) and increase the frequency and intensity of El Niño events. Recently it has been shown that for many seabird species breeding on the GBR the amount of food adults can gather during a single foraging trip is directly related to the SST over the same period. During periods of increased SST food availability to these species declines dramatically, decreasing meal sizes, feeding frequencies and rates of chick growth.

Total food availability to these same species decreases even further during intense El Niño events. For example, throughout the 2002 El Niño/coral bleaching episode in the southern GBR (see Chapter 10), food availability to shearwaters breeding on Heron Island dropped to one third normal levels. Adults were required to forage three to four times longer for equivalent meal sizes. This lack of food resulted in the complete reproductive failure of the Heron Island colony with almost 100% mortality of chicks and the loss of at least 2000 adults. It is likely that similar reproductive failures occurred throughout the Capricorn-Bunker island group.

Such findings demonstrate that predicted increases in both SST and the intensity or frequency of El Niño events will have serious detrimental impacts on tropical seabird populations throughout the GBR and in adjacent areas. Importantly, current evidence also suggests that at least some species, and possibly the majority of species, at all significant GBR breeding colonies are already in decline due to climate change related phenomena and/or show no recovery from recent ENSO impacts.

The eastern reef egret or eastern reef heron is common throughout the GBR and adjacent mainland. The species has an unusual, non-sexual colour dimorphism, with different individuals having either entirely white or slate grey plumage. The dark form is more common in temperate areas while the white form is more abundant in the tropics. The reason for the colour dimorphism is unknown. The two colour forms randomly interbreed and both white and dark offspring can be found in the same brood. When nesting on coral cays, reef egrets hunt both day and night at low tide using small territories they establish and maintain on the reef flat.

The buff-banded rail is widespread and secretive on the Australian mainland, but is also a highly dispersive and an obvious reef island coloniser, reaching many of the heavily vegetated cays. It is an omnivorous scavenger that feeds on small vertebrates and invertebrates as well as seeds, fallen fruit and other vegetable matter that it collects from within a defended territory. The species is highly susceptible to the impacts of introduced predators or competitors and often disappears from islands where these threats occur.

The pied imperial pigeon is one of the most numerous terrestrial birds on the GBR. It is a migratory species that visits coastal reef islands and the adjacent mainland south to about Mackay during summer. It then returns to Papua New Guinea during the winter. Greatest numbers of pigeons occur between Cooktown and Cape York with some colonies in this area reaching greater than ten thousand pairs.

Other less numerous but important terrestrial species on the GBR includes the capricorn white-eye (*Zosterops lateralis chlorocephalus*) and the yellow chat (*Epthianura crocea macgregori*). The Capricorn white-eye is the only endemic bird found within the GBR region and is restricted to the wooded cays of the southern Capricorn-Bunker region. Recent research suggests that it may be a distinct species. The yellow chat is the most vulnerable terrestrial bird species

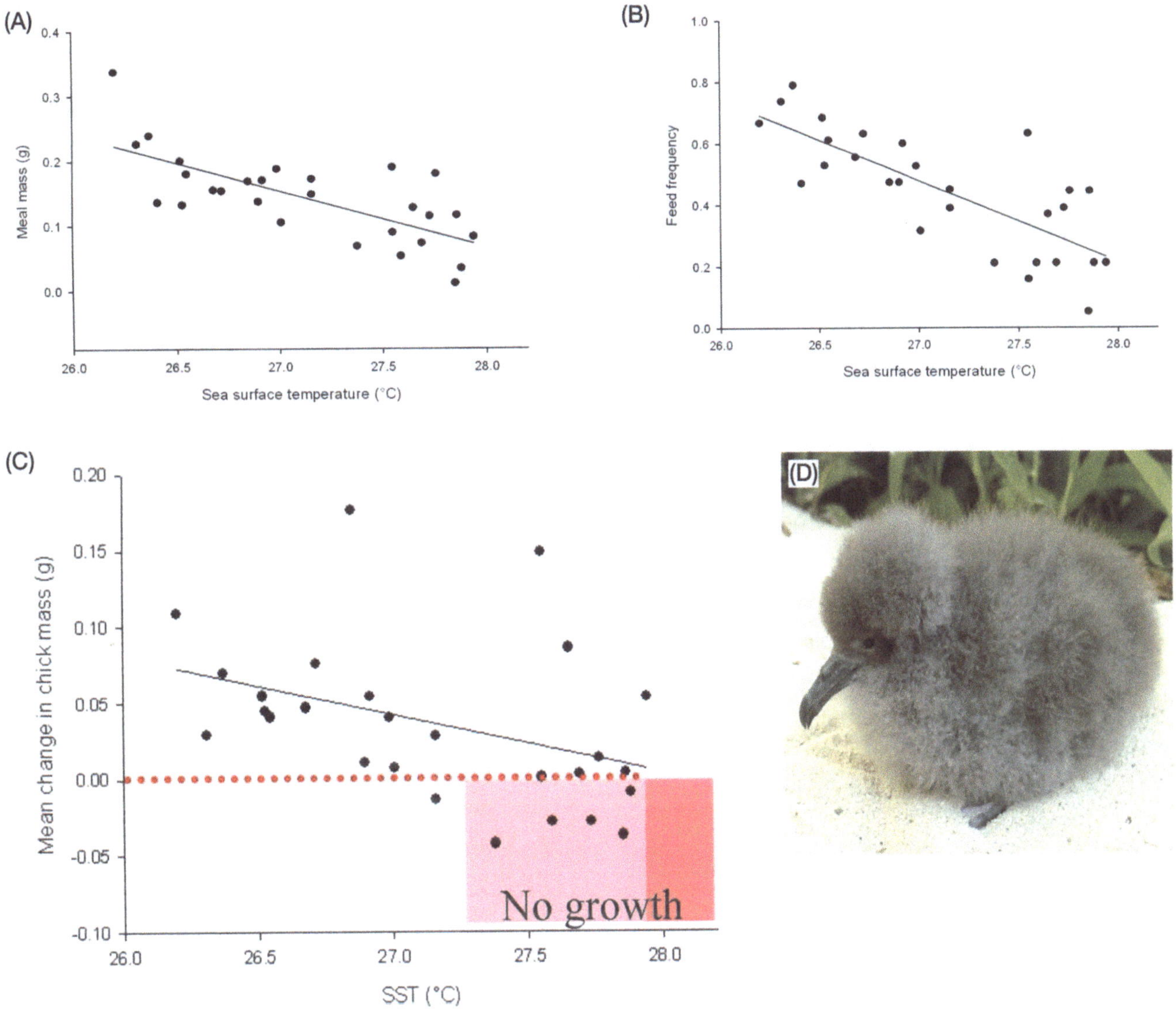

Figure 31.4 Relationships between day-to-day variation in sea surface temperatures and (*A*) meal size provided to chicks, (*B*) frequency of feeding, and (*C*) chick growth rates, for wedge-tailed shearwaters breeding on Heron I. February 2003. *D*, wedge-tailed shearwater chick. (Photos: D. Peck.)

breeding in the GBR. The Australian east coast population was believed to be extinct until a small number were discovered on a southern GBR island in the 1990's. The species currently has an extremely restricted coastal distribution and total population numbers remain low. It is listed as 'critically endangered'.

Large numbers of migratory wader species also spend their non-breeding season among the islands of the GBR. These birds breed in northern Asia and Alaska and migrate thousands of kilometres along the East Asian–Australasian Flyway to non-breeding sites in Australia and New Zealand. They are most commonly found in coastal estuaries and tidal mudflats but many species also form large groupings on coral cays and other islands. Common species include: greater (*Charadrius leschenaultii*) and lesser (*Charadrius mongolus*) sand plovers, turnstones (*Arenaria interpres*), golden plover (*Pluvialis apricaria*), grey-tailed tattler (*Tringa brevipes*), red-necked stint (*Calidris ruficollis*) and bar-tailed godwit (*Limosa lapponica*).

ADDITIONAL READING

Batianoff, G. N., and Cornelius, N. J. (2005). Birds of Raine Island: population trends, breeding behaviour and nesting habitats. *Proceedings of the Royal Society of Queensland* **112**, 1–29.

Congdon, B. C., Krockenberger, A. K., and Smithers, B. V. (2005). Dual-foraging and co-ordinated provisioning in a tropical Procellariiform, the wedge-tailed shearwater. *Marine Ecology Progress Series* **301**, 293–301.

Dyer, P. K., O'Neill, P., and Hulsman, K. (2005). Breeding numbers and population trends of Wedge-tailed Shearwater (*Puffinus pacificus*) and Black Noddy (*Anous minutus*) in the Capricornia Cays, southern Great Barrier Reef. *Emu* **105**, 249–257.

Erwin, C. A., and Congdon, B. C. (2007). Day-to-day variation in sea-surface temperature negatively impacts sooty tern (*Sterna fuscata*) foraging success on the Great Barrier Reef, Australia. *Marine Ecology Progress Series* **331**, 255–266.

Higgins, P. J., and Davies, S. J. J. F. (1996). 'Handbook of Australian New Zealand and Antarctic Birds. Vol. 3.' (Oxford University Press: Melbourne.)

Hulsman, K., O'Neill, P., and Stokes, T. (1997). Current status and trends of seabirds on the Great Barrier Reef. In 'State of the Great Barrier Reef World Heritage Area Workshop, Townsville, Queensland, Australia, 27–29 November 1995'. (Eds. D. Wachenfeld, J. Oliver and K. Davis.) pp. 259–282. Available at http://www.gbrmpa.gov.au/corp_site/info_services/publications/workshop_series [Verified 21 March 2008].

O'Neill, P., Heatwole, H., Preker, M., and Jones M. (1996). Populations, movements and site fidelity of brown and masked boobies on the Swain Reefs, Great Barrier Reef, as shown by banding recoveries. CRC Reef Research Centre, Technical Report, **11**, Townsville.

O'Neill, P., Minton, C., Ozaki, K., and White, R. (2005). Three populations of non-breeding Roseate Terns (*Sterna dougallii*) in the Swain Reefs, southern Great Barrier Reef. *Emu* **105**, 57–76.

Peck, D. R., Smithers, B. V., Krockenberger, A. K., and Congdon, B. C. (2004). Sea surface temperature constrains wedge-tailed shearwater foraging success within breeding seasons. *Marine Ecology Progress Series* **281**, 259–266.

Turner, M., Green, R., and Chin, A. (2006). Birds. In 'The State of the Great Barrier Reef On-line'. (Ed. A. Chin.) Great Barrier Reef Marine Park Authority, Townsville. Available at www.gbrmpa.gov.au/corp_site/info_services/publications/sotr/latest_updates/birds

Epilogue

P. A. Hutchings, M. J. Kingsford & O. Hoegh-Guldberg

The Great Barrier Reef has been an important part of Australia's identity and has become an international icon. It is, therefore, not surprising that Australia has the oldest coral reef society in the world, namely the Australian Coral Reef Society founded in 1922. This Society began as a Great Barrier Reef Committee. This committee instigated the first major expedition on the GBR, 1927–1928, which was the beginnings of the quest to enhance our knowledge of, and hence better manage the Great Barrier Reef. Eighty years later, the society has again supported this effort and has sponsored this book. It is also significant that most of the contributing experts have been, or are, members. As a result, all proceeds from the sale of this book go to the ACRS for the promotion and support of further efforts to understand and manage coral reefs not only in Australia but worldwide, especially through postgraduate students. The details for how to join the ACRS can be found at its website available at http://www.australiancoralreefsociety.org/ [Verified 9 March 2008].

There are many people who have contributed to this book by providing text or images. The Australian Museum, James Cook University and the University of Queensland have supported the editors in undertaking the task of bringing the book to completion and support research programs on the GBR. The Great Barrier Reef Foundation and its sponsors have also provided invaluable financial support and assistance to the courses that inspired this book. Over 50 experts provided their time for free to compile a fascinating series of chapters—the editors are enormously grateful for their contributions. We also thank Kate Loynes and CSIRO Publishing for their attention to detail in the final stages. It is our hope that the book will be a useful starting point to any study of coral reefs both in Australia and abroad.

Coral reefs are one of the natural wonders of the world. The Great Barrier Reef illustrates the intricacy of nature and its interactions. Only by understanding these interactions, do we have a chance of meeting the challenges of the future, most of which are rising from the extraordinary growth in human populations along tropical and subtropical coastlines. Adding to this, we now have fundamental changes arising from global warming and ocean acidification. We hope we have demonstrated that knowledge of key biological and physical components as well as the processes that influence them are essential if we are to manage marine resources such as coral reefs and reverse the current decline of coral reefs around the world.

Shaun Ahyong gratefully acknowledges Glenn Ahyong (Sydney), Claudia Arango (Queensland Museum), Roy Caldwell (University of California, Berkeley), Roger Springthorpe and Ian Loch (both Australian Museum), N. Dean Pentcheff (Los Angeles), Chris Tudge (American University, Washington DC) and the Great Barrier Reef Marine Park Authority for the use of photographs.

Mike Kingsford would like to thank Craig Steinberg and Iain Suthers for constructive comments on drafts of chapters and Richard Brickman for an original figure.

Maria Bryne gives thanks to many colleagues for permission to use their images. Thomas Prowse and Paulina Selvakumaraswamy assisted with the images.

Jon Day acknowledges that the RAP and the rezoning process resulted from the involvement and support of virtually the entire GBRMPA staff, as well as many external researchers, other experts, many thousands of local users, and the wider public, who were all concerned for the future of the GBR; their collective efforts therefore need to be acknowledged. Particular thanks also to Mick Bishop, Max Day, Kirstin Dobbs, Leanne Fernandes and Andrew Skeat for comments on drafts of his chapter for the book.

Guillermo Diaz-Pulido gives thanks to the Pew Program in Marine Conservation and to Laurence McCook for continuous support. Thanks also to Tyrone Ridgway for valuable comments.

Ove Hoegh-Guldberg thanks artist Diana Kleine, as well as members of the Coral Reef Targeted Research Project (www.gefcoral.org).

John Hooper would like to thank Natural Products Discovery Griffith University, and the GBR Seabed Biodiversity Project consortium (CRC Reef, AIMS, QDPI, QM and CSIRO Marine and Atmospheric Research) for their significant sponsorship of research on Porifera of the reefs and inter-reef regions, respectively, of the Great Barrier Reef. Without this funding we would still know very little about the tropical Australasian sponge faunas.

Terry Hughes acknowledges that his research was supported by the Australian Research Council's Centre of Excellence Program.

Pat Hutchings would like to thank Kathy Atkinson, Kate Attwood, David Bellwood, Karen Gowlett-Holmes, Jim Johnson, Brendan Kelaher, Tara Macdonald, Ashley Miskelly, Huy Nguyen, Greg Rouse, Roger Steene, Lyle Vail and Dave Wachenfeld for allowing her to use their superb photos, Tom Cribb for commenting on the section on platyhelminths, and Chris Glasby for commenting on the polychaetes. Mary Stafford-Smith provided information on coral genera distribution.

Dennis Gordon and Phillip Bock thank the New Zealand Foundation for Research, Science and Technology (Contract C01X0502) for funding (Dennis Gordon), and Karen Gowlett-Holmes (CSIRO) for permission to reproduce a photo.

Helene Marsh gives thanks to Rebecca Fong for assistance with the illustrations that were based on line drawings by Geoff Kelly and a concept diagram developed with the assistance of the late Peter Arnold. Isabel Beasely, Alistair Birtles, Mike Noad, Guido Parra, Steve van Dyck and anonymous referees provided valuable comments on various drafts of her manuscript. Photographs were provided by Guido Parra.

John Pandolfi and Russell Kelley acknowledge their thanks to Robin Beaman for use of the two photos he supplied, to David Hopley for his thoughtful review, to Jody Webster for discussions, and to the ARC Centre of Excellence in Coral Reef Studies for funding.

Roland Pitcher, Peter Doherty and Tara Anderson acknowledge that the Great Barrier Reef Seabed Biodiversity Project was a collaboration between the Australian Institute of Marine Science (AIMS), the Commonwealth Scientific and Industrial Research Organisation (CSIRO), Queensland Department of Primary Industries & Fisheries (QDPI&F), and the Queensland Museum (QM); funded by the CRC Reef Research Centre (CRC-Reef), the Fisheries Research and Development Corporation (FRDC), and the National Oceans Office (NOO) of the Department of Environment and Water Resources. We also thank the multi-agency teams and the crews of the RV Lady Basten (AIMS) and FRV Gwendoline May (QDPI&F) that contributed to the success of the fieldwork; the research agencies AIMS, CSIRO, QM, QDPI&F for providing support to the project; and all the project's team members without whose valuable efforts the project would not have been possible.

Richard Willan acknowledges Gary Cobb (Mooloolaba), Julie Jones (GBRMPA, Townsville), Allan Limpus (Bundaberg), Julie Marshall (Melbourne) and David Wachenfeld (GBRMPA, Townsville) for the use of an image. John Collins (James Cook University, Townsville) is acknowledged for helpful comments on an earlier draft of the mollusc chapter. Paul Southgate (James Cook University, Townsville) provided information on aquaculture of giant clams, and Uwe Weinreich (Cairns) is thanked for the use of images and for information on feeding in Bursidae.

ADDITIONAL READING

Hill, D. (1985). The Great Barrier Reef Committee, 1922–82. Part 11: The last three decades. *Historical Records of Australian Science* **6**(2), 195–221.

Index

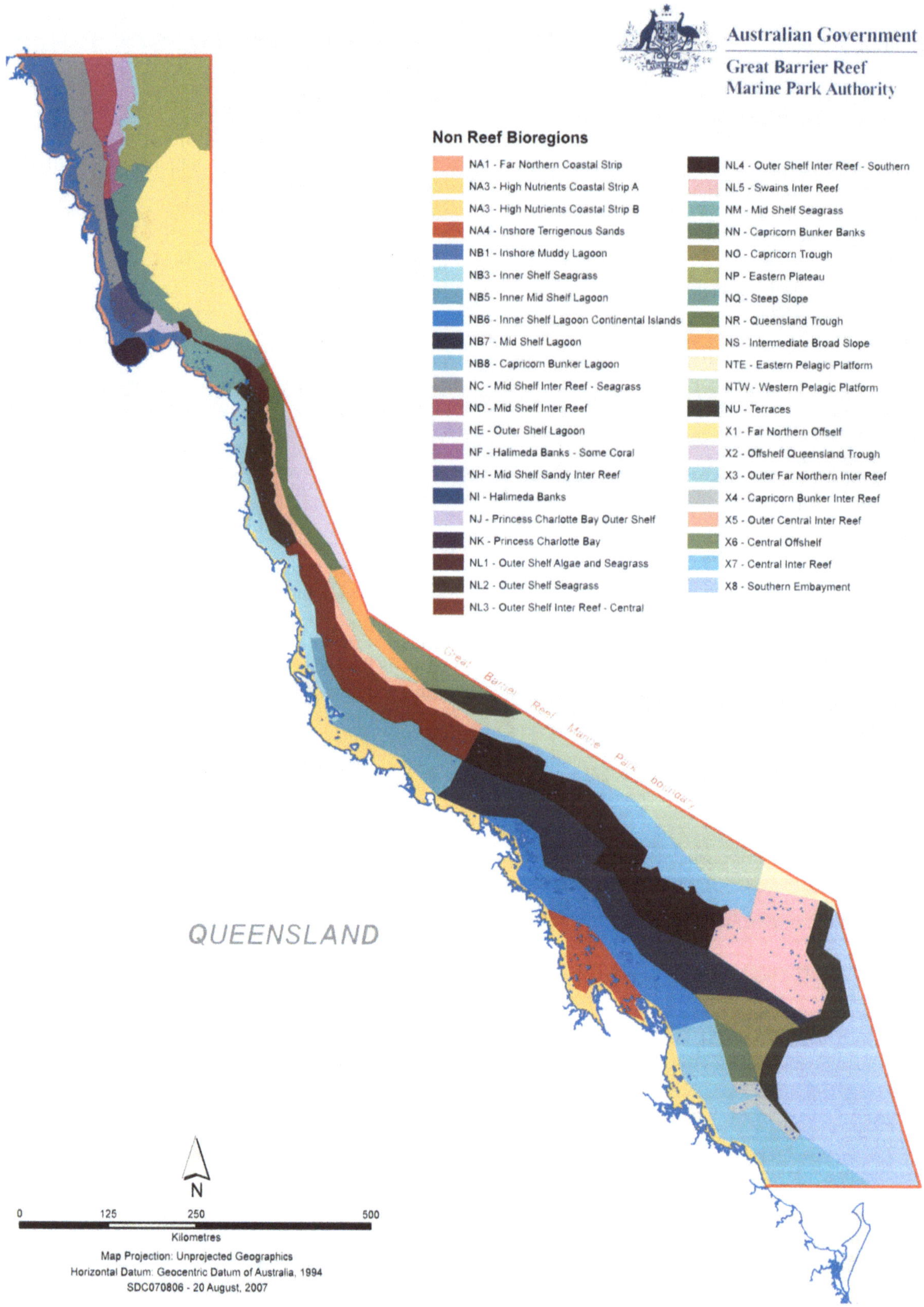

Non-reef bioregions in the Great Barrier Reef World Heritage Area. Each of these is characterised by a unique suite of biota and physical characteristics.

MIX
Papier aus verantwortungsvollen Quellen
Paper from responsible sources
FSC® C105338

If you have any concerns about our products,
you can contact us on
ProductSafety@springernature.com

In case Publisher is established outside the EU,
the EU authorized representative is:
Springer Nature Customer Service Center GmbH
Europaplatz 3, 69115 Heidelberg, Germany

Printed by Libri Plureos GmbH
in Hamburg, Germany